ASTROPHYSICS AND SPACE SCIENCE LIBRARY

VOLUME 340

PLASMA ASTROPHYSICS, PART I

PLASMA ASTROPHYSICS, PART I

Fundamentals and Practice

BORIS V. SOMOV
Moscow State University
Moscow, Russia

Boris V. Somov
Astronomical Institute
And Faculty of Physics
Moscow State University
Moscow, Russia
somov@sai.msu.ru

Cover illustration: Interaction of plasma with magnetic fields and light of stars creates many beautiful views of the night sky, like this one shown as the background – a part of the famous nebula IC434 located about 1600 light-years away from Earth and observed by the National Science Foundation's 0.9-meter telescope on Kitt Peak.

Photograph reproduced with kind permission from T.A.Rector (NOAO/AURA/NSF) and Hubble Heritage Team (STScI/AURA?NASA).

Library of Congress Control Number: 2006926924

ISBN-10: 0-387-34916-2
ISBN-13: 978-0387-34916-9

Printed on acid-free paper.

9 8 7 6 5 4 3 2 1

springer.com

Contents

About This Book

If you want to learn the most fundamental things about plasma astrophysics with the least amount of time – and who doesn't? – this text is for you. This book is addressed to young people, mainly to students, without a background in plasma physics; it grew from the lectures given many times in the Faculty of General and Applied Physics at the Moscow Institute of Physics and Technics (the well known 'fiz-tekh') since 1977. A similar full-year course was also offered to the students of the Astronomical Division of the Faculty of Physics at the Moscow State University over the years after 1990. A considerable amount of new material, related to modern astrophysics, has been added to the lectures. So the contents of the book can hardly be presented during a one-year lecture course, without additional seminars.

In fact, just the seminars with the topics '**how to make a cake**' were especially pleasant for the author and useful for students. In part, the text of the book retains the imprint of the seminar form, implying a more lively dialogue with the reader and more visual representation of individual notions and statements. At the same time, the author's desire was that these digressions from the academic language of the monograph will not harm the rigour of presentation of this textbook's subject – the physical and mathematical introduction to plasma astrophysics.

There is no unique simple model of a plasma, which encompasses all situations in space. We have to familiarize ourselves with many different models applied to different situations. We need clear guidelines when a model works and when it does not work. Hence **the best strategy** is to develop an intuition about plasma physics, but how to develop it?

The idea of the book is not typical for the majority of textbooks on plasma astrophysics. Its idea is

> the consecutive consideration of physical principles, starting from the most general ones, and of simplifying assumptions which give us a simpler description of plasma under cosmic conditions.

Thus I would recommend the students to read the book straight through each chapter to see the central line of the plasma astrophysics, its **classic fundamentals**. In so doing, the boundaries of the domain of applicability of the approximation at hand will be outlined from the viewpoint of physics rather than of many possible astronomical applications. After that, as an aid to detailed understanding, please return with pencil and paper to work out the missing steps (if any) in the formal mathematics.

On the basis of such an approach the student interested in modern astrophysics, its **current practice**, will find the answers to two key questions:

(1) what approximation is the best one (the simplest but sufficient) for description of a phenomenon in astrophysical plasma;

(2) how to build an adequate model for the phenomenon, for example, a solar flare or a flare in the corona of an accretion disk.

Practice is really important for the theory of astrophysical plasma. Related exercises (problems and answers supplemented to each chapter) to improve skill do not thwart the theory but serve to better understanding of plasma astrophysics.

As for the applications, preference evidently is given to physical processes in the solar plasma. Why? – Much attention to solar plasma physics is conditioned by the possibility of the all-round observational test of theoretical models. This statement primarily relates to the processes in the solar atmosphere. For instance, flares on the Sun, in contrast to those on other stars as well as a lot of other analogous phenomena in the Universe, *can be seen* in their development, i.e. we can obtain a sequence of images during the flare's evolution, not only in the optical and radio ranges but also in the ultraviolet, soft and hard X-ray, gamma-ray ranges.

This book is mainly intended for students who have mastered a course of general physics and have some initial knowledge of theoretical physics. For beginning students, who may not know in which subfields of astrophysics they wish to specialize,

> it is better to cover a lot of fundamental theories thoroughly than to dig deeply into any particular astrophysical subject or object,

even a very interesting one, for example black holes. Astronomers and astrophysicists of the future will need tools that allow them to explore in many different directions. Moreover astronomy of the future will be, more than hitherto, *precise science* similar to mathematics and physics.

The beginning graduate students are usually confronted with a confusing amount of work on plasma astrophysics published in a widely dispersed scientific literature. Knowing this difficulty, the author has tried as far as possible to represent the material in a self-contained form which does not require the reading of additional literature. However there is an extensive bibliography in the end of the book, allowing one to find the original works. In many cases, particularly where a paper in Russian is involved, the author has aimed to give the full bibliographic description of the work, including its title, etc.

Furthermore the book contains recommendations as to an introductory (unavoidable) reading needed to refresh the memory about a particular fact, as well as to additional (further) reading to refine one's understanding of the subject. Separate **remarks of an historical character** are included in many places. It is sometimes simpler to explain the interrelation of discoveries by representing the subject in its development. It is the author's opinion that the outstanding discoveries in plasma astrophysics are by no means governed by chance. With the same thought in mind, the author gives preference to original papers on a topic under consideration; it happens in science, as in art, that an original is better than nice-looking modernizations. Anyway,

> knowledge of the history of science and especially of natural science is of great significance for its understanding and development.

The majority of the book's chapters begin from an 'elementary account' and illustrative simple examples but finish with the most modern results of scientific importance. New problems determine the most interesting perspectives of plasma astrophysics as a new developing science. The author hopes, in this context, that professionals in the field of plasma astrophysics and adjacent sciences will enjoy reading this book too. Open issues are the focus of our attention in many places where they are. In this way, **perspectives of the plasma astrophysics** with its many applications will be also of interest for readers. The book can be used as a textbook but has higher potential of modern scientific monograph.

The first volume of the book is unique in covering the basic principles and main practical tools required for understanding and work in plasma astrophysics. The second volume "Plasma Astrophysics. 2. Reconnection and Flares" (referred in the text as vol. 2) represents the basic physics of the magnetic reconnection phenomenon and the flares of electromagnetic origin in space plasmas in the solar system, relativistic objects, accretion disks, their coronae.

Acknowledgements

The author is grateful to his young colleagues Sergei I. Bezrodnykh, Sergei A. Bogachev, Sergei V. Diakonov, Irina A. Kovalenko, Yuri E. Litvinenko, Sergei A. Markovskii, Elena Yu. Merenkova, Anna V. Oreshina, Inna V. Oreshina, Alexandr I. Podgornii, Yuri I. Skrynnikov, Andrei R. Spektor, Vyacheslav S. Titov, Alexandr I. Verneta, and Vladimir I. Vlasov for the pleasure of working together, for generous help and valuable remarks. He is also happy to acknowledge helpful discussions with many of his colleagues and friends in the world.

Moscow, 2006 Boris V. Somov

Plasma Astrophysics
History and Neighbours

Plasma astrophysics studies electromagnetic processes and phenomena in space, mainly the role of forces of an electromagnetic nature in the dynamics of cosmic matter. Two factors are specific to the latter: its gaseous state and high conductivity. Such a combination is unlikely to be found under natural conditions on Earth; the matter is either a non-conducting gas (the case of gas dynamics or hydrodynamics) or a liquid or a solid conductor. By contrast, **plasma is the main state of cosmic matter**. It is precisely the poor knowledge of cosmic phenomena and cosmic plasma properties that explains the retarded development of plasma astrophysics. It has been distinguished as an independent branch of physics in the pioneering works of Alfvén (see Alfvén, 1950).

Soon after that, the problem of thermonuclear reactions initiated a great advance in plasma research (Simon, 1959; Glasstone and Loveberg, 1960; Leontovich, 1960). This branch has been developing rather independently, although being partly 'fed' by astrophysical ideas. They contributed to the growth of plasma physics, for example, the idea of stelarators. Presently, the reverse influence of laboratory plasma physics on astrophysics is also important.

From the physical viewpoint,

> plasma astrophysics is a part of plasma theory related in the first place to the dynamics of a low-resistivity plasma in space.

However it is this part that is the most poorly studied one under laboratory conditions. During the 1930s, scientists began to realize that the Sun and other stars are powered by nuclear fusion and they began to think of recreating the process in the laboratory. The ideas of astro- and geophysics dominate here, as before. At present time, they mainly come from many space experiments and fine astronomical ground-based observations. From this viewpoint, plasma astrophysics belongs to experimental science.

Electric currents and, therefore, magnetic fields are easily generated in the astrophysical plasma owing to its low resistivity. The energy of magnetic fields is accumulated in plasma, and the sudden release of this energy – an original electrodynamical 'burst' or 'explosion' – takes place under definite but quite general conditions. It is accompanied by fast directed plasma ejections (jets), powerful flows of heat and radiation and impulsive acceleration of particles to high energies.

This phenomenon is quite a widespread one. It can be observed in flares on the Sun and other stars, in the Earth's magnetosphere as magnetic storms and substorms, in coronae of accretion disks of cosmic X-ray sources, in nuclei of active galaxies and quasars. The second volume of this book is devouted to the physics of *magnetic reconnection and flares* generated by reconnection in plasma in the solar system, single and double stars, relativistic objects, and other astrophysical objects.

The subject of the first volume of present book is the systematic description of the most important topics of plasma astrophysics. However the aim of the book is not the strict substantiation of the main principals and basic equations of plasma physics; this can be found in many wonderful monographs (Klimontovich, 1986; Schram, 1991; Liboff, 2003). There are also many nice textbooks (Goldston and Rutherford, 1995; Choudhuri, 1998; Parks, 2004) to learn general plasma physics without or with some astrophysical applications.

The primary aim of the book in your hands is rather the solution of a much more modest but still important problem, namely to help the students of astrophysics to understand the interrelation and limits of applicability of different approximations which are used in plasma astrophysics. If, on his/her way, the reader will continously try, following the author, to reproduce all mathematical transformation, he/she finally will soon find the pleasant feeling of real knowledge of the subject and the real desire for constructive work in plasma astrophysics.

The book will help the young reader to master the modern methods of plasma astrophysics and will teach the application of these methods while solving concrete problems in the physics of the Sun and many other astronomical objects. A good working knowledge of plasma astrophysics is essential for the modern astrophysicist.

Chapter 1

Particles and Fields: Exact Self-Consistent Description

There exist two different ways to describe *exactly* the behaviour of a system of charged particles in electromagnetic and gravitational fields. The first description, the Newton set of motion equations, is convenient for a small number of interacting particles. For systems of large numbers of particles, it is more advantageous to deal with the single Liouville equation for an *exact* distribution function.

1.1 Interacting particles and Liouville's theorem

1.1.1 Continuity in phase space

Let us consider a system of N interacting particle. Without much justification (which will be given in Chapter 2), let us introduce the distribution function

$$f = f(\mathbf{r}, \mathbf{v}, t) \tag{1.1}$$

for particles as follows. We consider the six-dimensional (6D) space called *phase space* $X = \{\mathbf{r}, \mathbf{v}\}$. The number of particles present in a small volume $dX = d^3\mathbf{r}\, d^3\mathbf{v}$ at a point X (see Figure 1.1) at a moment of time t is defined to be

$$dN(X, t) = f(X, t)\, dX. \tag{1.2}$$

Accordingly, the total number of the particles at this moment is

$$N(t) = \int f(X, t)\, dX \equiv \iint f(\mathbf{r}, \mathbf{v}, t)\, d^3\mathbf{r}\, d^3\mathbf{v}\,. \tag{1.3}$$

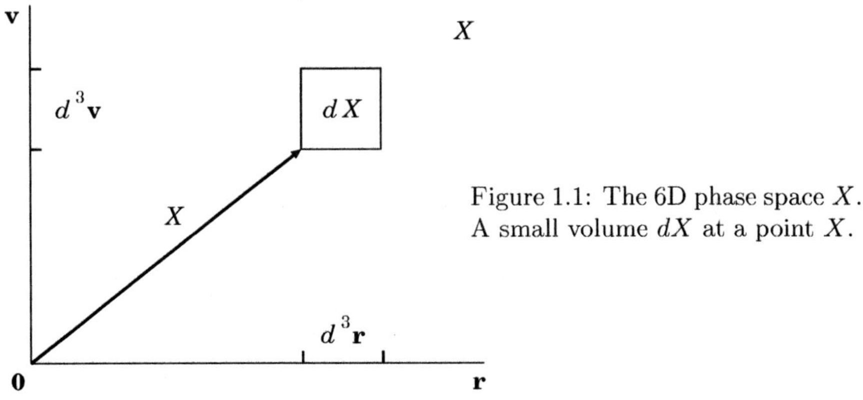

Figure 1.1: The 6D phase space X. A small volume dX at a point X.

If, for definiteness, we use the Cartesian coordinates, then

$$X = \{\, x,\, y,\, z,\, v_x,\, v_y,\, v_z \,\}$$

is a point of the phase space (Figure 1.2) and

$$\dot{X} = \{\, v_x,\, v_y,\, v_z,\, \dot{v}_x,\, \dot{v}_y,\, \dot{v}_z \,\} \tag{1.4}$$

is the velocity of this point in the phase space.

Let us suppose the coordinates and velocities of the particles are changing *continuously* – 'from point to point'. This corresponds to a continuous motion of the particles in phase space and can be expressed by the *continuity equation*:

$$\boxed{\frac{\partial f}{\partial t} + \text{div}_X\,(\dot{X} f) = 0} \tag{1.5}$$

or

$$\frac{\partial f}{\partial t} + \text{div}_{\mathbf{r}}\,(\mathbf{v} f) + \text{div}_{\mathbf{v}}\,(\dot{\mathbf{v}} f) = 0\,.$$

Equation (1.5) expresses the *conservation law* for the particles, since the integration of (1.5) over a volume U enclosed by the surface S in Figure 1.2 gives

$$\frac{\partial}{\partial t}\int\limits_U f\, dX + \int\limits_U \text{div}_X\,(\dot{X} f)\, dX =$$

by virtue of definition (1.2) and the Ostrogradskii-Gauss theorem

$$= \frac{\partial}{\partial t}\, N(t)\Big|_U + \int\limits_S (\dot{X} f)\, dS = \frac{\partial}{\partial t}\, N(t)\Big|_U + \int\limits_S \mathbf{J}\cdot d\mathbf{S} = 0\,. \tag{1.6}$$

Here a surface element $d\mathbf{S}$, normal to the boundary S, is oriented towards its outside, so that imports are counted as negative (e.g., Smirnov, 1965, Section 126). $\mathbf{J} = \dot{X} f$ is the *particle flux density* in phase space. Thus

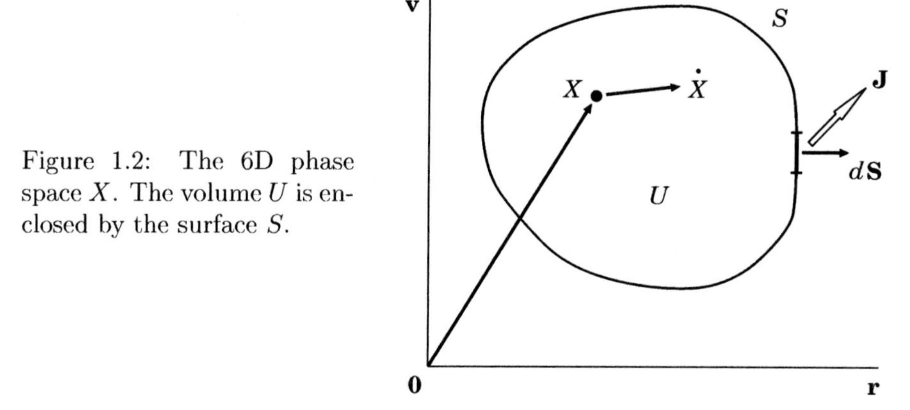

Figure 1.2: The 6D phase space X. The volume U is enclosed by the surface S.

a change of the particle number in a given phase space volume U is defined by the particle flux through the boundary surface S only.

The reason is clear. There are no sources or sinks for the particles inside the volume. Otherwise the source and sink terms must be added to the right-hand side of Equation (1.5).

1.1.2 The character of particle interactions

Let us rewrite Equation (1.5) in another form in order to understand the meaning of divergent terms. The first of them is

$$\operatorname{div}_{\mathbf{r}} (\mathbf{v} f) = f \operatorname{div}_{\mathbf{r}} \mathbf{v} + (\mathbf{v} \cdot \nabla_{\mathbf{r}}) f = 0 + (\mathbf{v} \cdot \nabla_{\mathbf{r}}) f ,$$

since $\mathbf{r}$ and $\mathbf{v}$ are independent variables in phase space X. The second divergent term is

$$\operatorname{div}_{\mathbf{v}} (\dot{\mathbf{v}} f) = f \operatorname{div}_{\mathbf{v}} \dot{\mathbf{v}} + \dot{\mathbf{v}} \cdot \nabla_{\mathbf{v}} f .$$

So far no assumption has been made as to the character of particle interactions. It is worth doing here. Let us restrict our consideration to the interactions with

$$\boxed{\operatorname{div}_{\mathbf{v}} \dot{\mathbf{v}} = 0 ,} \tag{1.7}$$

then Equation (1.5) can be rewritten in the equivalent form:

$$\frac{\partial f}{\partial t} + \mathbf{v} \cdot \nabla_{\mathbf{r}} f + \frac{\mathbf{F}}{m} \cdot \nabla_{\mathbf{v}} f = 0$$

or

$$\frac{\partial f}{\partial t} + \dot{X} \, \nabla_X f = 0 , \tag{1.8}$$

where

$$\dot{X} = \left\{ v_x, v_y, v_z, \frac{F_x}{m}, \frac{F_y}{m}, \frac{F_z}{m} \right\}.$$

Having written that, we 'trace' the particle phase trajectories. Thus Liouville's theorem is found to have the following formulation:

$$\frac{\partial f}{\partial t} + \mathbf{v} \cdot \nabla_{\mathbf{r}} f + \frac{\mathbf{F}}{m} \cdot \nabla_{\mathbf{v}} f = 0\,. \tag{1.9}$$

Liouville's theorem: *The distribution function remains constant on the particle phase trajectories* if condition (1.7) is satisfied.

We shall call Equation (1.9) the *Liouville equation*. Let us define also the Liouville operator

$$\frac{D}{Dt} \equiv \frac{\partial}{\partial t} + \dot{X} \frac{\partial}{\partial X} \equiv \frac{\partial}{\partial t} + \mathbf{v} \cdot \nabla_{\mathbf{r}} + \frac{\mathbf{F}}{m} \cdot \nabla_{\mathbf{v}}\,. \tag{1.10}$$

This operator is just the total time derivative following a particle motion in the phase space X. By using definition (1.10), we rewrite Liouville's theorem as follows:

$$\boxed{\frac{Df}{Dt} = 0\,.} \tag{1.11}$$

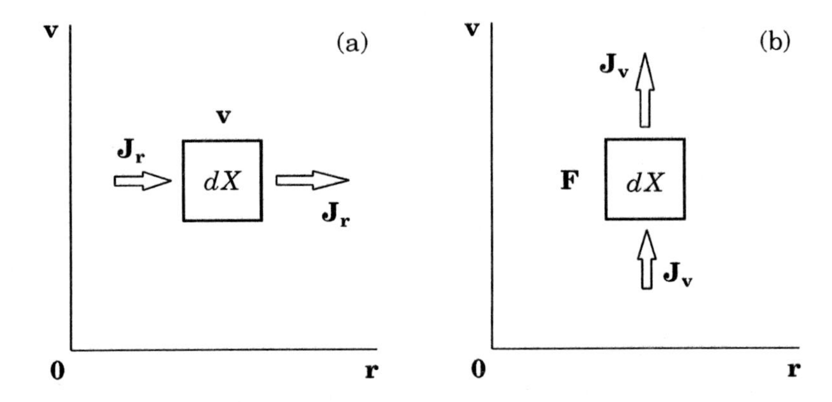

Figure 1.3: Action of the two different terms of the Liouville operator in the 6D phase space X.

What factors lead to the changes in the distribution function?

Let dX be a small volume in the phase space X. The second term in Equation (1.9), $\mathbf{v} \cdot \nabla_{\mathbf{r}} f$, means that the particles go into and out of the phase volume element considered, because their velocities are not zero (Figure 1.3a). So this term describes a simple kinematic effect. The third term, $(\mathbf{F}/m) \cdot \nabla_{\mathbf{v}} f$, means that the particles escape from the phase volume element dX or come into this element due to their acceleration or deceleration under the influence of forces (Figure 1.3b).

Some important properties of the Liouville equation are considered in Exercises 1.1–1.4.

1.1.3 The Lorentz force, gravity

Let us recall that the forces have to satisfy condition (1.7). We rewrite it as follows:

$$\frac{\partial \dot{v}_\alpha}{\partial v_\alpha} = \frac{1}{m} \frac{\partial F_\alpha}{\partial v_\alpha} = 0$$

or

$$\frac{\partial F_\alpha}{\partial v_\alpha} = 0 \,, \qquad \alpha = 1, 2, 3 \,. \tag{1.12}$$

In other words,

> the component F_α of the force vector $\mathbf{F}$ does not depend upon the velocity component v_α.

This is a sufficient condition.

The classical Lorentz force

$$F_\alpha = e \left[E_\alpha + \frac{1}{c} \left(\mathbf{v} \times \mathbf{B} \right)_\alpha \right] \tag{1.13}$$

obviously has that property. The gravitational force in the classical approximation is entirely independent of velocity.

Other forces may be considered, depending on the situation, for example the forces resulting from the emission and/or absorption of radiation by astrophysical plasma, which is electromagnetic in nature, though maybe quantum. These forces when they are important should be considered with account of their relative significance, conservative or dissipative character, and other physical properties taken.

1.1.4 Collisional friction in plasma

As a contrary example we consider the friction force (cf. formula (8.66) for the collisional drag force in plasma):

$$\mathbf{F} = -\, k \, \mathbf{v} \,, \tag{1.14}$$

where the constant $k > 0$. In this case the right-hand side of Liouville's equation is not zero:

$$-f \operatorname{div}_{\mathbf{v}} \dot{\mathbf{v}} = -f \operatorname{div}_{\mathbf{v}} \frac{\mathbf{F}}{m} = \frac{3k}{m} \, f \,,$$

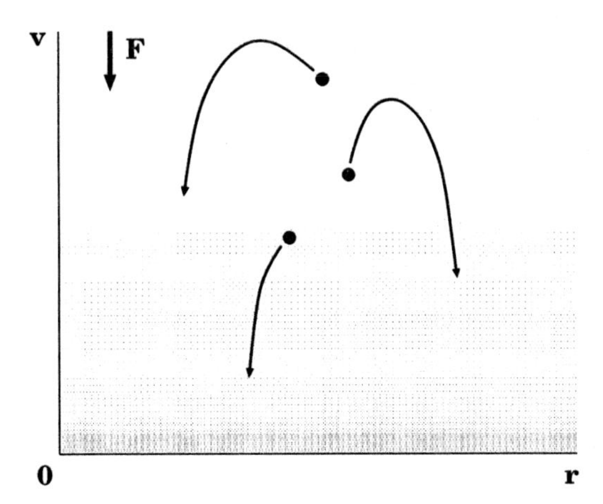

Figure 1.4: Particle density increase in phase space as a result of the action of the friction force **F**.

because

$$\frac{\partial v_\alpha}{\partial v_\alpha} = \delta_{\alpha\alpha} = 3\,.$$

Instead of Liouville's equation we have

$$\frac{Df}{Dt} = \frac{3k}{m} f > 0\,. \tag{1.15}$$

The distribution function (that is the particle density) does not remain constant on particle trajectories but increases as the time elapses. Along the phase trajectories, it increases exponentially:

$$f(t, \mathbf{r}, \mathbf{v}) \sim f(0, \mathbf{r}, \mathbf{v}) \exp\left(\frac{3k}{m}\, t\right). \tag{1.16}$$

The physical sense of this phenomenon is obvious. As the particles are decelerated by the friction force, they move down in Figure 1.4. By so doing, they are concentrated in the constantly diminishing region of phase space situated in the vicinity of the axis $\mathbf{v} = \mathbf{0}$.

There is a viewpoint that the Liouville theorem is valid for the forces that *do not disperse* particle velocities (Shkarofsky et al., 1966, Chapter 2). Why? It is usually implied that particle *collisions* enlarge such a dispersion: $\mathrm{div}_{\mathbf{v}}\, \dot{\mathbf{v}} > 0$. So

$$\frac{Df}{Dt} = \left(\frac{\partial f}{\partial t}\right)_{\mathrm{c}} = -f\, \mathrm{div}_{\mathbf{v}}\, \dot{\mathbf{v}} < 0\,. \tag{1.17}$$

In this case the right-hand side of Equation (1.17) is called the *collisional* integral (see Sections 2.1 and 2.2). In contrast to the right-hand side of (1.15), that of Equation (1.17) is usually negative.

The above example of the friction force is instructive in that it shows how the forces that are diminishing the velocity dispersion ($\mathrm{div}_{\mathbf{v}}\, \dot{\mathbf{v}} < 0$) lead to the violation of Liouville's theorem; in other words, how they lead to a change of the distribution function along the particle trajectories. For the validity of Liouville's theorem only the condition (1.7) is important; in the velocity space, the divergence of the forces has to equal zero. The sign of this divergence is unimportant.

1.1.5 The exact distribution function

Let us consider another property of the Liouville theorem. We introduce the N-particle distribution function of the form

$$\hat{f}(t, \mathbf{r}, \mathbf{v}) = \sum_{i=1}^{N} \delta\left(\mathbf{r} - \mathbf{r}_i(t)\right) \delta\left(\mathbf{v} - \mathbf{v}_i(t)\right). \tag{1.18}$$

We shall call such a distribution function the *exact* one. It is illustrated by schematic Figure 1.5.

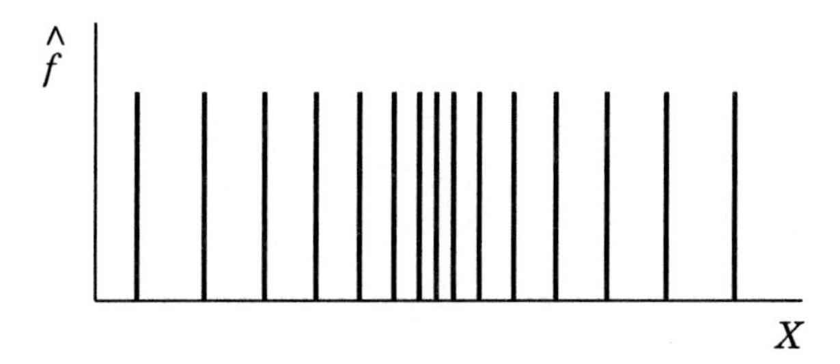

Figure 1.5: The one-dimensional analogy of the exact distribution function.

Let us substitute this expression for the distribution function in Equation (1.9). The resulting three terms are

$$\frac{\partial \hat{f}}{\partial t} = \sum_i (-1)\, \delta'_\alpha\left(\mathbf{r} - \mathbf{r}_i(t)\right) \dot{r}^{\,i}_\alpha\, \delta\left(\mathbf{v} - \mathbf{v}_i(t)\right) +$$

$$+ \sum_i (-1)\, \delta\left(\mathbf{r} - \mathbf{r}_i(t)\right) \delta'_\alpha\left(\mathbf{v} - \mathbf{v}_i(t)\right) \dot{v}^{\,i}_\alpha\,, \tag{1.19}$$

$$\mathbf{v} \cdot \nabla_{\mathbf{r}} \hat{f} \equiv v_\alpha \frac{\partial \hat{f}}{\partial r_\alpha} = \sum_i v_\alpha\, \delta'_\alpha\left(\mathbf{r} - \mathbf{r}_i(t)\right) \delta\left(\mathbf{v} - \mathbf{v}_i(t)\right), \tag{1.20}$$

$$\frac{\mathbf{F}}{m} \cdot \nabla_{\mathbf{v}} \hat{f} \equiv \frac{F_\alpha}{m} \frac{\partial \hat{f}}{\partial v_\alpha} = \sum_i \frac{F_\alpha}{m_i}\, \delta\left(\mathbf{r} - \mathbf{r}_i(t)\right) \delta'_\alpha\left(\mathbf{v} - \mathbf{v}_i(t)\right). \tag{1.21}$$

Here the index $\alpha = 1, 2, 3$ or (x, y, z). The prime denotes the derivative with respect to the argument of a function; for the delta function, see definition of the derivative in Vladimirov (1971). The overdot denotes differentiation with respect to time t. Summation over the repeated index α (contraction) is implied:

$$\delta'_\alpha \dot{r}^i_\alpha = \delta'_x \dot{r}^i_x + \delta'_y \dot{r}^i_y + \delta'_z \dot{r}^i_z .$$

The sum of terms (1.19)–(1.21) equals zero. Let us rewrite it as follows

$$0 = \sum_i \left(-\dot{r}^i_\alpha + v^i_\alpha\right) \delta'_\alpha \left(\mathbf{r} - \mathbf{r}_i(t)\right) \delta \left(\mathbf{v} - \mathbf{v}_i(t)\right) +$$

$$+ \sum_i \left(-\dot{v}^i_\alpha + \frac{F_\alpha}{m_i}\right) \delta \left(\mathbf{r} - \mathbf{r}_i(t)\right) \delta'_\alpha \left(\mathbf{v} - \mathbf{v}_i(t)\right) .$$

This can occur just then that all the coefficients of different combinations of delta functions with their derivatives equal zero as well. Therefore we find

$$\frac{d r^i_\alpha}{dt} = v^i_\alpha(t) , \qquad \frac{d v^i_\alpha}{dt} = \frac{1}{m_i} F_\alpha \left(\mathbf{r}_i(t), \mathbf{v}_i(t)\right) . \tag{1.22}$$

Thus

> the Liouville equation for an exact distribution function is *equivalent* to the Newton set of equations for a particle motion, both describing a purely dynamic behaviour of the particles.

It is natural since this distribution function is exact. No statistical averaging has been done so far. It is for this reason that both descriptions – namely, the Newton set and the Liouville theorem for the exact distribution function – are dynamic (as well as reversible, of course) and equivalent. Statistics will appear in the next Chapter when, instead of the exact description of a system, we begin to use some mean characteristics such as temperature, density etc. This is the statistical description that is valid for systems containing a large number of particles.

We have shown that finding a solution of the Liouville equation for an exact distribution function

$$\boxed{\frac{D\hat{f}}{Dt} = 0} \tag{1.23}$$

is the same as the integration of the motion equations. Therefore

> for systems of a large number of interacting particles, it is much more advantageous to deal with the single Liouville equation for the exact distribution function which describes the entire system.

Recommended Reading: Landau and Lifshitz, *Mechanics* (1976), Chapters 2 and 7; Landau and Lifshitz, *Statistical Physics* (1959b), Chapter 1, § 1–3.

1.2 Charged particles in the electromagnetic field

1.2.1 General formulation of the problem

Let us start from recalling basic physics notations and establishing a common basis. Maxwell's equations for the electric field $\mathbf{E}$ and magnetic field $\mathbf{B}$ are well known to have the form (see Landau and Lifshitz, *Classical Theory of Field*, 1975, Chapter 4, § 26):

$$\operatorname{curl} \mathbf{B} = \frac{4\pi}{c} \mathbf{j} + \frac{1}{c} \frac{\partial \mathbf{E}}{\partial t} , \tag{1.24}$$

$$\operatorname{curl} \mathbf{E} = -\frac{1}{c} \frac{\partial \mathbf{B}}{\partial t} , \tag{1.25}$$

$$\operatorname{div} \mathbf{B} = 0 , \tag{1.26}$$

$$\operatorname{div} \mathbf{E} = 4\pi \rho^{\mathrm{q}} . \tag{1.27}$$

The fields are completely determined by electric charges and electric currents. Note that, in general, Maxwell's equations imply the continuity equation for electric charge (see Exercise 1.5) as well as the conservation law for electromagnetic field energy (Exercise 1.6).

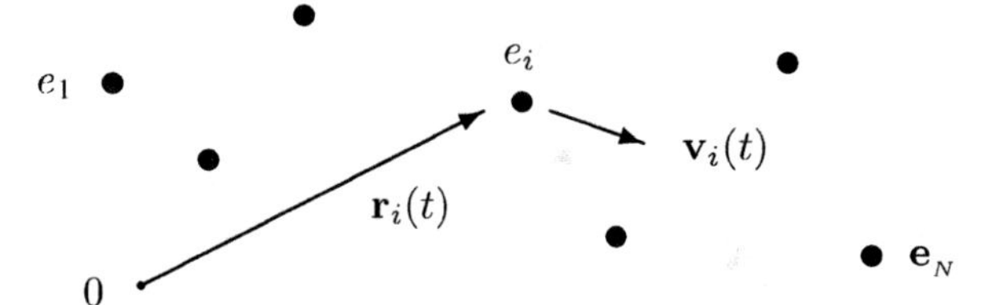

Figure 1.6: A system of N charged particles.

Let there be N particles with charges e_1, e_2, ... e_i, ... e_N, coordinates $\mathbf{r}_i(t)$ and velocities $\mathbf{v}_i(t)$, see Figure 1.6. By definition, the electric charge density

$$\rho^{\mathrm{q}}(\mathbf{r}, t) = \sum_{i=1}^{N} e_i \, \delta(\mathbf{r} - \mathbf{r}_i(t)) \tag{1.28}$$

and the density of electric current

$$\mathbf{j}(\mathbf{r}, t) = \sum_{i=1}^{N} e_i \, \mathbf{v}_i(t) \, \delta(\mathbf{r} - \mathbf{r}_i(t)) . \tag{1.29}$$

The delta function of the vector-argument is defined as usually:

$$\delta(\mathbf{r} - \mathbf{r}_i(t)) = \prod_{\alpha=1}^{3} \delta_\alpha = \delta\left(r_x - r_x^i(t)\right) \delta\left(r_y - r_y^i(t)\right) \delta\left(r_z - r_z^i(t)\right) . \tag{1.30}$$

The coordinates and velocities of particles can be found by integrating the equations of motion – the Newton equations:

$$\dot{\mathbf{r}}_i \equiv \frac{d\,\mathbf{r}_i}{dt} = \mathbf{v}_i(t)\,, \tag{1.31}$$

$$\dot{\mathbf{v}}_i \equiv \frac{d\,\mathbf{v}_i}{dt} = \frac{1}{m_i}\,e_i \left[\mathbf{E}\left(\mathbf{r}_i(t)\right) + \frac{1}{c}\,\mathbf{v}_i \times \mathbf{B}\left(\mathbf{r}_i(t)\right)\right]. \tag{1.32}$$

Let us count the number of unknown quantities: the vectors $\mathbf{B}$, $\mathbf{E}$, $\mathbf{r}_i$, and $\mathbf{v}_i$. We obtain: $3+3+3N+3N = 6\,(N+1)$. The number of equations is equal to $8 + 6N = 6\,(N + 1) + 2$. Therefore two equations seem to be unnecessary. Why is this so?

1.2.2 The continuity equation for electric charge

Let us make sure that the definitions (1.28) and (1.29) conform to the conservation law for electric charge. Differentiating (1.28) with respect to time gives (see Exercise 1.7)

$$\frac{\partial \rho^{\mathrm{q}}}{\partial t} = -\sum_i e_i\,\delta'_\alpha\,\dot{r}^{\,i}_\alpha\,. \tag{1.33}$$

Here the index $\alpha = 1, 2, 3$. The prime denotes the derivative with respect to the argument of the delta function, see Vladimirov (1971). The overdot denotes differentiation with respect to time t.

For the electric current density (1.29) we have the divergence

$$\operatorname{div}\mathbf{j} = \frac{\partial}{\partial r_\alpha}\,j_\alpha = \sum_i e_i\,v^i_\alpha\,\delta'_\alpha\,. \tag{1.34}$$

Comparing formula (1.33) with (1.34) we see that

$$\boxed{\frac{\partial \rho^{\mathrm{q}}}{\partial t} + \operatorname{div}\mathbf{j} = 0\,.} \tag{1.35}$$

Therefore the definitions for ρ^{q} and $\mathbf{j}$ conform to the continuity Equation (1.35).

As we shall see it in Exercise 1.5, conservation of electric charge follows also directly from the Maxwell Equations (1.24) and (1.27). The difference is that above we have not used Equation (1.27).

1.2.3 Initial equations and initial conditions

Operating with the divergence on Equation (1.24) and using the continuity Equation (1.35), we obtain

$$0 = \frac{4\pi}{c}\left(-\frac{\partial \rho^{\mathrm{q}}}{\partial t}\right) + \frac{1}{c}\,\frac{\partial}{\partial t}\operatorname{div}\mathbf{E}\,.$$

Thus, by postulating the definitions (1.28) and (1.29), by virtue of the continuity Equation (1.35) and without using the Maxwell Equation (1.27), we find that

$$\frac{\partial}{\partial t} \left(\operatorname{div} \mathbf{E} - 4\pi \rho^{\,\mathrm{q}} \right) = 0 \, . \tag{1.36}$$

Hence Equation (1.27) will be valid at any moment of time, provided it is true at the initial moment.

Let us operate with the divergence on Equation (1.25):

$$\frac{\partial}{\partial t} \operatorname{div} \mathbf{B} = 0 \, . \tag{1.37}$$

We come to the conclusion that the Equations (1.26) and (1.27) play the role of *initial conditions* for the time-dependent equations

$$\frac{\partial}{\partial t} \mathbf{B} = - c \operatorname{curl} \mathbf{E} \tag{1.38}$$

and

$$\frac{\partial}{\partial t} \mathbf{E} = + c \operatorname{curl} \mathbf{B} - 4\pi \mathbf{j} \, . \tag{1.39}$$

Equation (1.26) implies the absence of magnetic charges or, which is the same, the *solenoidal* character of the magnetic field.

Thus, in order to describe the gas consisting of N charged particles, we consider the time-dependent problem of N bodies with a given interaction law.

> The electromagnetic part of the interaction is described by Maxwell's equations, the time-independent scalar equations playing the role of initial conditions for the time-dependent problem.

Therefore the set consisting of eight Maxwell's equations and $6N$ Newton's equations is neither over- nor underdetermined. It is *closed* with respect to the time-dependent problem, i.e. it consists of $6\,(N+1)$ equations for $6\,(N+1)$ variables, once the initial and boundary conditions are given.

1.2.4 Astrophysical plasma applications

The set of equations described above can be treated analytically in just three cases:

1. $N = 1$, the motion of a charged particle in a given electromagnetic field, for example, drift motions and the so-called adiabatic invariants, wave-particle interaction and the problem of particle acceleration in astrophysical plasma; e.g., Chapters 7 and 18.

2. $N = 2$, Coulomb collisions of two charged particles. This is important for the kinetic description of physical processes, for example, the kinetic

effects under propagation of accelerated particles in plasma, collisional heating of plasma by a beam of fast electrons or/and ions, see Chapters 4 and 8.

3. $N \to \infty$, a very large number of particles. This case is the frequently considered one in plasma astrophysics, because it allows us to introduce macroscopic descriptions of plasma, the widely-used magnetohydrodynamic (MHD) approximation; Chapters 9, and 12.

Numerical integration of Equations (1.24)–(1.32) in the case of large but finite N, like $N \approx 3 \times 10^6$, is possible by using powerful modern computers. Such computations called 'particle simulations' have proved to be increasingly useful for understanding properties of astrophysical plasma. One important example of a simulation is *magnetic reconnection* in a collisionless plasma (Horiuchi and Sato, 1994; Cai and Lee, 1997). This process often leads to fast energy conversion from field energy to particle energy, flares in astrophysical plasma (see vol. 2).

Note also that the set of equations described above can be generalized to include consideration of neutral particles. This is necessary, for instance, in the study of the generalized Ohm's law (Chapter 9) which can be applied in the investigation of physical processes in *weakly-ionized* plasmas, for example in the solar photosphere and prominences.

Dusty and *self-gravitational* plasmas in space are interesting in view of the diverse and often surprising facts about planetary rings and comet environments, interstellar dark space (Bliokh et al., 1995; Kikuchi, 2001). Two effects are often of basic importance, gravitational and electric, since charged or polarized dust grains involved in such environments are much heavier than electrons and ions. So a variety of electric rather than magnetic phenomena are taking place predominantly; and gravitational forces acting on dust particles can become appreciable.

1.3 Gravitational systems

Gravity plays a central role in the dynamics of many astrophysical systems – from stars to the Universe as a whole (Lahav et al., 1996; Rose, 1998; Bertin, 1999; Dadhich and Kembhavi, 2000). It is important for many astrophysical applications that a *gravitational* force (as well as an electromagnetic force) acts on the particles:

$$m_i \, \dot{\mathbf{v}}_i = -m_i \, \nabla \phi \, . \tag{1.40}$$

Here the gravitational potential

$$\phi(t, \mathbf{r}) = - \sum_{n=1}^{N} \frac{G \, m_n}{|\, \mathbf{r}_n(t) - \mathbf{r} \,|} \, , \qquad n \neq i \, , \tag{1.41}$$

G is the gravitational constant. We shall return to this subject many times, for example, while studying the *virial theorem* in MHD (Chapter 19). This theorem is widely used in astrophysics.

At first sight, it may seem that a gravitational system, like stars in a galaxy, will be easier to study than a plasma, because there is gravitational charge (i.e. mass) of only one sign compared to the electric charges of two opposite signs. However the reality is the other way round. Though the potential (1.41) of the gravitational interaction looks similar to the Coulomb potential of charged particles (see formula (8.1)),

> physical properties of gravitational systems differ so much from properties of astrophysical plasma.

We shall see this fundamental difference, for example, in Section 3.3 and many times in what follows. A deep unifying theme which underlies many astrophysical results is that self-gravity is incompatible with thermodynamic equilibrium (see Section 9.6).

1.4 Practice: Exercises and Answers

Exercise 1.1 [Section 1.1.2] Show that any distribution function that is a function of the constants of the motion – the invariants of motion – satisfies Liouville's equation (1.11).

Answer. A general solution of the equations of motion (1.22) depends on $2N$ constants C_i where $i = 1, 2, \dots 2N$. If we assume that the distribution function is a function of these constants of the motion

$$f = f\,(C_1, \dots C_i, \dots C_{2N}\,)\,, \tag{1.42}$$

we can rewrite the left-hand side of Equation (1.11) as

$$\frac{Df}{Dt} = \sum_{i=1}^{2N} \left(\frac{DC_i}{Dt} \right) \left(\frac{\partial f}{\partial C_i} \right). \tag{1.43}$$

Because C_i are constants of the motion, $DC_i/Dt = 0$. Therefore the right-hand side of Equation (1.43) is also zero, and the distribution function (1.42) satisfies the Liouville equation. This is the so-called *Jeans theorem*. It will be used, for example, in the theory of wave-particle interaction in astrophysical plasma (Section 7.1).

Exercise 1.2 [Section 1.1.2] Rewrite the Liouville theorem by using the Hamilton equations instead of the Newton equations.

Answer. Rewrite the Newton set of the motion Equations (1.22) in the Hamilton form (see Landau and Lifshitz, *Mechanics*, 1976, Chapter 7, § 40):

$$\dot{q_\alpha} = \frac{\partial H}{\partial P_\alpha}\,, \quad \dot{P_\alpha} = -\frac{\partial H}{\partial q_\alpha} \quad (\alpha = 1, 2, 3)\,, \tag{1.44}$$

where $H(P, q)$ is the Hamiltonian of the system under consideration, q_α and P_α are the generalized coordinates and momemta, respectively.

Let us substitute the variables $\mathbf{r}$ and $\mathbf{v}$ in the Liouville equation (1.9) by the generalized variables $\mathbf{q}$ and $\mathbf{P}$. By doing so and using Equations (1.44), we obtain the following form of the Liouville equation

$$\frac{\partial f}{\partial t} + \nabla_{\mathbf{P}} H \cdot \nabla_{\mathbf{q}} f - \nabla_{\mathbf{q}} H \cdot \nabla_{\mathbf{P}} f = 0 \,. \tag{1.45}$$

Because of symmetry of the last equation, it is convenient here to use the Poisson brackets (see Landau and Lifshitz, *Mechanics*, 1976, Chapter 7, § 42). Recall that the Poisson brackets for arbitrary quantities A and B are defined to be

$$[\, A \,,\, B \,] = \sum_{\alpha=1}^{3} \left(\frac{\partial A}{\partial q_\alpha} \frac{\partial B}{\partial P_\alpha} - \frac{\partial A}{\partial P_\alpha} \frac{\partial B}{\partial q_\alpha} \right) . \tag{1.46}$$

Appling definition (1.46) to Equation (1.45), we find the final form of the Liouville theorem

$$\boxed{\frac{\partial f}{\partial t} + [\, f \,,\, H \,] = 0 \,.} \tag{1.47}$$

Q.e.d. Note that for a system in equilibrium

$$[\, f \,,\, H \,] = 0 \,. \tag{1.48}$$

Exercise 1.3 [Section 1.1.2] Discuss what to do with the Liouville theorem, if it is impossible to disregard quantum indeterminacy and assume that the classical description of a system is justified. Consider the case of dense fluids inside stars, for example, white dwarfs.

Comment. Inside a white dwarf star the temperature $T \sim 10^5$ K, but the density is very high: $n \sim 10^{28} - 10^{30}$ cm^{-3} (e.g., de Martino et al., 2003). The electrons cannot be regarded as classical particles. We have to consider them as a quantum system with a Fermi-Dirac distribution (see § 57 in Landau and Lifshitz, *Statistical Physics*, 1959b; Kittel, 1995).

Exercise 1.4 [Section 1.1.2] Recall the Liouville theorem in a course of mechanics – the phase volume is independent of t, i.e. it is the invariant of motion (e.g., Landau and Lifshitz, *Mechanics*, 1976, Chapter 7, § 46). Show that this formulation is equivalent to Equation (1.11).

Exercise 1.5 [Section 1.2.1] Show that Maxwell's equations imply the continuity equation for the electric charge.

Answer. Operating with the divergence on Equation (1.24), we have

$$0 = \frac{4\pi}{c} \, \mathrm{div}\, \mathbf{j} + \frac{1}{c} \frac{\partial}{\partial t} \, \mathrm{div}\, \mathbf{E} \,.$$

Substituting (1.27) in this equation gives us the continuity equation for the electric charge

$$\frac{\partial}{\partial t}\rho^{q} + \operatorname{div}\mathbf{j} = 0\,. \tag{1.49}$$

Thus Maxwell's equations conform to the charge continuity equation.

Exercise 1.6 [Section 1.2.1] Starting from Maxwell's equation, derive the energy conservation law for an electromagnetic field.

Answer. Let us multiply Equation (1.24) by the electric field vector $\mathbf{E}$ and add it to Equation (1.25) multiplied by the magnetic field vector $\mathbf{B}$. The result is

$$\frac{1}{c}\,\mathbf{E}\,\frac{\partial\mathbf{E}}{\partial t} + \frac{1}{c}\,\mathbf{B}\,\frac{\partial\mathbf{B}}{\partial t} = -\frac{4\pi}{c}\,\mathbf{j}\mathbf{E} - (\,\mathbf{B}\operatorname{curl}\mathbf{E} - \mathbf{E}\operatorname{curl}\mathbf{B}\,)\,.$$

By using the known formula from vector analysis

$$\operatorname{div}\,[\,\mathbf{a}\times\mathbf{b}\,] = \mathbf{b}\operatorname{curl}\mathbf{a} - \mathbf{a}\operatorname{curl}\mathbf{b}\,,$$

we rewrite the last equation as follows

$$\frac{1}{2c}\,\frac{\partial}{\partial t}\left(E^2 + B^2\right) = -\frac{4\pi}{c}\,\mathbf{j}\mathbf{E} - \operatorname{div}\,[\,\mathbf{E}\times\mathbf{B}\,]$$

or

$$\boxed{\frac{\partial}{\partial t}\,W = -\mathbf{j}\mathbf{E} - \operatorname{div}\mathbf{G}\,.} \tag{1.50}$$

Here

$$W = \frac{E^2 + B^2}{8\pi} \tag{1.51}$$

is the energy of electromagnetic field in a unit volume of space;

$$\mathbf{G} = \frac{c}{4\pi}\,[\,\mathbf{E}\times\mathbf{B}\,] \tag{1.52}$$

is the flux of electromagnetic field energy through a unit surface in space, i.e. the energy flux density for electromagnetic field. This is called the Poynting vector.

The first term on the right-hand side of Equation (1.50) is the power of work done by the electric field on all the charged particles in the unit volume of space. In the simplest approximation

$$e\,\mathbf{v}\,\mathbf{E} = \frac{d}{dt}\,\mathcal{E}\,, \tag{1.53}$$

where $\mathcal{E}$ is the particle kinetic energy (see Equation (5.6)). Hence instead of Equation (1.50) we write the following form of the energy conservation law:

$$\frac{\partial}{\partial t}\left(\frac{E^2 + B^2}{8\pi} + \frac{\rho v^2}{2}\right) + \operatorname{div}\left(\frac{c}{4\pi}\,[\,\mathbf{E}\times\mathbf{B}\,]\right) = 0\,. \tag{1.54}$$

Compare this simple approach to the energy conservation law for charged particles and an electromagnetic field with the more general situation considered in Section 12.1.3.

Exercise 1.7 [Section 1.2.2] Clarify the meaning of the right-hand side of Equation (1.33).

Answer. Substitute definition (1.30) of the delta-function in definition (1.28) of the electric charge density and differentiate the result over time t:

$$\frac{\partial \rho^{q}}{\partial t} = \sum_{i=1}^{N} e_i \sum_{\alpha=1}^{3} \left[\frac{\partial}{\partial \left(r_\alpha - r_\alpha^i(t)\right)} \prod_{\beta=1}^{3} \delta\left(r_\beta - r_\beta^i(t)\right) \right] \frac{\partial}{\partial t} \left(r_\alpha - r_\alpha^i(t)\right) =$$

$$= -\sum_{i=1}^{N} e_i \sum_{\alpha=1}^{3} \left[\frac{\partial}{\partial \left(r_\alpha - r_\alpha^i(t)\right)} \prod_{\beta=1}^{3} \delta\left(r_\beta - r_\beta^i(t)\right) \right] \frac{dr_\alpha^i(t)}{dt}. \qquad (1.55)$$

This is the right-hand side of Equation (1.33).

Chapter 2

Statistical Description of Interacting Particle Systems

In a system which consists of many interacting particles, the statistical mechanism of 'mixing' in phase space works and makes the system's behaviour *on average* more simple.

2.1 The averaging of Liouville's equation

2.1.1 Averaging over phase space

As was shown in the first Chapter, the exact state of a system consisting of N interacting particles can be given by the *exact* distribution function (see definition (1.18)) in six-dimensional (6D) phase space $X = \{\mathbf{r}, \mathbf{v}\}$. This is defined as the sum of δ-functions in N points of phase space:

$$\hat{f}(\mathbf{r}, \mathbf{v}, t) = \sum_{i=1}^{N} \delta(\mathbf{r} - \mathbf{r}_i(t))\, \delta(\mathbf{v} - \mathbf{v}_i(t)). \tag{2.1}$$

Instead of the equations of motion, we use Liouville's equation to describe the change of the system state (Section 1.1.5):

$$\frac{\partial \hat{f}}{\partial t} + \mathbf{v} \cdot \nabla_{\mathbf{r}} \hat{f} + \frac{\mathbf{F}}{m} \cdot \nabla_{\mathbf{v}} \hat{f} = 0. \tag{2.2}$$

Once the exact initial state of all the particles is known, it can be represented by N points in the phase space X (Figure 2.1). The motion of these

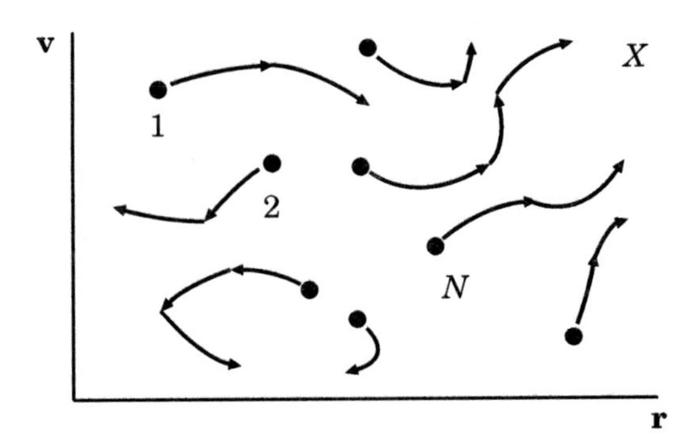

Figure 2.1: Particle trajectories in the 6D phase space X.

points is described by Liouville's equation (1.9) or by the $6N$ equations of motion (1.22).

In fact we usually know only some average characteristics of the system's state, such as the temperature, density, etc. Moreover the behaviour of each single particle is in general of no interest. For this reason, instead of the exact distribution function (2.1), let us introduce the distribution function averaged over a small volume ΔX of phase space, i.e. over a small interval of coordinates $\Delta \mathbf{r}$ and velocities $\Delta \mathbf{v}$ centered at the point $(\mathbf{r}, \mathbf{v})$, at a moment of time t:

$$\langle\, \hat{f}(\mathbf{r}, \mathbf{v}, t)\, \rangle_X = \frac{1}{\Delta X} \int\limits_{\Delta X} \hat{f}(X, t)\, dX =$$

$$= \frac{1}{\Delta \mathbf{r}\, \Delta \mathbf{v}} \int\limits_{\Delta \mathbf{r}\, \Delta \mathbf{v}} \hat{f}(\mathbf{r}, \mathbf{v}, t)\, d^3\mathbf{r}\, d^3\mathbf{v}\,. \tag{2.3}$$

Here $d^3\mathbf{r} = dx\, dy\, dz$ and $d^3\mathbf{v} = dv_x\, dv_y\, dv_z$, if use is made of Cartesian coordinates.

To put the same in another way, the mean number of particles present at a moment of time t in the element of phase volume ΔX is

$$\langle\, \hat{f}(\mathbf{r}, \mathbf{v}, t)\, \rangle_X \cdot \Delta X = \int\limits_{\Delta X} \hat{f}(\mathbf{r}, \mathbf{v}, t)\, dX\,.$$

The total number N of particles in the system is the integral over the whole phase space X.

Obviously the distribution function averaged over phase volume differs from the exact one as shown in Figure 2.2.

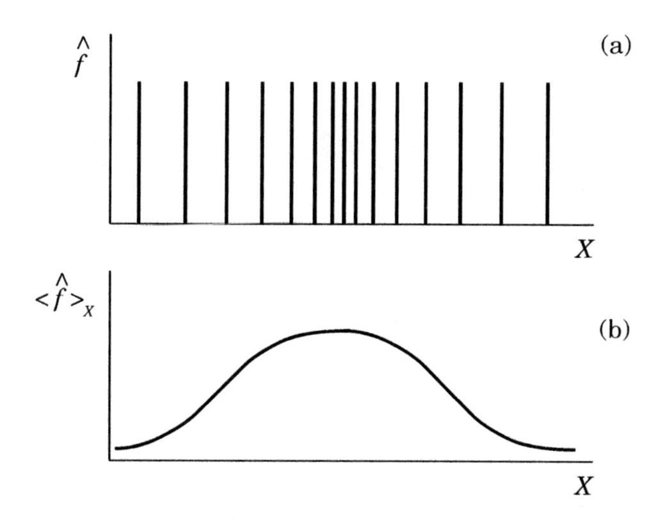

Figure 2.2: The one-dimensional analogy of the distribution function averaging over phase space X: (a) the exact distribution function (2.1), (b) the averaged function (2.3).

2.1.2 Two statistical postulates

Let us average the same exact distribution function (2.1) over a small time interval Δt centred at a moment of time t:

$$\langle \hat{f}(\mathbf{r}, \mathbf{v}, t) \rangle_t = \frac{1}{\Delta t} \int\limits_{\Delta t} \hat{f}(\mathbf{r}, \mathbf{v}, t)\, dt \,. \tag{2.4}$$

Here Δt is small in comparison with the characteristic time of the system's evolution:

$$\Delta t \ll \tau_{ev} \,. \tag{2.5}$$

We assume that the following *two statistical postulates* concerning systems containing a large number of particles are applicable to the system considered.

The first postulate. The mean values $\langle \hat{f} \rangle_X$ and $\langle \hat{f} \rangle_t$ exist for sufficiently small ΔX and Δt and are *independent* of the averaging scales ΔX and Δt.

Clearly the first postulate implies that the number of particles should be large. For a small number of particles the mean value depends upon the averaging scale: if, for instance, $N = 1$ then the exact distribution function (2.1) is simply a δ-function, and the average over the variable X is $\langle \hat{f} \rangle_X = 1/\Delta X$. For illustration, the case $(\Delta X)_1 > \Delta X$ is shown in Figure 2.3.

The second postulate is

$$\langle \hat{f}(X, t) \rangle_X = \langle \hat{f}(X, t) \rangle_t = f(X, t) \,. \tag{2.6}$$

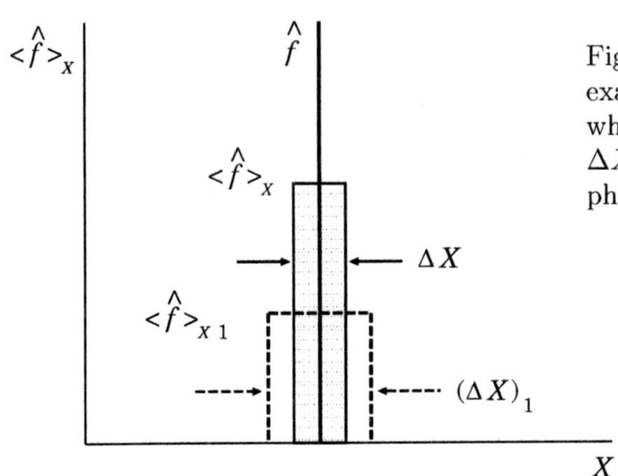

Figure 2.3: Averaging of the exact distribution function $\hat{f}$ which is equal to a δ-function. ΔX is a small volume of phase space X.

In other words, the averaging of the distribution function over phase space is *equivalent* to the averaging over time.

While speaking of the small ΔX and Δt, we assume that they are not too small: ΔX must contain a reasonably large number of particles while Δt must be large in comparison with the duration of drastic changes of the exact distribution function, such as the duration of the particle Coulomb collisions:

$$\Delta t \gg \tau_c \, . \tag{2.7}$$

It is in this case that the statistical mechanism of particle 'mixing' in phase space is at work and

> the averaging of the exact distribution function over the time Δt is equivalent to the averaging over the phase volume ΔX.

2.1.3 A statistical mechanism of mixing in phase space

Let us understand qualitatively how the mixing mechanism works in phase space. We start from the dynamical description of the N-particle system in $6N$-dimensional phase space in which

$$\Gamma = \{ \mathbf{r}_i, \, \mathbf{v}_i \} \, , \qquad i = 1, \, 2, \dots \, N,$$

a point is determined ($t = 0$ in Figure 2.4) by the initial conditions of all the particles. The motion of this point, that is the dynamical evolution of the system, can be described by Liouville's equation or equations of motion. The point moves along a complicated *dynamical trajectory* because the interactions in a many-particle system are extremely intricate and complicated.

The dynamical trajectory has a remarkable property which we shall illustrate by the following example. Imagine a glass vessel containing a gas consisting of a large number N of particles (molecules or charged particles). The state of this gas at any moment of time is depicted by a single point in the phase space Γ.

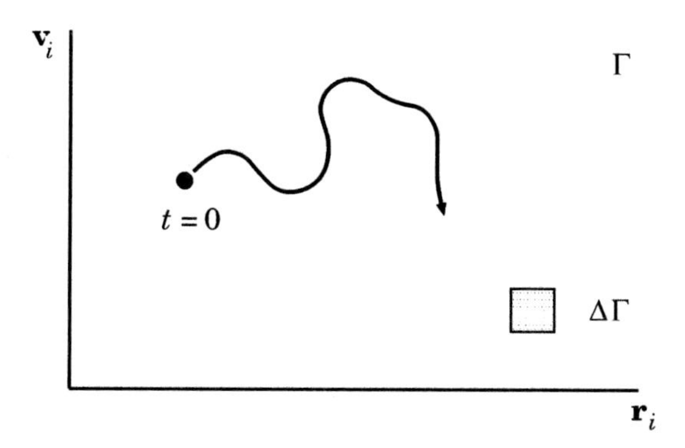

Figure 2.4: The dynamical trajectory of a system of N particles in the $6N$-dimensional phase space Γ.

Let us imagine another vessel which is identical to the first one, with one exception, being that at any moment of time the gas state in the second vessel is different from that in the first one. These states are depicted by two different points in the space Γ. For example, at $t = 0$, they are points 1 and 2 in Figure 2.5.

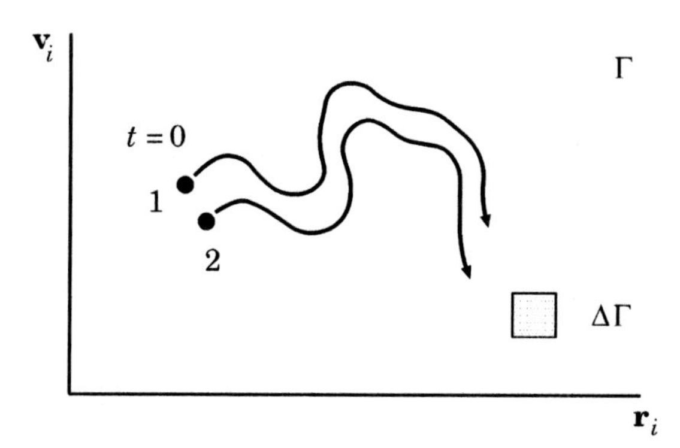

Figure 2.5: The trajectories of two systems never cross each other.

With the passage of time, the gas states in both vessels change, whereas the two points in the space Γ draw two different dynamical trajectories (Figure 2.5). These trajectories *do not* intersect. If they had intersected at just one point, then the state of the first gas, determined by $6N$ numbers $(\mathbf{r}_i, \mathbf{v}_i)$, would have coincided with the state of the second gas. These numbers could have been taken as the initial conditions which, in turn, would have uniquely determined the motion. The two trajectories would have merged into one. For the same reason the trajectory of a system cannot intersect itself. Thus we come to the conclusion that only one dynamical trajectory of a many particle system passes through each point of the phase space Γ.

Since the trajectories differ in initial conditions, we can introduce an infinite ensemble of systems (glass vessels) corresponding to the different initial conditions. In a finite time the ensemble of dynamical trajectories will closely fill the phase space Γ, without intersections. By averaging over the ensemble we can answer the question of what the probability is that, at a moment of time t, the system will be found in an element $\Delta\Gamma = \Delta\mathbf{r}_i \, \Delta\mathbf{v}_i$ of the phase space Γ:

$$dw = \langle \hat{f}(\mathbf{r}_i, \mathbf{v}_i) \rangle_{\Gamma} \, d\Gamma . \tag{2.8}$$

Here $\langle \hat{f}(\mathbf{r}_i, \mathbf{v}_i) \rangle_{\Gamma}$ is a function of all the coordinates and velocities. It plays the role of the *probability distribution density* in the phase space Γ and is called the statistical distribution function or simply the distribution function. It is obtained by way of statistical averaging over the ensemble and evidently corresponds to definition (2.3).

* * *

It is rather obvious that the same *probability density* can be obtained in another way – through the averaging over time. The dynamical trajectory of a system, given a sufficient time Δt, will closely cover phase space. There will be no self-intersections; but since the trajectory is very intricate it will repeatedly pass through the phase space element $\Delta\Gamma$. Let $(\Delta t)_{\Gamma}$ be the time during which the system locates in $\Delta\Gamma$. For a sufficiently large Δt, which is formally restricted by the characteristic time of slow evolution of the system as a whole, the ratio $(\Delta t)_{\Gamma}/\Delta t$ tends to the limit

$$\lim_{\Delta t \to \infty} \frac{(\Delta t)_{\Gamma}}{\Delta t} = \frac{dw}{d\Gamma} = \langle \hat{f}(\mathbf{r}_i, \mathbf{v}_i, t) \rangle_t . \tag{2.9}$$

By virtue of the role of the probability density, it is clear that

the statistical averaging over the ensemble (2.8) is equivalent to the averaging over time (2.9) as well as to the definition (2.4).

2.1.4 The derivation of a general kinetic equation

Now we have everything what we need to average the exact Liouville Equation (2.2). Since the equation contains the derivatives with respect to time t and phase-space coordinates $(\mathbf{r}, \mathbf{v})$ the procedure of averaging over the interval $\Delta X \, \Delta t$ is defined as follows:

$$f(X,t) = \frac{1}{\Delta X \, \Delta t} \int\limits_{\Delta X} \int\limits_{\Delta t} \hat{f}(X,t) \, dX \, dt . \tag{2.10}$$

Averaging the first term of the Liouville equation gives

$$\frac{1}{\Delta X\,\Delta t}\int\limits_{\Delta X}\int\limits_{\Delta t}\frac{\partial \hat{f}}{\partial t}\,dX\,dt = \frac{1}{\Delta t}\int\limits_{\Delta t}\frac{\partial}{\partial t}\left[\frac{1}{\Delta X}\int\limits_{\Delta X}\hat{f}\,dX\right]dt =$$

$$= \frac{1}{\Delta t}\int\limits_{\Delta t}\frac{\partial}{\partial t}f\,dt = \frac{\partial f}{\partial t}\,. \tag{2.11}$$

In the last equality the use is made of the fact that, by virtue of the second postulate of statistics (2.6), the averaging of the smooth averaged function does not change it.

Let us average the second term in Equation (2.2):

$$\frac{1}{\Delta X\,\Delta t}\int\limits_{\Delta X}\int\limits_{\Delta t}v_\alpha\frac{\partial \hat{f}}{\partial r_\alpha}\,dX\,dt = \frac{1}{\Delta X}\int\limits_{\Delta X}v_\alpha\frac{\partial}{\partial r_\alpha}\left[\frac{1}{\Delta t}\int\limits_{\Delta t}\hat{f}\,dt\right]dX =$$

$$= \frac{1}{\Delta X}\int\limits_{\Delta X}v_\alpha\frac{\partial}{\partial r_\alpha}f\,dX = v_\alpha\frac{\partial f}{\partial r_\alpha}\,. \tag{2.12}$$

Here the index $\alpha = 1, 2, 3$.

In order to average the term containing the force $\mathbf{F}$, let us represent it as a sum of a *mean force* $\langle \mathbf{F} \rangle$ and the force due to the difference of the real force field from the mean (smooth) one:

$$\mathbf{F} = \langle \mathbf{F} \rangle + \mathbf{F}'. \tag{2.13}$$

Substituting definition (2.13) in the third term in Equation (2.2) and averaging this term, we have

$$\frac{1}{\Delta X\,\Delta t}\int\limits_{\Delta X}\int\limits_{\Delta t}\frac{F_\alpha}{m}\frac{\partial \hat{f}}{\partial v_\alpha}\,dX\,dt =$$

$$= \frac{\langle F_\alpha \rangle}{m}\frac{1}{\Delta X}\int\limits_{\Delta X}\frac{\partial}{\partial v_\alpha}\left[\frac{1}{\Delta t}\int\limits_{\Delta t}\hat{f}\,dt\right]dX + \frac{1}{\Delta X\,\Delta t}\int\limits_{\Delta X}\int\limits_{\Delta t}\frac{F'_\alpha}{m}\frac{\partial \hat{f}}{\partial v_\alpha}\,dX\,dt =$$

$$= \frac{\langle F_\alpha \rangle}{m}\frac{\partial f}{\partial v_\alpha} + \frac{1}{\Delta X\,\Delta t}\int\limits_{\Delta X}\int\limits_{\Delta t}\frac{F'_\alpha}{m}\frac{\partial \hat{f}}{\partial v_\alpha}\,dX\,dt\,. \tag{2.14}$$

Gathering all three terms together, we write the averaged Liouville equation in the form

$$\boxed{\frac{\partial f}{\partial t} + \mathbf{v}\cdot\nabla_{\mathbf{r}} f + \frac{\langle \mathbf{F} \rangle}{m}\cdot\nabla_{\mathbf{v}} f = \left(\frac{\partial \hat{f}}{\partial t}\right)_{\mathrm{c}},} \tag{2.15}$$

where

$$\left(\frac{\partial \hat{f}}{\partial t} \right)_{\mathrm{c}} = - \frac{1}{\Delta X \, \Delta t} \int\limits_{\Delta X} \int\limits_{\Delta t} \frac{F'_{\alpha}}{m} \frac{\partial \hat{f}}{\partial v_{\alpha}} \, dX \, dt \, . \tag{2.16}$$

Equation (2.15) and its right-hand side (2.16) are called the *kinetic equation* and the *collisional integral* (cf. definition (1.17)), respectively.

Therefore we have found the *most general* form of the kinetic equation with a collisional integral, which is nice but cannot be directly used in plasma astrophysics, without making some additional simplifying assumptions. The main assumption, the binary character of collisions, will be taken into account in the next Section, see also Section 3.3.

2.2 A collisional integral and correlation functions

2.2.1 Binary interactions

We shall distinguish different kinds of particles, for example, electrons and protons, because their behaviours differ. Let $\hat{f}_k\left(\mathbf{r}, \mathbf{v}, t\right)$ be the exact distribution function (2.1) of particles of the *kind* k, i.e.

$$\hat{f}_k\left(\mathbf{r}, \mathbf{v}, t\right) = \sum_{i=1}^{N_k} \delta\left(\mathbf{r} - \mathbf{r}_{ki}(t)\right) \delta\left(\mathbf{v} - \mathbf{v}_{ki}(t)\right), \tag{2.17}$$

the index i denoting the ith particle of kind k, N_k being the number of particles of kind k. The Liouville Equation (2.2) for the particles of kind k takes a view

$$\frac{\partial \hat{f}_k}{\partial t} + \mathbf{v} \cdot \nabla_{\mathbf{r}} \hat{f}_k + \frac{\hat{\mathbf{F}}_k}{m_k} \cdot \nabla_{\mathbf{v}} \hat{f}_k = 0 \, , \tag{2.18}$$

m_k is the mass of a particle of kind k.

The force acting on a particle of kind k at a point $(\mathbf{r}, \mathbf{v})$ of the phase space X at a moment of time t, $\hat{F}_{k,\alpha}\left(\mathbf{r}, \mathbf{v}, t\right)$, is the sum of forces acting on this particle from all other particles:

$$\hat{F}_{k,\alpha}\left(\mathbf{r}, \mathbf{v}, t\right) = \sum_{l} \sum_{i=1}^{N_l} \hat{F}^{(i)}_{kl,\alpha}\left(\mathbf{r}, \mathbf{v}, \mathbf{r}_{li}(t), \mathbf{v}_{li}(t)\right) . \tag{2.19}$$

So the total force $\hat{F}_{k,\alpha}\left(\mathbf{r}, \mathbf{v}, t\right)$ depends upon the instant positions and velocities (generally with the time delay taken into account) of all the particles and can be written with the help of the exact distribution function as follows:

$$\hat{F}_{k,\alpha}\left(\mathbf{r}, \mathbf{v}, t\right) = \sum_{l} \int\limits_{X_1} \hat{F}_{kl,\alpha}\left(X, X_1\right) \hat{f}_l\left(X_1, t\right) dX_1 \, . \tag{2.20}$$

Here

$$\hat{f}_l(X,t) = \sum_{i=1}^{N_l} \delta(X - X_{li}(t))$$

is the exact distribution function of particles of kind l, the variable of integration is designated as $X_1 = \{\mathbf{r}_1, \mathbf{v}_1\}$ and $dX_1 = d^3\mathbf{r}_1\, d^3\mathbf{v}_1$.

Formula (2.20) takes into account that the forces considered are *binary* ones, i.e. they can be represented as a sum of interactions between two particles.

Making use of the representation (2.20), let us average the force term in the Liouville equation (2.2), as this has been done in (2.14). We have

$$\frac{1}{\Delta X\, \Delta t} \int\limits_{\Delta X} \int\limits_{\Delta t} \frac{1}{m_k} \hat{F}_{k,\alpha}(\mathbf{r}, \mathbf{v}, t)\, \frac{\partial \hat{f}_k}{\partial v_\alpha}\, dX\, dt =$$

$$= \frac{1}{\Delta X\, \Delta t} \int\limits_{\Delta X} \int\limits_{\Delta t} \sum_l \int\limits_{X_1} \frac{1}{m_k} \hat{F}_{kl,\alpha}(X, X_1)\, \hat{f}_l(X_1, t)\, \frac{\partial}{\partial v_\alpha} \hat{f}_k(X,t)\, dX\, dX_1\, dt =$$

$$= \frac{1}{\Delta X} \int\limits_{\Delta X} \sum_l \int\limits_{X_1} \frac{1}{m_k} \hat{F}_{kl,\alpha}(X, X_1) \times$$

$$\times \frac{\partial}{\partial v_\alpha} \left[\frac{1}{\Delta t} \int\limits_{\Delta t} \hat{f}_k(X,t)\, \hat{f}_l(X_1,t)\, dt \right] dX\, dX_1\,. \tag{2.21}$$

Here we have taken into account that the exact distribution function $\hat{f}_l(X_1,t)$ is independent of the velocity $\mathbf{v}$, which is a part of the variable $X = \{\mathbf{r}, \mathbf{v}\}$ related to the particles of the kind k, and that the interaction law $\hat{F}_{kl,\alpha}(X, X_1)$ is explicitly independent of time t.

Formula (2.21) contains the *pair products* of exact distribution functions of different particle kinds, as is natural for the case of *binary interactions*.

2.2.2 Binary correlation

Let us represent the exact distribution function $\hat{f}_k$ as

$$\hat{f}_k(X,t) = f_k(X,t) + \hat{\varphi}_k(X,t)\,, \tag{2.22}$$

where $f_k(X,t)$ is the *statistically averaged* distribution function, $\hat{\varphi}_k(X,t)$ is the deviation of the exact distribution function from the averaged one. In general the deviation is not small, of course. It is obvious that, according to definition (2.22),

$$\hat{\varphi}_k(X,t) = \hat{f}_k(X,t) - f_k(X,t)\,;$$

hence

$$\langle\, \hat{\varphi}_k(X,t) \,\rangle = 0\,. \tag{2.23}$$

Let us consider the integrals of pair products, appearing in the averaged force term (2.21). In view of definition (2.22), they can be rewritten as

$$\frac{1}{\Delta t} \int\limits_{\Delta t} \hat{f}_k (X, t) \, \hat{f}_l (X_1, t) \, dt = f_k (X, t) \, f_l (X_1, t) + f_{kl} (X, X_1, t) \,, \qquad (2.24)$$

where

$$f_{kl} (X, X_1, t) = \frac{1}{\Delta t} \int\limits_{\Delta t} \hat{\varphi}_k (X, t) \, \hat{\varphi}_l (X_1, t) \, dt \,. \qquad (2.25)$$

The function f_{kl} is referred to as the *correlation function* or, more exactly, the *binary* correlation function.

The physical meaning of the correlation function is clear from (2.24). The left-hand side of Equation (2.24) means the probability to find a particle of kind k at a point X of the phase space at a moment of time t *under condition* that a particle of kind l places at a point X_1 at the same time. In the right-hand side of (2.24) the distribution function $f_k (X, t)$ characterizes the probability that a particle of kind k stays at a point X at a moment of time t. The function $f_l (X_1, t)$ plays the analogous role for the particles of kind l.

If the particles of kind k did not interact with those of kind l, then their distributions would be independent, i.e. probability densities would simply multiply:

$$\langle \, \hat{f}_k (X, t) \;\, \hat{f}_l (X_1, t) \, \rangle = f_k (X, t) \;\, f_l (X_1, t) \,. \qquad (2.26)$$

So in the right-hand side of Equation (2.24) there should be

$$f_{kl} (X, X_1, t) = 0 \,. \qquad (2.27)$$

In other words there would be no correlation in the particle distribution.

With the proviso that the parameter characterizing the binary interaction, namely Coulomb collision considered below,

$$\zeta_{\rm i} \approx \frac{e^2}{\langle l \rangle} \Bigg/ \left\langle \frac{mv^2}{2} \right\rangle , \qquad (2.28)$$

is small under conditions in a wide range, the correlation function must be *relatively small*:

if the interaction is weak, the second term in the right-hand side of (2.24) must be small in comparison with the first one.

We shall come back to the discussion of this property in Section 3.1. This fundamental property allows us to construct a theory of plasma in many cases of astrophysical interest.

2.2.3 The collisional integral and binary correlation

Now let us substitute (2.24) in formula (2.21) for the averaged force term:

$$\frac{1}{\Delta X\,\Delta t}\int\limits_{\Delta X}\int\limits_{\Delta t}\frac{1}{m_k}\hat{F}_{k,\alpha}(X,t)\,\frac{\partial\hat{f}_k}{\partial v_\alpha}\,dX\,dt =$$

$$=\frac{1}{\Delta X}\int\limits_{\Delta X}\sum_l\int\limits_{X_1}\frac{1}{m_k}\hat{F}_{kl,\alpha}(X,X_1)\,\frac{\partial}{\partial v_\alpha}\,[\,f_k(X,t)\,f_l(X_1,t)\,+$$

$$+\,f_{kl}(X,X_1,t)\,]\,dX\,dX_1 =$$

since $f_k(X,t)$ is a smooth fuction, its derivative over v_α can be brought out of the averaging procedure:

$$=\left[\frac{\partial}{\partial v_\alpha}\,f_k(X,t)\right]\left\{\frac{1}{\Delta X}\int\limits_{\Delta X}\sum_l\int\limits_{X_1}\frac{1}{m_k}\hat{F}_{kl,\alpha}(X,X_1)\,f_l(X_1,t)\,dX\,dX_1\right\}+$$

$$+\frac{1}{\Delta X}\int\limits_{\Delta X}\sum_l\int\limits_{X_1}\frac{1}{m_k}\hat{F}_{kl,\alpha}(X,X_1)\,\frac{\partial}{\partial v_\alpha}\,f_{kl}(X,X_1,t)\,dX\,dX_1 =$$

$$=\frac{1}{m_k}\,F_{k,\alpha}(X,t)\,\frac{\partial f_k(X,t)}{\partial v_\alpha}+$$

$$+\sum_l\int\limits_{X_1}\frac{1}{m_k}\,F_{kl,\alpha}(X,X_1)\,\frac{\partial f_{kl}(X,X_1,t)}{\partial v_\alpha}\,dX_1\,. \qquad (2.29)$$

Here we have taken into account that the averaging of smooth functions does not change them, and the following definition of the *averaged force* is used:

$$F_{k,\alpha}(X,t)=\frac{1}{\Delta X}\int\limits_{\Delta X}\sum_l\int\limits_{X_1}\hat{F}_{kl,\alpha}(X,X_1)\;f_l(X_1,t)\;dX\,dX_1 =$$

$$=\sum_l\int\limits_{X_1}F_{kl,\alpha}(X,X_1)\,f_l(X_1,t)\,dX_1\,. \qquad (2.30)$$

This definition coincides with the previous definition (2.14) of the average force, since

> all the deviations of the real force $\hat{\mathbf{F}}_k$ from the mean (smooth) force $\mathbf{F}_k$ are taken care of in the deviations $\hat{\varphi}_k$ and $\hat{\varphi}_l$ of the real distribution functions $\hat{f}_k$ and $\hat{f}_l$ from their mean values f_k and f_l.

Thus the collisional integral can be represented in the form

$$\left(\frac{\partial \hat{f}_k}{\partial t}\right)_c = -\sum_l \int_{X_1} \frac{1}{m_k} F_{kl,\alpha}(X, X_1) \frac{\partial f_{kl}(X, X_1, t)}{\partial v_\alpha} dX_1 . \tag{2.31}$$

Moreover, if in the last term of (2.29) the binary interactions can be represented by smooth functions of the type $e_k e_l (|\mathbf{r}_k - \mathbf{r}_l|)^{-2}$ with account of the Debye shielding (Sections 3.2 and 8.2), then formally the velocity dependence may be neglected.

Let us recall an important particular case considered in Section 1.1. For the Lorentz force (1.13) as well as for the gravitational one (1.41), the condition (1.7) is satisfied. Let us require that in formula (2.31)

$$\frac{\partial}{\partial v_\alpha} F_{kl,\alpha}(X, X_1) = 0 . \tag{2.32}$$

In fact this condition was tacitly assumed from the early beginning, from Equation (2.2). Anyway, in the case (2.32), we obtain from formula (2.31) the following expession

$$\left(\frac{\partial \hat{f}_k}{\partial t}\right)_c = -\frac{\partial}{\partial v_\alpha} \sum_l \int_{X_1} \frac{1}{m_k} F_{kl,\alpha}(X, X_1)\, f_{kl}(X, X_1, t)\, dX_1 . \tag{2.33}$$

Hence the collisional integral, at least, for the Coulomb and gravity forces can be written in the divergent form in the velocity space $\mathbf{v}$:

$$\boxed{\left(\frac{\partial \hat{f}_k}{\partial t}\right)_c = -\frac{\partial}{\partial v_\alpha} J_{k,\alpha} ,} \tag{2.34}$$

where the flux of particles of kind k in the velocity space (cf. Figure 1.3b) is

$$J_{k,\alpha}(X, t) = \sum_l \int_{X_1} \frac{1}{m_k} F_{kl,\alpha}(X, X_1)\, f_{kl}(X, X_1, t)\, dX_1 . \tag{2.35}$$

Therefore we arrive to conclusion that the averaged Liouville equation or **the kinetic equation for particles of kind** k

$$\frac{\partial f_k(X,t)}{\partial t} + v_\alpha \frac{\partial f_k(X,t)}{\partial r_\alpha} + \frac{F_{k,\alpha}(X,t)}{m_k} \frac{\partial f_k(X,t)}{\partial v_\alpha} =$$

$$= -\frac{\partial}{\partial v_\alpha} \sum_l \int_{X_1} \frac{1}{m_k} F_{kl,\alpha}(X, X_1)\, f_{kl}(X, X_1, t)\, dX_1 \tag{2.36}$$

contains the *unknown* function f_{kl}. Hence the kinetic equation (2.36) for distribution function f_k is not closed. We have to find the equation for the correlation function f_{kl}. This will be done in the next Section.

2.3 Equations for correlation functions

To derive the equations for correlation functions (in the first place for the function of pair correlations f_{kl}), it is not necessary to introduce any new postulates or develop new formalisms. All the necessary equations and averaging procedures are at hand.

Looking at definition (2.25), we see that we need an equation which will describe the deviation of distribution function from its mean value, i.e. the function $\hat{\varphi}_k = \hat{f}_k - f_k$. In order to derive such equation, we simply have to subtract the averaged representation (2.36) from the exact Liouville equation (2.2). The result is

$$\frac{\partial \hat{\varphi}_k(X,t)}{\partial t} + v_\alpha \frac{\partial \hat{\varphi}_k(X,t)}{\partial r_\alpha} + \frac{\hat{F}_{k,\alpha}}{m_k}\frac{\partial \hat{f}_k}{\partial v_\alpha} - \frac{F_{k,\alpha}}{m_k}\frac{\partial f_k}{\partial v_\alpha} =$$

$$= \frac{\partial}{\partial v_\alpha}\sum_l \int_{X_1} \frac{1}{m_k} F_{kl,\alpha}(X,X_1)\, f_{kl}(X,X_1)\, dX_1 . \tag{2.37}$$

Here

$$\hat{F}_{k,\alpha}(X,t) = \sum_l \int_{X_1} F_{kl,\alpha}(X,X_1)\, \hat{f}_l(X_1,t)\, dX_1 \tag{2.38}$$

is the *exact* force (2.20) acting on a particle of the kind k at the point X of phase space, and

$$F_{k,\alpha}(X,t) = \sum_l \int_{X_1} F_{kl,\alpha}(X,X_1)\, f_l(X_1,t)\, dX_1 \tag{2.39}$$

is the statistically *averaged* force (2.30).

Thus the difference between the exact force and the averaged one is

$$\hat{F}_{k,\alpha} - F_{k,\alpha} = \sum_l \int_{X_1} F_{kl,\alpha}(X,X_1)\, \hat{\varphi}_l(X_1,t)\, dX_1 . \tag{2.40}$$

We substitute it in Equation (2.37) where the difference of force terms can be rewritten as follows:

$$\frac{\hat{F}_{k,\alpha}}{m_k}\frac{\partial \hat{f}_k}{\partial v_\alpha} - \frac{F_{k,\alpha}}{m_k}\frac{\partial f_k}{\partial v_\alpha} = \frac{\hat{F}_{k,\alpha} - F_{k,\alpha}}{m_k}\frac{\partial f_k}{\partial v_\alpha} + \frac{\hat{F}_{k,\alpha}}{m_k}\frac{\partial \hat{\varphi}_k}{\partial v_\alpha} .$$

The result of the substitution is

$$\frac{\hat{F}_{k,\alpha}}{m_k}\frac{\partial \hat{f}_k}{\partial v_\alpha} - \frac{F_{k,\alpha}}{m_k}\frac{\partial f_k}{\partial v_\alpha} =$$

$$= \sum_l \int_{X_1} \frac{1}{m_k} F_{kl,\alpha}(X,X_1)\, \hat{\varphi}_l(X_1,t)\, dX_1 \frac{\partial f_k}{\partial v_\alpha} + \frac{F_{k,\alpha}}{m_k}\frac{\partial \hat{\varphi}_k}{\partial v_\alpha} +$$

$$+ \sum_l \int_{X_1} \frac{1}{m_k} F_{kl,\alpha}(X,X_1)\, \hat{\varphi}_l(X_1,t)\, dX_1 \frac{\partial \hat{\varphi}_k}{\partial v_\alpha} . \tag{2.41}$$

On substituting (2.41) in Equation (2.37) we have the equation for the deviation $\hat{\varphi}_k$ of the exact distribution function $\hat{f}_k$ from its mean value f_k:

$$\frac{\partial \hat{\varphi}_k(X,t)}{\partial t} + v_\alpha \frac{\partial \hat{\varphi}_k(X,t)}{\partial r_\alpha} + \ldots = 0\,. \tag{2.42}$$

Considering that we have to derive the equation for the pair correlation function

$$f_{kl}(X_1, X_2, t) = \langle \hat{\varphi}_k(X_1,t)\, \hat{\varphi}_l(X_2,t) \rangle\,,$$

let us take two equations:
one for $\hat{\varphi}_k(X_1,t)$

$$\frac{\partial \hat{\varphi}_k(X_1,t)}{\partial t} + v_{1,\alpha} \frac{\partial \hat{\varphi}_k(X_1,t)}{\partial r_{1,\alpha}} + \frac{F_{k,\alpha}}{m_k} \frac{\partial \hat{\varphi}_k(X_1,t)}{\partial v_{1,\alpha}} + \ldots = 0 \tag{2.43}$$

and another for $\hat{\varphi}_l(X_2,t)$

$$\frac{\partial \hat{\varphi}_l(X_2,t)}{\partial t} + v_{2,\alpha} \frac{\partial \hat{\varphi}_l(X_2,t)}{\partial r_{2,\alpha}} + \frac{F_{l,\alpha}}{m_l} \frac{\partial \hat{\varphi}_l(X_2,t)}{\partial v_{2,\alpha}} + \ldots = 0\,. \tag{2.44}$$

Now we add the equations resulting from (2.43) multiplied by $\hat{\varphi}_l$ and (2.44) multiplied by $\hat{\varphi}_k$. We obtain

$$\hat{\varphi}_l \frac{\partial \hat{\varphi}_k}{\partial t} + \hat{\varphi}_k \frac{\partial \hat{\varphi}_l}{\partial t} + v_{1,\alpha} \frac{\partial \hat{\varphi}_k}{\partial r_{1,\alpha}} \hat{\varphi}_l + v_{2,\alpha} \frac{\partial \hat{\varphi}_l}{\partial r_{2,\alpha}} \hat{\varphi}_k + \ldots = 0$$

or

$$\frac{\partial (\hat{\varphi}_k \hat{\varphi}_l)}{\partial t} + v_{1,\alpha} \frac{\partial (\hat{\varphi}_k \hat{\varphi}_l)}{\partial r_{1,\alpha}} + v_{2,\alpha} \frac{\partial (\hat{\varphi}_k \hat{\varphi}_l)}{\partial r_{2,\alpha}} + \ldots = 0\,. \tag{2.45}$$

On averaging Equation (2.45) we finally have the equation for the *pair correlation* function in the following form:

$$\frac{\partial f_{kl}(X_1,X_2,t)}{\partial t} + v_{1,\alpha} \frac{\partial f_{kl}(X_1,X_2,t)}{\partial r_{1,\alpha}} + v_{2,\alpha} \frac{\partial f_{kl}(X_1,X_2,t)}{\partial r_{2,\alpha}} +$$

$$+ \frac{F_{k,\alpha}(X_1,t)}{m_k} \frac{\partial f_{kl}(X_1,X_2,t)}{\partial v_{1,\alpha}} + \frac{F_{l,\alpha}(X_2,t)}{m_l} \frac{\partial f_{kl}(X_1,X_2,t)}{\partial v_{2,\alpha}} +$$

$$+ \frac{\partial f_k(X_1,t)}{\partial v_{1,\alpha}} \sum_n \int_{X_3} \frac{1}{m_k} F_{kn,\alpha}(X_1,X_3)\, f_{nl}(X_3,X_2,t)\, dX_3 +$$

$$+ \frac{\partial f_l(X_2,t)}{\partial v_{2,\alpha}} \sum_n \int_{X_3} \frac{1}{m_l} F_{ln,\alpha}(X_2,X_3)\, f_{nk}(X_3,X_1,t)\, dX_3 =$$

$$= - \frac{\partial}{\partial v_{1,\alpha}} \sum_n \int_{X_3} \frac{1}{m_k} F_{kn,\alpha}(X_1,X_3)\, f_{kln}(X_1,X_2,X_3,t)\, dX_3 -$$

$$- \frac{\partial}{\partial v_{2,\alpha}} \sum_n \int_{X_3} \frac{1}{m_l} F_{ln,\alpha}(X_2,X_3)\, f_{kln}(X_1,X_2,X_3,t)\, dX_3\,. \tag{2.46}$$

Here

$$f_{kln}(X_1, X_2, X_3, t) = \frac{1}{\Delta t} \int_{\Delta t} \hat{\varphi}_k(X_1, t)\, \hat{\varphi}_l(X_2, t)\, \hat{\varphi}_n(X_3, t)\, dt \tag{2.47}$$

is the function of *triple correlations* (see also Exercise 2.1).

Thus Equation (2.46) for the pair correlation function contains the *unknown* function of triple correlations. In general,

> the chain of equations for correlation functions can be shown to be *unclosed*: the equation for the correlation function of sth order contains the function of the order $(s+1)$.

2.4 Practice: Exercises and Answers

Exercise 2.1 [Section 2.3] By analogy with formula (2.24), show that

$$\langle \hat{f}_k(X_1, t)\, \hat{f}_l(X_2, t)\, \hat{f}_n(X_3, t) \rangle = \tag{2.48}$$

$$= f_k(X_1, t)\, f_l(X_2, t)\, f_n(X_3, t) +$$

$$+ f_k(X_1, t)\, f_{ln}(X_2, X_3, t) + f_l(X_2, t)\, f_{kn}(X_1, X_3, t) +$$

$$+ f_n(X_3, t)\, f_{kl}(X_1, X_2, t) + f_{kln}(X_1, X_2, X_3, t)\,.$$

Exercise 2.2 Discuss a similarity and difference between the kinetic theory presented in this Chapter and the famous BBGKY hierarchy theory developed by Bogoliubov (1946), Born and Green (1949), Kirkwood (1946), and Yvon (1935).

Hint. Show that essential to both derivations is the weak-coupling assumption, according to which

> grazing encounters, involving small fractional energy and momentum exchange between colliding particles, dominate the evolution of the velocity distribution function.

The weak-coupling assumption provides justification of the widely appreciated practice which leads to a very significant simplification of the original collisional integral; for more detail see Klimontovich (1975, 1986).

Chapter 3

Weakly-Coupled Systems with Binary Collisions

In a system which consists of many interacting particles, the weak-coupling assumption allows us to introduce a well controlled approximation to consider the chain of the equations for correlation functions. This leads to a very significant simplification of the original collisional integral to describe collisional relaxation and transport in astrophysical plasma but not in self-gravitating systems.

3.1 Approximations for binary collisions

3.1.1 The small parameter of kinetic theory

The infinite chain of equations for the distribution function and correlation functions does not contain more information in itself than the initial Liouville equation for the exact distribution function. Actually, the statistical mixing of trajectories in phase space with subsequent statistical smoothing over the physically infinitesimal volume allows to lose 'useless information' – the information about the exact motion of particles. Just for this reason, description of the system's behaviour becomes irreversible.

The value of the chain is also that the chain allows a direct introduction of new physical assumptions which make it possible to break the chain off at some term (Figure 3.1) and to estimate the resulting error. We call this procedure a **well controlled approximation**.

There is no universal way of breaking the chain off. It is intimately related, in particular, to the physical state of a plasma. Different states (as well as different aims) require different approximations. In general, the physical state of a plasma can be characterized, at least partially, by **the ratio of the mean energy of two particle interaction to their mean kinetic**

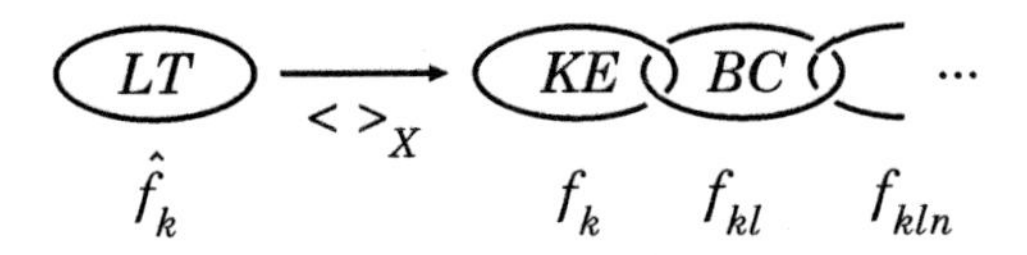

Figure 3.1: How to break the infinite chain of the equations for correlation functions? *LT* is the Liouville theorem (1.11) or Equation (2.18) for an exact distribution function $\hat{f}_k$. *KE* and *BC* are the kinetic Equation (2.36) for f_k and Equation (2.46) for the binary correlation function f_{kl}.

energy (parameter (2.28)). If the last one can be reasonably characterized by some temperature T (Section 9.1), then this ratio

$$\zeta_{\mathrm{i}} \approx \frac{e^2}{\langle l \rangle} \left(k_{\mathrm{B}} T \right)^{-1} . \tag{3.1}$$

As a mean distance between the particles we take $\langle l \rangle \approx n^{-1/3}$. Hence the ratio

$$\zeta_{\mathrm{i}} = \frac{e^2}{n^{-1/3}} \left(k_{\mathrm{B}} T \right)^{-1} = \frac{e^2}{k_{\mathrm{B}}} \times \frac{n^{1/3}}{T} \tag{3.2}$$

is termed the *interaction parameter*. It is small for a sufficiently *hot* and *rarefied* plasma.

In many astrophysical plasmas, for example in the solar corona (see Exercise 3.2), the interaction parameter is really very small. So the thermal kinetic energy of plasma particles is much larger than their interaction energy. **The particles are almost free** or moving on definite trajectories in the external fields if the later are present.

We shall call this case the approximation of *weak* Coulomb interaction. An existence of the small parameter allows us to have a complete description of this interaction by using the perturbation procedure. Moreover such a description is the simplest and the most exact one.

While constructing the kinetic theory, it is natural to use the perturbation theory with respect to the small parameter ζ_{i}. This means that

> the distribution function f_k must be taken to be of order unity, the pair correlation function f_{kl} of order ζ_{i}, the triple correlation function f_{kln} of order ζ_{i}^2, etc.

We shall see in what follows that this principle has a deep physical sense in kinetic theory. Such plasmas are said to be 'weakly coupled'.

An opposite case, when the interaction parameter takes values larger than unity, is very dense, relatively cold plasmas, for example in the interiors of white dwarf stars (Exercise 3.3). These plasmas are 'strongly coupled'.

3.1.2 The Vlasov kinetic equation

In the zeroth order with respect to the small parameter ζ_{i}, we obtain the Vlasov equation with the self-consistent electromagnetic field (Vlasov, 1938, 1945):

$$\frac{\partial f_k(X,t)}{\partial t} + v_\alpha \frac{\partial f_k(X,t)}{\partial r_\alpha} + \\ + \frac{e_k}{m_k}\left(\mathbf{E} + \frac{1}{c}\,\mathbf{v}\times\mathbf{B}\right)_\alpha \frac{\partial f_k(X,t)}{\partial v_\alpha} = 0\,. \tag{3.3}$$

Here $\mathbf{E}$ and $\mathbf{B}$ are the electric and magnetic fields obeying Maxwell's equations:

$$\operatorname{curl}\mathbf{E} = -\frac{1}{c}\frac{\partial \mathbf{B}}{\partial t}\,, \qquad \operatorname{div}\mathbf{E} = 4\pi\left(\rho^{\,0} + \rho^{\,\mathrm{q}}\right), \\ \operatorname{curl}\mathbf{B} = \frac{1}{c}\frac{\partial \mathbf{E}}{\partial t} + \frac{4\pi}{c}\left(\mathbf{j}^{\,0} + \mathbf{j}^{\,\mathrm{q}}\right), \qquad \operatorname{div}\mathbf{B} = 0\,. \tag{3.4}$$

$\rho^{\,0}$ and $\mathbf{j}^{\,0}$ are the densities of external charges and currents; they describe the external fields, for example, the uniform magnetic field $\mathbf{B}_0$. $\rho^{\,\mathrm{q}}$ and $\mathbf{j}^{\,\mathrm{q}}$ are the charge and current densities due to the plasma particles themselves:

$$\rho^{\,\mathrm{q}}(\mathbf{r},t) = \sum_k e_k \int_{\mathbf{v}} f_k(\mathbf{r},\mathbf{v},t)\, d^3\mathbf{v}\,, \tag{3.5}$$

$$\mathbf{j}^{\,\mathrm{q}}(\mathbf{r},t) = \sum_k e_k \int_{\mathbf{v}} \mathbf{v}\, f_k(\mathbf{r},\mathbf{v},t)\, d^3\mathbf{v}\,. \tag{3.6}$$

So, if we are considering processes which occur on a time scale much shorter than the time scale of collisions,

$$\tau_{ev} \ll \tau_c\,, \tag{3.7}$$

we may use a description which includes the electric and magnetic fields arising from the plasma charge density and current density, but **neglects the microfields responsible for binary collisions**. This means that $\mathbf{F}' = 0$ in formula (2.13), therefore the collisional integral (2.16) is also equal to zero.

The Vlasov kinetic Equation (3.3) together with the definitions (3.5) and (3.6), and with Maxwell's Equations (3.4) serve as a classic basis for the theory of oscillations and waves in a plasma (e.g., Silin, 1971; Schmidt, 1979; Benz, 2002) with the small parameter ζ_{i} and small correlational effects of higher orders. The Vlasov equation is also a proper basis for kinetic theory of wave-particle interactions in astrophysical plasma (Chapter 7) and shock waves in collisionless plasma (Section 16.4). The Vlasov equation was strongly criticized by Ginzburg et al. (1946).

One of the natural limitations of the Vlasov equation is that it will not make a plasma relax to a Maxwellian distribution (Section 9.5), since we effectively neglect collisions by neglecting the binary correlation function. Vlasov was the first to recognize that

> the electromagnetic interaction among plasma particles is qualitatively different from the interaction in an ordinary gas.

3.1.3 The Landau collisional integral

Using the perturbation theory with respect to the small interaction parameter ζ_i in the first order, and, therefore, neglecting the close Coulomb collisions (this will be justified in Section 8.1.5), we can find the kinetic equation with the collisional integral given by Landau (1937)

$$\left(\frac{\partial \hat{f}_k}{\partial t} \right)_c = - \frac{\partial}{\partial v_\alpha} J_{k,\alpha} \,, \tag{3.8}$$

where the flux of particles of kind k in the velocity space (cf. formula (2.35)) is

$$J_{k,\alpha} = \frac{\pi e_k^2 \ln \Lambda}{m_k} \sum_l e_l^2 \int_{\mathbf{v}_l} \left\{ f_k \frac{\partial f_l}{m_l \, \partial v_{l,\beta}} - f_l \frac{\partial f_k}{m_k \, \partial v_{k,\beta}} \right\} \times$$
$$\times \frac{(u^2 \, \delta_{\alpha\beta} - u_\alpha u_\beta)}{u^3} \, d^3 \mathbf{v}_l \,. \tag{3.9}$$

Here $\mathbf{u} = \mathbf{v} - \mathbf{v}_l$ is the relative velocity, $d^3 \mathbf{v}_l$ corresponds to the integration over the whole velocity space of 'field' particles l. $\ln \Lambda$ is the Coulomb logarithm which takes into account divergence of the Coulomb-collision cross-section (see Section 8.1.5). The full kinetic Equation (2.15) with the Landau collisional integral is a nonlinear integro-differential equation for the distribution function $f_k(\mathbf{r}, \mathbf{v}, t)$ of particles of the kind k.

The date of publication of the Landau (1937) paper may be considered as the date of birth of the kinetic theory of *collisional* fully-ionized plasma. The theory of *collisionless* plasma begins with the classical paper of Vlasov (1938). In fact, these two approaches correspond to **different limiting cases**.

> The Landau integral takes into account the part of the particle interaction which determines dissipation while the Vlasov equation allows for the average field, and is thus reversible.

For example, in the Vlasov theory the question of the role of collisions in the neighbourhood of resonances remains open. The famous paper by Landau (1946) was devoted to this problem. Landau used the reversible Vlasov equation as the basis to study the dynamics of a small perturbation of the Maxwell distribution function, $f^{(1)}(\mathbf{r}, \mathbf{v}, t)$. In order to solve the linearized

Vlasov equation (Section 7.1.1), he made use of the Laplace transformation, and defined the rule to avoid a pole in the divergent integral (see Section 7.1.2) by the replacement $\omega \to \omega + \mathrm{i}\,0$.

This technique for avoiding singularities may be formally replaced by a different procedure. Namely it is possible to add a small dissipative term $-\nu f^{(1)}(\mathbf{r}, \mathbf{v}, t)$ to the linearized Vlasov equation. In this way, the Fourier transform of the kinetic equation involves the complex frequency $\omega = \omega' + \mathrm{i}\,\nu$, leading with $\nu \to 0$ to the same expression for the *Landau damping*. Note, however, that

> the Landau damping is not by randomizing collisions but by a transfer of wave field energy into oscillations of resonant particles

(see Section 7.1.2).

Thus there are two different approaches to the description of plasma oscillation damping. The first is based on mathematical regularization of the Cauchy integral divergence. In this approach the physical nature of the damping seems to be not considered since the initial equation remains reversible. However the Landau method is really a beautiful example of complex analysis leading to an important new physical result.

The second approach reduces the reversible Vlasov equation to an irreversible one. Although the dissipation is assumed to be negligibly small, one cannot take the limit $\nu \to 0$ directly in the master equations: this can be done only in the final formulae. This second method of introducing the collisional damping is more natural. It shows that

> even very rare collisions play the principal role in the physics of collisionless plasma.

It is this approach that has been adopted in Klimontovich (1986). A more comprehensive solution of this principal question, however, can only be obtained on the basis of the dissipative kinetic equation.

The example of the Landau resonance and Landau damping demonstrates that some fundamental problems still remain unsolved in the kinetic theory of plasma. They arise from inconsistent descriptions of the transition from the reversible equations of the mechanics of charge particles and fields to the irreversible equations for statistically averaged distribution functions (Klimontovich, 1998).

In the first approximation with respect to the small interaction parameter ζ_{i} we find the Maxwellian distribution function and the effect of Debye shielding. This is the subject of the Section 3.2.

3.1.4 The Fokker-Planck equation

The smallness of the interaction parameter ζ_{i} signifies that, in the Landau collisional integral, the sufficiently distant Coulomb collisions are taken care of as the interactions with a **small momentum and energy transfer** (see

Section 8.1). For this reason, it comes as no surprise that the Landau integral can be considered as a particular case of a different approach which is the Fokker-Planck equation (Fokker, 1914; Planck, 1917). The latter generally describes systems of many particles that move under action of *stochastic* forces producing *small* changes in particle velocities (for a review see Chandrasekhar, 1943a).

Let us consider a distribution function independent of space so that $f = f(\mathbf{v}, t)$. The Fokker-Planck equation describes the distribution function evolution due to **nonstop overlapping weak collisions** resulting in particle diffusion in velocity space:

$$\left(\frac{\partial \hat{f}}{\partial t} \right)_{\mathrm{c}} = - \frac{\partial}{\partial v_\alpha} \left[a_\alpha f \right] + \frac{\partial^2}{\partial v_\alpha \, \partial v_\beta} \left[b_{\alpha\beta} \, f \right] . \tag{3.10}$$

The Fokker-Planck equation formally coincides with the diffusion-type equation (which is irreversible of course) for some admixture with concentration f, for example Brownian particles (or test particles) in a gas, on which stochastic forces are exerted by the molecules of the gas. The coefficient $b_{\alpha\beta}$ plays the role of the diffusion coefficient and is equal to

$$b_{\alpha\beta} = \frac{1}{2} \left(\delta v_{\alpha\beta} \right)_{\mathrm{av}} , \tag{3.11}$$

i.e. is expressed in terms of the averaged velocity changes in elementary acts – collisions:

$$\left(\delta v_{\alpha\beta} \right)_{\mathrm{av}} = \langle \, \delta v_\alpha \, \delta v_\beta \, \rangle \, . \tag{3.12}$$

The other coefficient is

$$a_\alpha = \left(\delta v_\alpha \right)_{\mathrm{av}} = \langle \, \delta v_\alpha \, \rangle \, . \tag{3.13}$$

It is known as the Fokker-Planck coefficient of dynamic friction. For example, a Brownian particle moving with velocity $\mathbf{v}$ through the gas experiences a drag opposing the motion (see Figure 1.4).

In order to find the mean values appearing in the Fokker-Planck equation, we have to make clear the physical and mathematical sense of expressions (3.12) and (3.13), see Exercise 3.4.

> The mean values of velocity changes are in fact statistically averaged and determined by the forces acting between a test particle and scatterers (field particles or waves).

Because of this, these averaged quantities have to be expressed by the collisional integral with the corresponding cross-sections (Exercises 3.5 and 3.6). The 'standard' derivation of the Fokker-Planck equation from the Boltzmann integral, with discussion of its particular features, can be found for example in Shoub (1987); however see Section 11.5 in Balescu (1975).

For electrons and ions in a plasma, such calculations can be made and give us the Landau integral; see Section 11.8 in Balescu (1975). The kinetic equation found in this way will allow us to study the Coulomb interaction of accelerated particle beams with astrophysical plasma (Chapter 4). The first term in the Fokker-Planck equation is a **friction** which slows down the particles of the beam and move them toward the zero velocity in the velocity space (Figure 3.2), the second term represents the three-dimensional **diffusion** of the beam particles in the velocity space.

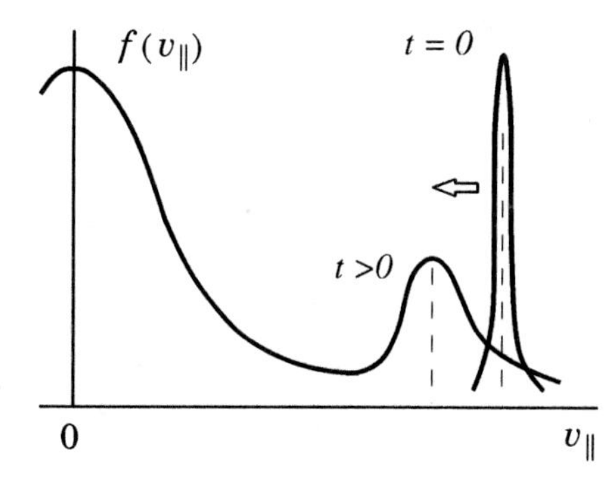

Figure 3.2: A beam of fast particles in plasma can generate the Langmuir waves due to the bump-on tail instability which will be shown in Chapter 7. Here we illustrate only the simplest effects of Coulomb collisions, that will be considered in Chapter 4.

During the motion of a beam of accelerated particles in a plasma a *reverse* current of thermal electrons is generated, which tends to compensate the electric current of accelerated particles – the *direct* current.

> The electric field driving the reverse current makes a great impact on the particle beam kinetics.

That is why, in order to solve the problem of accelerated particle propagation in, for example, the solar atmosphere, we inevitably have to apply a **combined approach**, which takes into account both the electric field influence on the accelerated particles (as in the Vlasov equation) and their scattering from the thermal particles of a plasma (as in the Landau equation; see Section 4.5).

The Landau collisional integral is effectively used in many problems of plasma astrophysics. It permits a considerable simplification of the calculations of many quantities determined by collisions of charged particles, such as the viscosity coefficient, thermal conductivity, electric conductivity, etc. (Section 9.5).

The Landau collisional integral does not take into account the close collisions since they are responsible for large exchange of the particle momentum (see Section 8.1). So the interaction parameter is not small, and the perturbation theory is not applicable (Exercise 3.6). The close Coulomb collisions of charged particles can play an important role in collective plasma phenomena (Klimontovich, 1986).

3.2 Correlation function and Debye shielding

We are going to understand the most fundamental property of the binary correlation function. With this aim in mind, we shall solve the second equation in the chain illustrated by Figure 3.1. To solve this equation we have to know two functions: the distribution function f_k from the first link in the chain and the triple correlation function f_{kln} from the third link.

3.2.1 The Maxwellian distribution function

Let us consider the stationary ($\partial/\partial t = 0$) solution to the equations for correlation functions, assuming the interaction parameter $\zeta_{\rm i}$ to be small and using the **method of successive approximations** in the following form. First, we set $f_{kl} = 0$ in the averaged Liouville equation (2.36) for the distribution function f_k, then we assume that the triple correlation function f_{kln} is zero in Equation (2.46) for the correlation function f_{kl} etc.

The plasma is supposed to be stationary, uniform and in the thermodynamic equilibrium state, i.e. the particle velocity distribution is assumed to be a Maxwellian function

$$f_k\,(X) = f_k\,(v^2) = c_k\,\exp\left(-\frac{m_k\,v^2}{2k_{\rm B}T_k}\right). \tag{3.14}$$

The constant c_k is determined by the normalizing condition and equals

$$c_k = n_k \left(\frac{m_k}{2\pi\,k_{\rm B}T_k}\right)^{3/2}.$$

It is obvious that the Maxwellian function (3.14) satisfies the kinetic equation (2.36) under assumption made above if the average force is equal to zero:

$$F_{k,\alpha}(X,t) = F_{k,\alpha}(X) = 0\,. \tag{3.15}$$

Since we will need the same assumption in the next Section, we shall justify it there.

3.2.2 The averaged force and electric neutrality

To a first approximation, i.e. with account of $f_{kl} \neq 0$, the distribution function is also uniform with respect to its space variables. Let us substitute the Maxwellian distribution function (3.14) in the pair-correlation function Equation (2.46), neglecting all the interactions except the Coulomb ones. For the latter, in circumstances where the averaged distribution functions for the components are *uniform*, we obtain the following expression for the averaged force (2.30):

$$F_{k,\alpha}\,(X_1) = \sum_l \int\limits_{X_2} F_{kl,\alpha}\,(X_1, X_2)\; f_l\,(X_2)\,dX_2 =$$

since plasma is uniform, f_l does not depend of $\mathbf{r}_2$

$$= \sum_l \int_{\mathbf{r}_2} F_{kl,\alpha}(\mathbf{r}_1, \mathbf{r}_2)\, d^3\mathbf{r}_2 \cdot \int_{\mathbf{v}_2} f_l(\mathbf{v}_2)\, d^3\mathbf{v}_2 =$$

$$= - \int_{\mathbf{r}_2} \sum_l \frac{\partial}{\partial r_{1,\alpha}} \left(\frac{e_k\, e_l}{|\, \mathbf{r}_1 - \mathbf{r}_2 \,|} \right) d^3\mathbf{r}_2 \cdot n_l =$$

$$= - \int_{\mathbf{r}_2} \frac{\partial}{\partial r_{1,\alpha}} \left(\frac{e_k}{|\, \mathbf{r}_1 - \mathbf{r}_2 \,|} \right) d^3\mathbf{r}_2 \cdot \sum_l n_l\, e_l\,. \tag{3.16}$$

Therefore

$$F_{k,\alpha} = 0\,, \tag{3.17}$$

if the plasma is assumed to be *electrically neutral*:

$$\boxed{\sum_l n_l\, e_l = 0\,,} \tag{3.18}$$

or *quasi-neutral* (see Section 8.2).

Balanced charges of ions and electrons determine the name *plasma* according Langmuir (1928). So the average force (2.30) is equal to zero in the electrically neutral plasma but is not equal to zero in a system of charged particles of the same charge sign: positive or negative, it does not matter. Such a system tends to expand.

There is no neutrality in gravitational systems. The large-scale gravitational field makes an overall thermodynamic equilibrium impossible (Section 9.6). Moreover, on the contrary to plasma, they tend to contract and collapse.

3.2.3 Pair correlations and the Debye radius

As a first approximation, on putting the triple correlation function $f_{kln} = 0$, we obtain from Equation (2.46), in view of condition (3.17), the following equation for the binary or pair correlation function f_{kl}:

$$v_{1,\alpha} \frac{\partial f_{kl}}{\partial r_{1,\alpha}} + v_{2,\alpha} \frac{\partial f_{kl}}{\partial r_{2,\alpha}} =$$

$$= - \sum_n \int_{X_3} \left\{ \frac{1}{m_k} F_{kn,\alpha}(X_1, X_3)\, f_{nl}(X_3, X_2)\, \frac{\partial f_k}{\partial v_{1,\alpha}} + \right.$$

$$\left. + \frac{1}{m_l} F_{ln,\alpha}(X_2, X_3)\, f_{nk}(X_3, X_1)\, \frac{\partial f_l}{\partial v_{2,\alpha}} \right\} dX_3\,. \tag{3.19}$$

Let us consider the particles of two kinds – electrons and ions, assuming the ions to be motionless and homogeneously distributed. Then the ions do not take part in any kinetic processes; hence $\hat{\varphi}_{\mathrm{i}} \equiv 0$ for ions and the correlation functions associated with $\hat{\varphi}_{\mathrm{i}}$ equal zero as well:

$$f_{\mathrm{ii}} = 0\,, \qquad f_{\mathrm{ei}} = 0 \qquad \text{etc.} \tag{3.20}$$

Among the pair correlation functions, only one has a non-zero magnitude

$$f_{\mathrm{ee}}\left(X_1, X_2\right) = f\left(X_1, X_2\right). \tag{3.21}$$

Taking into account (3.20), (3.21), and (3.14), rewrite Equation (3.19) as follows

$$\begin{aligned} \mathbf{v}_1 \frac{\partial f}{\partial \mathbf{r}_1} + \mathbf{v}_2 \frac{\partial f}{\partial \mathbf{r}_2} &= \\ = \frac{1}{k_{\mathrm{B}} T} \int\limits_{X_3} &\left[\mathbf{v}_1 \cdot \mathbf{F}\left(X_1, X_3\right) f\left(X_3, X_2\right) f_{\mathrm{e}}\left(\mathbf{v}_1\right) + \right. \\ &\left. + \mathbf{v}_2 \cdot \mathbf{F}\left(X_2, X_3\right) f\left(X_1, X_3\right) f_{\mathrm{e}}\left(\mathbf{v}_2\right)\right] dX_3\,. \end{aligned} \tag{3.22}$$

Since $\mathbf{v}_1$ and $\mathbf{v}_2$ are arbitrary and refer to the same kind of particles (electrons), Equation (3.22) takes the form

$$\frac{\partial f}{\partial \mathbf{r}_1} = \frac{1}{k_{\mathrm{B}} T} \int\limits_{X_3} \mathbf{F}\left(X_1, X_3\right) f\left(X_3, X_2\right) f_{\mathrm{e}}\left(\mathbf{v}_1\right) dX_3\,. \tag{3.23}$$

Taking into account the character of Coulomb force in the same approximation as in formula (3.17) and assuming the correlation to exist only between the positions of the particles in space (rather than between velocities), let us integrate both sides of Equation (3.23) over $d^3\mathbf{v}_1\, d^3\mathbf{v}_2$. The result is

$$\frac{\partial g\left(\mathbf{r}_1, \mathbf{r}_2\right)}{\partial \mathbf{r}_1} = -\frac{n e^2}{k_{\mathrm{B}} T} \int\limits_{\mathbf{r}_3} \nabla_{\mathbf{r}_1} \frac{1}{\left|\,\mathbf{r}_1 - \mathbf{r}_3\,\right|}\, g\left(\mathbf{r}_2, \mathbf{r}_3\right) d^3\mathbf{r}_3\,. \tag{3.24}$$

Here the function

$$g\left(\mathbf{r}_1, \mathbf{r}_2\right) = \int\limits_{\mathbf{v}_1} \int\limits_{\mathbf{v}_2} f\left(X_1, X_2\right) d^3\mathbf{v}_1\, d^3\mathbf{v}_2\,. \tag{3.25}$$

We integrate Equation (3.24) over $\mathbf{r}_1$ and designate the function

$$g\left(\mathbf{r}_1, \mathbf{r}_2\right) = g\left(r_{12}^2\right),$$

where $r_{12} = \left|\,\mathbf{r}_1 - \mathbf{r}_2\,\right|$. So we obtain the equation

$$g\left(r_{12}^2\right) = -\frac{n e^2}{k_{\mathrm{B}} T} \int\limits_{\mathbf{r}_3} \frac{g\left(r_{23}^2\right)}{r_{13}}\, d^3\mathbf{r}_3\,.$$

Its solution is

$$g(r) = \frac{c_0}{r} \exp\left(-\frac{r}{r_D}\right), \tag{3.26}$$

where

$$\boxed{r_D = \left(\frac{k_B T}{4\pi n e^2}\right)^{1/2}} \tag{3.27}$$

is the *Debye radius*. It will be defined in just this way (see formula (8.33)) for the case when the shielding is due to the particles of one kind – due to electrons. A more general formula for the Debye radius will be derived in Section 8.2.

The constant of integration

$$c_0 = -\frac{1}{4\pi r_D^2 n} \tag{3.28}$$

(see Exercise 3.8). Substituting (3.28) in solution (3.26) gives the sought-after pair correlation function, i.e. the velocity-integrated correlation function

$$g(r) = -\frac{1}{4\pi r_D^2 n} \frac{1}{r} \exp\left(-\frac{r}{r_D}\right) = -\frac{e^2}{k_B T} \frac{1}{r} \exp\left(-\frac{r}{r_D}\right). \tag{3.29}$$

Formula (3.29) shows that

> the Debye radius is a characteristic length scale of pair correlations in a fully-ionized equilibrium plasma:

$$\boxed{g(r) \sim \frac{1}{r} \exp\left(-\frac{r}{r_D}\right).} \tag{3.30}$$

This result proves to be fair in the context of Section 8.2 where the Debye shielding will be considered in another approach. Comparison of formula (3.30) with (8.32) shows that, as one might have anticipated,

> the binary correlation function reproduces the shape of the actual potential of interaction, i.e. the shielded Coulomb potential.

It is known that cosmic plasma can exhibit *collective phenomena* arising out of mutual interactions of many charged particles. Since the Debye radius r_D is a characteristic length scale of pair correlations, the number $n r_D^3$ gives us a measure of the number of particles which can interact simultaneously. The inverse of this number is the so-called *plasma parameter*

$$\zeta_p = \left(n r_D^3\right)^{-1}. \tag{3.31}$$

This is a small quantity as well as it can be expressed in terms of the small interaction parameter $\zeta_{\rm i}$ (Exercise 3.1). The fact that $\zeta_{\rm p} \ll 1$ implies **a large number of plasma particles in a volume enclosed by the sphere of the Debye radius**. In many astrophysical applications the plasma parameter (3.31) is really small (e.g., Exercise 3.2). So the collective phenomena can be really important in cosmic plasma.

3.3 Gravitational systems

There is a fundamental difference between plasma and the gravitational systems with potential (1.41), for example, the stars in a galaxy. This difference lies in the nature of the gravitational force: there is **no shielding** to vitiate this long-range $1/r^2$ force. The collisional integral formally equals infinity because the binary correlation function $g(r) \sim 1/r$.

The conventional wisdom of such system dynamics (see Binney and Tremaine, 1987) asserts that the structure and evolution of a collection of N self-gravitating point masses can be described by the collisionless kinetic equation, the gravitational analog of the Vlasov equation (Exercises 3.9 and 16.7). On the basis of what we have seen above,

> the collisionless appoach in gravitational systems, i.e. the entire neglect of particle pair correlations, constitutes an **uncontrolled approximation**.

Unlike the case of plasma, we cannot derive the next order correction to the collisionless kinetic equation in the context of a systematic perturbation expansion.

Physically, this is manifested by the fact that the $1/r$ potential yields an infinite cross-section, so that, when evaluating the effects of collisions in the usual way (Section 8.1.5) for an infinite homogeneous system, we encounter logarithmic divergences in the limit of large impact parameter (formula (8.18)), see however Exercise 3.9. We may hope to circumvent this difficulty, the problematic Coulomb logarithm of gravitational dynamics, by *first* identifying the bulk mean field force $\langle \mathbf{F} \rangle$ in definition (2.13), acting at any given point in space and *then* treating fluctuations $\mathbf{F}'$ away from the mean field force. This splitting into a mean field plus fluctuations can be introduced formally (Kandrup, 1998) and allows one to write down the collisional integral of the type (2.16). However, this is difficult to implement concretely because of the apparent absence of a clean separation of time scales.

For the N-body problem with $N \gg 1$ we might expect that these fluctuations are small, so that their effects do in fact constitute a small perturbation. So it is assumed that, on long time scales, one must allow for discreteness effects, described by the Fokker-Plank equation (3.10) or the kinetic equation with the Landau collisional integral (3.8); see Exercise 3.10.

Given that theoretical analyses have as yet proven inconclusive, one might instead seek resource to numerical experiments. This, however, is also difficult for gravitational systems not characterized by a high degree of symmetry. There is in fact only one concrete setting where detailed computations have been done, namely the toy model of one-dimensional gravity.

In summary, even though a **mean gravitational field theory** based on the Vlasov equation may seem well motivated physically, there is as yet no rigorous proof of its validity and, in particular, no rigorous estimate as to the time scale on which it might be expected to fail.

Hydrodynamic description of gravitational systems has a difficulty of the same origin. The gravitational attraction cannot be screened (Section 9.6).

3.4 Comments on numerical simulations

At present, astrophysical plasma processes are typically investigated in well developed and distinct approaches. One approach, described by the Vlasov equation, is the collisionless limit used when collective effects dominate. In cases where the plasma dynamics is determined by collisional processes in external fields and where the self-consistent fields can be neglected, the Fokker-Planck approach is used. At the same time, it is known that

> both collective kinetic effects and Coulomb collisions can play an essential role in a great variety of astrophysical phenomena

starting from the most simple one – propagation of fast particles in plasma (Chapter 4). Besides, as was mentioned in Section 3.1.3, **collisions play the principal role in the physics of collisionless plasma.** Taking collisions into account may lead not only to quantitative but also qualitative changes in the plasma behaviour, even if the collision frequency ν is much less than the electron plasma frequency.

It is known that, even in the collisionless limit, the kinetic equation is still too difficult for numerical simulations, and the 'macroparticle' methods are the most widely used algorithms. In these methods, instead of direct numerical solution of the kinetic equation, a set of ordinary differential equations for every macroparticle is solved. These equations are the characteristics of the Vlasov equation.

In the case of a collisional plasma, the position of a macroparticle satisfies the usual equation of the collisionless case

$$\dot{\mathbf{r}} \equiv \frac{d\,\mathbf{r}}{dt} = \mathbf{v}(t)\,, \tag{3.32}$$

but the momentum equation is modified owing to the Coulomb collisions. They are described by the Fokker-Planck operator (3.10) which introduces a friction (the coefficient a_α) and diffusion (the coefficient $b_{\alpha\beta}$) in velocity space. Thus it is necessary to find the effective collisional force $\mathbf{F}_{\mathrm{c}}$ which acts on the macroparticles:

$$\dot{\mathbf{v}} \equiv \frac{d\,\mathbf{v}}{dt} = \frac{1}{m}\left(\mathbf{F}_{\mathrm{L}} + \mathbf{F}_{\mathrm{c}}\right)\,. \tag{3.33}$$

The collisional force can be introduced phenomenologically (see Jones et al., 1996) but a more mathematically correct approach can be constructed using the stochastic equivalence of the Fokker-Planck and Langevin equations (see Cadjan and Ivanov, 1999). So **stochastic differential equations** can be regarded as an alternative to the description of astrophysical plasma in terms of distribution function.

> The Langevin approach allows one to overcome some difficulties related to the Fokker-Planck equation and to simulate actual plasma processes, taking account of both collective effects and Coulomb collisions.

Generally, if we want to construct an effective method for the simulation of complex nonlinear processes in astrophysical plasma, we have to satisfy the following obvious but conflicting conditions.

First, the method should be adequate for the task in hand. For a number of problems the application of simplified models of the collisional integral can provide a correct description and ensure good accuracy. The constructed model should describe collisional effects with the desired accuracy.

Second, the method should be computationally efficient. The algorithm should not be extremely time-consuming. In practice, some compromise between accuracy and complexity of the method should be achieved. Otherwise, we restrict ourselves either to a relatively simple setup of the problem or to a too-rough description of the phenomena.

A 'recipe': the choice of a particular collisional model (or a model of the collisional integral) is determined by the importance and particular features of the collisional processes in a given astrophysical problem.

3.5 Practice: Exercises and Answers

Exercise 3.1 [Section 3.1.1] Show that the interaction parameter

$$\zeta_{\rm i} = \frac{1}{4\pi}\, \zeta_{\rm p}^{2/3} , \tag{3.34}$$

if the Debye radius is given by formula (3.27). Discuss the difference between $\zeta_{\rm i}$ and $\zeta_{\rm p}$.

Exercise 3.2 [Section 3.1.1] How many particles are inside the Debye sphere in the solar corona?

Answer. From formula (8.31) for the Debye radius in two-component equilibrium plasma (see also formula (8.77) in Exercise 8.3) it follows that for electron-proton plasma with $T \approx 2 \times 10^6$ K and $n \approx 2 \times 10^8\ {\rm cm}^{-3}$ the Debye radius

$$r_{\rm D} = \left(\frac{kT}{8\pi\, e^2\, n} \right)^{1/2} \approx 4.9 \left(\frac{T}{n} \right)^{1/2} \approx 0.5\ {\rm cm}\,. \tag{3.35}$$

The number of particles inside the Debye sphere

$$N_{\mathrm{D}} = n\,\frac{4}{3}\,\pi r_{\mathrm{D}}^{3} \sim 10^{8} . \tag{3.36}$$

Hence the typical value of plasma parameter (3.31) in the corona is really small: $\zeta_{\mathrm{p}} \sim 10^{-8}$. The interaction parameter (3.2) is also small: $\zeta_{\mathrm{i}} \sim 10^{-6}$ (see formula (3.34)).

Exercise 3.3 [Section 3.1.1] Estimate the interaction parameter (3.2) in the interior of white dwarf stars (de Martino et al., 2003; see also Exercise 1.3).

Comment. It may seem at first sight that the mutual interactions between electrons would be very important inside a white dwarf star. However, in a system of fermions with most states filled up to the Fermi energy,

> collisions among nearby electrons are suppressed due to the fact that the electrons may not have free state available for occupation after the collision

(see Kittel, 1995). Hence electrons inside a white dwarf star are often approximated as a perfect gas made up of non-interacting fermions (see § 57 in Landau and Lifshitz, *Statistical Physics*, 1959b). For this reason, some results of plasma astrophysics are applicable to the electron gas inside white dwarfs.

Exercise 3.4 [Section 3.1.4] Let $w = w\,(\mathbf{v},\,\delta\mathbf{v})$ be the probability that a test particle changes its velocity $\mathbf{v}$ to $\mathbf{v} + \delta\mathbf{v}$ in the time interval δt. The velocity distribution at the time t can be written as

$$f(\mathbf{v},t) = \int f(\mathbf{v} - \delta\mathbf{v},\, t - \delta t)\, w\,(\mathbf{v} - \delta\mathbf{v},\, \delta\mathbf{v})\, d^{3}\delta\mathbf{v} . \tag{3.37}$$

Bearing in mind that the interaction parameter (3.1) is small and, therefore, $|\,\delta\mathbf{v}\,| \ll |\,\mathbf{v}\,|$, expand the product fw under the integral into a Taylor series.

Take the first three terms in the series and show that, in formulae (3.13) and (3.12), the average velocity change per time interval δt:

$$\langle\, \delta v_{\alpha} \,\rangle = \int \delta v_{\alpha}\, w\, d^{3}\delta\mathbf{v} , \tag{3.38}$$

$$\langle\, \delta v_{\alpha}\, \delta v_{\beta} \,\rangle = \int \delta v_{\alpha}\, \delta v_{\beta}\, w\, d^{3}\delta\mathbf{v} . \tag{3.39}$$

Show that the Fokker-Planck equation (3.10) follows from the Taylor series expansion of the function $f(\mathbf{v},t)$ given by formula (3.37).

Exercise 3.5 [Section 3.1.4] Express the collisional integral in terms of the differential cross-sections of interaction between particles (Smirnov, 1981).

Discussion. Boltzmann (1872) considered a delute neutral gas. Since the particles in a neutral gas do not have long-range interactions like the charged particles in a plasma, they are assumed to interact only when they collide, i.e. when the separation between two particles is not much larger than $2a$, where a is the 'radius of a particle'. A particle moves freely in a straight line between two collisions.

In a binary collision, let $\mathbf{v}_k$ and $\mathbf{v}_l$ be the velocities of particles k and l before the collision, $\mathbf{v}_k'$ and $\mathbf{v}_l'$ be the velocities of the same particles after the collision. There are two types of collisions: (a) one that increases the density of the particles at a given point of phase space by bringing in particles from other phase space locations, (b) the other that reduces the density of particles by taking particles away from this point to other phase space locations; these are the collisions $\mathbf{v}_k + \mathbf{v}_l \to \mathbf{v}_k' + \mathbf{v}_l'$.

By using notations taken into account that k and l can be different kinds of particles, we write the Boltzmann collisional integral in the form (cf. Boltzmann, 1956):

$$\left(\frac{\partial \hat{f}_k}{\partial t} \right)_c = \sum_l \int_{\mathbf{v}_l} \int_{\Omega} \left(f_k' \, f_l' - f_k \, f_l \right) v_{kl} \, d\sigma_{kl} \, d^3\mathbf{v}_l \, . \tag{3.40}$$

Here $\mathbf{v}_{kl} = \mathbf{v}_k - \mathbf{v}_l$ is the relative velocity, $d^3\mathbf{v}_l$ corresponds to the integration over the whole velocity space of 'field' particles l. $f_k = f_k\,(t, \mathbf{r}, \mathbf{v}_k)$ is the distribution function of particles of the kind k, $f_k' = f_k\,(t, \mathbf{r}, \mathbf{v}_k')$. The product $f_k' \, f_l'$ corresponds to the collisions $\mathbf{v}_k' + \mathbf{v}_l' \to \mathbf{v}_k + \mathbf{v}_l$ which inhance the particle density.

The precollision velocities $\mathbf{v}_k$ and $\mathbf{v}_l$ are related to the postcollision velocities $\mathbf{v}_k'$ and $\mathbf{v}_l'$ through the conservation laws of momentum and energy. These relations give us four scalar equations. However we need six equations to find two vectors $\mathbf{v}_k'$ and $\mathbf{v}_l'$.

A fifth condition comes from the fact the vectors $\mathbf{v}_k'$ and $\mathbf{v}_l'$ will have to lie in the plane of the vectors $\mathbf{v}_k$ and $\mathbf{v}_l$. This follows from the momentum conservation law and means that collisions are coplanar if the force of interaction between two particles is radial.

We need one more condition. We do not expect, of course, that the outcome of a collision is independent of the nature of interaction. If the impact parameter of the collision is given, we can calculate the defection produced by the collision from the interaction potential. The case of the Coulomb potential is considered in Chapter 8.

Since we are interested here in a statistical treatment, it is enough for us to know the probability of deflection in different direction or a differential scattering cross-section

$$d\sigma_{kl} = \frac{d\sigma_{kl}\,(v_{kl}, \chi)}{d\Omega} \, d\Omega \, , \tag{3.41}$$

where $d\Omega = 2\pi \, \sin\chi \, d\chi$ is a solid angle. If the particles are modelled as hard spheres undergoing two-body elastic collisions, the differential scattering cross-section is a function of the scattering angle χ alone. The Boltzmann gas model can be used for low-density neutral particles as well as for interactions of charged particles with neutral particles.

In plasma astrophysics, the Rutherford formula (8.8) is used to characterize the Coulomb collisions of charged particles. A general case is considered, for example, in Kogan (1967), Silin (1971), Lifshitz and Pitaevskii (1981).

Exercise 3.6 [Section 3.1.4] Show that the Fokker-Planck collisional model can be derived from the Boltzmann collisional integral (3.40) under the assumption that the change in the velocity of a particle due to a collision is rather small.

Exercise 3.7 [Section 3.1.4] The Landau collisional integral is generally thought to approximate the Boltzmann integral (3.40) for the 1/r potentials to 'dominant order', i.e. to within terms of order $1/\ln\Lambda$, where $\ln\Lambda$ is the Coulomb logarithm (see formula (8.23)). However this is not the whole truth. Show that the Landau integral approximates the Boltzmann integral to the dominant order only in parts of the velocity space.

Hint. This can be established by carring the Taylor series expansion of the Boltzmann integral to the fourth order. The first term in the series will be the familiar Landau-type collisional integral. The conclusion, drawn from the higher-order terms (Shoub, 1987), is that the large-angle scattering processes can play a more significant role in the evolution of the distribution function than currently believed. The normally 'nondominant' part of the diffusion tensor can make a contribution to the collisional term that decays more slowly with increasing velocity than do terms that are retained. In general, the approximations made are not uniformly valid in the velocity space, if the particle distribution functions are not sufficiently close to equilibrium distributions (Cercignani, 1969).

Exercise 3.8 [Section 3.2.3] Find the constant of integration c_0 in formula (3.26).

Answer. Let us solve the Poisson equation for the potential φ (more justification will be given in Section 8.2):

$$\Delta\varphi = -4\pi\, en \left\{ 1 - \left[1 + \frac{c_0}{r} \exp\left(-\frac{r}{r_{\rm D}} \right) \right] \right\} =$$

$$= n \, \frac{4\pi e\, c_0}{r} \exp\left(-\frac{r}{r_{\rm D}} \right). \tag{3.42}$$

Here it is taken into account that

$$\int\limits_{\mathbf{v}_1}\int\limits_{\mathbf{v}_2} \langle\, \hat{f}_k\,(X_1)\; \hat{f}_l\,(X_2)\,\rangle \; d^3\mathbf{v}_1\, d^3\mathbf{v}_2 = n_k\,(\mathbf{r}_1)\; n_l\,(\mathbf{r}_2) + g_{kl}\,(\mathbf{r}_1, \mathbf{r}_2)\,.$$

The general solution of Equation (3.42) in the spherically symmetric case, i.e. the solution of equation

$$\frac{1}{r}\frac{d^2}{dr^2}(r\varphi) = \frac{4\pi e\, c_0}{r}\exp\left(-\frac{r}{r_{\mathrm{D}}}\right)n\,,$$

is of the form

$$\varphi(r) = n\,\frac{4\pi e\, r_{\mathrm{D}}^2\, c_0}{r}\exp\left(-\frac{r}{r_{\mathrm{D}}}\right) + c_1 + \frac{c_2}{r}\,.$$

Since, as $r \to 0$, the potential φ takes the form $(-e)/r$, $c_1 = c_2 = 0$, and the only non-zero constant is

$$c_0 = -\frac{1}{4\pi\, r_{\mathrm{D}}^2\, n}\,. \tag{3.43}$$

Q.e.d.

Exercise 3.9 [Section 3.3] Following Section 3.1.2, write and discuss the gravitational analog of the Vlasov equation.

Answer. The basic assumption underlying the Vlasov equation is that the gravitational N-body system can be described probabilistically in terms of a statistically smooth distribution function $f(X,t)$. The Vlasov equation manifests the idea that this function will stream freely in the self-consistent gravitational potential $\phi(\mathbf{r},t)$ (cf. (1.41)) associated with $f(X,t)$, so that

$$\frac{\partial f(X,t)}{\partial t} + v_\alpha \frac{\partial f(X,t)}{\partial r_\alpha} - \frac{\partial \phi}{\partial r_\alpha}\frac{\partial f(X,t)}{\partial v_\alpha} = 0\,. \tag{3.44}$$

Here

$$\Delta\phi = -\,4\pi\, G\rho(\mathbf{r},t) \tag{3.45}$$

and

$$\rho(\mathbf{r},t) = \int f(\mathbf{r},\mathbf{v},t)\, d^3\mathbf{v}\,. \tag{3.46}$$

Note that, in the context of the mean field theory, a distribution of particles over their masses has no effect.

Applying for example to the system of stars in a galaxy, Equation (3.44) implies that the net gravitational force acting on a star is determined by the large-scale structure of the galaxy rather than by whether the star happens to lie close to some other star. The force on any star does not vary rapidly, and each star is supposed to accelerate smoothly through the force field generated by the galaxy as a whole.

In fact, encounters between stars may cause the acceleration $\dot{\mathbf{v}}$ to differ from the smoothed gravitational force $-\nabla\phi$ and therefore invalidate Equation (3.44). **Gravitational encounters are not screened**, they can be thought of as leading to an additional collisional term on the right side of the equation – a collisional integral. However very little is known mathematically about such possibility as we can see in Section 3.3.

Exercise 3.10 [Section 3.3] Following Section 3.1.3, discuss a gravitational analog of the Landau integral in the following form (e.g., Lancellotti and Kiessling, 2001):

$$\left(\frac{\partial \hat{f}}{\partial t}\right)_{\mathrm{c}} = \sigma \frac{\partial}{\partial \mathbf{v}} \int_{\mathbf{v}'} \frac{\partial^2 |\mathbf{v} - \mathbf{v}'|}{\partial \mathbf{v} \partial \mathbf{v}'} \cdot \left(\frac{\partial}{\partial \mathbf{v}} - \frac{\partial}{\partial \mathbf{v}'}\right) [f(\mathbf{r}, \mathbf{v}, t) f(\mathbf{r}, \mathbf{v}', t)] d^3\mathbf{v}'. \tag{3.47}$$

Here σ is a constant determined by the effective collision rate.

Chapter 4

Propagation of Fast Particles in Plasma

Among a variety of kinetic phenomena related to fast particles in astrophysical plasma, the simplest effect is Coulomb collisions under propagation of the particles in a plasma. An important role of the reverse-current electric field in this situation is demonstrated.

4.1 Derivation of the basic kinetic equation

4.1.1 Basic approximations

Among a rich variety of kinetic phenomena related to accelerated fast electrons and ions in astrophysical plasma (Kivelson and Russell, 1995) let us consider the simplest effect – Coulomb collisions under propagation of **fast particle beams in a fully-ionized thermal plasma**. We shall assume that there exists some external (background) magnetic field $\mathbf{B}_0$ which determines a way of fast particle propagation and which can be locally considered as a uniform one.

Electric and magnetic fields, $\mathbf{E}$ and $\mathbf{B}$, related to a beam of fast particles will be discussed in Section 4.5. Heating of plasma will be considered, for example, in Section 8.3. So, untill this will be necessary,

> accelerated particles will be considered as 'test' particles that do not influence the background plasma and magnetic field $\mathbf{B}_0$.

Let $f = f\,(t,\, \mathbf{r},\, \mathbf{v})$ be an unknown distribution function of test particles. In what follows, $q = Ze$ and $m = Am_{\rm p}$ are electric charge and mass of a test particle, respectively.

We restrict a problem by consideration of fast but non-relativistic particles interacting with background plasma which consists of thermal electrons ($m_1 =$

m_e and $e_1 = -e$) and thermal protons ($m_2 = m_p$ and $e_2 = +e$). Both components of a plasma are in thermodynamic equilibrium. Using the kinetic equation with the Landau collisional integral (3.8) we obtain

$$\frac{\partial f}{\partial t} + v_\alpha \frac{\partial f}{\partial r_\alpha} + \frac{q}{m} \left\{ E_\alpha + \frac{1}{c} \left[\mathbf{v} \times (\mathbf{B} + \mathbf{B}_0) \right]_\alpha \right\} \frac{\partial f}{\partial v_\alpha} = -\frac{\partial}{\partial v_\alpha} J_\alpha , \tag{4.1}$$

with $\mathbf{E} = 0$ and $\mathbf{B} = 0$,

$$J_\alpha = \frac{\pi q^2 \ln \Lambda}{m} \sum_{l=1}^{2} e_l^2 \int_{\mathbf{v}_l} \left\{ f \frac{\partial f_l}{m_l \partial v_{l,\beta}} - f_l \frac{\partial f}{m \partial v_\beta} \right\} \times$$

$$\times \frac{(u^2 \delta_{\alpha\beta} - u_\alpha u_\beta)}{u^3} \, d^3\mathbf{v}_l . \tag{4.2}$$

Here $\mathbf{u} = \mathbf{v} - \mathbf{v}_l$ is the relative velocity, $d^3\mathbf{v}_l$ corresponds to the integration over the whole velocity space of the plasma particles $l = 1, 2$. They are distributed by the Maxwellian function (3.14):

$$f_e(v) = n_e \left(\frac{m_e}{2\pi k_B T_e} \right)^{3/2} \exp \left(-\frac{m_e v^2}{2 k_B T_e} \right) \tag{4.3}$$

and

$$f_p(v) = n_p \left(\frac{m_p}{2\pi k_B T_p} \right)^{3/2} \exp \left(-\frac{m_p v^2}{2 k_B T_p} \right) . \tag{4.4}$$

For the sake of simplicity we assume $T_e = T_p = T$ (see, however, Section 8.3.2) as well as $n_e = n_p = n$. Also for the sake of simplicity we shall consider the stationary situation ($\partial / \partial t = 0$).

Moreover we shall assume that the distribution function f depends on one spatial variable – the coordinate z measured along the field $\mathbf{B}_0$, on the value of velocity v and the angle θ between the velocity vector $\mathbf{v}$ and the axis z. Therefore

$$f = f(z, v, \theta) . \tag{4.5}$$

In this case of the axial symmetry, the term containing the Lorentz force, related to the external field $\mathbf{B}_0$, in Equation (4.1) is equal to zero because the vector $\mathbf{v} \times \mathbf{B}_0$ is perpendicular to the plane ($\mathbf{v}$, $\mathbf{B}_0$) but the vector $\partial f / \partial \mathbf{v}$ is placed in this plane.

Under ussumptions made above, Equation (4.1) takes the following form:

$$v \cos\theta \frac{\partial f}{\partial z} = -\frac{1}{v^2} \frac{\partial}{\partial v} \left(v^2 J_v \right) - \frac{1}{v} \frac{1}{\sin\theta} \frac{\partial}{\partial \theta} \left(\sin\theta \, J_\theta \right) . \tag{4.6}$$

The distribution function f is *not* an isotropic one. So the angular component J_θ of the particle flux is not equal to zero.

4.1.2 Dimensionless kinetic equation in energy space

Let us introduce the dimensionless non-relativistic energy of the fast particles

$$x = \frac{mv^2}{2k_{\mathrm{B}}T} \left(\frac{m_{\mathrm{e}}}{m} \right) \tag{4.7}$$

and the dimensionless column depth along the magnetic field

$$\zeta = \xi / \tilde{\xi} \, . \tag{4.8}$$

Here

$$\xi = \int_0^z n(z)\, dz \, , \ \ \mathrm{cm}^{-2}, \tag{4.9}$$

is the dimensional column depth passed by the fast particles along the z axis; the unit of its measurement is

$$\tilde{\xi} = \frac{k_{\mathrm{B}}^2 T^2}{\pi e^2 q^2 \ln \Lambda} \left(\frac{m}{m_{\mathrm{e}}} \right)^2 , \ \ \mathrm{cm}^{-2}. \tag{4.10}$$

Equation (4.6) in the dimensionless variables (4.7) and (4.8) takes the following form (Somov, 1982):

$$\sqrt{x} \cos\theta \, \frac{\partial f}{\partial \zeta} = \frac{1}{\sqrt{x}} \frac{\partial}{\partial x} \left\{ \sqrt{x}\, D_{\gamma}(x) \left[\frac{\partial f}{\partial x} + \left(\frac{m}{m_{\mathrm{e}}} \right) f \right] \right\} + D_{\theta}(x)\, \Delta_{\theta} f. \tag{4.11}$$

Here

$$D_{\gamma}(x) = \left[\frac{\mathrm{erf}\,(\sqrt{x})}{\sqrt{x}} - \frac{2}{\sqrt{\pi}} \exp{(-x)} \right] + $$
$$+ \left(\frac{m_{\mathrm{e}}}{m_{\mathrm{p}}} \right)^{1/2} \left[\frac{\mathrm{erf}\,(\sqrt{\mathcal{X}})}{\sqrt{\mathcal{X}}} - \frac{2}{\sqrt{\pi}} \exp{(-\mathcal{X})} \right] \tag{4.12}$$

with

$$\mathcal{X} = \frac{m_{\mathrm{p}}}{m_{\mathrm{e}}}\, x$$

and

$$\mathrm{erf}\,(w) = \frac{2}{\sqrt{\pi}} \int_0^w \exp{(-t^2)}\, dt \, ,$$

which is the error function. The diffusion coefficient over the angle θ

$$D_{\theta}(x) = \frac{1}{8x^2} \left\{ \left[\frac{\mathrm{erf}\,(\sqrt{x})}{\sqrt{x}} (2x - 1) + \frac{2}{\sqrt{\pi}} \exp{(-x)} \right] + \right.$$
$$\left. + \left(\frac{m_{\mathrm{e}}}{m_{\mathrm{p}}} \right)^{1/2} \left[\frac{\mathrm{erf}\,(\sqrt{\mathcal{X}})}{\sqrt{\mathcal{X}}} (2\mathcal{X} - 1) + \frac{2}{\sqrt{\pi}} \exp{(-\mathcal{X})} \right] \right\} , \tag{4.13}$$

and

$$\Delta_\theta = \frac{1}{\sin\theta}\frac{\partial}{\partial\theta}\left(\sin\theta\,\frac{\partial}{\partial\theta}\right)$$

is the θ-dependent part of the Laplace operator.

To point out the similarity of the equation obtained with the Fokker-Planck equation (3.10), let us rewrite Equation (4.11) as follows:

$$\sqrt{x}\cos\theta\,\frac{\partial f}{\partial\zeta} = -\frac{\partial}{\partial x}\left[\,F(x)f\,\right] + \frac{\partial^2}{\partial x^2}\left[\,D(x)f\,\right] + D_\theta(x)\,\Delta_\theta f. \tag{4.14}$$

Here the first coefficient

$$F(x) = \frac{dD_\gamma}{dx} - \left(\frac{m}{m_{\rm e}} + \frac{1}{2x}\right) D_\gamma(x) \tag{4.15}$$

characterized the *regular losses* of energy when accelerated particles pass through the plasma. The second coefficient

$$D(x) = D_\gamma(x) \tag{4.16}$$

describes the *energy diffusion*. The third coefficient $D_\theta(x)$ corresponds to the fast particle diffusion over the angle θ.

Kudriavtsev (1958) derived the time-dependent equation which has the right-hand side similar to the one in our Equation (4.11) but for the isotropic distribution function $f = f(t,x)$ for fast ions in a thermal plasma. By using the Laplace transformation, Kudriavtsev solved the problem of maxwelization of fast ions that initially had the mono-energetic distribution $f(0,x) \sim \delta(x - x_0)$. The same problem has been solved numerically by MacDonald et al. (1957). (Note that in formula (8) by Kudriavtsev for the 'radial' component j_v of the fast ion flow in the velocity space, the factor $\sqrt{\pi}$ must be in the nominator but not in the denominator.) Both solutions (analytical and numerical) show, of course, that the higher the ion energy, the longer the maxwellization process.

In the particular case when all the particles are the same ($m = m_{\rm e} = m_{\rm p}$), the right-hand side of Equation (4.11) can be found, for example, by using the formulae for the Fokker-Planck coefficients (3.13) and (3.11) from Balesku (1963).

4.2 A kinetic equation at high speeds

Bearing in mind particles accelerated to high speeds in astrophysical plasma, let us consider some approximations and some solutions of the kinetic Equation (4.11) that correspond to these approximations. First of all, we shall assume that the dimensionless energy (4.7) of the fast particles

$$x \gg 1\,. \tag{4.17}$$

This means that speeds of the particles are much higher than the mean thermal velocity of plasma electrons (8.15). However, for the sake of simplicity, we restrict the problem by consideration of the fast but non-relativistic particles.

Under condition (4.17), we obtain from (4.12) and (4.13) the following simple formulae for the coefficients in the kinetic Equation (4.11):

$$D_{\gamma}(x) = \frac{1}{\sqrt{x}} \left(1 + \frac{m_{\rm e}}{m_{\rm p}} \right) , \tag{4.18}$$

$$D_{\theta}(x) = \frac{1}{2x\sqrt{x}} \, . \tag{4.19}$$

It is *not* taken into account here yet that $m_{\rm e} \ll m_{\rm p}$. The first term on the right-hand side of formula for D_{γ} (see the unit inside the brackets) is a contribution of collisions with the thermal electrons of a plasma, the second term (see the ratio $m_{\rm e}/m_{\rm p}$) comes from collisions with the thermal protons. However the electrons and protons give equal contributions to the angular diffusion coefficient D_{θ}. This is important to see when we derive formula (4.19) from (4.13).

Under the same assumption, the Fokker-Planck type equation (4.14) has the following coefficients:

$$D(x) = \frac{1}{\sqrt{x}} \left(1 + \frac{m_{\rm e}}{m_{\rm p}} \right) , \tag{4.20}$$

$$F(x) = - \frac{m}{m_{\rm e}} \, \frac{1}{\sqrt{x}} \left(1 + \frac{m_{\rm e}}{m} \, \frac{1}{x} \right) , \tag{4.21}$$

and the same coefficient of angular diffusion $D_{\theta}(x)$ of course.

Formulae (4.18) and (4.20) demonstrate that

> energy diffusion due to collisions with thermal electrons is faster in $m_{\rm p}/m_{\rm e}$ times than that due to collisions with thermal protons.

However the angular diffusion rate is equally determined by both electrons and protons in a plasma.

The second term on the right-hand side of the formula for $F(x)$ describes the regular losses of fast particle energy by collisions with thermal protons of plasma. Since $x \gg 1$ and $m \geq m_{\rm e}$, this term is always smaller than the first one. Taking into account that $m_{\rm e} \ll m_{\rm p}$ we also neglect the second term in formula for $D(x)$. Hence, in approximation under consideration,

$$F(x) = - \frac{m}{m_{\rm e}} \, \frac{1}{\sqrt{x}} \, , \qquad D(x) = \frac{1}{\sqrt{x}} \, , \qquad D_{\theta}(x) = \frac{1}{2x\sqrt{x}} \, . \tag{4.22}$$

Let us estimate a relative role of the first and second terms on the right-hand side of Equation (4.14). Dividing the former by the last with account of (4.22) taken gives the ratio

$$\frac{xF(x)}{D(x)} = \frac{m}{m_{\rm e}} \, x \, , \tag{4.23}$$

which is always much greater than unity. So, for fast particles with speeds much greater than the thermal velocity of plasma electrons,

> the regular losses of energy due to Coulomb collisions always dominate the energy diffusion.

However the energy diffusion may appear significant near the lower energy boundary $\mathcal{E}_1$ of the fast particle spectrum if $\mathcal{E}_1 \approx k_{\rm B} T$. This seems to be the case of electron acceleration in high-temperature turbulent-current layers in solar flares (see vol. 2, Sections 6.3 and 7.1). This simply means that, near the lower energy $\mathcal{E}_1 \approx 10$ keV, the initial assumption (4.17) becomes invalid. Instead of (4.17), $x \to 1$; so we have to solve exactly Equation (4.11).

Let us compare the first and third terms on the right-hand side of Equation (4.14). Dividing the former by the last with account of (4.22) taken gives the ratio

$$\frac{F(x)}{x D_\theta (x)} = 2 \, \frac{m}{m_{\rm e}} \, . \tag{4.24}$$

For fast protons and heavier ions, we can neglect angular scattering in comparison with the regular losses of energy.

Formula (4.24) shows, however, that

> for fast electons, it is impossible to neglect the angular diffusion in comparison with the regular losses of energy.

Since the case of fast electrons will be considered later on in more detail, let us rewrite the non-relativistic kinetic equation in the high-speed approximation as follows:

$$\boxed{\cos\theta \, \frac{\partial f}{\partial \zeta} = \frac{1}{x} \frac{\partial f}{\partial x} + \frac{1}{2x^2} \, \Delta_\theta f .} \tag{4.25}$$

Recall that the energy diffusion is neglected in (4.25) according to (4.23).

4.3 The classical thick-target model

We have just seen that, in the fast electron kinetic Equation (4.25), it is not reasonable to neglect the angular diffusion. Let us, however, consider the well-known and widely-used model of a *thick target*. From Equation (4.25), by *neglecting* the angular diffusion, we obtain the following equation

$$\cos\theta \, \frac{\partial f}{\partial \zeta} = \frac{1}{x} \frac{\partial f}{\partial x} \, . \tag{4.26}$$

With a new variable $y = \zeta / \mu$, where $\mu = \cos\theta$, this equation becomes especially simple:

$$\frac{1}{x} \frac{\partial f}{\partial x} - \frac{\partial f}{\partial y} = 0 \, . \tag{4.27}$$

General solution of this equation can be written as

$$f(x,y) = \mathcal{F}\left(\frac{x^2}{2} + y\right), \tag{4.28}$$

where $\mathcal{F}$ is an arbitrary function of its argument. Recall that μ = const, because we have neglected the angular diffusion; so the fast electrons move along straight lines θ = const without any scattering.

Let us consider the initial ($y = 0$) energy distribution of fast electrons – the *injection spectrum* – as a power law:

$$f(x,0) = c_0\, x^{-\gamma_0}\, \Theta(x - x_1)\, \Theta(x_2 - x)\, p_0(\mu)\,. \tag{4.29}$$

Here $\Theta(x)$ is the teta-function; $p_0(\mu)$ is the angular distribution of fast electrons, for example, for a beam of electrons injected parallel to the z axis

$$p_0(\mu) = \frac{1}{(1-\mu^2)^{1/2}}\; \delta\,(\mu - 1)\,. \tag{4.30}$$

According to (4.28) the general solution of the kinetic equation for the fast electrons at the column depth y has the following form:

$$f(x,y) = c_0\, 2^{-\gamma_0/2} \left(\frac{x^2}{2} + y\right)^{-\gamma_0/2} \Theta(x - x_1')\, \Theta(x_2' - x)\, p_0(\mu)\,, \tag{4.31}$$

where

$$x_{1,2}' = Re\,\left(x_{1,2}^2 - 2y\right)^{1/2}.$$

Let us consider the normalization condition for the distribution function, first, in the dimensional variables z, v, and θ (see definition (4.5)). If $n_b(z)$ is the **density of electrons in the beam** at distance z from the injection plane $z = 0$, then

$$n_b(z) = \int_0^\infty \int_0^\pi f\,(z,\, v,\, \theta)\, v^2 dv\, 2\pi\, \sin\theta\, d\theta, \;\; \text{cm}^{-3}. \tag{4.32}$$

It is taken into account here that we consider the case of a beam with the axial symmetry in velocity space.

Now we rewrite the same normalization condition in the dimensionless variable ζ, x, and μ:

$$n_b(\zeta) = \pi \left(\frac{2k_{\rm B}T}{m_{\rm e}}\right)^{3/2} \int_0^\infty \int_{-1}^1 f\,(\zeta,\, x,\, \mu)\, \sqrt{x}\, dx\, d\mu\,, \;\; \text{cm}^{-3}. \tag{4.33}$$

For initial energy distribution (4.29) and initial angular distribution (4.30), formula (4.33) gives

$$n_b(0) = \pi \left(\frac{2k_{\rm B}T}{m_{\rm e}}\right)^{3/2} c_0 \int_{x_1}^{x_2} x^{-\gamma_0 + 1/2}\, dx \equiv \int_{x_1}^{x_2} N(0,x)\, dx\,, \;\; \text{cm}^{-3}. \tag{4.34}$$

Here

$$N(0,x) = \pi \left(\frac{2k_{\rm B}T}{m_{\rm e}} \right)^{3/2} c_0 \, x^{-\gamma_0 + 1/2} \, \Theta(x - x_1) \, \Theta(x_2 - x) \tag{4.35}$$

is the differential spectrum of the fast electron density at the boundary $\zeta = 0$ where they are injected.

Let $\mathcal{E}$ be the kinetic energy of a fast electron measured in keV. Then we rewrite (4.35) as

$$N(0, \mathcal{E}) = K \, \mathcal{E}^{-(\gamma + 1/2)} \, \Theta(\mathcal{E} - \mathcal{E}_1) \, \Theta(\mathcal{E}_2 - \mathcal{E}) \,, \;\; {\rm cm}^{-3} \, {\rm keV}^{-1}, \tag{4.36}$$

where the coefficient

$$K = \pi \left(\frac{2k_{\rm B}T}{m_{\rm e}} \right)^{3/2} c_0 \left(\frac{k_{\rm B}T}{\rm keV} \right)^{\gamma + 1/2} , \;\; {\rm cm}^{-3} \, {\rm keV}^{\,\gamma - 1/2} , \tag{4.37}$$

and the spectral index

$$\gamma = \gamma_0 - 1 \,. \tag{4.38}$$

Hence the *injection* spectrum of fast electrons is determined by parameters (4.37) and (4.38).

Substituting c_0 and γ_0 from (4.37) and (4.38) in (4.31) allows us to obtain the differential spectrum of the number density of fast electrons passed the coulomn depth ξ measured in cm^{-2} (see definition (4.9)):

$$N(\xi, \mathcal{E}) = K \left(\mathcal{E}^2 + \mathcal{E}_0^2 \right)^{-(\gamma + 1/2)/2} \times \tag{4.39}$$

$$\times \, \Theta \left(\mathcal{E} - \mathcal{E}_1' \right) \, \Theta \left(\mathcal{E}_2' - \mathcal{E} \right) , \;\; {\rm cm}^{-3} \, {\rm keV}^{-1} .$$

Here

$$\mathcal{E}_0 = \left(2 a_0 \xi \right)^{1/2} \tag{4.40}$$

is the minimal energy of electrons that can pass the depth ξ, the 'constant' a_0 (a slow function of energy $\mathcal{E}$) originates from the Coulomb logarithm and equals

$$a_0 = 2\pi e^4 \, \ln \Lambda \approx \tag{4.41}$$

$$\approx 1.3 \times 10^{-19} \times \left[\ln \left(\frac{\mathcal{E}}{mc^2} \right) - \frac{1}{2} \, \ln n + 38.7 \right] , \;\; {\rm keV}^2 \, {\rm cm}^2 .$$

In formula (4.39)

$$\mathcal{E}_{1,2}'(\xi) = \left(\mathcal{E}_{1,2}^2 - \mathcal{E}_0^2(\xi) \right)^{1/2} \tag{4.42}$$

are the new boundaries of energetic spectrum, when the fast electrons have passed the column depth ξ.

Solution (4.39) shows that

> the regular losses of energy due to Coulomb collisions shift the spectrum of fast electrons to lower energies and make it harder

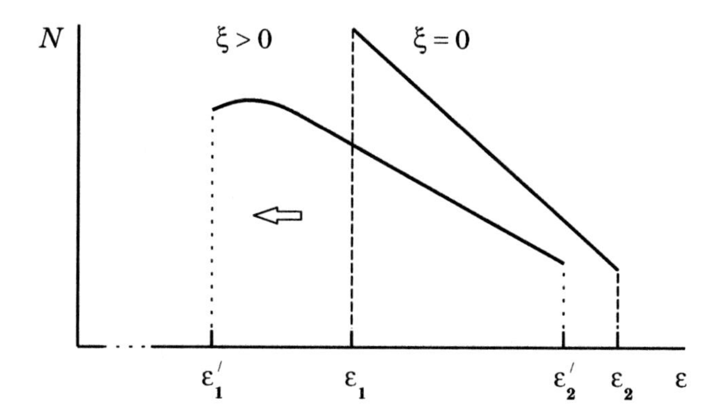

Figure 4.1: An injection spectrum ($\xi = 0$) and the spectrum of fast electrons that have passed the column depth ξ.

as illustrated by schematic Figure 4.1. Both effects follow from the fact that, in Equation (4.26), we have taken into account **only the regular losses** of energy (4.22). For non-relativistic electrons $F(x) = -1/\sqrt{x}$.

In the solar system, the Sun is the most energetic particle accelerator, producing electrons of up to tens of MeV and ions of up tens of Gev. The accelerated 20-100 keV electrons appear to contain a significant part of the total energy of a large solar flare (Lin and Hudson, 1971; Syrovatskii and Shmeleva, 1972), indicating that the particle acceleration and energy release processes are intimately linked. Flare accelerated electrons colliding with the ambient solar atmosphere produce the bremsstrahlung hard X-ray (HXR) emission.

Syrovatskii and Shmeleva (1972) used the solution (4.39) to calculate the HXR bremsstrahlung which arises during inelastic collisions of accelerated electrons with thermal ions in the solar atmosphere during flares (e.g., Strong et al., 1999). Brown (1971), in the same approximation but using a different method, has found a similar formula for HXR intensity but with the different numerical coefficient by factor π in Section 5 (see formulae (14) and (15)). Anyway, since that time,

> the simplest thick-target model is widely accepted as a likely mechanism and an appropriate mathematical tool to explain and describe the HXR emission observed during flares

on the Sun and other stars or generally in cosmic plasma (see, however, Sections 4.4 and 4.5). In the classical formulation of the thick-target model, beams of accelerated electrons stream along the magnetic field lines and loose their energy by Coulomb collisions in denser layers of the solar atmosphere, mainly in the chromosphere.

4.4 The role of angular diffusion

4.4.1 An approximate account of scattering

As we have seen in Section 4.2, for fast electrons, we cannot neglect the angular scattering in comparison with the regular losses of energy in kinetic Equation (4.14). Hence, in the classical thich-target model, we have to take the angular scattering into account at least approximately.

If, for example, the beam of fast electrons penetrates a plane parallel the stratified plasma such as the solar chromosphere, the scattering of an *average* beam of electrons may conveniently be described by the Chandrasekhar-Spitzer formulae (8.51) and (8.52) in terms of a coordinate z normal to the atmospheric strata and directed into the plasma. Then the *mean* electron energy $\mathcal{E}$ may be expressed as a function of z while the scattering is measured in terms of the angle $\theta(z)$ which the *mean* electron velocity $\mathbf{v}$ makes with the z axis at that point. So

$$v_{\parallel} \equiv v_z = v\mu \,, \quad \text{where} \quad \mu = \cos\theta . \tag{4.43}$$

The dimensional column depth passed by electrons along the z axis is

$$\xi = \int_0^z n(z)\, dz \,, \;\; \text{cm}^{-2} . \tag{4.44}$$

In terms of ξ, the Chandrasekhar-Spitzer formulae (8.51) and (8.52) are:

$$\frac{d\mathcal{E}}{d\xi} = -\frac{a_0}{\mathcal{E}}\,\frac{v}{v_z} \tag{4.45}$$

and

$$\frac{dv_z}{d\xi} = -\frac{3}{2}\,\frac{a_0}{\mathcal{E}^2}\, v \,, \tag{4.46}$$

where $a_0 = 2\pi e^4 \ln\Lambda$ (see definition (4.41)). Thus we have an ordinary differential equation

$$\frac{3}{2}\,\frac{1}{\mathcal{E}}\,\frac{d\mathcal{E}}{d\xi} = \frac{1}{v_z}\,\frac{dv_z}{d\xi}$$

with solution

$$\left(\frac{\mathcal{E}}{\mathcal{E}_0}\right)^{3/2} = \frac{v_z}{v_{z0}} \,, \tag{4.47}$$

where the suffix 0 refers to values at $\xi = 0$. Since $v_z/\mu = v$ and $v^2/v_0^2 = \mathcal{E}/\mathcal{E}_0$, we find that

$$\frac{v_z}{v_{z0}} = \frac{\mu}{\mu_0}\left(\frac{\mathcal{E}}{\mathcal{E}_0}\right)^{1/2} .$$

Therefore it follows from (4.47) that

$$\boxed{\frac{\mu}{\mu_0} = \frac{\mathcal{E}}{\mathcal{E}_0} .} \tag{4.48}$$

This nice formula (Brown, 1972) shows that on average when an electron has suffered a 60° deflection its energy has been reduced by 50 %.

Resubstituting (4.48) in (4.45) and (4.46) gives the solutions for μ and $\mathcal{E}$:

$$\frac{\mu}{\mu_0} = \frac{\mathcal{E}}{\mathcal{E}_0} = \left(1 - \frac{3\, a_0 \xi}{\mu_0\, \mathcal{E}_0^2} \right)^{1/3} . \tag{4.49}$$

For small depth ξ

$$\frac{\mu}{\mu_0} = \frac{\mathcal{E}}{\mathcal{E}_0} \approx 1 - \frac{a_0}{\mu_0\, \mathcal{E}_0^2}\, \xi . \tag{4.50}$$

Let us compare these results with the general solution (4.28) obtained without account taken of scattering in the classical thick-target model.

4.4.2 The thick-target model

According to (4.28)

$$\frac{x^2}{2} + y = \frac{x_0^2}{2} , \tag{4.51}$$

where x_0 is an initial energy of an electron. Hence

$$\frac{x}{x_0^2} = (1 - 2y)^{1/2} , \tag{4.52}$$

where $y = \zeta / \mu$ and $\mu = \text{const} = \mu_0$. Therefore for electrons with initial energy $\mathcal{E}_0$ solution (4.28) gives us:

$$\frac{\mathcal{E}}{\mathcal{E}_0} = \left(1 - \frac{2a_0}{\mu_0\, \mathcal{E}_0^2}\, \xi \right)^{1/2} . \tag{4.53}$$

If

$$\xi \ll \xi_0 = \frac{\mathcal{E}_0^2}{2a_0} ,$$

then

$$\frac{\mathcal{E}}{\mathcal{E}_0} \approx 1 - \frac{a_0}{\mu_0\, \mathcal{E}_0^2}\, \xi . \tag{4.54}$$

Formula (4.54) coincides with (4.50). The fast electrons in the thick-target model have the same behaviour at small depth ξ as that one predicted by the approximate Chandrasekhar-Spitzer formulae.

However, with increase of the column depth ξ, the approximate formula (4.49) predicts much faster losses of energy in comparison with the classical thick-target model which does not take collisional scattering into account.

In Figure 4.2, the dashed straight line (a) corresponds to the asymptotic formula (4.50) which is valid for small column depth ξ. Moreover here $\mu_0 = 0$, so

$$\frac{\mathcal{E}}{\mathcal{E}_0} \approx 1 - \frac{1}{2}\frac{\xi}{\xi_0}. \tag{4.55}$$

The solid curve (b) represents the **classical thick-target model**; it takes only the collisional losses of energy into account. With $\mu_0 = 0$, formula (4.53) is

$$\frac{\mathcal{E}}{\mathcal{E}_0} = \left(1 - \frac{\xi}{\xi_0}\right)^{1/2}. \tag{4.56}$$

An approximate scattering model described above is presented by the curve (c) which corresponds to formula (4.49) with $\mu_0 = 0$, so

$$\frac{\mathcal{E}}{\mathcal{E}_0} = \left(1 - \frac{3}{2}\frac{\xi}{\xi_0}\right)^{1/3}. \tag{4.57}$$

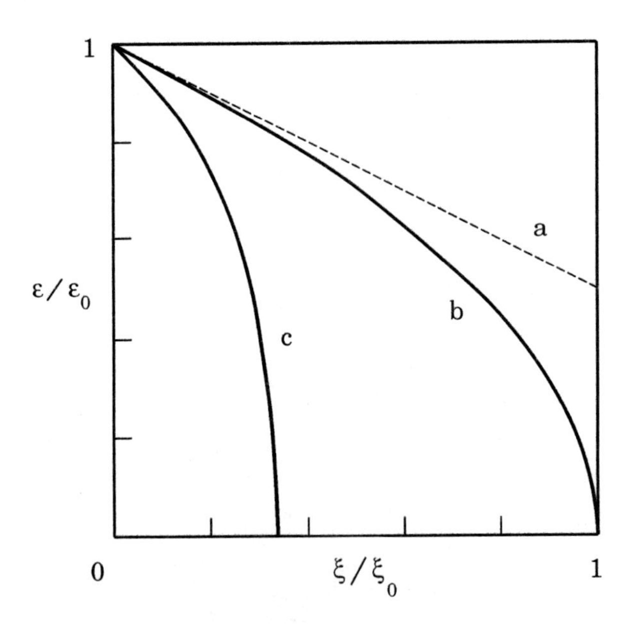

Figure 4.2: The mean energy $\mathcal{E}$ of fast electrons that have passed the column depth ξ (from Somov, 1982).

Figure 4.2 shows that

the collisional scattering and energy losses become very great in comparison with the classical thick-target model if the column depth ξ is not very small.

Brown (1972) used the approximate formula (4.49) to develop an approximate thick-target model in which accelerated electrons penetrate downward into the solar chromosphere during a flare. Here the electron distribution is greatly modified by collisions – not only by energy losses but also by scattering. **Directivity and polarization** of the hard X-ray bremsstrahlung emission have been calculated in such oversimplified thick-target model in which the guiding field $\mathbf{B}_0$ is vertical. The model predicted that the degree of polarization should rise from zero to around 30 % near the solar limb.

Unfortunately the accuracy of the model decreases when the collisional scattering and energy losses become not small. The reason is that the mean rates (4.45) and (4.46) represent well the modification of the electron velocity distribution only at small depth ξ. A more accurate formulation of the kinetic problem will be given in the next Section with account taken of the collisional scattering and one more mechanism of the electron beam anisotropization. Generally, it seems true that the total absorption of the accelerated electrons in a thick target might result in *negligible* directivity and polarization of the hard X-ray emission.

4.5 The reverse-current electric-field effect

4.5.1 The necessity for a beam-neutralizing current

We assume that some external magnetic field $\mathbf{B}_0$ channels a fast particle propagation and can be locally considered as uniform. The electric and magnetic fields $\mathbf{E}$ and $\mathbf{B}$ related to a beam of fast electrons are superposed on this field. In this way, the beam will be considered as a real electric current $\mathbf{J}$ which influences the background plasma and magnetic field $\mathbf{B}_0$. In order not to obscure the essential physical points related to the electromagnetic field of the beam, we shall neglect all other processes like the radiative and hydrodynamic response of the background plasma to a fast heating by the electron beam (Section 8.3.2).

In the classical thick-target model for hard X-ray bremsstrahlung emission during solar flares, if the fast electrons are supposed to have about the parallel velocities, then the number of injected beam particles per unit time has to be very large – in the order of $\gtrsim 10^{36}$ electrons s^{-1} above 25 keV during the impulsive phase of a flare (Hoyng et al., 1976). Given the large electron fluxes implied by the hard X-ray observations, various authors realized that the beam electric current must be enormous – $J \gtrsim 10^{17}$ Ampere.

This would imply the magnetic field of the beam $B \gtrsim 10^5$ G. So the magnetic energy contents of the coronal volume should be more than six orders of magnitude larger than the pre-flare contents for an average coronal field $B_0 \approx 100$ G. Such situation is not likely to occur because the electron beams are thought to be created by conversion of the magnetic energy available in the corona into kinetic energy.

Apart from this energy problem there is another difficulty related to beams of $\sim 10^{36}$ electrons s^{-1}; they create an enormous charge displacement. For a typical coronal volume of 10^{28} cm^3 and an electron density 10^9 cm^{-3}, the total number of electrons is 10^{37}. A stream of 10^{36} electrons s^{-1} would evacuate all the electrons out of the volume in about 10 s. As a result an enormous charge difference between the corona and the chromosphere would be build up.

In reality the above mentioned problems will not occur, because the beam propagates in a background **well-conducting plasma**. The charge displacement by the beam will quickly create an electric field $\mathbf{E}_1$ which causes the plasma electrons to redistribute in such a way as to *neutralize* the local charge built:

$$\operatorname{div} \mathbf{E}_1 = 4\pi \rho^q \, . \tag{4.58}$$

Because this electric field is caused by charge separation, it is frequently referred to as an *electrostatic* field.

The second effect is related to the inductive properties of a plasma. In a plasma the magnetic field will not vary considerably on a timescale shorter than the magnetic diffusion time. For beams with radii comparable to the radii of coronal flaring loops this scale is much longer than the duration of the impulsive phase. When the current varies in magnitude, immediately an *inductive* electric field $\mathbf{E}_2$ will be created. It drives a current $\mathbf{j}_2$ of plasma electrons in such a way to prevent magnetic field variations on a time scale shorter than the magnetic diffusion time. As a result the magnetic field will not vary much during the impulsive phase:

$$\operatorname{curl} \mathbf{B} \approx \text{const} \approx 0 \approx \frac{4\pi}{c} \mathbf{j}_2 + \frac{1}{c} \frac{\partial}{\partial t} \mathbf{E}_2 \, . \tag{4.59}$$

So the electrostatic effect allows the plasma to 'absorb' the excess charge imposed by the beam of fast electrons; and the inductive effect prevents the magnetic field from changing faster than the allowed diffusion time.

> Both the electrostatic and the inductive electric field will effectively result in an electron plasma current which is in opposite direction of the beam current **J**.

This electron plasma current is commonly referred to as the *reverse* or *return* current $\mathbf{J}_{rc}$.

Van den Oord (1990) has analyzed the electrostatic and inductive response of a plasma to a prescribed electron beam. By using the Maxwell equations together with the time-dependent Ohm's law (Section 11.2) and with the equation of motion for the plasma electrons in the hydrodynamic approximation (Section 9.4), he has shown that the non-linear terms are responsible for a coupling between the electrostatic (irrotational) and inductive (solenoidal) vector fields generated by the beam in a plasma. In order to obtain analytical solutions, van den Oord has decoupled the electrostatic and inductive fields, by ignoring the non-linear terms in the equation of motion, and has found solutions for a mono-energetic blunt beam.

An application of the model in conditions of the solar corona leads to the following results. Charge neutralization is accompanied by plasma oscillations (see formula (8.35)), that are present behind the beam front, and occurs on a time-scale of a few electron-ion collision times. This is also the time scale on which the plasma waves damp out. The net current in the system quickly becomes too low and therefore also the resulting magnetic field strength remains low ($B \ll B_0$).

Although the electric field near the beam front is locally strong, the oscillatory character prevents strong acceleration of the plasma electrons. According to the van den Oord model, all the beam energy is used initially to accelerate the plasma electrons from rest and later on to drive the reverse current against collisional losses. In what follows, we shall use these results and shall formulate an opposite problem in the kinetic approximation. We shall not consider the beam as prescribed. On the contrary, we shall consider **an influence of the electric field**, which drives the reverse current, **on the distribution function of fast electrons** in the thick-target plasma.

4.5.2 Formulation of a realistic kinetic problem

The *direct* electric current carried by the fast electrons is equal to

$$j_{dc}(z) = e \int_{\mathbf{v}} f(v,\, \theta,\, z)\, v\, \cos\theta\, d^3\mathbf{v}\,. \tag{4.60}$$

We shall consider this current to be fully balanced by the reverse current of the thermal electrons in the ambient plasma,

$$j_{dc}(z) = j_{rc}(z) \equiv j(z)\,. \tag{4.61}$$

This means that here we do not consider a very fast process of the reverse current generation. The time-dependent process of current neutralization, with account of both electrostatic and inductive effects taken (Section 4.5.1), has been investigated in linear approximation by van den Oord (1990). Instead of that we shall construct a self-consistent approach for solving the pure kinetic problem with a steady electric field $E = E(z)$ which drives the reverse current.

So, using Ohm's law, we determine the reverse-current electric field to be equal to

$$E(z) = \frac{j(z)}{\sigma}\,. \tag{4.62}$$

Here σ is conductivity of the plasma; we can assume that the conductivity is determined by, for example, Coulomb collisions (Section 11.1). This is the case of a cold dense astrophysical plasma.

On the other hand, the plasma turbulence effects are also important, for example, in the heat conductive front between the high-temperature source of energy and cold plasma of the thick-target. Anyway, even though we expect the wave-particle interactions to have some effects on the fast electrons (Chapter 7), it is unlikely that such effects can change significantly the distribution function of fast electrons with energies far exceeding the energies of the particles in a background cold plasma.

What is really important is the **reverse-current electric field**, it results in an essential change of the fast electron behaviour in the plasma. That is why, to solve the thick-target problem, we develop a combined approach which takes into account the electric field (4.62) as in the Vlasov equation and Coulomb collisions as in the Landau equation. So the distribution function for the fast electrons in the target is described by the following equation (Diakonov and Somov, 1988):

$$v \cos\theta \, \frac{\partial f}{\partial z} - \frac{eE(z)}{m_e} \cos\theta \, \frac{\partial f}{\partial v} - \frac{eE(z)}{m_e v} \sin^2\theta \, \frac{\partial f}{\partial \cos\theta} = \left(\frac{\partial f}{\partial t} \right)_{\rm c} . \tag{4.63}$$

Here the second and the third terms are the expression of the term

$$\frac{e_e}{m_e} \, \mathbf{E}\,(\mathbf{r}) \, \frac{\partial f}{\partial \mathbf{v}}$$

in the dimensional variables v and θ. On the right-hand side of Equation (4.63)

$$\left(\frac{\partial f}{\partial t} \right)_{\rm c} = \frac{1}{v^2} \frac{\partial}{\partial v} \left[v^2 \, \nu(v) \left(\frac{k_{\rm B} T_e}{m_e} \frac{\partial f}{\partial v} + v f \right) \right] +$$

$$+ \, \nu(v) \, \frac{\partial}{\partial \cos\theta} \left(\sin^2\theta \, \frac{\partial f}{\partial \cos\theta} \right) \tag{4.64}$$

is the linearized collisional integral; $\nu(v)$ is the collisional rate for fast electrons in the cold plasma.

To set the mathematical problem in the simplest form (see Figure 4.3), we assume that 'superhot' ($T_{e,0} = T_0 \gtrsim 10^8$ K) and 'cold' ($T_{e,1} = T_1 \sim 10^4 - 10^6$ K $\ll T_0$) plasmas occupy the two half-spaces separated by the plane turbulent front ($z = 0$). The superhot region represents the source of energy, for example, the high-temperature reconnecting current layer (RCL) in a solar flare. Let

$$f_s = f_s(v, \theta) \tag{4.65}$$

be the electron distribution function in the *source*. f_s is, for example, the Maxwellian function for the case of thermal electron runaway (Diakonov and Somov, 1988) or a superposition of thermal and nonthermal functions in the general case. To study the effect of the reverse-current electric field in the classical thick-target model, Litvinenko and Somov (1991b) considered only

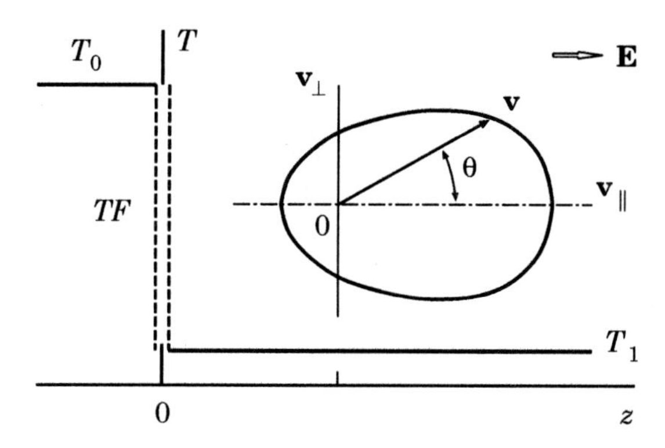

Figure 4.3: The fast electron propagation in a thick-target cold plasma. *TF* is the turbulent front between the superhot source of fast electrons and the cold plasma.

accelerated electrons with an energetic power-law spectrum. Anyway, the function f_s is normalized to the electron number density n_0 in the source:

$$\int f_s(v,\, \theta)\, d^3\mathbf{v} = n_0\,. \tag{4.66}$$

Because the electron runaway in a turbulent plasma (Gurevich and Zhivlyuk, 1966) is similar to the ordinary collisional runaway effect (Section 8.4.3), the electrons with velocities

$$v_e > v_{cr}\,, \tag{4.67}$$

where v_{cr} is some critical velocity, can freely penetrate through the turbulent front into the cold plasma. Electrons with lower velocities remain trapped in the source. In this Section, we are going to consider the distribution function for the fast electrons escaping into the cold plasma and propagating there. The boundary condition for the forward-flying (the suffix ff) fast electrons may be taken as

$$f_{ff}(v,\, \theta,\, 0) = f_s(v,\, \theta)\, \Theta(v - v_{cr})\,, \qquad 0 \le \theta \le \pi/2\,, \tag{4.68}$$

where Θ is the theta-function.

The distribution function for the back-flying electrons is determined from the solution of Equation (4.63) everywhere, including the boundary $z = 0$. Therefore the problem has been formulated. Note the obvious but important thing; Equation (4.63) contains **two unknown functions**: the fast electron distribution function $f(v,\, \theta,\, z)$ and the electric field $E(z)$. So the kinetic Equation (4.63) must be solved together with Equations (4.60) (4.62). This is the complete set of equations to be solved self-consistently.

4.5.3 Dimensionless parameters of the problem

In the dimensionless variables (4.7), (4.8) and $\mu = \cos\theta$, Equation (4.63) takes the form

$$\mu x^2 \frac{\partial f}{\partial \zeta} - 2\varepsilon \mu x^2 \frac{\partial f}{\partial x} - \varepsilon x \left(1 - \mu^2\right) \frac{\partial f}{\partial \mu} = x \frac{\partial f}{\partial x} + \tau x \frac{\partial^2 f}{\partial x^2} + \frac{1}{2} \Delta_\mu f. \quad (4.69)$$

Here the dimensionless electron energy

$$x = \frac{m_e v^2}{2 k_B T_0} \quad (4.70)$$

is normalized with the temperature T_0 of the superhot plasma; for example, $T_0 = T_{e,cl} \approx 100$ MK is an effective electron temperature of the high-temperature (super-hot) turbulent-current layer (see vol. 2, Section 6.3) The ratio of the cold-to-superhot plasma temperature

$$\tau = \frac{T_1}{T_0} \approx 10^{-4}, \quad (4.71)$$

if we consider as example the injection of fast electrons into the solar chromosphere. The dimensionless column depth ζ (see definition (4.8)) equals the dimensional column depth passed by fast electrons

$$\xi = \int_0^z n(z)\, dz\,, \;\; \text{cm}^{-2}, \quad (4.72)$$

divided by the unit of its measurement

$$\tilde{\xi} = \frac{k_B^2 T_0^2}{\pi e^4 \ln \Lambda}\,, \;\; \text{cm}^{-2}. \quad (4.73)$$

The dimensionless electric field

$$\varepsilon = \frac{E}{E_{D,1}} \frac{2}{\tau}\,, \quad (4.74)$$

where

$$E_{D,1} = \frac{4\pi e^3 \ln \Lambda}{k_B} \frac{n_1}{T_1} \quad (4.75)$$

is the Dreicer field in the cold plasma of the target (cf. definition (8.70)).

The parameter ε can be found from the self-consistent solution of the complete set of equations and the boundary conditions as desribed in Section 4.5.2. The parameter ε is not small in a general case and, in particular, in the solar flare problem $\varepsilon \approx 2 - 20$ (see Figure 4 in Diakonov and Somov, 1988). Therefore, from (4.74)

$$E = \varepsilon \frac{\tau}{2} E_{D,1} \approx \left(10^{-4} - 10^{-3}\right) E_{D,1}\,, \quad (4.76)$$

so Ohm's law (4.62) is well applicable in this case.

Let us set the specific form of the boundary distribution function (4.68). The processes of electron acceleration in astrophysical plasma and their heating are always closely related. However, for the sake of contrast of them to each other, we consider separately two different functions.

(a) We shall suppose that the electron distribution in the superhot plasma is near to the Maxwellian one. So the distribution function

$$f_s(x,\, \mu) = n_0\, c_0\, \exp\left(-x\right) h(\mu)\,, \qquad \mu \geq 0\,, \tag{4.77}$$

with the constant

$$c_0 = \left(\frac{m_{\rm e}}{2\pi k_{\rm B} T_0} \right)^{3/2} .$$

(b) For accelerated electrons we shall use the power-law spectum as the boundary distribution function for the forward-flying electrons

$$f_{ff}(x,\, \mu) = f_s(x,\, \mu)\, \Theta(\mu - 1) = n_0\, c_0\, x^{-\gamma}\, h(\mu)\,, \qquad \mu \geq 0\,, \tag{4.78}$$

with another normalization constant c_0. In principle, the function $h(\mu)$ is indefinite but should satisfy some additional conditions; at least the function $h(\mu)$ should be maximally smooth (Diakonov and Somov, 1988).

4.5.4 Coulomb losses of energy

4.5.4 (a) Electric current in the thick target

In Equation (4.69), the term $\tau x\, (\partial^2 f / \partial x^2)$ describes the energy diffusion. As we know from Section 4.2, for fast electrons with velocities much greater than the thermal velocity of plasma electrons, the regular losses of energy due to collisions always dominate the energy diffusion. So we neglect this term in comparison with the term $x\, (\partial f / \partial x)$.

However, as we also know from Section 4.2, we cannot neglect the term with the μ-dependent part $\Delta_\mu f$ of the differential operator Laplacian Δ. This term is responsible for the angular diffusion of electrons and is not small in comparison to the regular losses term $x\, (\partial f / \partial x)$.

Therefore we can ignore only the term with small parameter τ in Equation (4.69). After that we have

$$\mu x^2\, \frac{\partial f}{\partial \zeta} = 2\varepsilon\, \mu x^2\, \frac{\partial f}{\partial x} + \varepsilon x \left(1 - \mu^2\right) \frac{\partial f}{\partial \mu} + x\, \frac{\partial f}{\partial x} + \frac{1}{2}\, \frac{\partial}{\partial \mu} \left[\left(1 - \mu^2\right) \frac{\partial f}{\partial \mu} \right] . \tag{4.79}$$

By using this equation, we would like to obtain the equation which determines the behaviour of the **direct electric current** carried by fast electrons in the target. It follows from definition (4.60) that

$$j_{dc}(\zeta) = 2\pi e \left(\frac{2 k_{\rm B} T_0}{m_{\rm e}} \right)^2 \int\limits_0^\infty \int\limits_{-1}^{+1} f(x,\, \mu,\, \zeta)\, x \mu\, dx\, d\mu\, . \tag{4.80}$$

So we have to divide Equation (4.79) by x and to integrate it as in formula (4.80).

All terms on the right-hand side of Equation (4.79), except one, give zero contributions. The only term $x\,(\partial f / \partial x)$, describing the regular energy losses due to Coulomb collisions, determines the changes of electric current

$$j(\zeta) = j_{dc}(\zeta) = j_{rc}(\zeta) \tag{4.81}$$

along the coulomn depth ζ into the target. It gives the right-hand side of the equation:

$$\frac{dj}{d\zeta} = -\,c_j \int\limits_{-1}^{+1} f(x,\,\mu,\,\zeta)\,d\mu \tag{4.82}$$

with constant

$$c_j = \pi e \left(\frac{2k_{\rm B} T_0}{m_{\rm e}} \right)^2 . \tag{4.83}$$

The physical meaning of Equation (4.82) is that

> fast electrons lose their energy and mix with thermal particles of the ambient cold plasma due to Coulomb collisions.

Thus the self-consistent reverse-current problem demands to consider the term $x\,(\partial f / \partial x)$, describing the Coulomb energy losses.

4.5.4 (b) 2D versus 1D models for the thick target

Equation (4.82) shows that the electric current $j(\zeta)$ decreases along the coulomn depth ζ into the target because of the 'falling out' of 'completely' stopped ($x = 0$) electrons from the distribution function owing to collisional losses of energy. From the electric current continuity equation it follows that a current change is possible only when there are electron 'sources' and/or 'sinks' in the thick target.

In the energy region where Equation (4.69) is valid ($x \gg \tau$), the **collisional friction force** (Section 8.4.1) is inversely proportional to x. For this reason, the electrons with low energies quickly slow down to energies of the order of τ and thus mix with the thermal electrons in the ambient plasma. Since in Equation (4.79) formally $\tau = 0$, the 'falling out' takes place under $x = 0$ according to formula (4.82).

The models under consideration in this Chapter, except the classical thick-target model in Section 4.3, are *two-dimensional* (2D) in the velocity space (see definition (4.5)). This fact has an important consequence.

> Some electrons after injection into the thick target make a curve trajectory and cross the boundary in the reverse direction without significant losses of energy.

These electrons come back to the source (the place of acceleration) without being stopped in the target; they determine the boundary distribution function for *back-flying* electrons and constitute a *significant* part (possibly the bulk) of all injected electrons.

Such a process is impossible in one-dimensional (1D) models, like the classical thick-target model, because an electron cannot change the initial direction to the opposite one without being stopped to zeroth velocity and accelerated by the reverse-current electric field from the zeroth velocity in the reverse direction. So collisional losses of energy are involved twice in the 1D dynamics of all fast electrons stopped in the target. In general, the 1D kinetic models taking Coulomb collisions into account are non-physical approximations.

The other group of injected electrons considered in 2D models is composed of the fast electrons which, after moving in the target under electrostatic and friction forces, do not come back in the particle source. With suitable values of energy x and angle θ, they lose a lot of their initial energy and stop their motion in the target not far from the boundary. There seem to be small amounts of such particles. They determine the electric current change. Thus the current $j(\zeta)$ and, hence, the electric field $E(\zeta)$ can change slowly near the boundary.

Among the particles that determine the current, we may choose a small subgroup of fast electrons which penetrate to such a depth into the target where the electric field is very small ($\varepsilon \ll 1$) and further on they are moving affected only by collisions. Even for this small subgroup the 2D models are certainly more realistic in comparison with the 1D models which do not take into account the collisional scattering (Section 4.4).

4.5.5 New physical results

Usually to solve the 2D (in velocity space) kinetic equation one develops a complicated numerical method. Diakonov and Somov (1988) have developed a new technique to obtain an approximate analytical solution of Equation (4.63) taking the Coulomb collisions and the reverse-current field into account. They have applied this technique to the case of thermal runaway electrons in solar flares. It appears that the reverse-current electric field leads to a **significant reduction of the convective heat flux** carried by fast electrons escaping from the high-temperature plasma to the cold one.

> It is not justified to exclude the reverse-current electric-field effects in studies of convective heat transport by fast thermal electrons in astrophysical plasma, for example, in solar flares.

Litvinenko and Somov (1991b) have used the same technique to study the behaviour of the electrons accelerated inside a reconnecting current layer (RCL) in the solar atmosphere during flares. They have shown that the reverse-current electric field results in an essential change of the fast electron behaviour in the thick target.

> The reverse-current electric field leads to a *quicker* decrease of the distribution function with the column depth in comparison with the classical thick-target model.

It is worth mentioning here that both models (thermal and non-thermal) lead to practically the same value of the field near the boundary, ε_0, and this value is large: $\varepsilon_0 \gg 1$. So the effects of the reverse-current field are not small.

The distribution function appears to be an *almost isotropic* one. The main part of the injected electrons returns into the source. As a result, the hard X-ray polarization appears much smaller than in the collisional thick-target model without taking account of the reverse current. In calculations by Litvinenko and Somov (1991b), the maximum polarization was found to be of about 4 % only. So a major conclusion of this section is that

> in order to have a more precise insight into the problem of electron acceleration in solar flares, we inevitably have to take into account the reverse-current electric-field effects.

They make the accelerated electron distribution to be almost isotropic and leads to a significant decrease of expected hard X-ray bremsstrahlung polarization (Somov and Tindo, 1978).

4.5.6 To the future models

After all said above, it is rather surprising to conclude that the most of the above mentioned 2D models, which have been developed after the *classical* thick-target model (Section 4.3), are however not used to obtain a more realistic quantitative informaton on fast electrons in solar flares. The simplest classical thick-target model is still very popular. Up to now we do not have a realistic time-dependent self-consistent thick-target model (which must be simple enough to be easily used) to interpret and analyze the hard X-ray emission so frequently detected in space.

Future models will incorporate such fine effects like a nonuniform initial ionization of chromospheric plasma in the thick-target (Brown et al., 1998a; 2003), the time-of-flight effect (Aschwanden et al., 1998; Brown et al., 1998b; Aschwanden, 2002), with account taken of the effect of the **reverse-current electric field** as an effect of primary importance. Otherwise the accuracy of a model is lower that the accuracy of modern hard X-ray data obtained by *RHESSI* (Lin et al., 2002; 2003).

* * *

Now let us clarify our plans. Before transition to the hydrodynamic description that is valid for systems containing a large number of colliding particles, we have to study two particular but interesting cases.

First, $N = 1$, a particle in a given force field. This simplest approximation gives us clear approach to several fundamental issues of collisionless plasma.

In particular, it is necessary to outline the basis of kinetic theory for wave-particle interactions in astrophysical plasma (Chapter 7).

Second, $N = 2$, binary collisions of particles with the Coulomb potential of interaction. They are typical for collisional plasma. We have to know the Coulomb collisions well to justify the hydrodynamic description of astrophysical plasma (Chapter 9).

In the next Chapter we start from the former.

4.6 Practice: Exercises and Answers

Exercise 4.1. [Section 4.3] How deep can the accelerated electrons with the initial energy $\mathcal{E}_0 \approx 10$ keV penetrate from the solar corona into the chromosphere?

Answer. From formula (4.40) we find the simplest estimation for the column depth

$$\xi = \frac{\mathcal{E}_0^2}{2a_0}\,, \ \mathrm{cm}^{-2}. \tag{4.84}$$

Substituting $\mathcal{E}_0 \approx 10$ keV and $n \approx 10^{12}$ cm^{-3} in formula (4.41) gives $a_0 \approx 3 \times 10^{-18}$ keV^2 cm^2. With this value a_0 we find $\xi \approx 10^{19}$ cm^{-2}. At such depth in the chromosphere, the density of the plasma $n \approx 10^{12}$ cm^{-3} indeed.

Accelerated electrons with energies $\mathcal{E} > 10$ keV penetrate deeper and contribute significantly to impulsive heating of the optical part of a solar flare (see a temperature enhancement at $\xi \approx 10^{20}$ cm^{-2} in Figure 8.4).

Exercise 4.2. [Section 4.5] How strong is the reverse-current electric field in the chromosphere during a solar flare?

Answer. According to (4.76), the electric field

$$E = \varepsilon\,\frac{\mathcal{T}}{2}\,E_{\mathrm{D},1} \approx \left(10^{-4} - 10^{-3}\right) E_{\mathrm{D},1}\,. \tag{4.85}$$

In the chromosphere (Exercise 8.4), the Dreicer field $E_D > 0.1$ V cm^{-1}. So, under injection of accelerated electrons into the chromosphere during the impulsive phase of a flare, the reverse-current field $E > 10^{-5} - 10^{-4}$ V cm^{-1}. With the length scale $l \sim 10^3$ km, this electric field gives rise to a potential $\phi \approx E\,l \sim 1 - 10$ keV.

Exercise 4.3. [Section 4.5.4] Discuss expected properties of a solution of Equation (4.79) without the collisional energy losses term $x\,(\partial f / \partial x)$.

Chapter 5

Motion of a Charged Particle in Given Fields

Astrophysical plasma is often an extremely tenuous gas of charged particles, without net charge on average. If there are very few encounters between particles, we need only to consider the responses of a particle to the force fields in which it moves. The simplest situation, a single particle in given fields, allows us to understand the drift motions of different origin and electric currents in such collisionless plasma.

5.1 A particle in constant homogeneous fields

5.1.1 Relativistic equation of motion

In order to study the motion of a charged particle, let us consider the following basic equation:

$$\frac{d\,\mathbf{p}}{dt} = e\,\mathbf{E} + \frac{e}{c}\,\mathbf{v} \times \mathbf{B} + m\,\mathbf{g}\,. \tag{5.1}$$

In relativistic mechanics (see Landau and Lifshitz, *Classical Theory of Field*, 1975, Chapter 2, § 9) the particle momentum and energy are

$$\mathbf{p} = \frac{m\,\mathbf{v}}{\sqrt{1 - v^2/c^2}} \quad \text{and} \quad \mathcal{E} = \frac{mc^2}{\sqrt{1 - v^2/c^2}}\,, \tag{5.2}$$

respectively. By using the Lorentz factor

$$\gamma_{\mathrm{L}} = \frac{1}{\sqrt{1 - v^2/c^2}}\,, \tag{5.3}$$

we rewrite formulae (5.2) as

$$\mathbf{p} = \gamma_{\mathrm{L}}\, m\,\mathbf{v} \quad \text{and} \quad \mathcal{E} = \gamma_{\mathrm{L}}\, mc^2. \tag{5.4}$$

Hence

$$\mathbf{p} = \frac{\mathcal{E}}{c^2}\,\mathbf{v}\,. \tag{5.5}$$

By taking the scalar product of Equation (5.1) with the velocity vector $\mathbf{v}$ we obtain

$$\frac{d\mathcal{E}}{dt} = \mathbf{F}\cdot\mathbf{v}\,, \tag{5.6}$$

where

$$\mathbf{F} = e\,\mathbf{E} + m\,\mathbf{g}$$

is a *non-magnetic* force. The particle kinetic energy change during the time dt is $d\mathcal{E} = \mathbf{v}\cdot d\mathbf{p}$. Therefore, according to Equation (5.6), **the work on a particle is done by the non-magnetic force only**. In what follows we shall remember that magnetic fields are 'lazy' and do not work.

Let us consider the particle motion in *constant homogeneous* fields.

5.1.2 Constant non-magnetic forces

Now let a non-magnetic force be parallel to the y axis, $\mathbf{F} = F\,\mathbf{e}_y$, and let the initial momentum of the particle be parallel to the x axis, $\mathbf{p}_0 = p_0\,\mathbf{e}_x$.

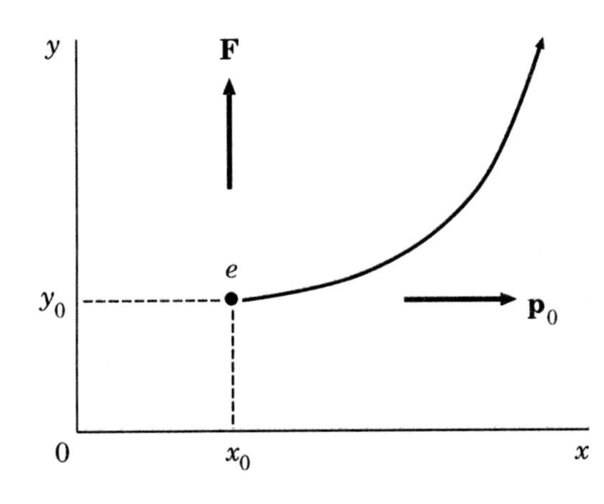

Figure 5.1: The trajectory of particle motion under the action of a constant non-magnetic force.

Then we integrate Equation (5.1) to find that the particle moves along the catenary shown in Figure 5.1:

$$y - y_0 = \frac{\mathcal{E}_0}{F}\left\{\cosh\left[\frac{F}{p_0 c}\,(x - x_0)\right] - 1\right\}. \tag{5.7}$$

Here $\mathcal{E}_0$ is an initial energy of the particle.

Formula (5.7) in the non-relativistic limit is that of a parabola:

$$y - y_0 = \frac{F}{2mv_0^2}\,(x - x_0)^2\,.$$

5.1.3 Constant homogeneous magnetic fields

Let the non-magnetic force $\mathbf{F} = 0$. The magnetic force in a constant and homogeneous field results in particle motions. Let us show that. From Equation (5.1) we have

$$\frac{d\mathbf{p}}{dt} = \frac{e}{c}\,\mathbf{v}\times\mathbf{B}\,. \tag{5.8}$$

We known by virtue of (5.6) that the particle kinetic energy $\mathcal{E} = \text{const}$. Therefore $|\,\mathbf{v}\,| = \text{const}$, and from Equation (5.8)

$$\dot{\mathbf{v}} = \omega_{\text{B}}\,\mathbf{v}\times\mathbf{n}\,. \tag{5.9}$$

Here the overdot denotes the derivative with respect to time t, $\mathbf{n}$ is the unit vector along the field $\mathbf{B} = B\,\mathbf{n}$, and the constant

$$\omega_{\text{B}} = \frac{ecB}{\mathcal{E}} \tag{5.10}$$

is the *gyrofrequency* or *cyclotron* frequency. We use sometimes, in what follows, the name Larmor frequency. The last is a slightly confusing terminology in view of the fact that there is the frequency of the Larmor precession (see § 45 in Landau and Lifshitz, *Classical Theory of Field*, 1975), ω_{L}, which turns out to be half of the gyrofrequency ω_{B}.

In the non-relativistic limit, the gyrofrequency

$$\boxed{\omega_{\text{B}} = \frac{eB}{mc}\,.} \tag{5.11}$$

By integrating Equation (5.9) we find the linear differential equation

$$\dot{\mathbf{r}} = \omega_{\text{B}}\,\mathbf{r}\times\mathbf{n} + \mathbf{C}\,, \tag{5.12}$$

where vector $\mathbf{C} = \mathbf{const}$.

By taking the scalar product of Equation (5.12) with the unit vector $\mathbf{n}$ we have

$$\mathbf{n}\cdot\dot{\mathbf{r}} = C_{\parallel} \equiv v_{\parallel}\,(t=0)\,.$$

The constant $\mathbf{C}_{\perp}$ can be removed from consideration by an appropriate choice of the moving reference system. $\mathbf{C}_{\perp} = 0$ in the reference system where $\mathbf{F} = 0$ (Section 5.1.4), and this choice is consistent with the initial Equation (5.8). Therefore

$$\dot{\mathbf{r}}_{\perp} = \omega_{\text{B}}\,\mathbf{r}_{\perp}\times\mathbf{n}\,. \tag{5.13}$$

The vector $\mathbf{r}_{\perp}$ is changing with the velocity $\mathbf{v}_{\perp}$ which is perpendicular to $\mathbf{r}_{\perp}$ itself. Hence the change of vector $\mathbf{r}_{\perp}$ is a *rotation with the constant frequency* $\boldsymbol{\omega} = \omega_{\text{B}}\,\mathbf{n}$. Thus we have

$$v_{\perp} = \omega_{\text{B}}\,r_{\perp} = \text{const} = v_{\perp}(0)\,,$$

and

$$r_{\perp} = \frac{v_{\perp}(0)}{\omega_{\mathrm{B}}} = \frac{\mathcal{E}\, v_{\perp}(0)}{ecB} = \frac{cp_{\perp}}{eB} \,,$$

since it follows from formula (5.5) that

$$\mathcal{E}\, v_{\perp} = c^2 p_{\perp} \,.$$

We have obtained the expression for the *gyroradius* or the Larmor radius

$$\boxed{\; r_{\mathrm{L}} = \frac{cp_{\perp}}{eB} \,. \;} \tag{5.14}$$

The term 'rigidity' is introduced in cosmic physics:

$$\mathcal{R} = \frac{c\mathbf{p}}{e} \,. \tag{5.15}$$

The rigidity of a particle is measured in Volts:

$$[\,\mathcal{R}\,] = \frac{[\,cp\,]}{[\,e\,]} = \frac{\mathrm{eV}}{\mathrm{e}} = \mathrm{V}.$$

Rigidity is usually used together with the term 'pitch-angle'

$$\theta = \left(\widehat{\mathbf{v}_0, \mathbf{B}} \right) . \tag{5.16}$$

From (5.14) and (5.15) it follows that the particle's gyroradius or Larmor radius is

$$r_{\mathrm{L}} = \frac{\mathcal{R}_{\perp}}{B} \,. \tag{5.17}$$

That is why

> the particles with the same rigidity and pitch-angle move along the same trajectories in a magnetic field.

This fact is used in the physics of the magnetospheres of the Earth and other planets, as well as in general physics of "cosmic rays" (Ginzburg and Syrovatskii, 1964; Schlickeiser, 2002).

The cosmic rays, high-energy (from 10^9 eV to somewhat above 10^{20} eV) particles of cosmological origin, were discovered almost a century ago but they are one of the very few means available to an Earth-based observer to study astrophysical or cosmological phenomena. The knowledge of their incoming direction and their energy spectrum are the bits and pieces of a complex puzzle which can give us information on the mechanism that produced them at the origin, unfortunately distorted by many effects they undergo during their journey over huge distances.

5.1.4 Non-magnetic force in a magnetic field

Let us consider the case when a non-magnetic force $\mathbf{F}$ is perpendicular to the homogeneous magnetic field $\mathbf{B}$ (see Figure 5.2). For the sake of simplicity, we shall consider the non-relativistic equation of motion:

$$m\,\dot{\mathbf{v}} = \mathbf{F} + \frac{e}{c}\,\mathbf{v}\times\mathbf{B}\,. \tag{5.18}$$

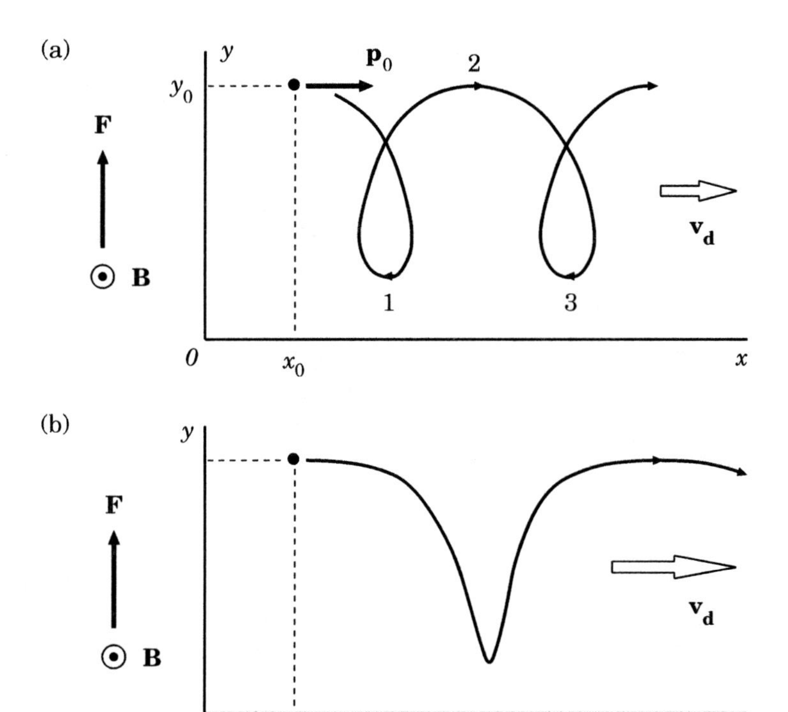

Figure 5.2: The trajectory of motion a positively charged particle in a uniform magnetic field under the action of a non-magnetic force. Slow (a) and fast (b) drifts.

Let us try to find the solution of this equation in the form

$$\mathbf{v} = \mathbf{v}_{\rm d} + \mathbf{u}\,. \tag{5.19}$$

Here $\mathbf{v}_{\rm d}$ is some constant velocity, so that substituting (5.19) in Equation (5.18) gives

$$m\,\dot{\mathbf{u}} + \mathbf{0} = \frac{e}{c}\,\mathbf{u}\times\mathbf{B} + \mathbf{F} + \frac{e}{c}\,\mathbf{v}_{\rm d}\times\mathbf{B}\,.$$

We choose $\mathbf{v}_{\rm d}$ in such a way that the two last terms vanish:

$$\mathbf{F} + \frac{e}{c}\,\mathbf{v}_{\rm d}\times\mathbf{B} = 0\,.$$

This is the case if the following expression is chosen:

$$\boxed{\mathbf{v}_{\mathrm{d}} = \frac{c}{e}\,\frac{\mathbf{F}\times\mathbf{B}}{B^2}\,.} \tag{5.20}$$

Actually, by using the known vector identity

$$\mathbf{a}\times(\mathbf{b}\times\mathbf{c}) = \mathbf{b}\,(\mathbf{a}\cdot\mathbf{c}) - \mathbf{c}\,(\mathbf{a}\cdot\mathbf{b})\,,$$

we infer

$$\frac{e}{c}\,\mathbf{v}_{\mathrm{d}}\times\mathbf{B} = \mathbf{n}\,(\mathbf{n}\cdot\mathbf{F}) - \mathbf{F} = -\,\mathbf{F}\,,$$

since $\mathbf{F}\perp\mathbf{n} = \mathbf{B}/B$. So formula (5.20) is correct.

Thus if a non-magnetic force $\mathbf{F}$ is perpendicular to the field $\mathbf{B}$, the particle motion is a sum of the *drift* with the velocity (5.20) called *drift velocity*, which is perpendicular to both $\mathbf{F}$ and $\mathbf{B}$, and the spiral motion round the magnetic field lines – the *gyromotion*:

$$m\,\dot{\mathbf{u}} = \frac{e}{c}\,\mathbf{u}\times\mathbf{B}\,. \tag{5.21}$$

Depending on a relative speed of these two motions, we distinguish *slow* ($v_{\mathrm{d}} < u$) and *fast* ($v_{\mathrm{d}} > u$) drifts, see (a) and (b) in Figure 5.2.

To understand the motion, let us think first about how the particle would move if only the magnetic field were present. It would gyrate in a circle, and the direction of motion around the circle would depend on the sign of the particle's charge. The radius of the circle, r_{L}, would vary with the particle's mass and would therefore much larger for an ion than for an electron if their velocities were the same (see formula 5.14).

The non-magnetic force $\mathbf{F}$ accelerates the particle during part of each orbit (see $1 \to 2$ in Figure 5.2a) and decelerates it during the remaining part of the orbit (see $2 \to 3$ in Figure 5.2a). The result is that the orbit is a distorted circle with a larger-than-average radius of curvature during half of the orbit and a smaller-than-average radius of curvature during the remaining half of the orbit. A net displacement is perpendicular to the force $\mathbf{F}$ and the magnetic field $\mathbf{B}$.

5.1.5 Electric and gravitational drifts

As we have seen above, in collisionless plasma, any force $\mathbf{F}$, that is capable of accelerating or decelerating particles as they gyrate about the magnetic field $\mathbf{B}$, will result in a drift perpendicular to both the field and the force.

(a) If $\mathbf{F} = e\,\mathbf{E}$, then the drift is called *electric* drift, its velocity

$$\mathbf{v}_{\mathrm{d}} = c\,\frac{\mathbf{E}\times\mathbf{B}}{B^2} \tag{5.22}$$

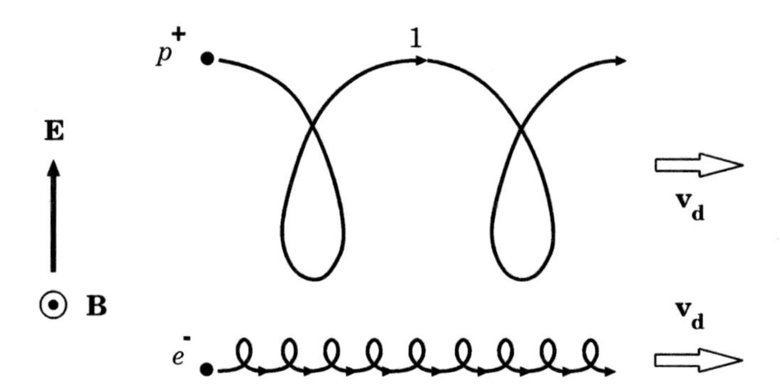

Figure 5.3: Electric drift. The kinetic energy $\mathcal{E}$ of a positively charged particle p^+ is a maximum at the upper point 1, hence the curvature radius r_L of the trajectory is a maximum at this point.

being independent of the particle charge and mass (Figure 5.3).

Since the drift velocity depends upon neither the charge nor the mass of the particle,

> the electric drift generates the motion of collisionless plasma as a whole with the velocity $\mathbf{v} = \mathbf{v}_{\rm d}$ relative to a magnetic field.

Being involved in the electric drift, the collisionless plasma tends: (a) to flow similar to a fluid, and (b) to be 'squeezed out' from direct action of the electric field **E** applied in a direction which is perpendicular to the magnetic field **B**. Formula (5.22) says that the drift velocity is perpendicular to both the electric and magnetic fields. This is sometimes referred to as an 'E-cross-B drift', but its magnitude is inversely proportional to the magnitude of **B**.

We should not forget that formula (5.22) was obtained in the non-relativistic limit. In fact, formula (5.22) would formally result in $v_{\rm d} \geq c$ for $E \geq B$.

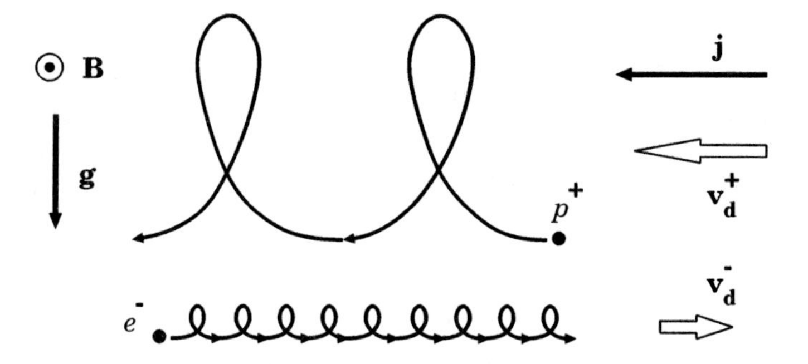

Figure 5.4: Gravitational drift. Initiation of an electric current by the action of the gravity force in a collisionless plasma with magnetic field.

(b) For the gravitational force $\mathbf{F} = m\,\mathbf{g}$ formula (5.20) gives the drift velocity

$$\mathbf{v}_{\mathrm{d}} = \frac{mc}{e}\,\frac{\mathbf{g}\times\mathbf{B}}{B^2}\,. \tag{5.23}$$

The *gravitational* drift velocity is seen to depend upon the particle mass and charge. Positively charged particles drift in the direction coinciding with that of the product $\mathbf{g}\times\mathbf{B}$, while negatively charged particles drift in the opposite direction as shown in Figure 5.4. Therefore

> a gravitational field is capable of generating an electric current in a magnetized collisionless plasma.

5.2 Weakly inhomogeneous slowly changing fields

5.2.1 Small parameters in the motion equation

Let us take the non-relativistic Equation (5.18) for the motion of a charged particle and rewrite it as follows:

$$\frac{m}{e}\,(\,\ddot{\mathbf{r}} - \mathbf{g}\,) = \mathbf{E} + \frac{1}{c}\,\dot{\mathbf{r}}\times\mathbf{B}\,. \tag{5.24}$$

On making this expression non-dimensional

$$\mathbf{r}^* = \frac{\mathbf{r}}{L}\,,\quad t^* = \frac{t}{\tau}\,,\quad \mathbf{v}^* = \frac{\mathbf{v}}{v_0}\,,\quad \mathbf{g}^* = \frac{\mathbf{g}}{g}\,,\quad \mathbf{B}^* = \frac{\mathbf{B}}{B_0}\,,\quad \mathbf{E}^* = \frac{\mathbf{E}}{E_0}\,,$$

we have the following equation

$$\frac{m}{e}\,\frac{L}{\tau^2}\left(\ddot{\mathbf{r}}^* - \frac{g\tau^2}{L}\,\mathbf{g}^*\right) = E_0\,\mathbf{E}^* + \frac{L}{c\tau}\,B_0\,\dot{\mathbf{r}}^*\times\mathbf{B}^*\,.$$

Normalize this equation with respect to the last term (the Lorentz force) by dividing the equation by $LB_0/c\tau$:

$$\frac{m}{e}\,\frac{c}{B_0}\,\frac{1}{\tau}\,(\,\ddot{\mathbf{r}}^* - \alpha_{\mathbf{g}}\,\mathbf{g}^*) = \frac{E_0}{B_0}\,\frac{c\tau}{L}\,\mathbf{E}^* + \dot{\mathbf{r}}^*\times\mathbf{B}^*\,.$$

Introduce the dimensionless parameter

$$\alpha_{\mathrm{B}} = \frac{m}{e}\,\frac{c}{B_0}\,\frac{1}{\tau}\,.$$

Two situations are conceivable.

(a) Spatially homogeneous magnetic and electric fields are slowly changing in time. The characteristic time $\tau = 1/\omega$, where ω is a characteristic field change frequency. Therefore the dimensionless parameter α_{B} is equal to

$$\alpha_{\mathrm{B}} = \frac{\omega}{\omega_{\mathrm{B}}}\,. \tag{a}$$

(b) For the fields that are constant in time but weakly inhomogeneous, the characteristic time is to be defined as $\tau = L/v_0$, L and v_0 being the characteristic values of the field dimensions and the particle velocity, respectively. In this case

$$\alpha_{\rm B} = \frac{r_{\rm L}}{L} \,. \tag{b}$$

Generally, a superposition of these two cases takes place. The field is called **weakly inhomogeneous slowly changing** field, if

$$\boxed{\alpha_{\rm B} \approx \frac{\omega}{\omega_{\rm B}} \approx \frac{r_{\rm L}}{L} \ll 1 \,.} \tag{5.25}$$

The second parameter of the problem,

$$\alpha_{\rm E} = \frac{E_0}{B_0} \frac{c\tau}{L} \,,$$

characterizes the relative role of the electric field. We assume $\alpha_{\rm E} = 1$, because, if this parameter is small, this can be taken into account in the final result.

The third dimensionless parameter $\alpha_{\mathbf{g}} = g\tau^2/L$ is not important for our consideration in this Section; so we put $\alpha_{\mathbf{g}} = 1$.

Thus we have

$$\alpha_{\rm B} \left(\ddot{\mathbf{r}}^* - \mathbf{g}^* \right) = \mathbf{E}^* + \dot{\mathbf{r}}^* \times \mathbf{B}^* \,, \tag{5.26}$$

the equation formally coinciding with the initial dimensional one. That is why it is possible to work with Equation (5.24), using as a *small parameter* the dimensional quantity m/e. This method is rather unusual but quite justified and widely used in plasma physics. The corresponding expansion in the Taylor series is termed the expansion in powers of m/e. We find such a solution of Equation (5.24).

5.2.2 Expansion in powers of m/e

Now let us represent the solution of Equation (5.24) as a sum of two terms,

$$\mathbf{r}(t) = \mathbf{R}(t) + \mathbf{r}_{\rm L}(t) \,. \tag{5.27}$$

The first term $\mathbf{R}(t)$ describes the motion of the *guiding center* of the Larmor circle, the second term $\mathbf{r}_{\rm L}(t)$ corresponds to the rotational motion or gyromotion of the particle. The case of an electron e^- is shown in Figure 5.5.

Recall that for the constant homogeneous magnetic field (see (5.14))

$$r_{\rm L} = \frac{cp_\perp}{eB} = \frac{m}{e} \frac{cv_\perp}{B} \,,$$

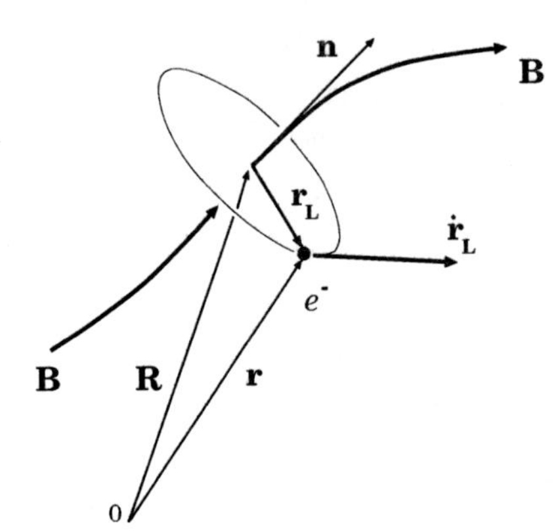

Figure 5.5: The Larmor motion of a negatively charged particle (an electron) in a weakly inhomogeneous slowly changing field.

i.e., the Larmor radius is proportional to the parameter m/e. It is natural to suppose that the dependence is the same for the weakly inhomogeneous slowly changing field, i.e.

$$|\,\mathbf{r}_{\mathrm{L}}\,| \sim \frac{m}{e}\,.$$

For example, if the magnetic field does not change in time and does not change much within the gyroradius, then the particle moves through a nearly uniform magnetic field while making a circular round. However the non-uniformities make the guiding center move in a way different from a simple translatory motion. We are going to find the equation describing the guiding center motion.

Let us substitute (5.27) in Equation (5.24) and expand the fields $\mathbf{g}$, $\mathbf{E}$, and $\mathbf{B}$ in the Taylor series about the point $\mathbf{r} = \mathbf{R}$:

$$\begin{aligned}
\mathbf{g}\,(\mathbf{r}) &= \mathbf{g}\,(\mathbf{R}) + (\mathbf{r}_{\mathrm{L}} \cdot \nabla)\,\mathbf{g}\,(\mathbf{R}) + \dots\,, \\
\mathbf{E}\,(\mathbf{r}) &= \mathbf{E}\,(\mathbf{R}) + (\mathbf{r}_{\mathrm{L}} \cdot \nabla)\,\mathbf{E}\,(\mathbf{R}) + \dots\,, \\
\mathbf{B}\,(\mathbf{r}) &= \mathbf{B}\,(\mathbf{R}) + (\mathbf{r}_{\mathrm{L}} \cdot \nabla)\,\mathbf{B}\,(\mathbf{R}) + \dots\,.
\end{aligned} \tag{5.28}$$

From Equation (5.24) we have

$$\ddot{\mathbf{r}} = \mathbf{g} + \left(\frac{m}{e}\right)^{-1} \left[\mathbf{E}\,(\mathbf{r}) + \frac{\mathbf{1}}{\mathbf{c}}\,\dot{\mathbf{r}} \times \mathbf{B}\,(\mathbf{r})\right].$$

Hence the basic equation contains the small parameter m/e to the power (-1). By substituting (5.27) and (5.28) in this equation we obtain

$$\begin{aligned}
\ddot{\mathbf{R}} + \ddot{\mathbf{r}}_{\mathrm{L}} &= \mathbf{g}\,(\mathbf{R}) + (\mathbf{r}_{\mathrm{L}} \cdot \nabla)\,\mathbf{g}\,(\mathbf{R}) + \\
&+ \left(\frac{m}{e}\right)^{-1} \{\,\mathbf{E}\,(\mathbf{R}) + (\mathbf{r}_{\mathrm{L}} \cdot \nabla)\,\mathbf{E}\,(\mathbf{R})\,\} + \\
&+ \left(\frac{m}{e}\right)^{-1} \left\{\frac{1}{c}\left(\dot{\mathbf{R}} + \dot{\mathbf{r}}_{\mathrm{L}}\right) \times [\,\mathbf{B}\,(\mathbf{R}) + (\mathbf{r}_{\mathrm{L}} \cdot \nabla)\,\mathbf{B}\,(\mathbf{R})\,]\right\} + \dots\,.
\end{aligned} \tag{5.29}$$

Note that we have to think carefully about smallness of different terms in Equation (5.29). For example, the magnitude of $\dot{\mathbf{r}}_{\rm L}$ is not small:

$$|\,\dot{\mathbf{r}}_{\rm L}\,| \sim \frac{|\,\mathbf{r}_{\rm L}\,|}{\tau} \sim r_{\rm L}\,\omega_{\rm B} \sim \alpha_{\rm B}\,\alpha_{\rm B}^{-1} \sim 1\,.$$

The particle velocity is not small, although the Larmor radius is small. That is the physical reason for the term

$$\left(\frac{m}{e}\right)^{-1} \frac{1}{c}\;[\;\dot{\mathbf{r}}_{\rm L} \times (\mathbf{r}_{\rm L}\cdot\nabla)\;\mathbf{B}\,(\mathbf{R})\,]$$

having zero order with respect to the small parameter m/e.

The acceleration term $\ddot{\mathbf{r}}_{\rm L}$ is not small either:

$$|\,\ddot{\mathbf{r}}_{\rm L}\,| \sim \frac{|\,\mathbf{r}_{\rm L}\,|}{\tau^2} \sim r_{\rm L}\,\omega_{\rm B}^2 \sim \alpha_{\rm B}^{-1} \sim \left(\frac{m}{e}\right)^{-1}.$$

In the expansion (5.29) let us retain only the terms with the order of smallness less than one, that is

$$\begin{aligned}\underbrace{\ddot{\mathbf{R}}}_{(0)} \; &= \underbrace{-\,\ddot{\mathbf{r}}_{\rm L}}_{(-1)} + \underbrace{\mathbf{g}\,(\mathbf{R})}_{(0)} + \underbrace{\left(\frac{m}{e}\right)^{-1}\left[\mathbf{E}\,(\mathbf{R}) + \frac{1}{c}\;\dot{\mathbf{R}} \times \mathbf{B}\,(\mathbf{R})\right]}_{(-1)} + \\ &+ \underbrace{\left(\frac{m}{e}\right)^{-1}(\mathbf{r}_{\rm L}\cdot\nabla\,)\,\mathbf{E}\,(\mathbf{R})}_{(0)} + \underbrace{\left(\frac{m}{e}\right)^{-1}\frac{1}{c}\;\dot{\mathbf{R}} \times [\,(\,\mathbf{r}_{\rm L}\cdot\nabla\,)\,\mathbf{B}\,(\mathbf{R})\,]}_{(0)} + \\ &+ \underbrace{\left(\frac{m}{e}\right)^{-1}\frac{1}{c}\;\dot{\mathbf{r}}_{\rm L} \times [\,(\,\mathbf{r}_{\rm L}\cdot\nabla\,)\,\mathbf{B}\,(\mathbf{R})\,]}_{(0)} + O\left(\frac{m}{e}\right). \qquad (5.30)\end{aligned}$$

Here the orders of smallness of the corresponding terms are given in brackets under the braces.

5.2.3 The averaging over gyromotion

In order to obtain the equation for guiding center motion let us average Equation (5.30) over a small period of the Larmor rotation,

$$T_{\rm B} = \frac{2\pi}{\omega_{\rm B}}\,.$$

Since $\langle\;\mathbf{r}_{\rm L}\;\rangle = \langle\;\dot{\mathbf{r}}_{\rm L}\;\rangle = \langle\;\ddot{\mathbf{r}}_{\rm L}\;\rangle = 0$, we infer the following equation

$$\begin{aligned}\ddot{\mathbf{R}} = \mathbf{g}\,(\mathbf{R}) + \frac{e}{m}\left[\mathbf{E}\,(\mathbf{R}) + \frac{1}{c}\;\dot{\mathbf{R}} \times \mathbf{B}\,(\mathbf{R})\right] + O\left(\frac{m}{e}\right) + \\ + \frac{e}{mc}\,\langle\;\dot{\mathbf{r}}_{\rm L} \times [\,(\,\mathbf{r}_{\rm L}\cdot\nabla\,)\,\mathbf{B}\,(\mathbf{R})\,]\;\rangle\,. \qquad (5.31)\end{aligned}$$

Let us consider the last term which also has to be averaged. Here we may put

$$\dot{\mathbf{r}}_{\rm L} = \omega_{\rm B}\, \mathbf{r}_{\rm L} \times \mathbf{n}\,.$$

On rearrangement (see Exercise 5.9), we obtain

$$\frac{e}{mc}\,\langle\, \dot{\mathbf{r}}_{\rm L} \times [\,(\,\mathbf{r}_{\rm L} \cdot \nabla\,)\,\mathbf{B}\,(\mathbf{R})\,]\,\rangle = -\,\frac{\mathcal{M}}{m}\,\nabla B\,. \tag{5.32}$$

Here

$$\mathcal{M} = \frac{1}{c}\,\frac{e\,\omega_{\rm B}}{2\pi}\,\left(\pi r_{\rm L}^2\right) = \frac{1}{c}\,JS \tag{5.33}$$

is the *magnetic moment* of a particle on the Larmor orbit (Figure 5.6). The case of electron e^- is shown here.

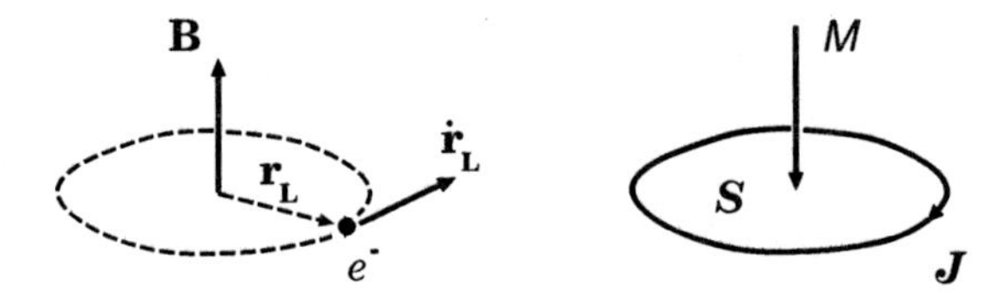

Figure 5.6: The motion of a negatively charged particle on the Larmor orbit and its magnetic moment. The moment is antiparallel to the magnetic field.

We interpret $-e\,(\omega_{\rm B}/2\pi)$ as the current $+J$ associated with the gyrating electron. That is why we call $\mathcal{M}$ a dipole magnetic moment as the name usually refers to a property of a *current loop* defined as the current J flowing through the loop times the area S of the loop (see Sivukhin, 1952). Hence it is clear from (5.33) that $\mathcal{M}$ is the magnetic moment of the gyrating particle.

So a single gyrating charge generates a magnetic dipole. Note that, for any charge of a particle, positive or negative,

> the direction of the dipole magnetic moment is opposite to the direction of the magnetic field.

Therefore the **diamagnetic effect** has to occur.

Substituting the non-relativistic formula $\omega_{\rm B} = eB/mc$ in (5.33) gives

$$\mathcal{M} = \frac{1}{2\pi}\,\frac{e^2}{mc^2}\,B\,\pi r_{\rm L}^2\,. \tag{5.34}$$

Therefore

> the magnetic moment is proportional to the magnetic field flux through the surface covering the particle's Larmor orbit.

It is also obvious from (5.32) that we can use the following formula for the force acting on the magnetic moment:

$$\boxed{\mathbf{F} = -\,\mathcal{M}\,\nabla B\,.} \tag{5.35}$$

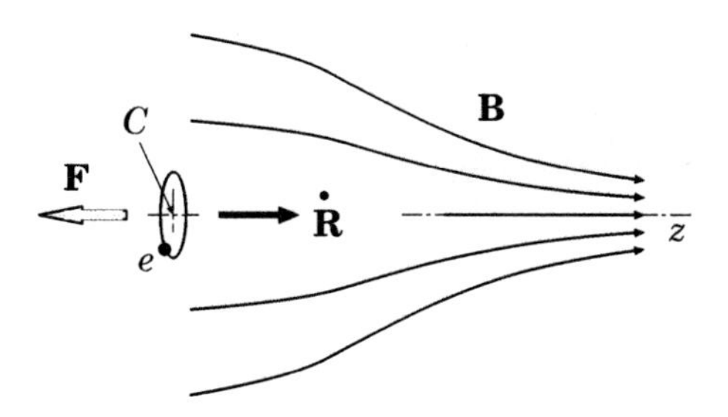

Figure 5.7: The diamagnetic force acts on the guiding center moving along the symmetry axis of a magnetic mirror configuration.

Let the field strength increase along the field direction. For the sake of simplicity, we consider a magnetic configuration symmetric around the central field line as shown in Figure 5.7. The strength of the magnetic field increases when the guiding center (not a particle) of a particle moves along the central line in the direction of the axis z. The force (5.35) is exerted along the field and **away from the direction of increase of the field**. As a consequence, the parallel component of the guiding center velocity $\dot{\mathbf{R}}$ decreases to zero at some maximum strength of the magnetic field and then changes sign. We say that the particle experiences a *mirror force*, and we shall call the place where it turns around a *magnetic mirror*. Note that a charged particle moving along the symmetry axis z is unaffected by magnetic force of course.

Finally, from Equation (5.31), we obtain the equation of the guiding center motion:

$$\ddot{\mathbf{R}} = \mathbf{g}\,(\mathbf{R}) + \frac{e}{m}\left[\,\mathbf{E}\,(\mathbf{R}) + \frac{1}{c}\,\dot{\mathbf{R}} \times \mathbf{B}\,(\mathbf{R})\right] - \frac{\mathcal{M}}{m}\,\nabla B\,(\mathbf{R}) + O\left(\frac{m}{e}\right). \quad (5.36)$$

The guiding center calculations involve considerably less amount of numerical work and produce trajectories in good agreement with detailed calculations if the non-uniformities of the magnetic and other fields are really small over the region through which the particle is making the circular motion. Moreover

> the guiding center theory helps us to develope an intuition about the motions of charged particles in magnetic field.

And this intuition turns out to be useful in solving many practical problems of plasma astrophysics, for example, in physics of the Earth magnetosphere.

5.2.4 Spiral motion of the guiding center

Even without regarding the terms $O(m/e)$, Equation (5.36) is more difficult in comparison with (5.24). The term $\mathbf{g}\,(\mathbf{R})$, the term with electric field $\mathbf{E}\,(\mathbf{R})$, and the two last terms in Equation (5.36) apart, it is seen that

$$\ddot{\mathbf{R}} = \frac{e}{mc}\,\dot{\mathbf{R}} \times \mathbf{B}\,. \quad (5.37)$$

Therefore the guiding center *spirals*, as does the particle (cf. Equation (5.8)).

By analogy with formula (5.14), the guiding center spiral radius can be found

$$R_{\perp} = \frac{mc\dot{R}_{\perp}}{eB} \,. \tag{5.38}$$

So it is a small quantity of order

$$\frac{R_{\perp}}{r_{\mathrm{L}}} = \frac{\dot{R}_{\perp}}{v_{\perp}} \sim \frac{r_{\mathrm{L}}}{L}$$

as compared with the particle Larmor radius (5.14).

The radius of the guiding center spiral is of the order of m/e as compared with the particle Larmor radius. Consequently, this spiral has a higher order with respect to the small parameter m/e and can be neglected in the approximation under study.

5.2.5 Gradient and inertial drifts

Let us neglect the term $O(m/e)$ in Equation (5.36) and take the vector product of Equation (5.36) with the unit vector $\mathbf{n} = \mathbf{B}/B$:

$$\ddot{\mathbf{R}} \times \mathbf{n} = \mathbf{g} \times \mathbf{n} + \frac{e}{m}\,\mathbf{E} \times \mathbf{n} + \frac{eB}{mc}\,(\,\dot{\mathbf{R}} \times \mathbf{n}\,) \times \mathbf{n} + \frac{\mathcal{M}}{m}\,\mathbf{n} \times \nabla B \,.$$

From this we find the drift velocity across the magnetic field

$$\dot{\mathbf{R}}_{\perp} \equiv \mathbf{n} \times (\,\dot{\mathbf{R}} \times \mathbf{n}\,) = c\,\frac{\mathbf{E} \times \mathbf{n}}{B} + \frac{mc}{eB}\,\mathbf{g} \times \mathbf{n} +$$

$$+ \frac{\mathcal{M}c}{eB}\,\mathbf{n} \times \nabla B - \frac{mc}{eB}\,\ddot{\mathbf{R}} \times \mathbf{n} \,. \tag{5.39}$$

The first term on the right-hand side of Equation (5.39) corresponds to the *electric* drift (5.22), the second one presents the *gravitational* drift (5.23). The third term is new for us in this Chapter; it describes the *gradient* drift arising due to the magnetic field inhomogeneity. The gradient drift velocity

$$\boxed{\mathbf{v}_{\mathrm{d}} = \frac{\mathcal{M}c}{eB}\,\mathbf{n} \times \nabla B \,.} \tag{5.40}$$

The same formula follows of course from (5.20) after substituting in it the formula (5.35) for the force acting on the magnetic moment $\mathcal{M}$ in the weakly inhomogeneous field.

So, if a particle gyrates in a magnetic field whose strength changes from one side of its gyration orbit to the other, the instantaneous radius of the curvature of the orbit will become alternately smaller and larger. Averaged over several gyrations, the particle drifts in a direction perpendicular to both the magnetic field and the direction in which the strength of the field changes.

The fourth term on the right-hand side of (5.39) corresponds to the *inertial* drift:

$$\mathbf{v}_{\mathrm{d}} = -\frac{mc}{eB}\,\ddot{\mathbf{R}} \times \mathbf{n}\,. \tag{5.41}$$

Let us consider it in some detail. For calculating the inertial drift velocity (5.41), we have to know the guiding center *acceleration* $\ddot{\mathbf{R}}$. It will suffice for the calculation of $\ddot{\mathbf{R}}$ to consider Equation (5.39) in the zeroth order, since the last term of (5.39) contains the small parameter m/e. In this order with respect to m/e, we have

$$\dot{\mathbf{R}}_{\perp} = c\,\frac{\mathbf{E} \times \mathbf{n}}{B}\,.$$

Hence the guiding center acceleration

$$\ddot{\mathbf{R}} = \frac{d}{dt}\,\dot{\mathbf{R}} = \frac{d}{dt}\,(\dot{\mathbf{R}}_{\parallel} + \dot{\mathbf{R}}_{\perp}) = \frac{d}{dt}\left(v_{\parallel}\,\mathbf{n} + c\,\frac{\mathbf{E} \times \mathbf{n}}{B}\right). \tag{5.42}$$

Because this aspect of particle motion is important in accounting for the special properties of a collisionless cosmic plasma, it is good to understand it not only mathematically but also in an intuitive manner.

5.2.5 (a) The centrifugal drift

At first, we consider the particular case assuming the electric field $\mathbf{E} = 0$ in formula (5.42), the magnetic field $\mathbf{B}$ being *time-independent* but *weakly inhomogeneous*. Under these conditions

$$\ddot{\mathbf{R}} = \frac{d}{dt}\,(v_{\parallel}\,\mathbf{n}) = \mathbf{n}\,\frac{dv_{\parallel}}{dt} + v_{\parallel}\,\frac{d\mathbf{n}}{dt}\,.$$

The first term on the right-hand side does not contribute to the drift velocity since $\mathbf{n} \times \mathbf{n} = \mathbf{0}$. Rewrite the second term as follows:

$$v_{\parallel}\,\frac{d\mathbf{n}}{dt} = v_{\parallel}\left(\frac{\partial \mathbf{n}}{\partial t} + v_{\parallel}\,(\mathbf{n} \cdot \nabla)\,\mathbf{n}\right). \tag{5.43}$$

In this formula, the first term on the right equals zero for the time-independent field. The second one is equal to

$$v_{\parallel}^{2}\,(\mathbf{n} \cdot \nabla)\,\mathbf{n} = -v_{\parallel}^{2}\left(\frac{\mathbf{e}_{\mathrm{c}}}{\mathcal{R}_{\mathrm{c}}}\right). \tag{5.44}$$

Here $\mathcal{R}_{\mathrm{c}}$ is a radius of *curvature* for the field line at a given point $\mathbf{R}$. At this point the unit vector $\mathbf{e}_{\mathrm{c}}$ is directed from the curvature center 0_{c} as shown in Figure 5.8.

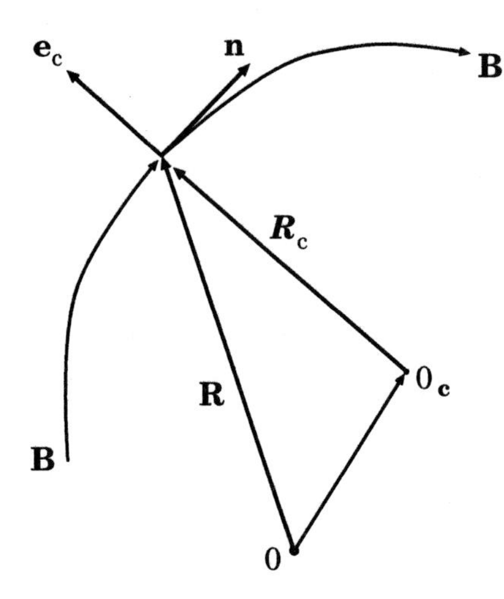

Figure 5.8: The frame of reference for derivation of the formula for the inertial drift in weakly inhomogeneous magnetic field.

Thus the dependence of the inertial drift velocity on the curvature of the weakly inhomogeneous magnetic field is found

$$\dot{\mathbf{R}}_{\perp}\Big|_{c} = \frac{1}{\mathcal{R}_c\,\omega_{_B}}\, v_{\parallel}^{2}\, \mathbf{e}_c \times \mathbf{n}\,. \tag{5.45}$$

This is the drift of a particle under action of the *centrifugal* force

$$\mathbf{F}_c = \frac{m v_{\parallel}^{2}}{\mathcal{R}_c}\, \mathbf{e}_c\,. \tag{5.46}$$

In formula (5.45), the centrifugal force produced by motion of a particle along the magnetic field appears explicitly. Therefore the *centrifugal* drift velocity can be seen to be a special case of the expression (5.20) obtained for drift produced by an arbitrary non-magnetic force **F**.

5.2.5 (b) The curvature-dependent drift

Let us come back to the gradient drift and consider a time-independent weakly-inhomogeneous magnetic field. Its gradient

$$\nabla B = \frac{1}{2B}\,\nabla\,(\mathbf{B}\cdot\mathbf{B}) = \frac{1}{B}\,[\,(\mathbf{B}\cdot\nabla\,)\,\mathbf{B} + \mathbf{B}\times \mathrm{curl}\,\mathbf{B}\,]\,.$$

In a current-free region $\mathrm{curl}\,\mathbf{B} = 0$, and hence

$$\nabla B = \frac{1}{B}\,(\mathbf{B}\cdot\nabla\,)\,\mathbf{B} = (\mathbf{n}\cdot\nabla)\,\mathbf{B} = (\mathbf{n}\cdot\nabla)\,B\,\mathbf{n} = B\,(\mathbf{n}\cdot\nabla)\,\mathbf{n} +$$

$$+\,\mathbf{n}\,(\mathbf{n}\cdot\nabla B\,) = -B\left(\frac{\mathbf{e}_c}{\mathcal{R}_c}\right) + \mathbf{n}\,(\mathbf{n}\cdot\nabla B\,)\,.$$

The last term does not contribute to the gradient drift velocity (5.40). The contribution of the first term to the drift velocity is

$$\dot{\mathbf{R}}_{\perp} = \frac{\mathcal{M}c}{eB}\,\mathbf{n}\times\left((-B)\,\frac{\mathbf{e}_{\rm c}}{\mathcal{R}_{\rm c}}\right) = -\frac{\mathcal{M}}{e\,\mathcal{R}_{\rm c}}\,\mathbf{n}\times\mathbf{e}_{\rm c} = \frac{\mathcal{M}}{e\,\mathcal{R}_{\rm c}}\,\mathbf{e}_{\rm c}\times\mathbf{n}\,. \qquad (5.47)$$

Here, according to definition (5.33) and formula (5.14), the magnetic moment

$$\mathcal{M} = \frac{1}{c}\,JS = \frac{e\,\omega_{\rm B}\,r_{\rm L}^2}{2c} = \frac{e\,v_{\perp}^2}{2c\,\omega_{\rm B}}\,. \qquad (5.48)$$

On substituting formula (5.48) into (5.47) we see that the gradient drift in a time-independent weakly-inhomogeneous magnetic field has a structure analogous to the centrifugal drift (5.45):

$$\dot{\mathbf{R}}_{\perp}\Big|_{\rm gr} = \frac{1}{\mathcal{R}_{\rm c}\,\omega_{\rm B}}\,\frac{1}{2}\,v_{\perp}^2\;\mathbf{e}_{\rm c}\times\mathbf{n}\,. \qquad (5.49)$$

Therefore we can add the curvature-dependent part of the gradient drift (5.49) to the centrifugal drift (5.45):

$$\dot{\mathbf{R}}_{\perp} = \frac{1}{\mathcal{R}_{\rm c}\,\omega_{\rm B}}\left(v_{\parallel}^2 + \frac{1}{2}\,v_{\perp}^2\right)\mathbf{e}_{\rm c}\times\mathbf{n}\,. \qquad (5.50)$$

This formula unites the two drifts that depend on the field line curvature of a weakly inhomogeneous magnetic field.

In a curved magnetic field, the gradient drift is present in combination with the centrifugal drift.

5.2.5 (c) The curvature-independent gradient drift

It is worth considering the part of the gradient drift, that is independent of the field line curvature. Let the field lines be straight ($\mathcal{R}_{\rm c} \to \infty$), their density increasing unidirectionally as shown in Figure 5.9. The field strength B_2 at a point 2 is greater than that one at a point 1. So, according to (5.17), the Larmor radius

$$r_{\rm L}\big|_2 < r_{\rm L}\big|_1\,.$$

The particle moves in the manner indicated in Figure 5.9.

For comparison purposes, it is worth remembering another illustration. This is related to, on the contrary, the non-magnetic force $\mathbf{F}$ (Section 5.1.4). Under action of the force, the particle velocity at a point 1 in Figure 5.10, v_1, is greater than at a point 2. Hence the Larmor radius $r_{\rm L} = cp_{\perp}/eB$ is greater at a point 1 than at a point 2 as well.

In other words, when the particle is at the point 2 at the top of its trajectory, the force $\mathbf{F}$ and the Lorentz force $(e/c)\,\mathbf{v}\times\mathbf{B}$ both act in the downward

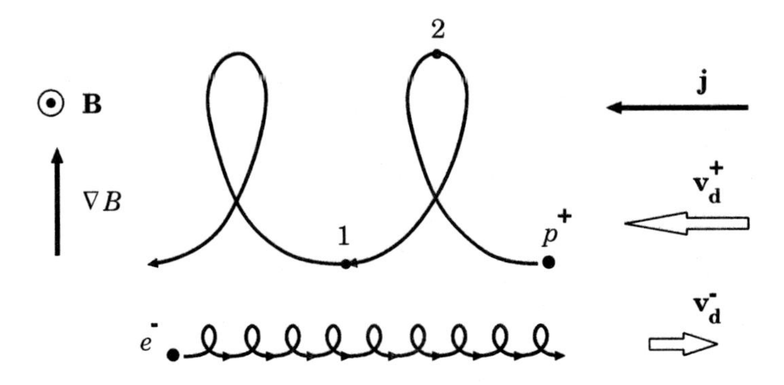

Figure 5.9: The simplest interpretation of the gradient drift. A gradient in the field strength, ∇B, in the direction perpendicular to **B** will produce a drift motion of ions and electrons.

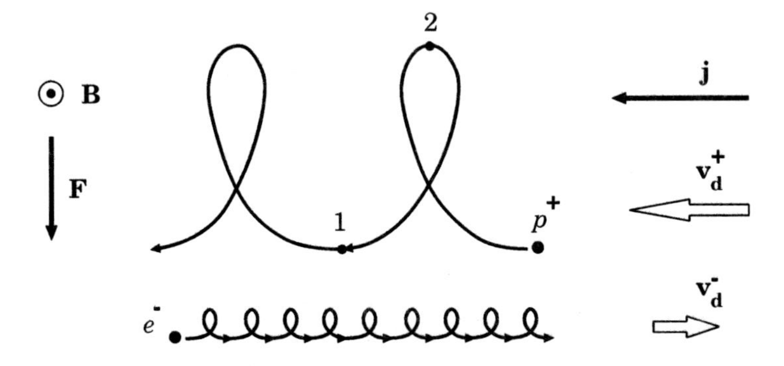

Figure 5.10: The physical nature of the drift under the action of a non-magnetic force **F** which is perpendicular to the uniform magnetic field **B**.

direction in Figure 5.10. This enhanced normal acceleration makes the trajectory more sharply bent than it would have been in the absence of the force **F**. On the other hand, when the particle is at the bottom point 1, the Lorentz force is diluted by **F**, thereby causing the trajectory to be less sharply bent. As a result, there is a drift of the guiding center in a direction perpendicular to both **B** and **F**.

Figures 5.9 and 5.10 also demonstrate the validity of formula (5.35).

The drifts with velocity which depends on the particle charge and mass, like the gradient drift, can give rise to a current by making the electrons and ions drift in opposite directions. Such drifts can also be important for the problem of element abundances or element fractionation (see the second volume of this book).

Recommended Reading: Sivukhin (1965), Morozov and Solov'ev (1966b)

5.3 Practice: Exercises and Answers

Exercise 5.1 [Section 5.1] Evaluate the gyrofrequency for thermal electrons and protons in the solar corona above a sunspot.

Answer. At typical temperature in the corona, $T \approx 2 \times 10^6$ K, from the non-relativistic formula (5.11), it follows that: the electron gyrofrequency

$$\omega_{\mathrm{B}}^{(e)} = 1.76 \times 10^7 \, B\,(\mathrm{G})\,, \quad \mathrm{rad\ s^{-1}}\,; \tag{5.51}$$

the proton gyrofrequency

$$\omega_{\mathrm{B}}^{(p)} = 9.58 \times 10^3 \, B\,(\mathrm{G})\,, \quad \mathrm{rad\ s^{-1}}\,. \tag{5.52}$$

The gyrofrequency of electrons is $m_{\mathrm{p}} \,/\, m_{\mathrm{e}} \approx 1.84 \times 10^3$ times larger than that one of protons. Just above a sunspot the field strength can be as high as $B \approx 3000\,\mathrm{G}$. Here $\omega_{\mathrm{B}}^{(e)} \approx 5 \times 10^{10}\,\mathrm{rad\ s^{-1}}$. The emission of thermal electrons at this height in the corona can be observed at wavelength $\lambda \approx 4$ cm.

Exercise 5.2 [Section 5.1] Under conditions of the corona (Exercise 5.1), evaluate the *mean thermal velocity* and the Larmor radius of thermal electrons and protons.

Answer. The thermal velocity of particles with mass m_{i} and temperature T_{i} is

$$V_{\mathrm{Ti}} = \left(\frac{3 k_{\mathrm{B}}\, T_{\mathrm{i}}}{m_{\mathrm{i}}} \right)^{1/2} . \tag{5.53}$$

Respectively, for electrons and protons:

$$V_{\mathrm{Te}} = 6.74 \times 10^5 \, \sqrt{T_{\mathrm{e}}\,(\mathrm{K})}\,, \quad \mathrm{cm\ s^{-1}}\,, \tag{5.54}$$

and

$$V_{\mathrm{Tp}} = 1.57 \times 10^4 \, \sqrt{T_{\mathrm{p}}\,(\mathrm{K})}\,, \quad \mathrm{cm\ s^{-1}}\,. \tag{5.55}$$

At the coronal temperature $V_{\mathrm{Te}} \approx 9.5 \times 10^3$ km s^{-1} $\sim 10^9$ cm s^{-1} and $V_{\mathrm{Tp}} \approx$ 220 km s^{-1}.

From (5.14) we find the following formulae for the Larmor radius:

$$r_{\mathrm{L}}^{(e)} = \frac{V_{\mathrm{Te}}}{\omega_{\mathrm{B}}^{(e)}} = 3.83 \times 10^{-2} \, \frac{\sqrt{T_{\mathrm{e}}\,(\mathrm{K})}}{B\,(\mathrm{G})}\,, \quad \mathrm{cm}\,, \tag{5.56}$$

and

$$r_{\mathrm{L}}^{(p)} = \frac{V_{\mathrm{Tp}}}{\omega_{\mathrm{B}}^{(p)}} = 1.64 \, \frac{\sqrt{T_{\mathrm{p}}\,(\mathrm{K})}}{B\,(\mathrm{G})}\,, \quad \mathrm{cm}\,. \tag{5.57}$$

At $T \approx 2 \times 10^6$ K and $B = 3000$ G we find $r_{\mathrm{L}}^{(e)} \approx 0.2$ mm and $r_{\mathrm{L}}^{(p)} \approx 1$ cm.

Exercise 5.3. [Section 5.1] During solar flares electrons are accelerated to energies higher than 20–30 keV. These electrons produce the bremsstrahlung emission. The lower boundary of the spectrum of accelerated electrons is not known because the thermal X-ray emission of the high-temperature (super-hot) plasma masks the lower boundary of the non-thermal X-ray spectrum. Assuming that the lower energy of accelerated electrons $\mathcal{K} \approx 30$ keV, find their velocity and the Larmor radius in the corona.

Answer. The kinetic energy of a particle

$$\mathcal{K} = \mathcal{E} - mc^2, \tag{5.58}$$

where $\mathcal{E}$ is the total energy (5.2), $mc^2 = 511$ keV for an electron. Since $\mathcal{K}/mc^2 \ll 1$, formula (5.58) can be used in the non-relativistic limit: $\mathcal{K} = mv^2/2$. From here the velocity of a 30 keV electron $v \approx 10^{10}$ cm s^{-1} $\approx 0.3\, c$.

The Larmor radius of a non-relativistic electron according to (5.14)

$$r_{\rm L}^{(e)} = 5.69 \times 10^{-8}\, \frac{v_\perp\,({\rm cm\ s^{-1}})}{B\,({\rm G})}\,. \tag{5.59}$$

For a 30 keV electron

$$r_{\rm L}^{(e)} \approx 5.6 \times 10^{2}\, \frac{1}{B\,({\rm G})}\,. \tag{5.60}$$

Above a sunspot with $B \approx 3000$ G the Larmor radius $r_{\rm L}^{(e)} \approx 2$ mm. Inside a coronal magnetic trap with a field $B \approx 100$ G the electrons with kinetic energy $\mathcal{K} \approx 30$ keV have the Larmor radius $r_{\rm L}^{(e)} \approx 6$ cm.

Exercise 5.4 [Section 5.1] Under conditions of the previous Exercise estimate the Larmor radius of a proton moving with the same velocity as a 30 keV electron.

Answer. For a non-relativistic proton it follows from formula (5.14) that

$$r_{\rm L}^{(p)} = 1.04 \times 10^{-4}\, \frac{v_\perp\,({\rm cm\ s^{-1}})}{B\,({\rm G})}\,,\quad {\rm cm}\,. \tag{5.61}$$

Above a sunspot a proton with velocity $\approx 0.3\, c$ has the Larmor radius ≈ 3 m. Inside a coronal trap with magnetic field ≈ 100 G the Larmor radius $\approx 10^4$ cm. So

> non-relativistic protons (and other ions) can be well trapped in coronal magnetic traps including collapsing ones

(see vol. 2, Chapter 7). This is important for the problem of ion acceleration in solar flares.

Exercise 5.5 [Section 5.1] The stronger magnetic field, the smaller is the Larmor radius $r_{\rm L}$ of an electron. Find the condition when $r_{\rm L}$ is so small as the *de Broglie* wavelength of the electron

$$\lambda_{\rm B} = \frac{h}{m_e v} = 1.22 \times 10^{-7}\, \frac{1}{\sqrt{\mathcal{K}({\rm eV})}}\,. \tag{5.62}$$

Here h is Planck's constant, $\mathcal{K}$ is the kinetic energy (5.58) of the electron. If $\mathcal{K} = 1$ eV, the de Broglie wavelength $\lambda_{\text{B}} \approx 10^{-7}$ cm ≈ 10 Angström.

Answer. In the non-relativistic limit, the electron with kinetic energy $\mathcal{K}$ has the Larmor radius

$$r_{\text{L}} = 3.37\, \frac{\sqrt{\mathcal{K}\,(\text{eV})}}{B\,(\text{G})}\,, \quad \text{cm}\,. \tag{5.63}$$

When the energy of the electron is 1 eV and the field has a strength of 1 G, the Larmor radius $r_{\text{L}} \approx 3$ cm. However for a field of 3×10^7 G, the Larmor radius is diminished to the de Broglie wavelength $\approx 10^{-7}$ cm. So for white dwarfs which have $B > 10^7$ G, and especially for neutron stars, we have to take into account

> the *quantization* effect of the magnetic field: the Larmor radius is no longer arbitrary but can take only certain definite values.

We call a magnetic field the *superstrong* one, if $r_{\text{L}} < \lambda_{\text{B}}$. Substituting (5.63) and (5.62) into this condition, we rewrite it as follows

$$B > 3 \times 10^7\, \mathcal{K}\,(\text{eV})\,, \quad \text{G}\,. \tag{5.64}$$

In superstrong fields the classic theory of particle motion, developed above, is no longer valid and certain *quantum effects* appear.

The energy difference between the levels of a non-relativistic electron in a superstrong field is

$$\delta \mathcal{E}_{B} \approx \frac{eB}{mc}\,\frac{h}{2\pi} \sim 10^{-8}\, B\,, \quad \text{eV}\,. \tag{5.65}$$

On the other hand, the difference between energy levels in an atom, for example a hydrogen atom, is of about 10 eV; this is comparable with $\delta \mathcal{E}_{B}$ in a superstrong field $B > 10^8 - 10^9$ G. In ordinary conditions B is not so large and does not affect the internal structure of atoms.

Inside and near neutron stars $B > 10^{11} - 10^{12}$ G. In such fields a lot of abnormal phenomena come into existence due to the profound influence of the external field on the interior of atoms. For example, the electron orbits around nuclei become very oblate. Two heavy atoms, e.g. iron atoms, combine into a molecule (Fe_2) and, moreover, these molecules form polymolecular substances, which are constituents of the hard surface of neutron stars (Ruderman, 1971; Rose, 1998).

Exceedingly superstrong (ultrastrong) fields, $\gtrsim 10^{14}$ G, are suggested in the so-called *magnetars*, the highly-magnetized, newly-born neutron stars (see Section 19.1.3).

Exercise 5.6 [Section 5.1] Is it justified to neglect the radiation reaction in the motion Equation (5.8) while considering the gyromotion of electrons in astrophysical plasmas?

Answer. In the non-relativistic limit $v^2 \ll c^2$, the total energy radiated per unit time by a charge e moving with acceleration $\ddot{\mathbf{r}}$ can be calculated in the dipolar approximation (see Landau and Lifshitz, *Classical Theory of Field*, 1975, Chapter 9, § 67):

$$I = \frac{2}{3c^3}\, \ddot{\mathbf{d}}^{\,2} . \tag{5.66}$$

Here $\mathbf{d} = e\,\mathbf{r}$ and $\ddot{\mathbf{d}} = e\,\ddot{\mathbf{r}}$.

In a uniform magnetic field $\mathbf{B}$, an electron moves in a helical trajectory. For the transversal motion in the Larmor orbit $r = r_{_L}$, the total power radiated by the electron

$$I = \frac{2}{3c^3}\, e^2\, r_{_L}^2\, \omega_{_B}^4 = \frac{2e^2}{3c^3}\, v^2\, \omega_{_B}^2 . \tag{5.67}$$

Here $v = \omega_{_B}\, r_{_L}$ is the velocity of the electron in the Larmor orbit.

Let us estimate the strength of the magnetic field such that an electron with kinetic energy $\mathcal{K} = mv^2/2$ would radiate an appreciable amount of energy during one period of gyration, $\tau_{_B} = 2\pi/\omega_{_B}$. Consider a ratio

$$\gamma_r = \frac{\tau_{_B}}{\tau_r} = \frac{1}{\mathcal{K}}\, \frac{d\mathcal{K}}{dt}\, \frac{2\pi}{\omega_{_B}} . \tag{5.68}$$

Substituting (5.67) in (5.68) gives

$$\gamma_r = \frac{8\pi}{3}\, \frac{e^3}{(mc^2)^2}\, B \approx 1.4 \times 10^{-15}\, B\,(\mathrm{G}) . \tag{5.69}$$

Therefore, while considering the gyromotion of non-relativistic electrons in cosmic plasmas, the radiation reaction could be important in the motion Equation (5.8) only in ultrastrong magnetic fields with

$$B \gtrsim \frac{3}{8\pi}\, \frac{(mc^2)^2}{e^3} \approx 7 \times 10^{14}\ \mathrm{G} . \tag{5.70}$$

However other physical processes already dominate under such conditions; see discussion in Exercise 5.5.

Recall that formula (5.67) is not valid for a relativistic electron moving in the Larmor orbit; see next Exercise.

Exercise 5.7 [Section 5.1] For a relativistic electron moving in the Larmor orbit with a speed $v = \beta c$, the total power of radiation is given by formula (see Landau and Lifshitz, *Classical Theory of Field*, 1975, Chapter 9, § 74):

$$I = \frac{2}{3c^3}\, \frac{e^4}{m^2}\, \frac{\beta^2}{1 - \beta^2}\, B^2 . \tag{5.71}$$

Therefore, in contrast to the non-relativistic formula (5.67), $I \to \infty$ when $\beta \to 1$. Find the rate of energy loss for such an electron.

Answer. According to (5.4), for a relativistic particle

$$\mathcal{E}^2 = (pc)^2 + (mc^2)^2 . \tag{5.72}$$

By using this expression we rewrite formula (5.71) as follows

$$\frac{d\mathcal{E}}{dt} = -I = \frac{2e^4 B^2}{3m^4 c^7} \left((mc^2)^2 - \mathcal{E}^2 \right) . \tag{5.73}$$

From here we find

$$\frac{\mathcal{E}}{mc^2} = \text{cth} \left(\frac{2e^4 B^2}{3m^3 c^5} \, t + \text{const} \right) . \tag{5.74}$$

With an increase of time t, the particle's energy monotonuouly decreases to the value $\mathcal{E} = mc^2$ with the characteristic time

$$\tau_r = \frac{3m^3 c^5}{2e^4 B^2} \, . \tag{5.75}$$

Comparing between this time and $2\pi \, \omega_{\text{B}}^{-1}$ gives us the characteristic value of magnetic field

$$B = \frac{3m^2 c^4}{4\pi \, e^3} \left(1 - \beta^2 \right)^{1/2} . \tag{5.76}$$

We see that $B \to 0$ when $\beta \to 1$. So, for relativistic electrons, there is no need in strong magnetic fields to radiate efficiently unless they become non-relativistic particles (see Exercise 5.6). This means that

> for relativistic electrons, the radiative losses of energy can be important even in relatively weak magnetic fields.

That is why the synchrotron radiation is very widespread in astrophysical conditions (e.g., Ginzburg and Syrovatskii, 1965). It was the first radio-astronomical radiation mechanism which had been successfully used by classical astrophysics to interpret the continuum spectrum of the Crab nebula. The synchrotron mechanism of radio emission works in any source which contains relativistic electrons in a magnetic field: in the solar corona during flares, in the Jovian magnetosphere, interstellar medium, supernova remnants etc.

Exercise 5.8 [Section 5.2.3] Consider an actual force acting on a particle gyrating around the central field line in the magnetic mirror configuration shown in Figure 5.7.

Answer. Let us use the cylindrical coordinates (r, z, φ) with the axis z along the central field line as shown in Figure 5.7. In the weakly inhomogeneous magnetic field, the predominant component is B_z but there is a small component B_r which produces the z component of the Lorentz force:

$$F_z = -\frac{q}{c} \, v_\varphi B_r \, . \tag{5.77}$$

Here the φ-component is the gyromotion velocity $\mathbf{v}_\perp$; for a negatively (positively) charged particle, it is directed in the positive (negative) φ-direction (see Figure 5.6).

The component B_z of the magnetic field can be found from condition $\operatorname{div} \mathbf{B} = 0$ as follows:

$$B_r = -\frac{1}{2}\, r\, \frac{\partial B_z}{\partial z}\,. \tag{5.78}$$

Substituting (5.78) into (5.77) gives

$$F_z = -\mathcal{M}\, \frac{\partial B_z}{\partial z}\,, \tag{5.79}$$

where $\mathcal{M}$ is the magnetic moment (5.33) of the gyrating particle.

Exercise 5.9 [Section 5.2.3] Derive formula (5.32) for the last term in the averaged Equation (5.31).

Answer. We have to write down the following expression explicitly

$$(\,\mathbf{r}_{\text{L}} \times \mathbf{n}\,) \times [\,(\,\mathbf{r}_{\text{L}} \cdot \nabla\,)\,\mathbf{B}\,(\mathbf{R})\,]$$

and then to average it. It is a matter to do that, once we make use of the following tensor identities:

$$(\mathbf{a} \times \mathbf{b})_{\,\alpha} = e_{\alpha\beta\gamma}\, a_\beta\, b_\gamma\,.$$

Here $e_{\alpha\beta\gamma}$ is the unit antisymmetric tensor, and

$$e_{\alpha\beta\gamma}\, e_{\mu\nu\gamma} = \delta_{\alpha\mu}\, \delta_{\beta\nu} - \delta_{\alpha\nu}\, \delta_{\beta\mu}\,.$$

On rearrangement, we average the last term in Equation (5.31) to obtain

$$\frac{e}{mc}\, \langle\, \dot{\mathbf{r}}_{\text{L}} \times [\,(\,\mathbf{r}_{\text{L}} \cdot \nabla\,)\,\mathbf{B}\,(\mathbf{R})\,]\,\rangle = -\,\frac{\mathcal{M}}{m}\, \nabla B\,, \tag{5.80}$$

where

$$\mathcal{M} = \frac{1}{c}\, \frac{e\,\omega_{\text{B}}}{2\pi}\, \left(\pi r_{\text{L}}^2\right) = \frac{1}{c}\, JS \tag{5.81}$$

is the *magnetic moment* of a particle on the Larmor orbit.

Chapter 6

Adiabatic Invariants in Astrophysical Plasma

Adiabatic invariants are useful to understand many interesting properties of collisionless plasma in cosmic magnetic fields: trapping and acceleration of charged particles in collapsing magnetic traps, the Fermi acceleration, "cosmic rays" origin.

6.1 General definitions

As is known from mechanics (see Landau and Lifshitz, *Mechanics*, 1976, Chapter 7, § 49), the so-called *adiabatic invariants* remain constant under changing conditions of motion, if these changes are slow. Recall that the system executing a *finite one-dimensional* motion is assumed to be characterized by a parameter λ that is slowly – adiabatically – changing with time:

$$\lambda \,/\, \dot{\lambda} \gg T \, . \tag{6.1}$$

Here T is a characteristic time for the system (e.g., a particle in given fields) motion.

More precisely, if the parameter λ did not change, the system would be closed and would execute a strictly periodic motion with the period T like a simple pendulum in gravitational field. In this case the energy of the system, $\mathcal{E}$, would be invariant.

Under the slowly changing parameter λ, if $\dot{\mathcal{E}} \sim \dot{\lambda}$, then the integral

$$I = \oint P \, dq \, , \tag{6.2}$$

rather than the energy $\mathcal{E}$, is conserved. Here P and q are the *generalized* momentum and coordinate, respectively. The integral is taken along the trajectory of motion under given $\mathcal{E}$ and λ. The integral I is referred to as the adiabatic invariant.

6.2 Two main invariants

6.2.1 Motion in the Larmor plane

The motion of a charged particle in slowly changing weakly inhomogeneous fields has been considered in the previous section. Several types of periodic motion can be found. In particular, the particle's motion in the plane perpendicular to the magnetic field – the Larmor motion – is periodic. Let $\mathbf{P}$ be the generalized momentum. According to definition (6.2) for such a motion the adiabatic invariants are the integrals

$$I_1 = \oint P_1 \, dq_1 = \text{const} \quad \text{and} \quad I_2 = \oint P_2 \, dq_2 = \text{const} \,,$$

taken over a period of the motion of coordinates q_1 and q_2 in the plane of the Larmor orbit.

It is convenient to combine these integrals, that is simply to add them together:

$$I = \oint \mathbf{P}_{\perp} \cdot d\mathbf{q} = \text{const} \,. \tag{6.3}$$

(This is the same, of course, as $q = r_{\mathrm{L}} \phi$ in definition (6.2) with $0 \leq \phi \leq 2\pi$.) Here

$$\mathbf{P}_{\perp} = \mathbf{p}_{\perp} + \frac{e}{c} \, \mathbf{A}$$

is the generalized momentum (see Landau and Lifshitz, *Classical Theory of Field*, 1975, Chapter 3, § 16) projection onto the plane mentioned above. In this plane $\mathbf{q} = \mathbf{r}_{\mathrm{L}}$. The vector potential $\mathbf{A}$ is perpendicular to the vector $\mathbf{B}$ since $\mathbf{B} = \text{curl}\, \mathbf{A}$, and $\mathbf{p}$ is the ordinary kinetic momentum of a particle.

Now perform the integration in formula (6.3)

$$I = \oint \mathbf{P}_{\perp} \cdot d\mathbf{r}_{\mathrm{L}} = \oint \mathbf{p}_{\perp} \cdot d\mathbf{r}_{\mathrm{L}} + \frac{e}{c} \oint \mathbf{A} \cdot d\mathbf{r}_{\mathrm{L}} =$$

$$= 2\pi r_{\mathrm{L}} \, p_{\perp} - \frac{e}{c} \int\limits_S \text{curl}\, \mathbf{A} \cdot d\mathbf{S} =$$

by virtue of the Stokes theorem

$$= 2\pi r_{\mathrm{L}} \, p_{\perp} - \frac{e}{c} \int\limits_S \mathbf{B} \cdot d\mathbf{S} = 2\pi r_{\mathrm{L}} \, p_{\perp} - \frac{e}{c} \, B \, \pi r_{\mathrm{L}}^2 \,. \tag{6.4}$$

Substituting $r_{\mathrm{L}} = c p_{\perp} / eB$ (cf. formula (5.17)) into (6.4) gives

$$I = \frac{\pi c}{e} \frac{p_{\perp}^2}{B} = \text{const} \,.$$

Thus we come to the conclusion that the conserving quantity is

$$\boxed{\frac{p_\perp^2}{B} = \text{const}\,.} \tag{6.5}$$

This quantity is called the *first* or *transversal* adiabatic invariant.

According to definition (5.33), the particle magnetic moment for the Larmor orbit is

$$\mathcal{M} = \frac{1}{c}\, JS = \frac{p_\perp^2}{2mB} = \frac{\mathcal{K}_\perp}{B}\,. \tag{6.6}$$

Here use is made of the non-relativistic formula for the Larmor frequency (5.11) and the non-relativistic kinetic energy of the particle transversal motion is designated as

$$\mathcal{K}_\perp = \frac{p_\perp^2}{2m}\,.$$

When (6.5) is compared with (6.6), it is apparent that the particle *magnetic moment is conserved in the non-relativistic approximation*.

In the relativistic limit the particle magnetic moment (6.6) does not remain constant; however, the first adiabatic invariant can be interpreted to represent the magnetic field flux through the surface covering the particle Larmor orbit,

$$\Phi = B\,\pi r_{\mathrm{L}}^2 = \frac{\pi c^2}{e^2}\,\frac{p_\perp^2}{B} = \text{const}\,. \tag{6.7}$$

This also follows directly from (6.4), when we substitute the relativistic formula

$$p_\perp = r_{\mathrm{L}}\,\frac{eB}{c} \tag{6.8}$$

into the first term on the right-hand side of formula (6.4). We obtain

$$I = \frac{e}{c}\left(B\,\pi r_{\mathrm{L}}^2\right) = \frac{e}{c}\,\Phi\,. \tag{6.9}$$

Therefore

> in the *relativistic* case, the magnetic field flux Φ through the surface S covering the particle Larmor orbit is conserved.

6.2.2 Magnetic mirrors and traps

Let us imagine the time-independent magnetic field, the field lines forming the convergent flux. As a rule, the field takes such a form in the vicinity of its sources, for instance, a sunspot S in the photosphere Ph in Figure 6.1.

The particle transversal momentum is

$$p_\perp = p\,\sin\theta\,, \tag{6.10}$$

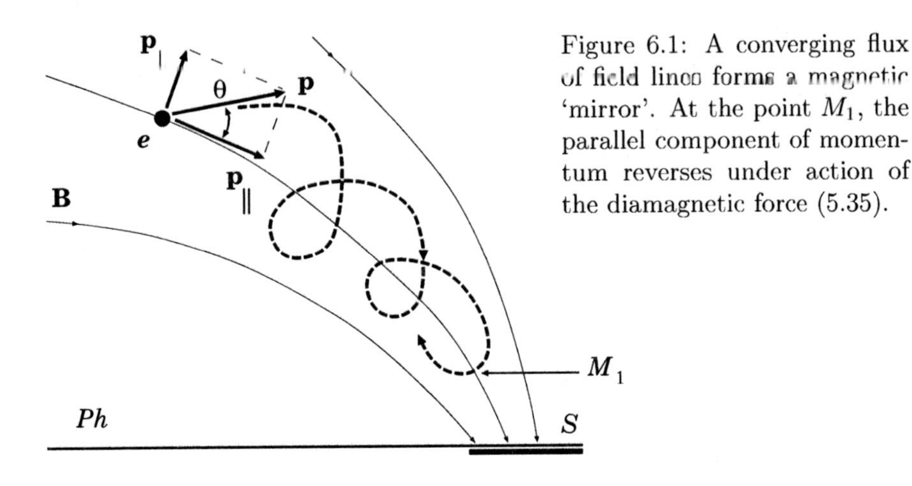

Figure 6.1: A converging flux of field lines forms a magnetic 'mirror'. At the point M_1, the parallel component of momentum reverses under action of the diamagnetic force (5.35).

it being known that p = const, since by virtue of (5.6) we have $\mathcal{E}$ = const. Substituting (6.10) into (6.5) gives

$$\frac{\sin^2\theta}{B} = \text{const} = \frac{\sin^2\theta_0}{B_0}$$

or

$$\sin^2\theta = \frac{B}{B_0}\,\sin^2\theta_0\,. \tag{6.11}$$

This formula shows that, for the increasing B, a point M_1 must appear in which $\sin^2\theta_1 = 1$. The corresponding value of the field is equal to

$$B_1 = B_0 \,/\, \sin^2\theta_0\,. \tag{6.12}$$

At this point the particle 'reflection' takes place:

$$p_{\parallel} = p\,\cos\theta_1 = 0\,.$$

The regions of convergent field lines are frequently referred to as magnetic 'mirrors'.

So, if there is a field-aligned gradient of the magnetic-field strength, the component of velocity parallel to the field decreases as the particle moves into a region of increasing field magnitude, although the total velocity is conserved. Eventually, under action of the diamagnetic force (5.35), the parallel velocity reverses (see the point M_1 in Figure 6.1). Such reflections constitute the principle of a *magnetic trap*. For example, magnetic fields create traps for fast particles in the solar atmosphere as shown in Figure 6.2. The particles are injected into the coronal magnetic tubes called flaring loops, during a flare. Let us suppose that this injection occurs at the loop apex.

Let us also suppose that, having hit the chromosphere Ch, the particles 'die' because of collisions. The particles do not return to the coronal part of

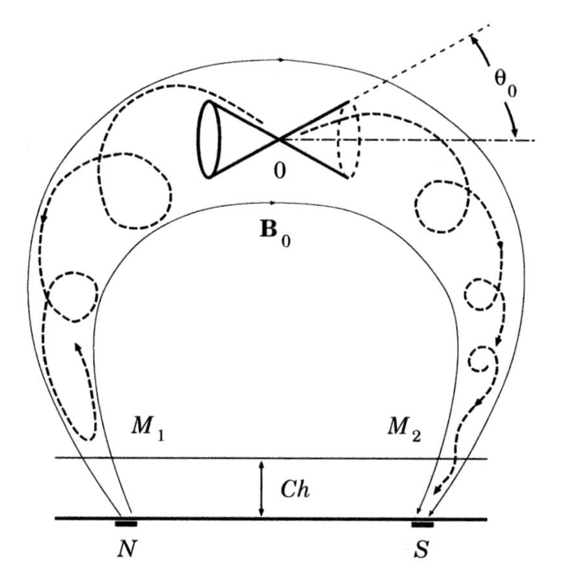

Figure 6.2: A coronal magnetic tube as a trap for particles accelerated in a solar flare. $\theta < \theta_0$ is the loss cone. Motion between the mirror points M_1 and M_2 is called bounce motion.

the trap, their energy being transferred to the chromospheric plasma, leading to its heating. Such particles are termed *precipitating* ones. Their pitch-angles have to be less than θ_0:

$$\theta < \theta_0 \tag{6.13}$$

with

$$\theta_0 = \arcsin \sqrt{B_0 \,/\, B_1} \tag{6.14}$$

in accordance with (6.12). Here B_0 is the magnetic field at the trap apex, B_1 is the field at the upper chromosphere level at the mirror points M_1 and M_2 as shown in Figure 6.2. The quantity $B_1 \,/\, B_0$ is called the *cork ratio*.

The angle region (6.13) is termed the *loss cone*. The particles with the initial momenta inside the loss cone precipitate from the trap. By contrast, the particles with $\theta > \theta_0$ at the loop apex experience reflection and do not reach the chromosphere. Such particles are termed *trapped* ones.

An interesting situation arises if the diffusion of the trapped particles into the loss cone is slower than their precipitation from the trap into the chromosphere. Then the distribution function becomes anisotropic, since the loss cone is 'eaten away', and non-equilibrium. The situation is quite analogous to the case of the distribution function formation with the positive derivative in some velocity region, like the *bump-on-tail* distribution (Figure 7.2). As a result, some *kinetic instabilities* (e.g., Silin, 1971; Schram, 1991; Shu, 1992) can be excited which lead to such plasma processes as wave excitation, anomalous particle transfer owing to the particles scattering off the waves, and *anomalous diffusion* into the loss cone (see also Chapter 7).

6.2.3 Bounce motion

Let us consider another example of a particle motion in a magnetic trap, namely that of a motion between two magnetic corks, the transversal drift being small during the period of longitudinal motion. In other words, the conditions of periodic longitudinal motion are changing adiabatically slowly. Then the *second* adiabatic invariant, referred to as the *longitudinal* one, is conserved:

$$I = \oint P_{\parallel} \, dl = p \oint \sqrt{1 - \sin^2 \theta} \, dl = p \oint \sqrt{1 - \frac{B}{B_1}} \, dl \, . \tag{6.15}$$

Here account is taken of the facts that the vector **A** is perpendicular to the vector **B** and $p = |\, \mathbf{p} \,| = \text{const}$ since $\mathcal{E} = \text{const}$; the formula (6.11) for the first adiabatic invariant is used in the last equality.

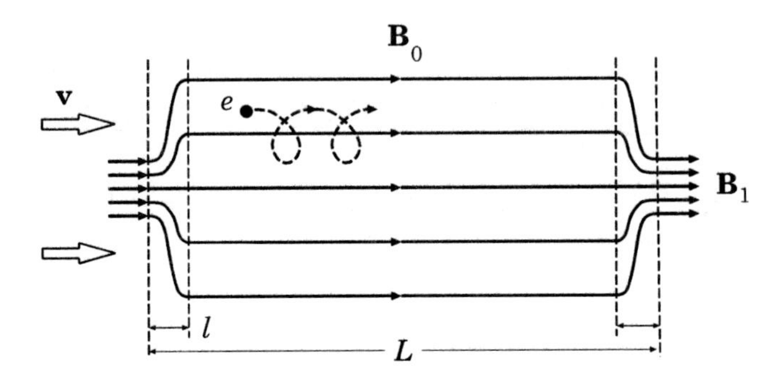

Figure 6.3: An idealized model of a long trap with a short moving cork. Unless a charged particle has its velocity vector very close to the axis of the trap, it is reflected back and forth between the mirrors, thereby remaining trapped.

Let us apply formula (6.15) to the case of a long trap with short corks: $l \ll L$ in Figure 6.3. The longitudinal invariant for such a trap is

$$I = \oint p_{\parallel} \, dl \approx 2 p_{\parallel} L = \text{const} \, .$$

Therefore the second adiabatic invariant is associated with the cyclical bounce motion between two mirrors or corks and is equal to

$$\boxed{p_{\parallel} L = \text{const} \, .} \tag{6.16}$$

Let us suppose now that the distance between the corks is changing, that is the trap length $L = L(t)$. Then from (6.16) it follows that

$$p_{\parallel}(t) = p_{\parallel}(0) \, \frac{L(0)}{L(t)} \, . \tag{6.17}$$

It is evident from (6.17) that (a) increasing the distance between the corks decreases the longitudinal momentum and, consequently, the particle energy, and (b) particle acceleration takes place in the trap if two magnetic corks are approaching each other as is shown by vector **v** in Figure 6.3.

The former case can describe the so-called 'adiabatic cooling' of accelerated particles, for example, in a magnetic trap which is captured by the solar wind and is expanding into interplanetary space. The latter case is more interesting. It corresponds to the Fermi mechanism considered in the next Section.

6.2.4 The Fermi acceleration

The famous theory of Fermi (1949) discussed the so-called interstellar 'clouds' that carry magnetic fields and could reflect charged particles. The same role could be played for instance by magnetic inhomogeneities in the solar wind or interplanetary medium. Fermi visualized that charged particles can be accelerated by being repeatedly hit by the moving magnetic clouds.

The energy of a particle, $\mathcal{E}$, will increase or decrease according to whether a cloud (an inhomogeneity of magnetic field) that causes the reflection moves toward the particle (head-on collision) or away from it (overtaken collision). The particle gains energy in a head-on collision but there can be also 'trailing' collisions in which energy is lost. It was shown by Fermi (1949, 1954) that

> on the average, the energy increases because the head-on collisions are *more probable* than the overtaking collisions

(see a non-relativistic treatment of the problem in Exercise 6.1). Through this *stochastic* mechanism

> the energy of the particle increases at a rate that, for relativistic particles, is proportional to their energy

(Exercise 6.2):

$$\frac{d\mathcal{E}}{dt} \propto \mathcal{E} \, . \tag{6.18}$$

That is why such a mechanism is often called the *first-order* (in energy $\mathcal{E}$) Fermi acceleration. **The higher the energy $\mathcal{E}$, the faster acceleration.** This is the most important feature of the Fermi mechanism. However we shall call it the *stochastic* Fermi acceleration to avoid a slightly confusing terminology in view of the fact that there is another parameter (a relative velocity of magnetic clouds) which characterizes the coefficient of proportionality in the problem under consideration (see Exercise 6.2).

From formula (6.18) follows that the energy $\mathcal{E}$ increases exponentially with time:

$$\mathcal{E}(t) = \mathcal{E}_0 \, \exp \frac{t}{t_a} \, , \tag{6.19}$$

where $\mathcal{E}_0$ is the initial energy, t_a is the acceleration time scale.

Large-scale MHD turbulence is generally considered as a source of magnetic inhomogeneities accelerating particles in astrophysical plasma. Acceleration of particles by MHD turbulence has long been recognized as a possible mechanism for solar and galactic cosmic rays (Davis, 1956).

Though the Fermi acceleration has been popular, it appears to be neither efficient nor selective. A mirror reflects particles on a *nonselective* basis: thermal particles may be reflected as well as suprathermal ones. Therefore one is faced with the conclusion (Eichler, 1979) that **most of the energy in the MHD turbulence goes into bulk heating** of the plasma rather than the selective acceleration of only a minority of particles. We shall come back to this question in Chapter 7.

If we somehow arrange that only head-on collisions take place, then the acceleration process will be much more efficient. We should call the acceleration resulting from such a situation the *regular* Fermi acceleration. More often, however, this mechanism is called the first-order (in the small parameter v_m/c, where v_m is the velocity of the moving magnetic clouds; see Exercise 6.1). The simplest example of this type mechanism is a pair of converging shock waves (Wentzel, 1964). In this case, there is no deceleration by trailing collisions (see formula (6.22) in Exercise 6.1) that reduce the net efficiency to the second order in the parameter v_m/c (Exercise 6.2).

One of several well-known examples of this type of the Fermi acceleration is the impulsive (with high rate of energy gain) acceleration between two approaching shocks S_{up} in the model of a flaring loop as shown in Figure 6.4. To explain the hard X-ray and gamma-ray time profiles in solar flares, Bai et al. (1983) assumed that pre-accelerated electrons penetrate into the flare loop and heat the upper chromosphere to high-temperatures rapidly. As a consequence of the fast expansion of a high-temperature plasma into the corona – the process of chromospheric 'evaporation', two shock waves S_{up} move upward from both footpoints.

Energetic particles are to be reflected only by colliding with the shock fronts. In such a way, the regular Fermi acceleration of particles between two shocks was suggested as a mechanism for the second-step acceleration of protons and electrons in flares. A similar example of the regular Fermi-type acceleration also related to a collapsing ($L(t) \to 0$) magnetic trap in solar flares is considered in vol. 2, Chapter 7.

The cosmic rays (see Section 5.1.3) were assumed to be accelerated by crossing shock fronts generated in explosive phenomena such as supernovae. However a very simple dimensional argument shows the kind of difficulties encountered even by the most violent phenomena in the Universe.

> The more energetic are the particles, the larger are their Larmor radius and/or the higher are the magnetic fields B necessary to confine them within the limits of a cosmic accelerator.

The size of a accelerator R must be larger than the Larmor radius of a particle. The product BR large enough to suit the 10^{20} eV energy range exists in no

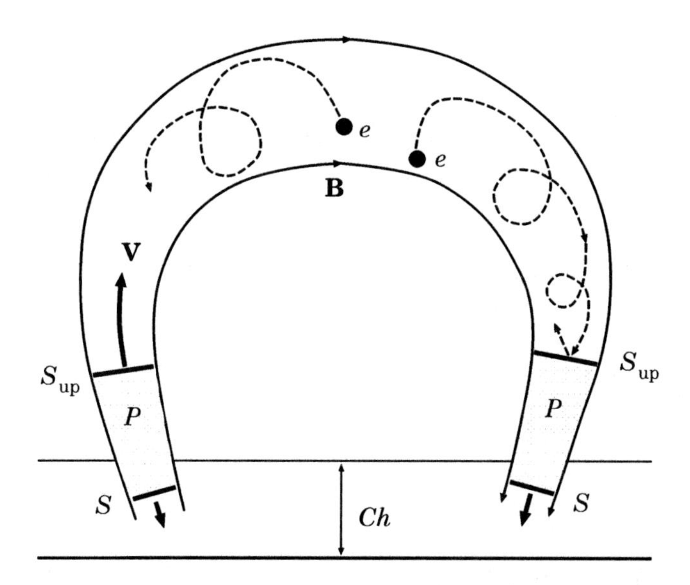

Figure 6.4: The flare-heated chromospheric plasma P rapidly expands into the corona. Particle acceleration of the first order Fermi type may occur in a magnetic loop between two converging shock waves S_{up}.

known standard astrophysical object.

6.3 The flux invariant

Let us consider the axisymmetric trap which is modelled on, for example, the Earth's magnetic field. Three types of the particle's motion are shown in Figure 6.5.

First, on the time scales of Larmor period, the particle spirals about a field line. Second, since there is a field-aligned gradient of the field strength, the particle oscillates between two mirrors M_1 and M_2. Third, if the guiding center does not lie on the trap's symmetry axis then **the radial gradient of field** (cf. Figure 5.9) **causes the drift** around this axis. This drift (formula (5.40)) is superimposed on the particle's oscillatory of rotation.

As the particle bounces between the mirrors and also drifts from one field line to another one, it traces some magnetic surface $S_{\rm d}$. The latter is called the *drift shell*. Let $T_{\rm s}$ be the period of particle motion on this surface.

If the magnetic field $\mathbf{B} = \mathbf{B}\,(t)$ is changing so slowly that $B \,/\, \dot{B} \gg T_{\rm s}\,$, then a *third* adiabatic invariant, referred to as a flux one, is conserved:

$$\Phi = \int\limits_{S} \mathbf{B} \cdot d\,\mathbf{S} = \text{const}\,. \tag{6.20}$$

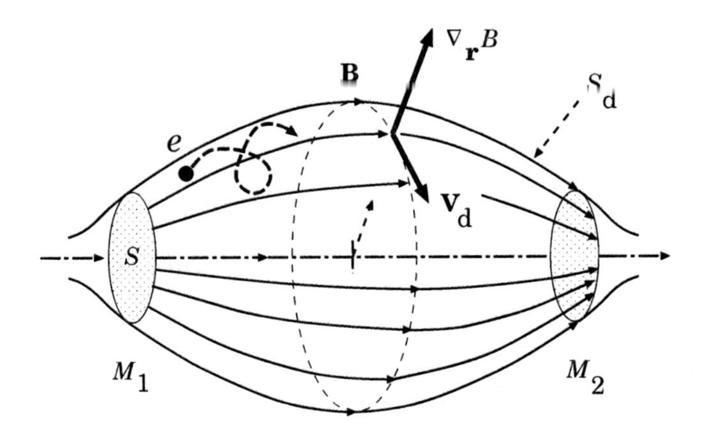

Figure 6.5: Particle drift in a trap, due to the radial gradient of field.

Thus the first adiabatic invariant implies conservation of the magnetic flux through the Larmor orbit, $B\,\pi r_{\rm L}^2$, whereas

> the flux invariant implies conservation of the magnetic flux through the closed orbit of guiding center motion,

that is the flux through the shaded surface S in Figure 6.5.

6.4 Approximation accuracy. Exact solutions

Adiabatic invariants have been obtained in the approximation of weakly inhomogeneous slowly changing magnetic fields. The invariants are *approximate* integrals of motion, widely used in plasma astrophysics. However we should not forget two important facts. First, the adiabatic theory has a limited, though *exponential*, accuracy. Second, this theory has a limited, though wide, area of applicability. The second vo;ume of this book will be devoted to the effect of magnetic reconnection and will present a situation when the adiabatic theory a priory does not apply.

Exact solutions to the equations of charged particle motion usually require numerical integration. The motion in the field of a magnetic dipole is a simple case that, nevertheless, is of practical significance. The reason for that is the possibility to approximate the Earth's magnetic field at moderately large distances by the dipole field. It was Störmer (1955) who contributed significantly to the solution of this problem.

Two types of trajectories are considered.

(a) The ones coming from infinity and returning there. These have been calculated in order to find out whether a particle can reach a given point along a given direction. An answer to this question is important for cosmic ray theories. For each point on the Earth and for each direction the so-called 'threshold rigidity' has been calculated. If a rigidity is greater than the threshold one, then the particle can reach the point. The vertical threshold rigidity is the most universally used one. This characterizes particle arrival in the direction of the smallest column depth of the Earth atmosphere.

(b) The orbits of trapped particles. Two radiation belts of the Earth, the inner and the outer, have been shown to exist. The mechanisms which generate trapped particles are not yet fully understood. They are presumably related to geomagnetic storms (Tverskoy, 1969; Walt, 1994).

Both gradient drift and curvature drift cause the positive particles in the radiation belt to drift westward in the Earth dipole magnetic field. Thus the radiation belt forms a ring of westward current circulating the Earth. This current tends to decrease the strength of the basic northward magnetic field observed at low latitudes on the Earth surface. There is a simple theoretical relationship between the depression of the magnetic field at the surface of the Earth and the total energy in the trapped particles. This relation allows us to use the observed change of the magnetic field as an indication of the amount of the energy in ring-current particles.

Recommended Reading: Northrop (1963), Kivelson and Russell (1995).

6.5 Practice: Exercises and Answers

Exercise 6.1 [Section 6.2.4] Show that a non-relativistic particle on average gains energy in collisions with moving magnetic clouds.

Answer. Let us consider the simplest model of one-dimensional motions of clouds: half of the clouds are moving in one direction and the other half moving in the opposite direction with the same velocity v_m. Let a particle of initial velocity V_0 undergo a head-on collision. The initial velocity seen from the rest frame of the cloud is $V_0 + v_m$. If the collision is elastic, the particle bounces back in the opposite direction with the same magnitude of velocity $V_0 + v_m$ in this rest frame. In the observer's frame, the reflected velocity appears to be $V_0 + 2v_m$. Hence the gain of kinetic energy $\mathcal{K}$ according to the observer equals

$$\delta \mathcal{K}_{+} = \frac{1}{2}\, m \left(V_0 + 2v_m \right)^2 - \frac{1}{2}\, m V_0^2 = 2m\, v_m \left(V_0 + v_m \right) . \tag{6.21}$$

Similarly, the energy loss in a trailing collision

$$\delta \mathcal{K}_{-} = -\, 2m\, v_m \left(V_0 - v_m \right) . \tag{6.22}$$

The probability of head-on collisions is proportional to the relative velocity $V_0 + v_m$, whereas the probability of trailing collisions is proportional to $V_0 - v_m$. Therefore the average gain of kinetic energy is equal to

$$\delta \mathcal{K}_{av} = \delta \mathcal{K}_{+} \frac{V_0 + v_m}{2v_m} + \delta \mathcal{K}_{-} \frac{V_0 - v_m}{2v_m} = 4m\, v_m^2 . \tag{6.23}$$

So a particle is accelerated.

Exercise 6.2 [Section 6.2.4] Prove the Fermi formula (6.18) for a relativistic particle.

Answer. Make the same procedure as that one in Exercise 6.1 by using the corresponding expressions in special relativity to see that the average energy gain

$$\delta \mathcal{E}_{av} = 4 \left(\frac{v_m}{c} \right)^2 \mathcal{E} \, . \tag{6.24}$$

Formula (6.24) obviously reduces to (6.23) in the non-relativistic limit on putting $\mathcal{E} = mc^2$.

So the average energy gain is proportional to the energy. Therefore the energy of a relativistic particle suffering repeated collisions with moving nagnetic clouds increases according to formula

$$\frac{d\mathcal{E}}{dt} = \alpha_{_F} \, \mathcal{E} \, , \tag{6.25}$$

where $\alpha_{_F}$ is a constant. Q.e.d.

Note also that the average energy gain (6.24) is propotional to the dimensionless parameter $(v_m/c)^2$. Since actual clouds are moving at non-relativistic velocities, this parameter should be a very small number. Hence the acceleration process is quite inefficient. Because of this quadratic dependence on v_m, this process is referred as the *second-order* Fermi acceleration.

If only head-on collisions take place, then the acceleration is much more efficient. It follows from formula (6.21) that, for $V_0 \gg v_m$, the energy gain will depend linearly on v_m. So the acceleration resulting from such conditions is called the *first-order* Fermi acceleration. Such conditions are well possible, for example, in collapsing magnetic traps created by the magnetic reconnection process in solar flares (see vol. 2, Chapter 7).

Powerful shock waves in a plasma with magnetic field (like the solar wind) may well provide sites for the first-order Fermi acceleration. Magnetic inhomogeneities are expected on both sides of the shock front. It is possible that a charged particle is trapped near the front and repeatedly reflected from magnetic inhomogeneities on both sides. Such collisions may lead to more efficient acceleration (see Chapter 18) compared to original Fermi's acceleration by moving interstellar clouds.

Chapter 7

Wave-Particle Interaction in Astrophysical Plasma

The growth or damping of the waves, the emission of radiation, the scattering and acceleration of particles – all these phenomena may result from wave-particle interaction, a process in which a wave exchanges energy with the particles in astrophysical plasma.

7.1 The basis of kinetic theory

7.1.1 The linearized Vlasov equation

In this Chapter we shall only outline the physics and main methods used to describe the wave-particle interaction in collisionless astrophysical plasmas as well as in Maxwellian plasmas where fast particles interact with electromagnetic waves. In the simplest – *linear* – approach, the idea is in the following.

We assume the unperturbed plasma to be uniform and characterized by the distribution functions $f_k^{(0)}$ of its components k: electrons and ions. The unperturbed plasma is also assumed to be steady. So

$$f_k^{(0)} = f_k^{(0)} (\mathbf{v}) . \tag{7.1}$$

Let $\mathbf{B}^{(0)}$ be the unperturbed uniform magnetic field inside the plasma. We further assume that the only zero-order force is the Lorentz force with $\mathbf{E}^{(0)} = 0$.

The dynamics of individual particles is determined by the first-order forces related to the wave electric field $\mathbf{E}^{(1)}$ and wave magnetic field $\mathbf{B}^{(1)}$. To describe these particles we shall use the perturbation function $f_k^{(1)}$, which is linear in $\mathbf{E}^{(1)}$ and $\mathbf{B}^{(1)}$. Under the assumptions made, we see that the Vlasov equation (Section 3.1.2) can be a proper basis for the kinetic theory of wave-particle interaction. For this reason we shall realize the following procedure

(a) We linearize the Vlasov equation (3.3) together with the Maxwell equations (3.4) for the self-consistent wave field. Equation (3.3) becomes

$$\frac{\partial f_k^{(1)}(X,t)}{\partial t} + v_\alpha \frac{\partial f_k^{(1)}(X,t)}{\partial r_\alpha} + $$
$$+ \frac{e_k}{m_k}\left(\frac{1}{c}\,\mathbf{v}\times\mathbf{B}^{(0)}\right)_\alpha \frac{\partial f_k^{(1)}(X,t)}{\partial v_\alpha} =$$
$$- \frac{e_k}{m_k}\left(\mathbf{E}^{(1)} + \frac{1}{c}\,\mathbf{v}\times\mathbf{B}^{(1)}\right)_\alpha \frac{\partial f_k^{(0)}(\mathbf{v})}{\partial v_\alpha}\,. \tag{7.2}$$

The left-hand side of the linear Equation (7.2) is the Liouville operator (1.10) acting on the first-order distribution function for particles following *unperturbed* trajectories in phase space $X = \{\mathbf{r}, \mathbf{v}\}$:

$$\frac{D}{Dt}\, f_k^{(1)} = -\frac{F_{k,\alpha}^{(1)}}{m_k}\,\frac{\partial f_k^{(0)}}{\partial v_\alpha}\,. \tag{7.3}$$

This fact (together with the linear Lorentz force in the right-hand side of (7.3) and the linearized Maxwell equations) can be used to find the general solution of the problem. We are not going to do this here (see Exercise 7.1). Instead, we shall make several simplifying assumptions to demonstrate the most important features of kinetic theory on the basis of Equation (7.3).

(b) Let us consider a small harmonic perturbation varying as

$$f_k^{(1)}(t,\mathbf{r},\mathbf{v}) = \tilde{f}_k(\mathbf{v})\,\exp\left[-\mathrm{i}\,(\omega t - \mathbf{k}\cdot\mathbf{r})\right]. \tag{7.4}$$

Substituting the plane wave expression (7.4) with a similar presentation of the perturbed electromagnetic field in Equation (7.2) gives us the following linear equation:

$$\mathrm{i}\,(\omega - \mathbf{k}\cdot\mathbf{v})\,\tilde{f}_k(\mathbf{v}) - \frac{e_k}{m_k}\left(\frac{1}{c}\,\mathbf{v}\times\mathbf{B}^{(0)}\right)_\alpha \frac{\partial \tilde{f}_k(\mathbf{v})}{\partial v_\alpha} =$$
$$= \frac{e_k}{m_k}\left[\tilde{\mathbf{E}}\left(1 - \frac{\mathbf{k}\cdot\mathbf{v}}{\omega}\right) + \mathbf{k}\left(\frac{\mathbf{v}\cdot\tilde{\mathbf{E}}}{\omega}\right)\right]_\alpha \frac{\partial f_k^{(0)}(\mathbf{v})}{\partial v_\alpha}\,. \tag{7.5}$$

Here the Faraday law (1.25) has been used to substitute for the wave magnetic field.

(c) We shall assume that the waves propagate parallel to the ambient field $\mathbf{B}^{(0)}$ which defines the z axis of a Cartesian system. From Section 5.1 it follows that in a uniform magnetic field there exist two constants of a particle's motion: the parallel velocity $\mathbf{v}_\parallel$ and the magnitude of the perpendicular velocity

$$v_\perp = |\,\mathbf{v}_\perp| = \left(v_x^2 + v_y^2\right)^{1/2}.$$

Hence the unperturbed distribution function

$$f_k^{(0)} = f_k^{(0)} \left(v_{\parallel}, v_{\perp} \right), \tag{7.6}$$

as required by Jeans's theorem (Exercise 1.1). Therefore in what follows we can consider two cases of resonance, corresponding two variables in the distribution function (7.6).

7.1.2 The Landau resonance and Landau damping

Let us consider the so-called *electrostatic* waves which have only a parallel electric field $\mathbf{E}^{(1)} = \mathbf{E}_{\parallel}$ under the assumption of parallel propagation:

$$\mathbf{k} \times \mathbf{B}^{(0)} = 0 . \tag{7.7}$$

In this case the linearized Vlasov Equation (7.5) reduces to

$$\mathrm{i} \left(\omega - k_{\parallel} v_{\parallel} \right) \tilde{f}_k - \frac{e_k}{m_k} \left(\frac{1}{c} \, \mathbf{v} \times \mathbf{B}^{(0)} \right)_{\alpha} \frac{\partial \tilde{f}_k}{\partial v_{\alpha}} = \frac{e_k}{m_k} \, \tilde{E}_{\parallel} \, \frac{\partial f_k^{(0)}}{\partial v_{\alpha}} \, . \tag{7.8}$$

Now let us find the perturbation of charge density according to definition (3.5):

$$\rho^{q\,(1)} (\mathbf{r}, t) = \sum_k e_k \int_{\mathbf{v}} f_k^{(1)} (\mathbf{r}, \mathbf{v}, t) \, d^3\mathbf{v} \, . \tag{7.9}$$

Hence the amplitude

$$\tilde{\rho}^{q} = \sum_k e_k \int_{\mathbf{v}} \tilde{f}_k (\mathbf{v}) \, d^3\mathbf{v} \, . \tag{7.10}$$

When we calculate the charge density by using Equation (7.8), the second term on the left-hand side of this equation vanishes on integration over perpendicular velocity.

Therefore, for parallel propagating electrostatic waves, the harmonic perturbation of charge density is given by

$$\tilde{\rho}^{q} = - \mathrm{i} \, \tilde{\mathbf{E}}_{\parallel} \sum_k \frac{e_k^2}{m_k} \int_{v_{\parallel}} \frac{1}{\left(\omega - k_{\parallel} v_{\parallel} \right)} \, \frac{\partial f_k^{(0)}}{\partial v_{\parallel}} \, d v_{\parallel} \, . \tag{7.11}$$

Formula (7.11) shows that there is a *resonance* which occurs when

$$\boxed{\omega - k_{\parallel} v_{\parallel} = 0} \tag{7.12}$$

or when the particle velocity equals the parallel phase velocity of the wave, $\omega / k_{\parallel}$. This is the *Landau resonance*.

A physical picture of Landau resonance is simple.

When the resonance condition (7.12) is satisfied the particle 'sees' the electric field of the wave as a *static* electric field in the particle's rest system

(see Exercise 7.3).

Particles in resonance moving slightly faster than the wave will lose energy, while those moving slightly slower will gain energy. Since the Maxwellian distribution is decreasing with velocity,

in a Maxwellian plasma, near the Landau resonance, there are more particles at lower velocities than at higher velocities. That is why **the plasma gains energy at the expense of the wave.**

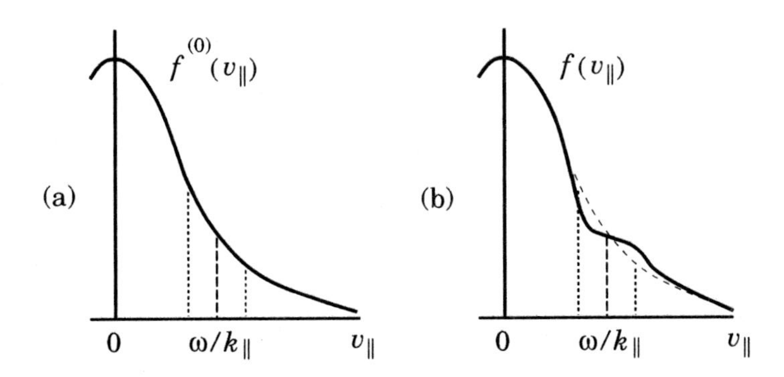

Figure 7.1: The Landau damping. (a) The initial distribution function of thermal electrons with some narrow region centered at the resonance with the wave. (b) The distribution function after an evolution due to interaction of the electrons with the wave.

This effect, illustrated by Figure 7.1 (see also Exercise 7.6), is called the *Landau damping* (Landau, 1946) or collisionless damping. Normally we think of damping as a dissipative process and hence expect it to be present only in systems where collisions can convert a part of the wave energy into thermal energy. At first sight, damping in a collisionless system seems mystifying since we ask the question where could the energy have gone. For a negative slope of the distribution function at the phase velocity ω/k, there are more particle which are accelerated than which are decelerated. For this reason the wave puts a net amount of energy in the particles so that there is a loss of wave energy. Therefore the Landau damping is not by randomizing collisions but by a transfer of wave field energy into oscillations of resonant particles.

Landau damping is often the dominant **damping mechanism for waves**, such as ion-acoustic waves and Langmuir waves, in thermal plasma without a magnetic field.

The absorption of longitudinal waves in plasma in the thermal equilibrium is often determined by collisionless damping (e.g., Zheleznyakov, 1996).

On the other hand, if a distribution function has more particles at higher velocities than at lower velocities in some region of phase space as shown in Figure 7.2, this distribution will be unstable to waves that are in resonance with the particles. This is the known 'bump-on-tail' instability. Due to this type of instability, a beam of fast electrons (with velocities much higher than the thermal speed of electrons in the plasma) causes Langmuir waves to grow. Langmuir waves generated through the bump-on-tail instability play an essential role, for example, in solar radio bursts.

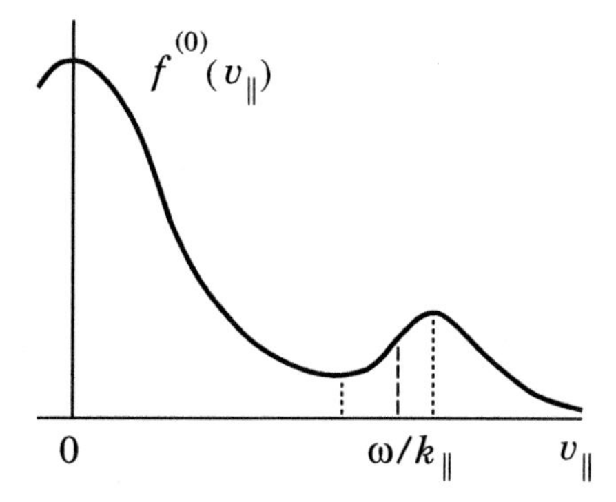

Figure 7.2: The bump-on-tail distribution function with the resonance condition in the region of a positive slope.

There are many examples in plasma astrophysics in which one species (e.g., electrons) moves relative to another. Solar flares produce a significant flux of fast electrons moving through the plasma in interplanetary space. Fast electrons move away from a planetary shock through the solar wind. Aurorae are produced by fast electrons moving along Earth's magnetic-field lines. If we consider a stream of plasma with an average velocity impinding on another plasma at rest, we have just the same situation. The system has an instability such that

> the kinetic energy of the relative motion between the plasma steams is fed into a plasma wave of the appropriate phase velocity.

So all the two-stream instabilities have, in fact, the same origin.

The above derivation emphasizes the close relation of the Landau damping with the **Cherenkov effect** (see Exercises 7.2–5). It has been definitely pointed out by Ginzburg and Zheleznyakov (1958) that

> the Landau damping and the Cherenkov absorption of plasma waves, the inverse Cherenkov effect, are the same phenomenon

initially described in two different ways.

The discussion hitherto has focused on the linear Landau damping, i.e. the behaviour of a small perturbation which satisfies the linearized Vlasov equation. However this picture can be extended to **finite amplitude** perturbations (Kadomtsev, 1976, Chapter 4). In the context of plasma astrophysics, this means considering *nonlinear* Landau damping, which generalized the linear theory by incorporating the possibility of mode-mode couplings that allow energy transfer between different modes.

In fact, the linear theory illuminates only a narrow window out of the wealth of all effects related to wave-particle interactions. Mathematically, the **linear theory uses a well-developed algorithm**. Few analytical methods are known to treat the much wider field of nonlinear effects, and most of these methods rely on approximations and lowest-order perturbation theory. The theory of *weak* wave-particle interaction or *weak turbulence* as well as the *quasi-linear* theory for different types of waves are still today the most important parts in astrophysical applications (e.g., Treumann and Baumjohann, 1997; Benz, 2002).

7.1.3 Gyroresonance

As for the Landau resonance, we shall use the linear Equation (7.5) as a basis, assuming that a wave is propagating parallel to the ambient field $\mathbf{B}^{(0)}$. However, this time, we shall further assume that the wave electric field $\mathbf{E}^{(1)}$ and hence the wave magnetic field $\mathbf{B}^{(1)}$ are perpendicular to the ambient magnetic field.

Under the assumption of a harmonic perturbation (7.4) we shall make use of the so-called *polarized coordinates*:

$$\tilde{E}_l = \frac{\tilde{E}_x + \mathrm{i}\tilde{E}_y}{\sqrt{2}}, \qquad \tilde{E}_r = \frac{\tilde{E}_x - \mathrm{i}\tilde{E}_y}{\sqrt{2}}. \tag{7.13}$$

Subscripts l and r correspond to the waves with left- and right-hand circular polarizations, respectively.

By definition, the wave is right-hand circular polarized if $\tilde{E}_x$ leads $\tilde{E}_y$ by a quarter of a wave period. If, for such a wave, we multiply Equation (7.5) by velocity

$$\mathbf{v}_r = \frac{v_x - \mathrm{i}\, v_y}{\sqrt{2}} \tag{7.14}$$

and integrate over velocity space, making use of (7.6) and the fact that the unperturbed distriburion function $f^{(0)}$ is a symmetric function of $v_\perp$, we find the equation which determines (see definition (3.6)) the current density in the harmonic perturbation:

$$\tilde{\mathbf{j}}_r^{\,\mathrm{q}} = -\mathrm{i} \sum_{\mathrm{k}} \frac{e_{\mathrm{k}}^2}{m_{\mathrm{k}}} \tilde{E}_r \times \tag{7.15}$$

$$\times \int_{\mathbf{v}} \frac{1}{\left(\omega - k_\| v_\| - s\,\omega_{\mathrm{B}}^{(\mathrm{k})}\right)} \left[\left(1 - \frac{k_\| v_\|}{\omega}\right) \frac{\partial f_k^{(0)}}{\partial v_\perp} + \frac{k_\| v_\perp}{\omega} \frac{\partial f_k^{(0)}}{\partial v_\|} \right] v_\perp \, d^3\mathbf{v}\,.$$

Here $\omega_{\rm B}^{(k)}$ is the Larmor frequency of a particle of a kind k, the integer s can be positive or negative. The resonance condition in formula (7.15) for current density is the *gyroresonance*:

$$\boxed{\omega - k_{\parallel} v_{\parallel} - s\,\omega_{\rm B}^{(k)} = 0\,.} \tag{7.16}$$

We see that a gyroresonant interaction occurs when the Doppler-shifted wave frequency

$$\omega_{\rm D} = \omega - k_{\parallel}\, v_{\parallel}\,, \tag{7.17}$$

as observed by a particle moving with the parallel velocity $v_{\parallel}$, is an integer multiple s of the Larmor frequency in the guiding center frame, i.e.

$$\omega_{\rm D} = s\,\omega_{\rm B}^{(k)}\,. \tag{7.18}$$

Depending upon the initial relative phase of the wave and particle, the particle will corotate with either an accelerating or decelerating electric field over a significant portion of its Larmor motion,

resulting in an appreciable gain or loss of energy, respectively.

If the particle and transversal electric field rotate in the same sense, the integer $s > 0$, whereas an opposite sense of rotation requires $s < 0$. However the strongest interaction usually occurs when the Doppler-shifted frequency exactly matches the particle Larmor frequency.

The gyroresonance is important for generating waves such as the *wistler* mode, which is polarized predominantly perpendicular to the ambient field.

For a wave to grow from gyroresonance, there should be a net decrease in particle energy as the particle diffuses down the phase-space density gradient defined by the numerator in formula (7.15),

i.e. by the expression enclosed in large square brackets under the integral in formula (7.15).

For the parallel propagation of a wave in plasma, the Landau resonance is associated with parallel electric fields. For perpendicular electric fields, particles and fields can be in gyroresonance. It is clear that the Landau resonance diffuses particles parallel to the ambient magnetic field, whereas **gyroresonance causes diffusion in the pitch angle**. This can be seen in the wave frame, i.e. the frame in which the parallel phase velocity of the wave is zero. If we transform the expression enclosed in large square brackets in formula (7.15) to the wave frame, we find that in this frame the gradient in velocity space is gradient with respect to pitch angle θ. Hence

the main effect of gyroresonance is to cause particles to change pitch angle in the wave frame.

This is contrasted with the Landau resonance, where the diffusion is in the parallel velocity $v_\parallel$ due to the term $\partial f^{(0)}/\partial v_\parallel$ and therefore mainly in energy, rather than pitch angle.

As such, then the Landau-resonant instabilities are often driven by bump-on-tail distributions of particles, whereas gyroresonant instabilities are driven by *pitch-angle* anisotropy. Thus the gyroresonance-type instabilities can appear as soon as a 'tail' or beam is formed in the direction parallel to the background field $\mathbf{B}^{(0)}$. They excite waves that scatter the particles back to a nearly isotropic state.

7.2 Stochastic acceleration of particles by waves

7.2.1 The principles of particle acceleration by waves

In Section 7.1 we considered the resonant interaction between particles and one wave propagating parallel to the uniform magnetic field $\mathbf{B}^{(0)}$ in a uniform plasma without an external electric field: $\mathbf{E}^{(0)} = 0$. The dynamics of individual particles was determined by the first-order forces related to the wave electric field $\mathbf{E}^{(1)}$ and wave magnetic field $\mathbf{B}^{(1)}$. We described behavior of these particles by the linearized Vlasov equation (7.2) for the perturbation function $f_k^{(1)}$, which is linear in $\mathbf{E}^{(1)}$ and $\mathbf{B}^{(1)}$.

Under simplifing assumptions made, we saw that, in addition to the Landau resonance (7.12):

$$\omega_{\mathrm{D}} = 0\,, \tag{7.19}$$

other resonances (7.16) arise in wave-particle interaction. These are the gyroresonances which occur when the Doppler-shifted frequency

$$\omega_{\mathrm{D}} = \omega - k_\parallel v_\parallel \tag{7.20}$$

(as observed by a particle moving with parallel velocity $v_\parallel$) is some integer multiple s of the particle Larmor frequency $s\,\omega_{\mathrm{B}}^{(k)}$:

$$\boxed{\omega_{\mathrm{D}} = s\,\omega_{\mathrm{B}}^{(\mathrm{k})}\,.} \tag{7.21}$$

If a wave is, in general, oblique, its electric field has components transversal and parallel to $\mathbf{B}^{(0)}$, whereas if the wave is parallel, its electric field is transversal. Since the transversal field typically consists of left- and right-hand polarized components, the integer s may be either positive or negative. Anyway

> the energy gain is severely limited due to the particle losing resonance with the wave.

Large gains of energy are possible, in principle, if a spectrum of waves is present. In this case, the resonant interaction of a particle with one wave can result in an energy change that brings this particle into resonance with a neighboring wave, which then changes the energy so as to allow the particle to resonate with another wave, and so on. Such an energy change can be diffusive, but over long time scales there is a net gain of energy, resulting in *stochastic* acceleration.

A traditional problem of the process under discussion is the so-called *injection* energy. The problem arises since for many waves in plasma their phase velocity along the ambient magnetic field, $\omega / k_{\parallel}$, is much greater than the mean thermal velocity of particles. Let us re-write the gyroresonance condition (7.21) as

$$\gamma_{\mathrm{L}} \left(\frac{\omega}{k_{\parallel}} - v_{\parallel} \right) = \frac{s\, \omega_{\mathrm{B}}^{(\mathrm{k})}}{k_{\parallel}} \, . \tag{7.22}$$

Here the relativistic Lorentz factor γ_{L} has been taken into account (see Exercise 7.3). Consider two opposite cases.

(a) For low thermal velocities we can neglect $v_{\parallel}$ in Equation (7.22) and see that, in order to resonate with a thermal particle, the waves must have very high frequencies $\omega \approx \omega_{\mathrm{B}}^{(\mathrm{k})}$ or very small $k_{\parallel}$.

For the case of thermal electrons and protons in the solar corona, their Larmor frequencies are very high (Exercise 5.1). If we try to choose a minimal value of $k_{\parallel}$, we are strongly restricted by a maximal value of wavelenghts, which must be certainly smaller than the maximal size of an acceleration region. These difficulties naturally lead to much doubt about the viability of stochastic acceleration and to a search for *preacceleration* mechanisms.

(b) On the other hand, high energy particles need, according to the resonance condition (7.22), waves with very low frequencies: $\omega \ll \omega_{\mathrm{B}}^{(\mathrm{k})}$. Therefore

> a very *broad-band* spectrum of waves (extending from $\approx \omega_{\mathrm{B}}^{(\mathrm{k})}$ to very low frequencies) is necessary to accelerate particles from thermal to relativistic energies.

In principle, the so-called *wave cascading* from low to high frequencies can be a way of producing the necessary broad-band spectrum. The idea comes from the Kolmogorov theory of hydrodynamic turbulence (Kolmogorov, 1941). Here the evolution of turbulence can be described by the **Kolmogorov-style dimensional analysis** or by a **diffusion of energy in wavenumber space**. The last idea was subsequently introduced to MHD by Zhou and Matthaeus (1990). They presented a general transfer equation for the wave spectral density. In Section 7.2.2, we shall discuss briefly both approaches and their applications; see also Goldreich and Sridhar (1997).

The stochastic acceleration of particles by waves is essentially the resonant form of Fermi acceleration (see Section 6.2 (c)). An important feature of stochastic acceleration is an isotropization process because the pitch-angle scattering increases the volume of wave phase space that can be sampled by the resonant particles (7.22).

In general, if isotropization exists and keeps the distribution isotropic during an acceleration time, it increases the acceleration efficiency. For example, Alfvén (1949) considered the betatron acceleration in an uniform magnetic field $\mathbf{B}^{(0)}(t)$ which changes periodically in time and has local nonuniformities $\mathbf{B}^{(1)}$ characterized by significant variations at distances smaller than the Larmor radius of accelerated particles.

When a particle passes through such nonuniformities its motion becomes random, with the momenta tending to be uniformly distributed between the three degrees of freedom. For this reason, when the field $\mathbf{B}^{(0)}(t)$ contracts, a fraction of the energy acquired due to betatron acceleration is transferred to the parallel component of the particle motion. As a consequence, the decrease in the energy of the transverse motion with decreasing magnetic field is smaller than its increase in the growth time. Thus the particle acquires an additional energy on completion of the full cycle. Therefore the total particle energy can systematically increase even if the fluctuating magnetic field does not grow. This phenomenon is known as the *Alfvén pumping*.

Tverskoi (1967, 1968) showed that in a turbulent cosmic plasma, the Fermi acceleration related to the reflection from long strong waves is efficient only in the presence of fast particle scattering by short waves whose length is comparable to the particle Larmor radius.

7.2.2 The Kolmogorov theory of turbulence

In general terms, a hydrodynamic flow tends to become turbulent if the ratio of inertial to viscous terms in the equation of motion, as described by the Reynolds number (see Chapter 12), is sufficiently large. In order not to obscure the essential physical point made in this section, we assume that a turbulence is *isotropic* and *homogeneous*. So we define a one-dimensional spectral density $W(k)$, which is the wave energy density per unit volume in the wave vector space $\mathbf{k}$.

First, we remind the Kolmogorov (1941) treatment of *stationary* turbulence of *incompressible* fluid. The steady state assumption implies that the energy flux F through a sphere of radius k is independent of time. In the *inertial* range of wave numbers, for which supply and dissipation of energy are neglected, the flux F is also independent of the wave vector k. If $\mathcal{P}$ denotes the total rate of energy dissipation at the short wave ($k = k_{\,\text{max}}$) edge of the inertial range, which equals the rate of energy supply at the long wave ($k = k_{\,\text{min}}$) edge, then $F = \mathcal{P}$ and $dF/dk = 0$ in the inertial range in Figure 7.3.

Kolmogorov's theory adopts the hypothesis that with the above assumptions the flux F through a sphere of radius k in the *inertial* range depends only upon the energy in that sphere and upon the wave number. Thus by

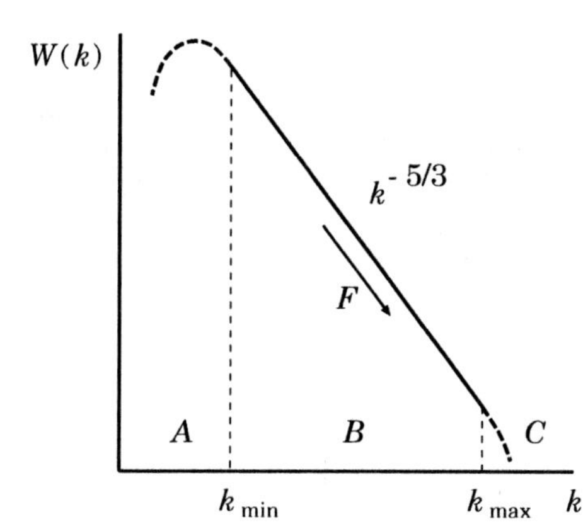

Figure 7.3: The energy per unit wave number in Kolmogorov's turbulence is plotted as a function of wavenumber in the inertial range B between the source A at small k and the sink C at large k.

dimensional analysis we arrive at

$$F = \mathcal{P} \sim W^{3/2}\, k^{5/2} . \tag{7.23}$$

From here it follows that the one-dimensional spectral density

$$\boxed{W(k) = C_k\, \mathcal{P}^{2/3}\, k^{-5/3} .} \tag{7.24}$$

This is the famous *Kolmogorov spectrum* for the fluid isotropic turbulence, involving the Kolmogorov constant C_k.

The turbulent velocity field in fluid can be thought of as being made of many eddies of different sizes. The input energy is usually fed into the system in a way to produce the largest eddies. Kolmogorov had realized that these large eddies can feed energy to the smaller eddies and these in turn feed the still smaller eddies, resulting in a cascade of energy from the larger eddies to the smaller ones.

If we anticipate the viscosity ν (see Section 12.2.2) to be not important for this process, we neglect dissipation of energy. However we cannot have eddies of indefinitely small size. For sufficiently small eddies of size $l_{\rm min}$ and velocity $v_{\rm min}$, the Reynolds number is of order unity, i.e.

$$l_{\rm min}\, v_{\rm min} \sim \nu . \tag{7.25}$$

So the energy in these small eddies is dissipated by viscosity.

Let the energy be fed into the turbulence at some rate $\mathcal{P}$ per unit mass per unit time at the larges eddies of size $l_{\rm max}$ and velocity $v_{\rm max}$, for which the Reynolds number

$$\mathrm{Re} = \frac{l_{\rm max}\, v_{\rm max}}{\nu} \gg 1 . \tag{7.26}$$

Then this energy cascades to smaller and smaller eddies untill it reaches the smallest eddies satisfying condition (7.25).

The intermediate eddies merely transmit the energy to the smaller eddies. Let characterize these intermediate eddies only by their size l and velocity v. Since they are able to transmit the energy at the required rate $\mathcal{P}$, Kolmogorov postulated that it must be possible to express $\mathcal{P}$ in terms of l and v. On dimensional grounds, there is only one way of writing $\mathcal{P}$ in terms of l and v:

$$\mathcal{P} \sim \frac{v^3}{l}. \tag{7.27}$$

From here

$$v \sim (\mathcal{P}l)^{1/3}. \tag{7.28}$$

So

> the velocity associated with the turbulent eddies of a particular size is proportional to the cube root of this size.

This result is known as the **Kolmogorov scaling law**. The scaling law (7.28) expresses the same thing as (7.24). This is shown in Exercise 7.10.

The Kolmogorov scaling law (7.28) was verified by doing experiments on a turbulent fluid with a sufficiently large inertial range. In laboratory it is very difficult to reach high enough Reynolds numbers to produce a sufficiently broad inertial range. One of the first confirmations of it was reported by Grant et al. (1962) by conducting experiments in a tidal channel between Vancouver Island and mainland Canada (see also Stewart and Grant, 1969).

The Kolmogorov power spectrum (7.24) is observed in the turbulent boundary layer on the ground and in some other turbulent flows in astrophysical plasma (for example, in the solar wind), in spite of the fact that, in all these cases, the original assumptions of incompressibility and isotropy are not fulfilled.

7.2.3 MHD turbulent cascading

The Kolmogorov concept of independence of widely separated wave numbers in the inertial range of fluid turbulence was modified for the MHD case by Iroshnikov (1963) and Kraichnan (1965). When the magnetic energy in subinertial wave numbers exceeds the total energy in the inertial range, the predicted inertial range spectrum is proportional to $k^{-3/2}$, instead of $k^{-5/3}$. Note that the Kolmogorov spectrum is steeper than the Kraichnan spectrum ($5/3 > 3/2$).

Leith (1967) introduced a diffusion approximation for spectral transfer of energy in isotropic hydrodynamic turbulence. This approach may be viewed as an alternative to the straight-forward dimensional analysis discussed above. However it is a natural extension since this approach approximates the spectral transfer as a local process in wave number space, i.e. in accordance with the spirit of the Kolmogorov hypotheses that the total energy is conserved with respect to couplings between waves. Therefore just diffusion is a physically appealing framework for the simplest model to describe this kind of *local conservative* transfer.

If some waves, propagating parallel to the uniform field $\mathbf{B}^{(0)}$, are injected at the longest wavelength $\lambda = \lambda_{\text{max}}$ and if a Kolmogorov-like nonlinear cascade transfers the wave energy to smaller scales, then the diffusion equation in wave number space

$$\frac{\partial W}{\partial t} = \frac{\partial}{\partial k_{\|}} \left(D_{\|\|} \frac{\partial W}{\partial k_{\|}} \right) - \gamma \left(k_{\|} \right) W + S \tag{7.29}$$

can describe injection, cascading, and damping of the waves. Here $D_{\|\|}$ is a diffusion coefficient that depends on W and can be determined for Kolmogorov-type cascading. $\gamma \left(k_{\|} \right)$ is the damping rate usually due to particle acceleration in high-temperature low-density astrophysical plasma. The wave energy is dissipated by accelerating particles in smallest scales $\lambda \sim \lambda_{\text{min}}$.

The source term S in Equation (7.29) is proportional to the injection rate Q of the wave energy. A mechanism by which the waves are generated is typically unknown but easily postulated. For example, MHD waves can be formed by a large-scale restructuring of the magnetic field in astrophysical plasma, which presumably occurs in nonstationary phenomena with flare-like energy releases due to magnetic reconnection.

In summary, the wave cascading and particle acceleration are described by one wave-diffusion equation, in which the damping depends on the accelerating particle spectra, and by diffusion equations (one for each kind k of particles: electrons, protons and other ions) for accelerating particles. The system is therefore highly coupled and generally nonlinear or *quasilinear* in the case of small-amplitude waves.

7.3 The relativistic electron-positron plasma

According to present views, in a number of astrophysical objects there is a relativistic plasma that mainly consists of electrons and positrons. Among these objects are pulsar magnetospheres (Ruderman and Sutherland, 1975; Michel, 1991), accretion disks in close binary systems (Takahara and Kusunose, 1985; Rose, 1998), relativistic jets from active galactic nuclei (Begelman et al., 1984; Peacock, 1999), and magnetospheres of rotating black holes in active galactic nuclei (Hirotani and Okamoto, 1998).

Because of synchrotron losses, the relativistic collisionless plasma in a strong magnetic field should be strongly anisotropic: its particle momenta should have a virtually one-dimensional distribution distended along the field. The transversal (with respect to the field) momentum of a particle is small compared with the longitudinal momentum. In accordance with Ruderman and Sutherland (1975), such a particle distribution is formed near the pulsar surface under the action of a strong longitudinal electric field and synchrotron radiation. What equations can be used as starting ones for a description of the electron-positron plasma? – The answer depends upon a property of the plasma, which we would like to describe.

It is known that the anisotropy can result in various types of instabilities, for example, the fire-hose instability of the relativistic electron-positron plasma (Mikhailovskii, 1979). Behaviour of Alfvén waves in the isotropic and anisotropic plasmas can be essentially different (Mikhailovskii et al., 1985).

We suppose that the anisotropic relativistic approach of a type of the CGL approximation (Section 11.5) can be used to consider the problem of Alfvén waves of finite amplitude. However the dispersion effects are important for such waves and are not taken into account in the CGL approximation. The problem can be analysed on the basis of the standard kinetic approach with use of the Vlasov equation (Section 3.1.2). As we saw above, such a procedure is sufficiently effective in the case of linear problems but is complicated in study of nonlinear processes when one must deal with parts of the distribution function square and cubic to the wave amplitude.

More effective kinetic approaches are demonstrated in Mikhailovskii et al. (1985). One of them is based on expansion in the series of the inverse power of the background magnetic field (Section 5.2) and allowance for the cyclotron effects as a small corrections. Using this approach, Mikhailovskii et al. consider the nonlinear Alfvén waves both in the case of an almost one-dimensional momentum particle distribution (the case of a pulsar plasma) and in the case of an isotropic plasma. The later case is interesting, in particular, for the reason that it has been also analysed by means of the MHD equations (Section 20.1.4). Two types of Alfvén solitons (the moving-wave type and the nonlinear wave-packet type) can exist in relativistic collisionless electron-positron plasma.

Magnetic reconnection in a collisionless relativistic electron-positron plasma is considered as a mechanism of electron and lepton acceleration in large-scale extragalactic jets, pulsar outflows like the Crab Nebular and core regions of active galactic nuclei (AGN) as the respectiv jet origin (see Larrabee et al., 2003; Jaroschek et al., 2004).

Recommended Reading: Lifshitz and Pitaevskii, *Physical Kinetics* (1981) Chapters 3 and 6.

7.4 Practice: Exercises and Answers

Exercise 7.1 [Section 7.1.1] Write the general solution of the linear Equation (7.2).

Answer. Since the left-hand side of (7.2) is the time derivative (more exactly, the Liouville operator (1.10) acting on the first-order distribution function for particles following *unperturbed* trajectories), the solution of (7.2) is formally the integral over time

$$f_k^{(1)}(\mathbf{r},\mathbf{v},t) = -\frac{e_k}{m_k}\int\limits_{-\infty}^{t}\left(\mathbf{E}^{(1)} + \frac{1}{c}\,\mathbf{v}\times\mathbf{B}^{(1)}\right)_\alpha \frac{\partial f_k^{(0)}(\mathbf{r},\mathbf{v},\tau)}{\partial v_\alpha}\,d\tau. \tag{7.30}$$

Here the integration follows an unperturbed-particle trajectory to the point $(\mathbf{r},\mathbf{v})$ in phase space X.

In principle, substitution of (7.30) into the Poisson law for electrostatic waves gives a perturbation of electric charge density (3.5). Similarly, one can determine a perturbation of current density (3.6) by substitution of (7.30) into the Ampére law in the case of electromagnetic waves. In practice, solving (7.30) is fairly complicated.

Exercise 7.2 [Section 7.1.2] Show that, for a particle with velocity $\mathbf{v}$ in a plasma without magnetic field, the resonance condition correspondes to:

$$\omega - \mathbf{k}\cdot\mathbf{v} = 0\,. \tag{7.31}$$

This is usually called the *Cherenkov condition.*

Exercise 7.3 [Sections 7.1.2, 7.2.1] Consider a wave that has frequency ω and wave vector $\mathbf{k}$ in the laboratory frame. Show that in the rest frame of the particle the frequency of the wave is

$$\omega_0 = \gamma_{\rm L}\,(\omega - \mathbf{k}\cdot\mathbf{v})\,, \tag{7.32}$$

where

$$\gamma_{\rm L} = \left(1 - \frac{v^2}{c^2}\right)^{-1/2} \tag{7.33}$$

is the Lorentz factor of the particle. Therefore the Cherenkov resonance condition (7.31) corresponds to $\omega_0 = 0$, which means that the fields appear static in the rest frame of the particle.

Answer. Apply the Lorentz transformation to the four-vector $\{\,\mathbf{k},\, i\,\omega/c\,\}$ (see Landau and Lifshitz, *Classical Theory of Field*, 1975, Chapter 6, § 48).

Exercise 7.4 [Section 7.1.2] In a transparent medium with a refraction index n, greater than unity, the Cherenkov condition (7.31) can be satisfied for fast particles with

$$\beta = \frac{v}{c} \geq \frac{1}{n}\,. \tag{7.34}$$

Let χ be the angle between the particle's velocity $\mathbf{v}$ and the wave vector $\mathbf{k}$ of appearing emission which is called *Cherenkov emission* (Cherenkov, 1934, 1937).

As we know, a charged particle must move non-uniformly to radiate in vacuum. As an example we may recall the formula (5.66) for dipole emission.

In a medium, however, condition (7.34) allows the uniformly moving paricle to radiate.

Show that Cherenkov emission is confined to the surface of a cone with the cone half-angle (as shown in Figure 7.4)

$$\chi = \arccos \frac{1}{n} \,. \tag{7.35}$$

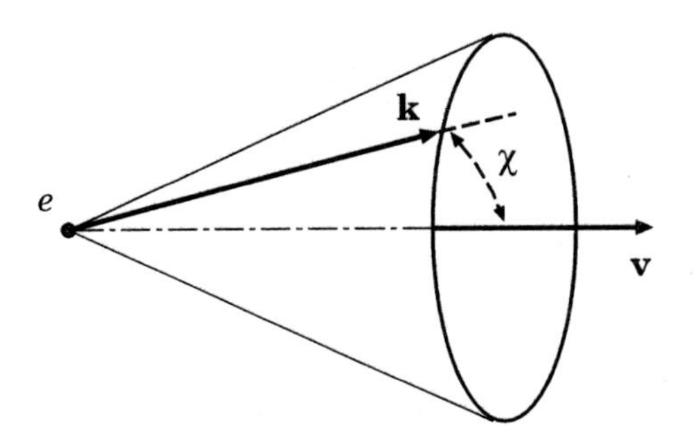

Figure 7.4: The wave-vector cone of the Cherenkov emission.

Radiation with wave vectors along the conic surface (7.35) is generated as a result of the Cherenkov emission. Discuss an analogy between the Cherenkov emission pattern and the bow wave of a ship or a supersonic aircraft.

Exercise 7.5 [Section 7.1.2] Consider the one-dimensional motion of an electron in the electric field of a Langmuir wave of a small but finite amplitude.

Answer. Let the electric field potential of the wave be of the form

$$\varphi = \varphi_0 \cos\left(\omega_{pl}^{(e)}\, t - kx\right). \tag{7.36}$$

In the reference frame moving with the wave (see Section 10.2.2), the field is static:

$$\varphi = \varphi_0 \cos kx \,. \tag{7.37}$$

This potential is shown in Figure 7.5a.

For an electron having a small velocity near $x = 0$, we have the following equation of motion:

$$m_e\, \ddot{x} = e\, \frac{\partial \varphi}{\partial x} = -e\varphi_0\, k \,\sin kx \approx -e\varphi_0\, k^2\, x \,. \tag{7.38}$$

So such a trapped electron is oscillating with frequency

$$\omega_{tr}^{(e)} = k \left(\frac{e\varphi_0}{m_e}\right)^{1/2} . \tag{7.39}$$

This is illustrated by particle trajectories in the two-dimensional phase space (Figure 7.5b).

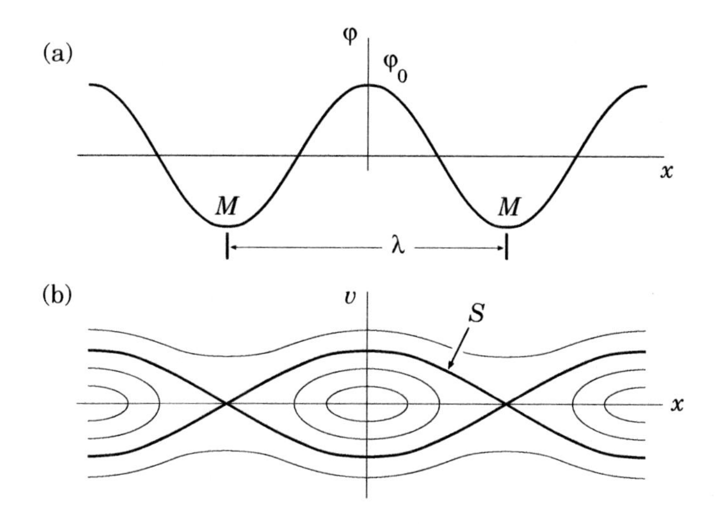

Figure 7.5: (a) The the electric field potential in a Langmuir wave of a small but finite amplitude. (b) The phase trajectories of an electron in the wave.

The potential energy $-e\varphi$ of the trapped electron has maximum at the minimum of the potential φ, at points M which determine the separatrix S.

Exercise 7.6 [Section 7.1.2] Consider the Landau resonance for electrons in a Maxwellian plasma. It is clear that electrons moving much slower or much faster than the wave tend to see the electric field that averages to zero. So we have to consider only the particles in some small part of velocity space close to the phase velocity as shown in Figure 7.1.

Since the slope of the initial distribution function is negative, there are more electrons at lower velocity than at higher velocity near the resonance (7.12). Estimate a difference.

Exercise 7.7 [Section 7.1.2] Show that the Landau damping prevents plasma waves from escaping the region where $\omega = \omega_{pl}^{(e)}$ (see definition (8.78)) into rarefied plasma, for example, from the solar corona to interplanetary medium (see Zheleznyakov, 1996).

Hint. Consider the dispersion equation for electromagnetic waves in a homogeneous equilibrium plasma without a magnetic field.

Exercise 7.8 [Section 7.1.2] In the fire-hose instability, the driving force is the beam pressure parallel to the magnetic field. Show that this pressure increases the amplitude of an electromagnetic transverse wave in a way analogous to that of a water flowing through a hose.

Hint. Consider low-frequency transverse waves in a homogeneous equilibrium plasma with a magnetic field. Such waves are called the kinetic Alfven waves. They extend to frequencies higher than that are valid for MHD. Let a beam of protons or electrons travel parallel to the magnetic field. An analysis of linear disturbances similar to the MHD waves will introduce an additional term into the dispersion equation of the Alfven wave. Note that an instability occurs for beams of protons or electrons. Consider the threshold condition in both cases.

Exercise 7.9 [Section 7.1.3] Show that fast ions can generate whistler-mode waves when the resonant particles are traveling faster than the wave. Show that, in this case, the effect of Doppler shift is to change the sense of rotation of the wave electric field in the resonant-particle frame from right-handed to left-handed.

Exercise 7.10 [Section 7.2.2] Show that the Kolmogorov spectrum formula (7.24) follows from the Kolmogorov scaling law (7.28).

Answer. The kinetic energy density associated with some wavenumber k is $W(k)\,dk$, which can be roughly written as

$$W(k)\,k \sim v^2 \,. \tag{7.40}$$

Substituting for v from formula (7.28) with $l \sim 1/k$, we have

$$W(k)\,k \sim \mathcal{P}^{2/3}\,k^{-2/3} \,. \tag{7.41}$$

From here the Kolmogorov spectrum (7.24) readily follows.

Chapter 8

Coulomb Collisions in Astrophysical Plasma

Binary collisions of particles with the Coulomb potential of interaction are typical for physics of collisional plasmas in space and especially for gravitational systems. Coulomb collisions of fast particles with plasma particles determine momentum and energy losses of fast particles, the relaxation processes in astrophysical plasma.

8.1 Close and distant collisions

8.1.1 The collision parameters

Binary interactions of particles, described by the Coulomb potential

$$\varphi(r) = \frac{e}{r}\,, \tag{8.1}$$

have been studied in mechanics (see Landau and Lifshitz, *Mechanics*, 1976, Chapter 4, § 19). Considering binary interactions as *collisions*, we are interested only in their final result, the duration of the interaction and the actual form of particle trajectories being neglected. Thus in the centre-of-mass system, each particle is deflected through an angle χ defined by the relation

$$\tan\frac{\chi}{2} = \frac{e_1 e_2}{mv^2\, l} \tag{8.2}$$

or

$$l\,(\chi) = \frac{e_1 e_2}{mv^2}\cot\frac{\chi}{2}\,. \tag{8.3}$$

Here

$$m = \frac{m_1 m_2}{m_1 + m_2} \tag{8.4}$$

is the reduced mass, v is the relative particle velocity at infinity, l is the '*impact parameter*'. The last is the closest distance of the particle's approach, were it not for their interaction as shown in Figure 8.1.

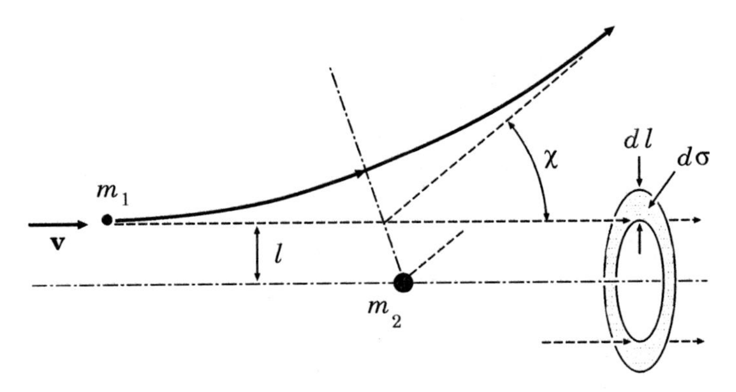

Figure 8.1: The trajectory of a light particle with mass m_1 near a heavy particle with mass m_2.

For particles deflected through a right angle

$$l\left(\frac{\pi}{2}\right) \equiv l_{\perp} = \frac{e_1 e_2}{m v^2}\,, \tag{8.5}$$

so the initial formula (8.2) is conveniently rewritten as

$$\boxed{\tan\frac{\chi}{2} = \frac{l_{\perp}}{l}\,.} \tag{8.6}$$

The collisions are called *close* if

$$\pi/2 \le \chi \le \pi\,, \qquad \text{i.e.} \qquad 0 \le l \le l_{\perp}\,. \tag{8.7}$$

Correspondingly, for *distant* collisions $l > l_{\perp}$ and $0 \le \chi < \pi/2$. Both cases are shown in Figure 8.2.

8.1.2 The Rutherford formula

The average characteristics of the Coulomb collisions are obtained with the aid of the formula for the *differential* cross-section. It is called the Rutherford formula and is derived from (8.3) as follows:

$$d\sigma = 2\pi\, l(\chi)\, dl = 2\pi\, l(\chi) \left| \frac{dl}{d\chi} \right| d\chi =$$

$$= \frac{\pi e_1^2 e_2^2}{m^2 v^4}\, \frac{\cos(\chi/2)}{\sin^3(\chi/2)}\, d\chi = \left(\frac{e_1 e_2}{2 m v^2}\right)^2 \frac{d\Omega}{\sin^4(\chi/2)}\,. \tag{8.8}$$

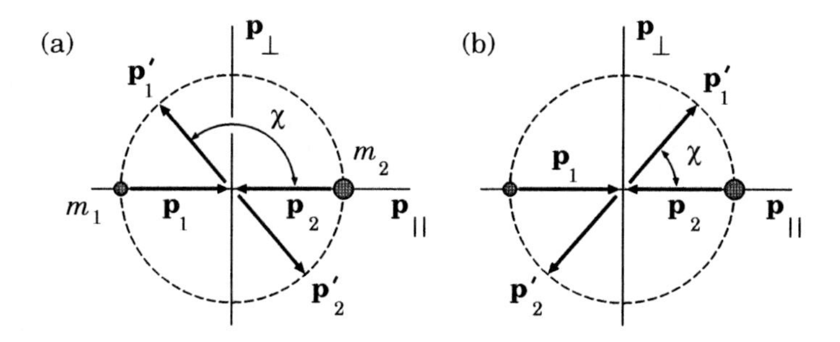

Figure 8.2: Close (a) and distant (b) collisions of particles in the momentum space in the centre-of-mass system.

Here the modulus bars indicate the absolute value of the derivative $dl/d\chi$ because it has a negative sign: with increase of the impact parameter l, the scattering angle χ decreases; the solid angle $d\Omega = 2\pi \sin\chi \, d\chi$.

By integrating (8.8) over the back hemisphere (8.7), we find the *total* cross-section of close collisions

$$\sigma_{cl} = \frac{\pi e_1^2 e_2^2}{m^2 v^4} = \pi l_\perp^2 \,. \tag{8.9}$$

This formula follows directly from definition (8.5), of course, without integrating the differential cross-section (8.8).

8.1.3 The test particle concept

By analogy with the usual gas, the concept of a 'test' particle is introduced to analyse the collisions in plasma. For instance the frequency of test particle (m_1, e_1) collisions with 'field' particles (m_2, e_2) is introduced:

$$\nu_{cl} = n_2 \, v_1 \, \sigma_{cl} = \frac{\pi e_1^2 e_2^2 \, n_2}{m_1^2 v_1^3} \,. \tag{8.10}$$

Here, for simplicity's sake, it is assumed that $m_2 \gg m_1 \approx m$ (see formula (8.4)) and $v_2 \ll v_1$. So this is, for example, the case of an electron colliding with 'cold' ions.

The length of *mean free path* λ of a test particle in a gas consisting of field particles is, by definition, the distance along which the particle suffers one collision,

$$\lambda = v_1 \, \nu^{-1} \,. \tag{8.11}$$

From (8.10) and (8.11) it follows for *close* collisions that

$$\lambda_{cl} = \frac{1}{n_2 \, \sigma_{cl}} \,. \tag{8.12}$$

Hence the time between two consecutive collisions is

$$\tau_{cl} = \frac{\lambda_{cl}}{v_1} = \frac{m_1^2 v_1^3}{\pi e_1^2 e_2^2 n_2} \sim \frac{v_1^3}{n_2}, \tag{8.13}$$

or the frequency of close collisions

$$\nu_{cl} = \frac{1}{\tau_{cl}} = \frac{\pi e_1^2 e_2^2 n_2}{m_1^2 v_1^3} \sim \frac{n_2}{v_1^3}, \tag{8.14}$$

which is the same as formula (8.10) of course.

8.1.4 Particles in a magnetic trap

Formulae (8.10) and (8.13) are frequently used in order to find out what approximation we have to use to consider the astrophysical plasma. For example, if the length of mean free path λ of the test particles inside a magnetic trap (Section 6.2) is greater than the trap's size, then such particles can be considered in the collisionless approximation. Here *charge separation* may be found to be essential, as well as the electric field resulting from it (Alfvén and Fälthammar, 1963; Persson, 1963).

> While the magnetic mirror is the primary trapping mechanism, the electrostatic potential also traps electrons

with energies low to overcome the electrostatic potential.

In the solar atmosphere, the electrostatic potential produced, in solar-flare magnetic traps, has an energy equivalent of the average energy of accelerated electrons. The number and energy fluxes of the electrons that escape from the trap can be reduced by as much as ~ 50 or more depending on the magnetic mirror ratio of the flare loop and the ratio of the ion and electron anisotropy factors (Spicer and Emslie, 1988).

Some other effects due to non-collisional particles in the so-called *collapsing* magnetic traps are mentioned in Section 6.2; they will be considered in Section 18.3 and vol. 2, Chapter 7. For example, the electric potential mentioned above increases the efficiency of confinement and acceleration of electrons in solar flares (Kovalev and Somov, 2002).

On the other hand, if the length of the mean free path of the test particles is much less than the trap's size, the collisions play an important role. As a rule they maxwellise the plasma (the gas of test particles), making it an equilibrium one. In such a plasma the notion of *temperature* is meaningful, as we shall see in Chapter 9. For example, while considering thermal electrons (having the density $n_{\rm e}$ and the temperature $T_{\rm e}$) in the trap, an electron with the *mean thermal velocity* (see definition (5.53))

$$V_{\rm Te} = \sqrt{\frac{3k_{\rm B}T_{\rm e}}{m_{\rm e}}} \tag{8.15}$$

should be taken as the test particle. Then we obtain the known 'T to the 3/2 power' law for the time of the Coulomb collisions (8.13):

$$\boxed{\tau \sim \frac{T_{\rm e}^{3/2}}{n_{\rm e}} .} \tag{8.16}$$

> The hotter the astrophysical plasma is, the more non-collisional is it with respect to some physical phenomenon or another.

The characteristic time τ of the Coulomb collisions has to be compared with the characteristic times of other physical processes: the time of particle motion between magnetic corks in the trap, the period of the Larmor rotation, the time of heating or cooling, etc.

8.1.5 The role of distant collisions

Because for small angles χ the differential cross-section (8.8) is

$$d\sigma \sim \frac{d\chi}{\chi^3} , \tag{8.17}$$

the total cross-section diverges.

> Such divergence of the collisional cross-section always occurs, once the interaction potential has no restricting factor,

or, to put the same in another way, if the interaction forces do not break off at some distance, as in the case of hard balls. This fact is of fundamental importance, for example, in *stellar dynamics* (Jeans, 1929; Chandrasekhar, 1943a) or, more exactly, in any astrophysical system governed by gravitational force (say a gravitational system), see Sections 3.3 and 9.6.

Although each distant collision causes only a small deflection of the test particle trajectory, they are present in such large numbers that their total action upon the particle is *greater* or much greater than that of relatively rare close collisions. Let us convince ourselves that this is true.

Each collision causes a small change in momentum perpendicular to the initial direction of the particle's motion:

$$\delta p_{\perp} = p \sin \chi = m_1 v_1 \, \frac{2 \tan (\chi/2)}{1 + \tan^2 (\chi/2)} = \frac{2 m_1 v_1 \, (l_{\perp}/l)}{1 + (l_{\perp}/l)^2} = 2 m_1 v_1 \, \frac{x}{1 + x^2} \, .$$

Here $x = l_{\perp}/l$, and $0 \leq x \leq 1$.

Since distant collisions occur chaotically, we are usually interested in the mean rate of change in the quantity $p_{\perp}^2$:

$$\frac{d}{dt} \, p_{\perp}^2 = \int\limits_{\chi = \pi/2}^{\chi = 0} (\delta p_{\perp})^2 \; n_2 \, v_1 \, d\sigma =$$

$$- 8\pi n_2 m_1^2 v_1^3 l_\perp^2 \int_1^0 \frac{dx}{(1+x^2)^2 x} \sim \ln x \Big|_1^0 . \tag{8.18}$$

The integral diverges logarithmically on the upper limit. Let us restrict it to some maximal value of the impact parameter

$$\Lambda = l_{\max}/l_\perp . \tag{8.19}$$

Then the integral is approximately equal to

$$\frac{d}{dt} p_\perp^2 = 8\pi n_2 m_1^2 v_1^3 l_\perp^2 \ln \Lambda = \pi e_1^2 e_2^2 \frac{n_2}{v_1} 8 \ln \Lambda . \tag{8.20}$$

The factor $\ln \Lambda$ is referred to as the Coulomb logarithm.

Introduce the characteristic time $\tau_\perp$ during which the perpendicular component of the momentum acquires a value equal to the initial momentum $m_1 v_1$:

$$\tau_\perp = (m_1 v_1)^2 \left(\frac{d}{dt} p_\perp^2 \right)^{-1} = \frac{m_1^2 v_1^3}{\pi e_1^2 e_2^2 n_2 \, (8 \ln \Lambda)} . \tag{8.21}$$

In other words, the mean resulting deflection becomes comparable with the quantity $\pi/2$ in a time $\tau_\perp$. Recall that this deflection through a large angle is a result of many distant collisions.

The effective frequency of distant collisions that corresponds to the time $\tau_\perp$ is

$$\nu_\perp = \frac{1}{\tau_\perp} = \frac{\pi e_1^2 e_2^2 n_2}{m_1^2 v_1^3} \, 8 \ln \Lambda , \tag{8.22}$$

which is $8 \ln \Lambda$ larger than the close collisions frequency (8.14):

$$\boxed{\nu_\perp = 8 \ln \Lambda \cdot \nu_{cl} .} \tag{8.23}$$

The factor $8 \ln \Lambda$ is usually much greater than unity; its typical value is $\gtrsim 10^2$ under physical definition of $\ln \Lambda$ given in Section 8.2.

> The influence of the close Coulomb collisions on kinetic processes in astrophysical plasma is, as a rule, negligibly small in comparison to the action of distant collisions.

For example, the distant collisions determine an evolution of the distribution function of fast electrons injected into the thermal plasma in the solar atmosphere diring solar flares. However **this does not mean that the close collisions do never play any role** in plasma astrophysics. Just in the same example, the close collisions of fast electrons with thermal ions create hard X-ray bremsstrahlung emission in the range 10–100 keV, because the close collisions are responsible for large exchange of the particle momentum. For typical flare parameters ($h\nu \approx 20$ keV, $\ln \Lambda \approx 20$) the efficiency of the bremsstrahlung process is $\sim 3 \times 10^{-6}$ (Brown, 1971; Korchak, 1971).

8.2 Debye shielding and plasma oscillations

8.2.1 Simple illustrations of the shielding effect

While considering the distant collisions, we have removed the divergence of the integral (8.18) which describes the mean rate of change of the test particle transversal momentum, purely formally – by artificially restricting the radius of action of the Coulomb forces at some maximal distance l_{max}. Meanwhile this maximal distance may be chosen quite justifiably, based on the following reasoning. In a plasma,

> each charged particle attracts oppositely charged particles and, at the same time, repels the particles of the same charge.

As a consequence, the oppositely charged particles tend to gather around the particle, thus weakening its Coulomb field. As a result of such 'shielding' the action of the field extends over a distance no greater than some quantity r_{D} called *Debye radius*.

The concept of Debye shielding has a clear meaning. Let us assume that a plasma contains an immovable charge which then creates the electrostatic field in its vicinity. As a final result of shielding interactions mentioned above, some *equilibrium* distribution of *two components*: positive and negative plasma particles is established in this field. Its electrostatic potential φ is related to the densities of ions n_{i} and electrons n_{e} via the Poisson equation

$$\Delta\varphi = -4\pi e\,(Zn_{\text{i}} - n_{\text{e}})\,, \tag{8.24}$$

where Ze is the ion charge.

In the thermodynamic equilibrium state the ion and electron densities in the electrostatic field with potential $\varphi\,(r)$ are to be distributed according to Boltzmann's law

$$n_{\text{i}} = n_{\text{i}}^{0}\,\exp\left(-\frac{Ze\varphi}{k_{\text{B}}T_{\text{i}}}\right), \qquad n_{\text{e}} = n_{\text{e}}^{0}\,\exp\left(\frac{e\varphi}{k_{\text{B}}T_{\text{e}}}\right). \tag{8.25}$$

The constant coefficients are set equal to the mean densities n_{i}^{0} and n_{e}^{0} of plasma particles, since $\varphi \to 0$ far from the particle considered.

Supposing that the Coulomb interaction is so weak that

$$Ze\varphi \ll k_{\text{B}}T_{\text{i}} \quad \text{and} \quad e\varphi \ll k_{\text{B}}T_{\text{e}}\,, \tag{8.26}$$

or restricting our consideration to the approximate solutions applicable at large distances from the shielded charge, we expand both exponents (8.25) in a series and substitute in Equation (8.24). We obtain the following equation:

$$\frac{1}{r^2}\frac{d}{dr}\left(r^2\frac{d\varphi}{dr}\right) = -\,4\pi e\left[Zn_{\text{i}}^{0}\left(1 - \frac{Ze\varphi}{k_{\text{B}}T_{\text{i}}}\right) - n_{\text{e}}^{0}\left(1 + \frac{e\varphi}{k_{\text{B}}T_{\text{e}}}\right)\right] =$$

$$= 4\pi e \left[\left(n_{\mathrm{e}}^{0} - Z n_{\mathrm{i}}^{0} \right) + \frac{e}{k_{\mathrm{B}}} \left(\left(Z n_{\mathrm{i}}^{0} \right) \frac{Z}{T_{\mathrm{i}}} + \left(n_{\mathrm{e}}^{0} \right) \frac{1}{T_{\mathrm{e}}} \right) \varphi \right] . \tag{8.27}$$

As usual the actual plasma is *quasi-neutral on average* (see the next Section); instead of this let us assume here (like in Sections 3.2.2 and 3.2.3) that the plasma is *ideally neutral*:

$$Z n_{\mathrm{i}}^{0} = n_{\mathrm{e}}^{0} . \tag{8.28}$$

Thus we have an equation

$$\frac{1}{r^2} \frac{d}{dr} \left(r^2 \frac{d\varphi}{dr} \right) = \frac{4\pi e^2 n_{\mathrm{e}}^{0}}{k_{\mathrm{B}}} \left(\frac{Z}{T_{\mathrm{i}}} + \frac{1}{T_{\mathrm{e}}} \right) \varphi = \frac{\varphi}{r_{\mathrm{D}}^2} . \tag{8.29}$$

On the right-hand side of Equation (8.29) we have two terms for a two-component plasma. We divide them by φ, then

$$\frac{1}{r_{\mathrm{D}}^2} = \frac{1}{r_{\mathrm{D}}^{(\mathrm{i})\,2}} + \frac{1}{r_{\mathrm{D}}^{(\mathrm{e})\,2}} = \frac{4\pi e^2 n_{\mathrm{e}}^{0}}{k_{\mathrm{B}} T_{\mathrm{e}}} \left(1 + Z \frac{T_{\mathrm{e}}}{T_{\mathrm{i}}} \right) . \tag{8.30}$$

Therefore

$$r_{\mathrm{D}} = \left(\frac{k_{\mathrm{B}}}{4\pi e^2 n_{\mathrm{e}}^{0}} \frac{T_{\mathrm{e}} T_{\mathrm{i}}}{Z T_{\mathrm{e}} + T_{\mathrm{i}}} \right)^{1/2} \tag{8.31}$$

is known as the *Debye radius*, being first derived by Debye and Hückel (1923) in the theory of electrolytes.

The solution of Equation (8.27) corresponding to the charge e situated at the origin of the coordinates is the potential

$$\boxed{\varphi = \frac{e}{r} \exp \left(- \frac{r}{r_{\mathrm{D}}} \right) .} \tag{8.32}$$

At distances greater than r_{D}, the electrostatic interaction is exponentially small.

> The Debye length is an effective range for collisions, the potential between charged particles being the *shielded* Coulomb potential (8.32) rather than the Coulomb one (8.1) which would apply in a vacuum.

That is why:

(a) the binary correlation function (3.30) reproduces the shape of the shielded Coulomb potential (8.32),

(b) the Debye radius r_{D} is substituted in the Coulomb logarithm (8.20) in place of l_{max}.

A formula that is simpler than (8.31) is frequently used for the Debye radius, namely

$$r_{\mathrm{D}}^{(\mathrm{e})} = \left(\frac{k_{\mathrm{B}} T}{4\pi e^2 n_{\mathrm{e}}} \right)^{1/2} . \tag{8.33}$$

This variant of the formula for the Debye radius implies that the shielding is due to just the particles of one sign, more exactly, *electrons*, i.e. in the formulae (8.25) we have $T_{\rm i} = 0$ (the approximation of *cold ions*) and $T_{\rm e} = T$ (see Exercise 9.3). This is the *electron* Debye radius. The corresponding formula for the Coulomb logarithm is

$$\ln \Lambda = \ln \frac{3}{2e^3} \left(\frac{k_{\rm B}^3 \, T^3}{\pi n_{\rm e}} \right)^{1/2} . \tag{8.34}$$

Its values typical of the solar atmosphere are around 20 (Exercise 8.1).

Formula (8.33) shows that the electron Debye radius increases with an increase of temperature, since electrons with higher kinetic energy can withstand the attraction of the positive ion charge Ze up to larger distances. It decreases with an increase of density n_0, since a larger number of electrons and ions can be accommodated in shorter distances to screen the electric field of charge Ze.

8.2.2 Charge neutrality and oscillations in plasma

The Debye shielding length is fundamental to the nature of a plasma. That is why this important characteristic appears again and again in plasma astrophysics, starting from the binary correlation function (3.30).

The first point to note is that a plasma maintains *approximate charge neutrality* (Sections 11.5.2 and 3.2.2). The reason for this is simply that any significant imbalance of positive and negative charge could only be maintained by a huge electric field. The movement of electrons to neutralize a charge inhomogeneity would be followed by an oscillatory motion (e.g., Alfvén and Fälthammar, 1963, Chapter 4).

This brings us to a second characteristic of plasmas called the *plasma frequency* or, more exactly, the electron plasma frequency:

$$\boxed{\omega_{pl}^{(\rm e)} = \left(\frac{4\pi e^2 n_{\rm e}}{m_{\rm e}} \right)^{1/2} .} \tag{8.35}$$

A charge density disturbance oscillates with this frequency (see Section 10.2.1). These oscillations are called *Langmuir waves* or *plasma waves*. Therefore, under most circumstances,

> plasma cannot sustain electric fields for lengths in excess of the Debye radius or times in excess of a plasma period $T_{pl}^{(\rm e)} = 2\pi / \omega_{pl}^{(\rm e)}$.

However one cannot talk of plasma oscillations unless a large number of thermal particles are involved in the motion. It is the Debye shielding length which determines the spatial range of the field set up by the charge inequality:

$$r_{\rm D} = \frac{1}{\sqrt{3}} \frac{V_{\rm Te}}{\omega_{pl}^{(e)}} . \tag{8.36}$$

Here $V_{\rm Te}$ is the mean thermal velocity of electrons. Therefore the Debye length

$$\boxed{r_{\rm D} \approx \frac{V_{\rm Te}}{\omega_{pl}^{(e)}} .} \tag{8.37}$$

So a fully-ionized plasma in the termodynamic equilibrium is a quasi-neutral medium. The *space* and *time* scales of charge separation in such plasma are the Debye radius and the inverse plasma frequency. Therefore the plasma oscillations are a typical example of **collective phenomena** (Section 3.2.3).

The Coulomb collisions, of course, damp the amplitude of the plasma oscillations with the rate which is proportional to the frequency $\nu_{\rm ei}$ of electron-ion collisions (see Exercise 10.3).

8.3 Collisional relaxations in cosmic plasma

8.3.1 Some exact solutions

It was shown in Section 8.1 that, as a result of the Coulomb collisions, a particle deflects through an angle comparable with $\pi/2$ in a characteristic time given by formula (8.21). More exact calculations of the Coulomb collisions times, that take into account the thermal motion of field particles, have been carried out by Spitzer (1940) and Chandrasekhar (1943). These calculations are cumbersome, so we give only their final results.

Let us consider the electron component of a plasma. Suppose that the test particles likewise are electrons moving with mean thermal velocity. Then the exact calculation gives instead of the formula (8.21) the time

$$\tau_{\rm ee} = \frac{m_{\rm e}^2 \, (3k_{\rm B} T_{\rm e}/m_{\rm e})^{3/2}}{\pi e_{\rm e}^4 \, n_{\rm e} \, (8 \ln \Lambda)} \cdot \frac{1}{0.714} . \tag{8.38}$$

This is called the time of mutual electron collisions or simply the *electron collisional time*. Comparison of formula (8.38) with (8.21) shows that the difference (the last factor in (8.38)) is not large. So the consideration of binary collisions in the approximation used in Section 8.1 is accurate enough, at least for astrophysical applications.

The analogous time of mutual collisions for ions, having mass $m_{\rm i}$, charge $e_{\rm i}$, temperature $T_{\rm i}$ and density $n_{\rm i}$, is equal to

$$\tau_{\rm ii} = \frac{m_{\rm i}^2 \, (3k_{\rm B} T_{\rm i}/m_{\rm i})^{3/2}}{\pi e_{\rm i}^4 \, n_{\rm i} \, (8 \ln \Lambda)} \cdot \frac{1}{0.714} . \tag{8.39}$$

If a plasma is quasi-neutral: $e_{\rm i}\, n_{\rm i} \approx -e_{\rm e}\, n_{\rm e} = en$, where $e_{\rm i} = -Ze_{\rm e}$, and if $T_{\rm e} \approx T_{\rm i}$, then the ratio

$$\boxed{\frac{\tau_{\rm ii}}{\tau_{\rm ee}} \approx \left(\frac{m_{\rm i}}{m_{\rm e}} \right)^{1/2} \frac{1}{Z^3} \, .} \tag{8.40}$$

Coulomb collisions between thermal ions occur much more rarely than those between thermal electrons.

However it is not the time of collisions between ions $\tau_{\rm ii}$ – the *ion collisional time*, but rather the time of electron-ion collisions that is the greatest. This characterizes, in particular, the process of temperature equalizing between the electron and ion components in a plasma. The rate of temperature equalizing can be determined from the equation

$$\frac{dT_{\rm e}}{dt} = \frac{T_{\rm i} - T_{\rm e}}{\tau_{\rm ei}\,(\mathcal{E})}\,, \tag{8.41}$$

where $\tau_{\rm ei}\,(\mathcal{E})$ is the time of equilibrium establishment between the electron and ion plasma components. It characterizes the rate of exchange of energy $\mathcal{E}$ between the components and equals (Spitzer, 1940, 1962; see also Sivukhin, 1966, § 9 and § 17; cf. formulae (42.5) in Lifshitz and Pitaevskii, 1981, § 42)

$$\tau_{\rm ei}\,(\mathcal{E}) = \frac{m_{\rm e} m_{\rm i} \left[\, 3k_{\rm B} \left(T_{\rm e}/m_{\rm e} + T_{\rm i}/m_{\rm i} \right) \right]^{3/2}}{e_{\rm e}^2\, e_{\rm i}^2\, (6\pi)^{1/2}\, (8\, \ln \Lambda)}\,. \tag{8.42}$$

For comparison with formula (8.40) let us put $T_{\rm i} = T_{\rm e}$. Then

$$\tau_{\rm ei}\,(\mathcal{E}) = 0.517\, \frac{e_{\rm i}^2}{e_{\rm e}^2} \left(\frac{m_{\rm i}}{m_{\rm e}} \right)^{1/2} \tau_{\rm ii}\,. \tag{8.43}$$

Thus the time of energy exchange between electrons and ions is much greater than the time of mutual ion collisions.

In a plasma consisting of electrons and protons with equal temperatures we have

$$\tau_{\rm ep}(\mathcal{E}) \approx 22\, \tau_{\rm pp} \approx 950\, \tau_{\rm ee}\,. \tag{8.44}$$

The energy exchange between electron and ion components occurs so slowly that for each component a distribution may be set up which is close to Maxwellian with the proper temperature.

That is the reason why we often deal with a *two-temperature* plasma. Moreover the so-called adiabatic model for two-temperature plasma (Section 8.3.3) is often used in astrophysics.

8.3.2 Two-temperature plasma in solar flares

8.3.2 (a) Impulsive heating by accelerated electrons

Let us illustrate the situation, discussed above, by two examples from the physics of flares. The first is the impulsive heating of the solar atmosphere by a powerful beam of accelerated electrons. The beam impinges on the chromosphere from the coronal part of a flare along the magnetic field tubes. The maximal energy flux is $F_{\rm max} \gtrsim 10^{11}$ erg cm^{-2} s^{-1}. The time profile with the maximum at $t \lesssim 5$ s of the energy flux at the upper boundary of the chromosphere has been used for numerical solution of the two-temperature dissipative hydrodynamic equations (Chapter 2 in Somov, 1992).

Yohkoh observations, made using three of the instruments on board – the Hard X-ray Telescope (HXT), the Soft X-ray Telescope (SXT), and the Bragg Crystal Spectrometer (BCS) – show that the nonthermal electron energy flux can be even larger, for example, in the flare of 16 December 1991 (see Figure 6a in McDonald et al., 1999), the maximal energy flux is

$$F_{\rm max} \approx 2.5 \times 10^{29} \text{ erg s}^{-1} / \, 2 \times 10^{17} \text{ cm}^2 \sim 10^{12} \text{ erg cm}^{-2}\text{s}^{-1} .$$

Weak beams do not produce a significant response of the chromosphere (see Figure 6b in McDonald et al., 1999), of course, just hard X-ray bremsstrahlung.

In the chromosphere, beam electrons lose their energy by mainly Coulomb collisions.

> The fastest process is the primary one, namely that of energy transfer from the beam electrons to the thermal electrons

of chromospheric plasma (Figure 8.3).

As a result, plasma electrons are rapidly heated to high temperatures: in a matter of seconds the electron temperature reaches values of the order of ten million degrees. At the same time, the ion temperature lags considerably, by one order of magnitude, behind the electron temperature (Figure 8.4). Here the Lagrange variable

$$\xi = - \int_{z_{\rm max}}^{z} n(z) \, dz + \xi_{\rm min} \, , \;\; \text{cm}^{-2} , \qquad (8.45)$$

z is the height above the photosphere, $z_{\rm max}$ corresponds to the transition layer between the chromosphere and corona before an impulsive heating. Therefore ξ is the column depth – the number of atoms and ions in a column (of the unit cross-section) measured down into the chromosphere from its upper boundary, the transition layer.

The column depth $\xi_{\rm min} = n_c l_r$ is the number of ions inside a flaring loop which is the coronal part of a reconnected magnetic-field-line tube (see vol. 2, Section 3.2.1); l_r is the length of the reconnected field line, n_c is the plasma

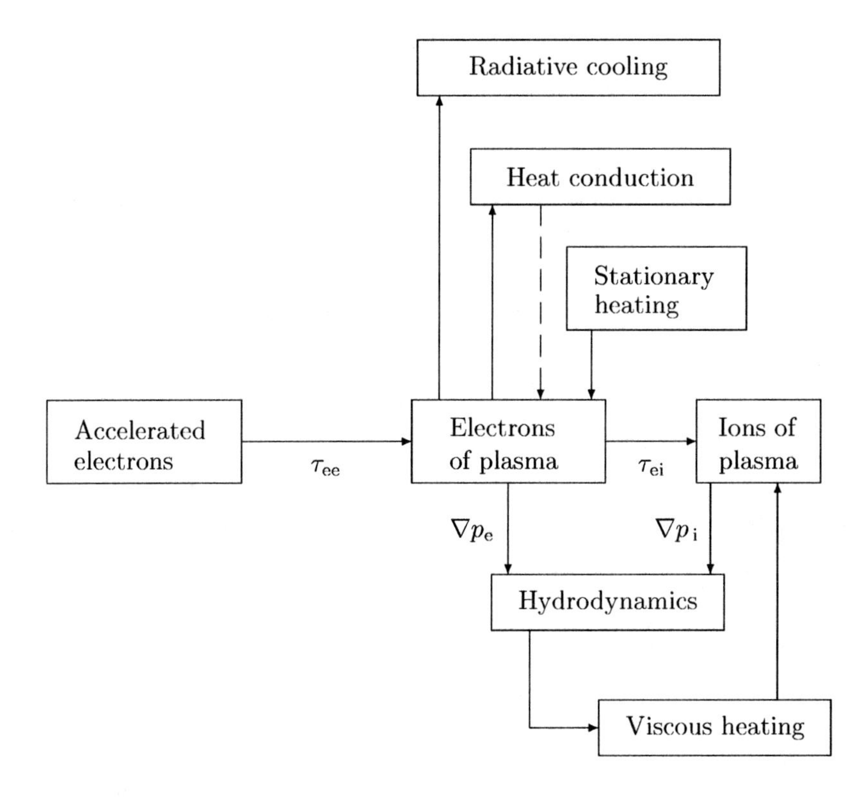

Figure 8.3: A scheme of the energy exchange in the two-temperature model of hydrodynamic response of the solar atmosphere to impulsive heating by an electron beam.

density inside the tube above the transition layer between the chromosphere and corona before an impulsive heating. Let us assume, for simplicity, that

$$\xi_{\,\mathrm{min}} \ll \xi_1 = \frac{\mathcal{E}_1^2}{2a_1}\,, \ \mathrm{cm}^{-2}. \tag{8.46}$$

Here ξ_1 is the column thickness that the accelerated electrons with the minimal energy $\mathcal{E}_1$ measured in keV can pass in a plasma before they stop (see formula (4.40)). The assumption (8.46) means that we neglect the energy losses of the electrons in the coronal part of the loop. In this way, we consider direct impulsive heating of the chromosphere by an electron beam. Accelerated electrons penetrate into the chromosphere to significant depth; for this reason a significant fraction of the beam energy is lost as radiation in optical and EUV lines. The column depth of evaporated plasma $\xi \approx 2 \times 10^{19}$ cm^{-2} but its temperature does not exceed $T_{\mathrm{max}} \approx 10^7$ K.

> The difference between the electron and ion temperatures is essential, at first, for the dynamics of high-temperature plasma

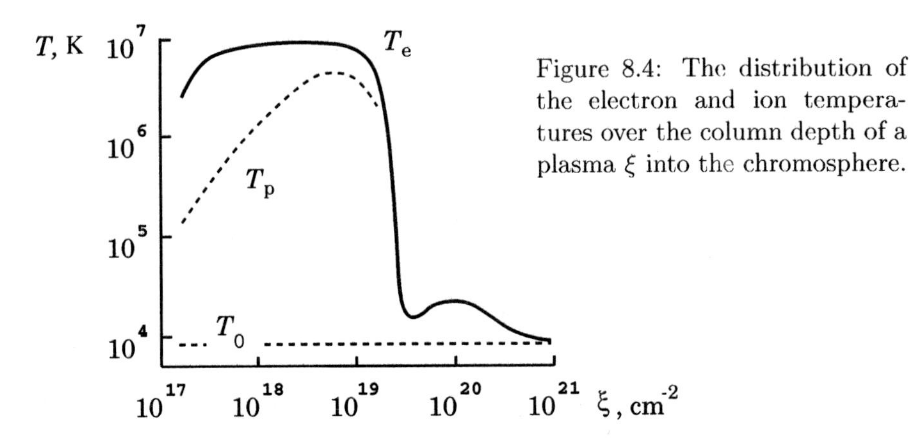

Figure 8.4: The distribution of the electron and ion temperatures over the column depth of a plasma ξ into the chromosphere.

which absorbs the main part ($\geq 90\,\%$) of the beam energy flux. Let us imagine that only the electrons are heated, while the ion heating can be neglected. In this case the electron temperature is twice as large as it would be in the case of equal heating of the electrons and ions,

$$(\,T_{\rm e}\,)_1 \simeq 2\,(\,T_{\rm e}\,)_2\,.$$

The rate of high-temperature plasma cooling is mainly determined by heat fluxes into colder plasma. These can be evaluated by the formula for the classical heat flux

$$F_c = -\,\kappa_{\rm e}\,\nabla T_{\rm e} \tag{8.47}$$

under conditions when this formula is applicable, of course (see Somov et al., 1981). Here $\kappa_{\rm e} = \kappa_0\,T_{\rm e}^{5/2}$ is the classical heat conductivity due to the Coulomb collisions of plasma electrons. From formula (8.47) we see that the heat flux is proportional to $T_{\rm e}^{7/2}$. Therefore the real heat flux

$$F_c\,(\,T_{\rm e}\,)_1 \simeq 2^{7/2}\,F_c\,(\,T_{\rm e}\,)_2 \tag{8.48}$$

can be an order of magnitude ($2^{7/2} \sim 10$) larger than the flux calculated in one-temperature ($\,T_{\rm e} = T_{\rm i}\,$) models. Because of this, the one-temperature models are much less dynamic than one would expect.

The effect becomes even more important if the accelerated electrons heat a preliminary (before a flare) evaporated 'hot' plasma. This formally means that, in formula (8.45), the column depth $\xi_{\,\rm min} = n_c l_r$ is not small in comparison with ξ_1. So we have to take into account the direct impulsive heating of the plasma inside the coronal part of the flaring loop. Such process (Duijveman et al., 1983; MacNeice et al., 1984) can very efficiently produce a 'superhot' plasma which has an electron temperature $T_{\rm e}$ much higher than the maximal temperature in the case of chromospheric heating considered above.

8.3.2 (b) Heating by high-temperature current layers

The difference between the electron and ion temperatures is known to be critical for a wide variety of kinetic effects, in particular for the generation of some turbulence (for example, ion-acoustic or ion-cyclotron) in the impulsively heated plasma. The turbulence, in its turn, has a great impact on the efficiency of heating and particle acceleration in a plasma.

The second example, when the electron component of a plasma has a temperature that is considerably different from the ion temperature, is supplied by the high-temperature turbulent-current layers (Somov, 1981 and 1986; Somov and Titov, 1983) in the regions of reconnection. Since the layer thickness $2a$ is small in comparison with its width $2b$ (see vol. 2, Figure 6.1), the plasma inflow quickly enters the region of the Joule dissipation of reconnecting magnetic field components. Here the impulsively **fast heating of the electrons and ions takes place, resulting in considerably different temperatures**. The conditions in a reconnecting current layer (RCL) in the solar corona, especially, in flares (vol. 2, Section 6.3) are such that

> the Coulomb exchange of energy between the impulsively heated electrons and ions inside the RCL can be entirely neglected.

One of distinctive features of fast reconnection in RCLs, proposed as the primary energy source in solar flares, is the presence of fast plasma outflows, or jets, whose velocities are nearly equal to the Alfvén velocity, see definition (15.30). Outflows can give origin to plasma velocity distributions with equal and opposite components along the x axis in Figure 8.5 and, as a consequence, along the line-of-sight (l.o.s.) to an observer. Therefore, in this way, they can create a **symmetric supra-thermal broadening** in the soft X-ray and EUV lines observed during solar flares. The broadening mainly depends on the electron and ion temperatures inside the RCL (Antonucci and Somov, 1992).

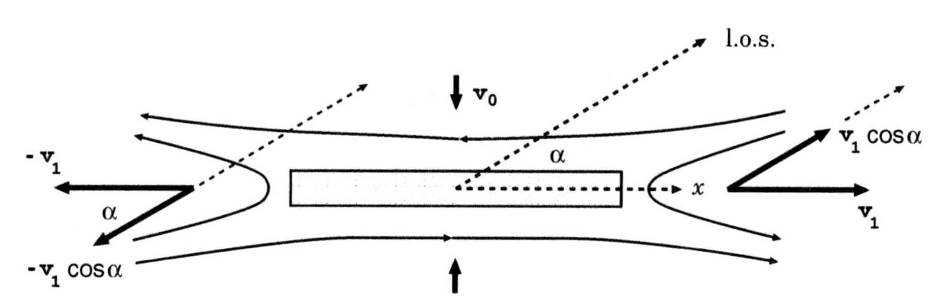

Figure 8.5: High-temperature plasma velocities near a reconnecting current layer.

A comparison of the supra-thermal profiles of the Fe XXV emission lines observed at flare onset with the predictions of the high-temperature turbulent-current layer model suggests that the observed supra-thermal broadenings are consistent with the presence in the flare region of several small-scale or one (a few) curved large-scale RCLs (Antonucci et al., 1996).

The energy release by reconnection has been invoked to explain both large-scale events, such as solar flares and coronal mass ejections (CMEs), and small-scale phenomena, such as the coronal and chromospheric microflares that probably heat the corona (vol. 2, Section 12.4) and accelerate the solar wind. Ultraviolet observations of the so-called explosive events in the solar chromosphere by SUMER (the Solar Ultraviolet Measurements of Emitted Radiation instrument) on the spacecraft *SOHO* (the Solar and Heliospheric Observatory) reveal the presence of **bi-directional plasma jets** ejected from small sites above the solar surface (Innes et al., 1997; cf. Antonucci and Somov, 1992). The structure of these jets evolves in the manner predicted by theoretical models of reconnection (see Figure 1 in Somov and Syrovatskii, 1976a), thereby leading strong support to the view that reconnection is the fundamental process for accelerating plasma on the Sun.

8.3.3 An adiabatic model for two-temperature plasma

As we saw in Section 8.3.1, equilibrium in an electron-proton plasma is achieved in three stages. First, the electrons reach a Maxwellian distribution with temperature T_e on a time τ_{ee}. Then, on a longer time,

$$\tau_{pp} \approx (m_p/m_e)^{1/2}\, \tau_{ee} \, ,$$

the protons reach a Maxwellian distribution with temperature T_p. Finally, the two temperatures equalize on the longest time of order

$$\tau_{ep} \sim (m_p/m_e)\, \tau_{ee} \, .$$

Let us suppose that a two-temperature plasma is created by a strong shock wave in an electron-proton plasma. **The shock primarily heats ions** because the kinetic energy of a particle is proportional to the particle mass. In the postshock region, the protons reach thermal equilibrium on a time τ_{pp} after they are heated through the shock (Zel'dovich and Raizer, 1966, 2002). Within this time the proton temperature is significantly higher than the electron one. Subsequently the protons share their thermal energy with the electrons through Coulomb collisions.

> In astrophysical plasma, sometimes, a difference between electron and ion temperatures can be observed at huge linear scales.

For example, the so-called X-ray clusters, or clusters of galaxies, with the X-ray temperatures $(4 - 10) \times 10^7$ K show noticeable differences between their electron and ion temperatures at radii greater than 2 Mpc.

The clusters of galaxies are the largest objects in the Universe, containing galaxies and dark matter, collisionless particles and a diffuse gas component. The last one is called the *intracluster medium* and has a temperature of about 10^8 K, thus emitting hard X-rays (HXR) mainly through the thermal bremsstrahlung of the electrons. In the outer parts of the clusters, the free-free cooling time is much longer than the Hubble time. So we neglect radiative cooling in such plasma which is supposed to be heated by the shock in the accretion flow (see Takizawa, 1998).

If we could also neglect heat conduction (for example, by assuming that the thermal conductivity of the intracluster medium is strongly reduced by a temperature gradient-driven kinetic instability, see Hattori and Umetsu, 2000), then the electrons would be considered as an adiabatic gas. It would be very convenient to calculate the electron and ion temperature profiles by using the *adiabatic model* of a two-temperature plasma by Fox and Loeb (1997). This is also the case if tangled magnetic fields, for example of turbulent origin, can suppress heat conduction in high-temperature plasma. So we assume that there exists

> a chaotic magnetic field that is sufficiently strong to suppress heat conduction in high-temperature astrophysical plasma, *yet small enough* to have negligible dynamical and dissipative effects including Joule heating.

These conditions seem to be approximatelly satiesfied in cluster environments; for more detail see Fox and Loeb (1997).

The general case of a strong shock in a fully ionized plasma with heat conduction is complicated by the fact that the electron thermal speed exceeds the shock speed, allowing the electrons to preheat the plasma ahead of the shock (Zel'dovich and Raizer, 1966). Usually **heat conduction determines internal scales of the problem** being in competition with the thermal instability driven by radiative cooling (Field, 1965; see also Somov and Syrovatskii, 1976a). Radiation emitted by the high-temperature plasma behind the shock also may heat a preshock region. Fast particles, escaping from the high-tempertature plasma (see Section 8.4.3), may contribute the preshock heating too. So we have to be very careful when we apply the adiabatic model of two-temperature plasma to astrophysical conditions.

If come back to HXR tails observed in the X-ray spectra of some clusters, one suggestion is that all or part of this emission might be nonthermal bremsstrahlung from suprathermal electrons with energies of $\sim 10-100$ keV. This nonthermal electrons would form a population in excess of the normal thermal gas, which is the bulk of the intracluster medium. The most natural explanation of this suprathermal population would be that they are particles currently being accelerated to high energies by turbulence in the intracluster medium. Sarazin and Kempner (2000) have calculated models for the nonthermal HXR bremsstrahlung in the clasters of galaxies.

The high-Mach-number shocks in young supernova remnants (SNRs) do not produce electron-ion temperature equilibration either. The heating process in these collisionless shocks is not well understood, but the Coulomb collisions times are too long to provide the required heating. Presumably the plasma collective processes should be responsible for the heating; see discussion and references in Section 16.4. This raises the question of whether the heating process leads to temperature equilibration or not. It appears that the observed electron temperature ($T_e \sim 1$ keV) remains very low compared to the observed ion temperature ($T_i \sim 500$ keV for ions O VII) behind the shock.

8.3.4 Two-temperature accretion flows

Magnetized accretion disks have become the most convincing physical paradigm to explain a low emission from the central engines of active galactic nuclei (AGN) and X-ray binary sources (see also Section 13.2). The observed radiation comes from the energy dissipation required to maintain steady accretion of plasma on to the central object. In the standard model of the optically-thin accreation disk, the heat energy released by viscous dissipation is radiated almost immediatelly by the accreating plasma. So

> the net luminosity must be equal to ($\approx$ one-half) the gravitational energy released as the mass falls onto the central object.

In a few of binary stellar systems, the mass of the primary star has been measured and found to be consistent with the mass of a neutron star, $\sim 1.4\, M_\odot$. In several other systems, however, the mass of the primary is found to be greater than 3 $M_\odot$, which makes these stars too massive to be neutron stars. These are considered as black hole candidates.

Although neutron stars and black holes have been distinguished on the basis of their masses, the real physical distinction between the two is that black holes must have a horizon (a surface through which the matter and energy fall in but from which nothing escapes) while neutron stars are normal stars with surfaces. This basic difference provides an opportunity to test the reality of black holes (see Narayan et al., 1997).

Two-temperature advection-dominated accretion flows (ADAFs) have received much attention in an effort to explain low-luminosity stellar and galactic accreting sources (Blackman, 1999; Wiita, 1999; Manmoto, 2000). Here the ions are assumed to receive the energy dissipated by the steady accretion **without having enough time to transfer their energy to the cooler electrons** before falling on to the central object.

While the electrons can almost always radiate efficiently, the protons will not, as long as Coulomb processes are the only thing that share energy between electrons and protons. So some or most of the dissipated energy is *advected* (Section 13.2.3), not radiated, as it would have been if the electrons received all of the dissipated energy. In the ADAF model, the heat generated via viscosity is advected inward rather than radiated away locally like a standard accretion disk (Novikov and Torn, 1973; Shakura and Sunyaev, 1973).

> When the central object is a black hole, the advected energy is lost forever rather than reradiated as it would be for a neutron star.

Precisely such observed differences between corresponding X-ray binary systems have been purported to provide evidence for black hole horizons (Narayan et al., 1997; see also Chakrabarti, 1999); see, however, discussion of the ADAF model in Section 9.3.3.

8.4 Dynamic friction in astrophysical plasma

8.4.1 The collisional drag force and energy losses

8.4.1 (a) Chandrasekhar-Spitzer's formulae

As in Sections 8.1 and 8.3, we use the concept of a test particle to illustrate the effects of the collisional drag force in astrophysical plasma. A test particle of mass m_1 and charge e_1 is incident with velocity $\mathbf{v}$ in a gas containing field particles of mass m_2, charge e_2 and density n_2. In what follows, $v_{\parallel}$ will be the component of the test particle velocity parallel to the original direction of its motion.

First, for the sake of simplicity, let us consider the field particles *at rest*. As in Section 8.1.5, integration over all possible values of the impact parameter up to the upper cut-off at $l = l_{\max}$ yields the following formulae describing the **mean rates** of energy losses and of scattering for the incident particle (Spitzer, 1962):

$$\frac{d\mathcal{E}}{dt} = -\frac{2\pi e_1^2 e_2^2 \ln \Lambda}{\mathcal{E}} \frac{m_1}{m_2} n_2 v \tag{8.49}$$

and

$$\frac{d}{dt} v_{\parallel} = -\frac{\pi e_1^2 e_2^2 \ln \Lambda}{\mathcal{E}^2} \left(1 + \frac{m_1}{m_2} \right) n_2 v^2 . \tag{8.50}$$

Here $\mathcal{E}$ is the energy of the incident particle (see definition (5.2)).

If we consider a beam of accelerated electrons in astrophysical ionized plasma, the most important are interactions with electrons and protons. So

$$\frac{d\mathcal{E}}{dt} = -\frac{2\pi e^4 \ln \Lambda}{\mathcal{E}} \left(1 + \frac{m_{\rm e}}{m_{\rm p}} \right) n_{\rm e} v \tag{8.51}$$

and

$$\frac{d}{dt} v_{\parallel} = -\frac{\pi e^4 \ln \Lambda}{\mathcal{E}^2} \left(3 + \frac{m_{\rm e}}{m_{\rm p}} \right) n_{\rm e} v^2 . \tag{8.52}$$

Thus

> both ambient electrons and protons produce scattering (8.52) of the incident electrons but **only ambient electrons contribute significantly to the energy losses**;

the contribution of protons in the rate of energy losses (8.51) is proportional to the small ratio $m_{\rm e}/m_{\rm p}$. This is consistent, of course, with what we have concluded in Section 4.2 for fast particles propagating in thermal plasma.

We neglect collective effects due to interaction of the plasma and the electron beam as a whole without any justification here. It must be emphasized also at this point that formulae (8.51) and (8.52) describe the *mean* rates of change of $\mathcal{E}$ and $v_{\parallel}$ for the electrons of an incident beam but neglect the dispersions about these means. The accuracy of such procedure decreases as the scattering and energy losses become not small. These ristrictions have been discussed in Section 4.4. Now we recall that we have neglected the proper motions of the plasma particles. Let us take them into account.

8.4.1 (b) Energy losses in plasma

The most general non-relativistic formula for Coulomb losses in the many-component thermal plasma is given, for example, in Trubnikov (1965), Sivukhin (1966) and can be expressed as follows:

$$P \equiv \frac{d\mathcal{E}}{dt} = \sum_k \left(\frac{d\mathcal{E}}{dt} \right)_k = - \sum_k \frac{4\pi e^4 \ln \Lambda}{m_k} \, \frac{Z^2 Z_k^2 \, n_k}{v_k} \, \mathcal{P}_k \left(\frac{v}{v_k}, \frac{m_k}{M} \right) . \tag{8.53}$$

Here Z_k, m_k, n_k and v_k are the charge, mass, density and thermal velocity of the plasma particles of the kind k; they have a temperature T_k. Z, $M = Am_{\text{p}}$ and v are the charge, mass and velocity of the incident particles; their kinetic energy $\mathcal{E} = Mv^2/2$. Contrary to definition (8.15) of the mean thermal velocity, in formula (8.53) the thermal velocity is equal to the *most probable* velocity of thermal particles (Sivukhin, 1966):

$$v_k = \left(\frac{2k_{\text{B}} T_k}{m_k} \right)^{1/2} . \tag{8.54}$$

It is convenient to determine the dimensionless variable

$$x_k = \frac{v}{v_k} = \left(\frac{m_k}{M} \, \frac{\mathcal{E}}{k_{\text{B}} T_k} \right)^{1/2} \tag{8.55}$$

and to rewrite the dimensionless function $\mathcal{P}_k$ as follows

$$\mathcal{P}_k \left(x_k, \frac{m_k}{M} \right) = \frac{1}{x_k} \, \text{erf} \left(x_k \right) - \left(1 + \frac{m_k}{M} \right) \frac{2}{\sqrt{\pi}} \, \exp \left(-x_k^2 \right) . \tag{8.56}$$

Here

$$\text{erf} \left(x_k \right) = \frac{2}{\sqrt{\pi}} \int_0^{x_k} \exp \left(-t^2 \right) dt \tag{8.57}$$

is the probability integral.

Let us consider the low-energy limit. Note that

$$\mathcal{P}_k \left(x_k, \frac{m_k}{M} \right) \approx \frac{2}{\sqrt{\pi}} \left[- \frac{m_k}{M} + \frac{2}{3} \left(1 + \frac{m_k}{M} \right) x_k^2 \right] \quad \text{if} \quad x_k \ll 1 . \tag{8.58}$$

Hence the dimensionless function

$$\mathcal{P}_k\left(0, \frac{m_k}{M}\right) = -\frac{2}{\sqrt{\pi}}\frac{m_k}{M} < 0 \tag{8.59}$$

and, according to formula (8.53), the energy losses rate

$$P_k \equiv \left(\frac{d\mathcal{E}}{dt}\right)_k = \frac{8\sqrt{\pi}e^4 \ln\Lambda}{M}\,\frac{Z^2 Z_k^2 n_k}{v_k} > 0\,. \tag{8.60}$$

This means that a test particle with zeroth (or very small) velocity takes energy from the field particles having the temperature T_k. **The hot field particles heat a cold test particle.**

Consider an opposite limiting case. If $x_k \gg 1$, then, being positive, the function

$$\mathcal{P}_k\left(x_k, \frac{m_k}{M}\right) \sim \frac{1}{x_k} \to 0 \quad \text{when} \quad x_k \gg 1\,. \tag{8.61}$$

So the higher the energy of a test particle, the smaller are the Coulomb losses.

The maximum of the dimensionless function $\mathcal{P}_k$ is reached at $x_{k,\,\max} \approx 1.52$, see schematical Figure 8.6.

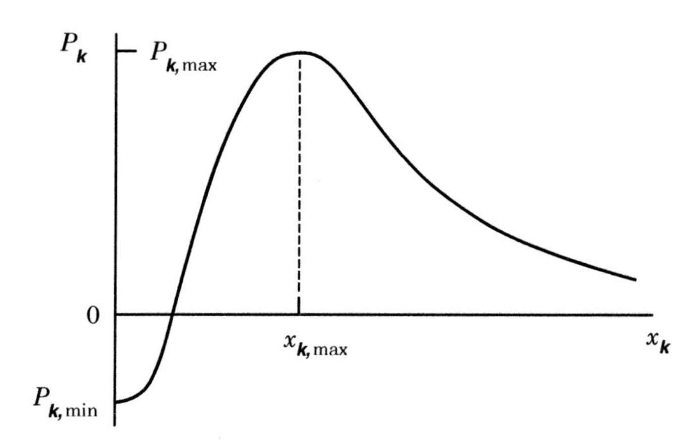

Figure 8.6: The Coulomb losses (with the sign *minus* in formula (8.53)) of energy of a test particle as a function of its velocity measured in the most probable velocity of the field thermal particles of the kind k.

Astrophysical plasma consists of many components. To obtain the total losses it is necessary to sum over all of them in formula (8.53). However two components – electrons and protons – give the largest contribution. In a plasma consisting of electrons and protons with $n_e = n_p = n$ and temperatures T_e and T_p we have (Korchak, 1980):

$$P = -c_{\varepsilon}\,\frac{Z^2}{A}\,\frac{n\,\ln\Lambda}{\sqrt{k_B T_e}}\left[\mathcal{P}_e\left(x_e, \frac{m_e}{M}\right) + \left(\frac{m_e T_e}{m_p T_p}\right)^{1/2} \mathcal{P}_p\left(x_p, \frac{m_p}{M}\right)\right], \tag{8.62}$$

where the constant $c_\varepsilon \approx 1.6 \times 10^{-23}$.

The location of both maxima of the function (8.62) is determined by conditions:

$$x_1 = x_\mathrm{p} \approx 1.52 \quad \text{and} \quad x_2 = x_\mathrm{e} \approx 1.52 \,. \tag{8.63}$$

As follows from formula (8.62), the ratio of losses in the maxima

$$\frac{P_{\max,\,\mathrm{p}}}{P_{\max,\,\mathrm{e}}} = \left(\frac{m_\mathrm{e}\, T_\mathrm{e}}{m_\mathrm{p}\, T_\mathrm{p}} \right)^{1/2} \approx \frac{1}{43} \left(\frac{T_\mathrm{e}}{T_\mathrm{p}} \right)^{1/2} . \tag{8.64}$$

> The maximum of the electron Coulomb losses is the main energy threshold of the particle acceleration from low energies.

The proton barrier is considerably lower than the electron one.

The energy loss contribution of the proton component of astrophysical plasma does not seem to be important. This is not always true, however. First of all, formula (8.64) shows that the Coulomb losses on thermal protons increase with the growth of the ratio $T_\mathrm{e}/T_\mathrm{p}$. This may be an important case if particles of low energies are accelerated in super-hot turbulent-current layers (SHTCLs, see vol. 2, Section 6.3). The second argument comes from a consideration of very low energies of accelerated particles. In this region, the efficiency of acceleration is low for the majority of accelerating mechanisms. However, just in this region of low energies,

> the Coulomb losses can strongly influence the nuclear composition and the charge-state of accelerated particles in astrophysical plasma

(Korchak, 1980; see also Holman, 1995; Bodmer and Bochsler, 2000; Bykov et al., 2000).

When particular acceleration mechanisms in a astrophysical plasma are considered, the role of Coulomb collisions often reduces to the energy losses of the accelerated particles and, in particular, to the presence of the loss barrier at low velocities. As a result, Coulomb collisions decrease the efficiency of any acceleration mechanism. Contrary to this statement, we shall see that in many cases Coulomb collisions can play a much less trivial and not so passive role (e.g., vol. 2, Section 12.3.1). This makes plasma astrophysics more interesting.

8.4.1 (c) Dynamic friction in plasma

The collisional drag force F_f acts on a test particle (mass M, charge Ze) moving through the many-component plasma with the Maxwellian distribution of field particles:

$$M \frac{d}{dt} v_\parallel = -F_f = -\sum_k F_k \left(v_\parallel \right) . \tag{8.65}$$

Here the velocity component $v_\parallel$ is parallel to the vector of the initial velocity of an incident test particle.

For a test particle with a velocity v much below the thermal velocity (8.54) of the field particles with the mass m_k, temperature T_k, and number density n_k,

$$F_f \approx \sum_k \frac{4\pi e^4 \ln \Lambda}{k_{\text{B}}} \frac{Z^2 Z_k^2 n_k}{T_k} \left(1 + \frac{m_k}{M}\right) \frac{2}{3\sqrt{\pi}} \frac{v_{\parallel}}{v_k} \sim v_{\parallel} \,. \tag{8.66}$$

Therefore at small velocities the collisional drag force is proportional to the component $v_{\parallel}$ (cf. formula (1.14)).

When the test particle velocity exceeds the thermal velocity of the field particles, the drag force decreases with $v_{\parallel}$ as follows:

$$F_f = \sum_k F_k \approx \sum_k \frac{2\pi e^4 \ln \Lambda}{k_{\text{B}}} \frac{Z^2 Z_k^2 n_k}{T_k} \left(1 + \frac{m_k}{M}\right) \left(\frac{v_{\parallel}}{v_k}\right)^{-2} \sim v_{\parallel}^{-2} \,. \tag{8.67}$$

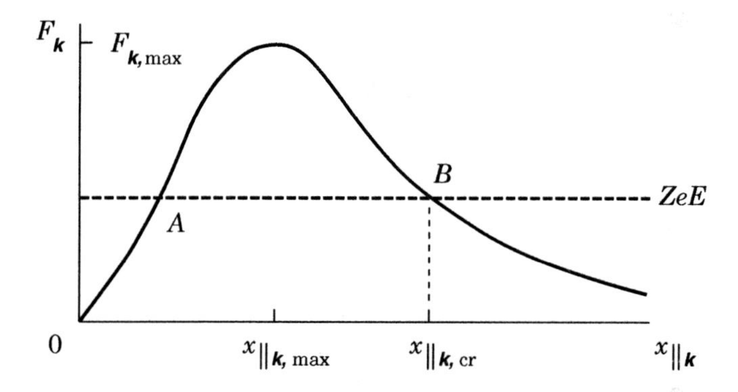

Figure 8.7: The collisional drag force F_k (with the sign *minus* in formula (8.65)) on a test particle as a function of its velocity $v_{\parallel}$ measured in the most probable velocity v_k of the field particles of the kind k.

The general formula for collisional drag force is given, for example, in Sivukhin (1966) and is illustrated by schematical Figure 8.7; here the dimensionless variable $x_{\parallel k} = v_{\parallel}/v_k$. The drag force vanishes when $x_{\parallel k} = 0$; it linearly increases with increasing $x_{\parallel k}$, becoming a maximum when

$$x_{\parallel k} = x_{\parallel k,\,\max} \approx 0.97, \tag{8.68}$$

and then falls off, approaching zero asymptotically as $x_{\parallel k} \to \infty$. This behaviour of the drag force has important consequences discussed below.

8.4.2 Electric runaway

It has been assumed above that the plasma is characterized by the Maxwellian distribution and that there are no external fields. Let us now assume that a uniform electric field **E** is switched on at some instant of time, the velocity distribution being assumed to be Maxwellian at this time. At least, at the beginning of the process when the velocity distribution has not yet changed appreciably, the time variation of the test-particle momentum $M\mathbf{v}$ due to Coulomb collisions with plasma particles will still be given by formulae (8.66) and (8.67) supplemented by the electric force $Ze\mathbf{E}$ in Equation (8.65).

Thus, considering the component $v_{\|}$ as a component of the test-particle velocity **v** which is parallel to the electric field **E**, we rewrite Equation (8.65) as follows:

$$M \frac{d}{dt} v_{\|} = -F_f + ZeE = -\sum_k F_k + ZeE \,. \tag{8.69}$$

If the test-particle velocity is not small in comparison with the thermal velocity v_k, then the collisional drag force on a test particle falls off with increasing velocity v, according to formula (8.67), while the electric force is velocity independent. Therefore

> for all particles with high enough velocities the electric force exceeds the collisional drag force, and the particles are able to *run away* from the thermal distribution.

Equating the electric and collisional drag forces allows us to see the critical velocity v_{cr} above which runaway will occur for a given electric field strength E, see point B in Figure 8.7. Runaway in astrophysical plasma can occur as long as there is a component of the electric field along the magnetic field. Before the acceleration of the heavy ions becomes significant, the acceleration of the light electrons gives rise to the *electron runaway* effect which was first predicted by Giovanelli (1949). He has shown that

- as the electric field applied to a highly ionized gas is increased, **the current**, which is initially limited by elastic collisions between electrons and positive ions, **increases** rapidly as the field strength reaches a critical value;
- this is due to a reduction in the cross-section of positive ions for scattering of electrons with increasing electron velocity.

In a strong electric field (or in a plasma of sufficiently low density and high temperature) all the electrons are accelerated by the field, i.e. become the runaway electrons. The Dreicer field (Dreicer, 1959):

$$E_{\rm Dr} = \frac{4\pi e^3 \ln \Lambda}{k_{\rm B}} \, \frac{n_{\rm e}}{T_{\rm e}} \tag{8.70}$$

approximately correspondes to the electric field strength for which $v_{cr} = v_{\rm e}$. Here $v_{\rm e}$ is the most probable velocity of thermal electrons (8.54).

In a weak field only very fast electrons will run away, i.e. those velocity $v_{\|} \gg v_{cr}$. The velocity v_{cr} depends in an essential manner on the magnitude of electric field. In a weak field, the velocity v_{cr} is naturally much larger than the thermal velocity of electrons in the plasma. Therefore the number of runaway electrons should be very small if their distribution would remain maxwellian for velocities $v_{\parallel} \lesssim v_{\rm cr}$. This is not true however.

In order to determine the flux of runaway electrons we must know the way in which the density of electrons having a velocity $v_{\parallel} \sim v_{\rm cr}$ varies under action of the runaway effect. This means that we must know the velocity distribution for the electrons for $v_{\parallel} \sim v_{\rm cr}$. To consider this problem self-consistently it is necessary to solve the kinetic equation taking both collisions and the electric field into account (Section 4.5). It appears that Coulomb collisions creat a power-law tail distribution between a region of thermal velocities and the region where $v_{\parallel} \approx v_{\rm cr}$ with a constant flux of electrons directed from low to high velocities. By so doing, **Coulomb collisions increase the flux of runaway electrons** (Gurevich, 1961).

To have an idea of the magnitude of the Dreicer field (see Exercise 8.4), let us substitute the definition of the Debye radius (8.31) in formula (8.70) and assume that $T_{\rm e} = T_{\rm p} = T$ and $n_{\rm e} = n_{\rm p} = n$. We find

$$E_{\rm Dr} = \frac{e}{r_{\rm D}^2} \, \frac{\ln \Lambda}{2} \sim \frac{e}{r_{\rm D}^2} \, . \qquad (8.71)$$

So the Dreicer field is approximatelly equal to the electric field of a positive charge at a distance slightly smaller than the Debye radius.

8.4.3 Thermal runaway in astrophysical plasma

Let us consider a plasma with a non-uniform distribution of electron temperature $T_{\rm e}$. Let $l_{_T}$ be the characteristic length of the temperature profile and $\lambda_{\rm e}$ be the mean free path of thermal electrons. For the classical heat conductivity to be applicable, it is necessary to satisfy a condition (Section 9.5):

$$\lambda_{\rm e} \ll l_{_T} \equiv \frac{T_{\rm e}}{|\,\nabla T_{\rm e}\,|} \, . \qquad (8.72)$$

The mean free path of a particle increases with its velocity. This can be seen from formula (8.13) which gives us the mean free path

$$\lambda = \tau v_1 \sim v_1^4 \, . \qquad (8.73)$$

That is why

> a number of fast electrons can penetrate from a hot plasma into cold one even if the gradient of temperature is very small.

In such a way, the hot plasma can lose some part of its thermal energy transferred by fast thermal escaping electrons. In addition to the usual heat flux (8.47), which is determined *locally* by the Coulomb collisions of plasma electrons, there appears a *non-local* energy flux carried by the fast electrons practically without collisions. A classical diffusive heat transfer and a convective one, determined by *thermal runaway* electrons, are always present in plasma.

It is interesting for astrophysical applications that, at not too small temperature gradients, the convective transfer of thermal energy can play a principal role. Gurevich and Istomin (1979) have examined the case of a **small temperature gradient**. By using a perturbation analysis for the high-speed kinetic equation (Section 4.2), they have shown that the fast growth of the mean free path with increasing velocity gives an abrupt growth of the number of fast electrons in the cold plasma.

The opposite case of a **large temperature gradient** in the narrow transition layer between a high-temperature plasma and a cold one was investigated by many authors with applications to the problem of energy transfer in the solar atmosphere. For example, Shoub (1983) has solved numerically the boundary-value problem for the Fokker-Planck equation in the model of the transition layer between the corona and the chromosphere in quiet conditions. An excess of fast electrons has been found in the low transion layer region. As for solar flares, the prevailing view is that

> the high-temperature plasma can lose energy efficiently by the convective heat transfer by the thermal runaway electrons

(see Somov, 1992).

In both cases, however, it is important to take into account that the **fast runaway electrons**, similar to any beam of fast particles, **generate the electric field which drives the reverse current** of thermal electrons. Diakonov and Somov (1988) have found an analytical solution to the self-consistent kinetic problem on the beam of escaping thermal electrons and its associated reverse current (Section 4.5). They have shown that the reverse-current electric field in solar flares leads to a significant reduction of the convective heat flux carried by fast electrons escaping from the high-temperature plasma to the cold one.

Recommended Reading: Sivukhin (1966), Somov (1992).

8.5 Practice: Exercises and Answers

Exercise 8.1 [Section 8.1] For an electron, which moves in the solar corona with a mean thermal velocity (Exercise 5.2), evaluate the characteristic time of *close* and *distant* collisions with thermal protons.

Answer. Characteristic time of close electron-proton collisions follows from formula (8.13) and is equal to

$$\tau_{cl,\,\mathrm{ep}} = \frac{m_{\mathrm{e}}^2}{\pi e^4} \frac{V_{\mathrm{Te}}^3}{n_{\mathrm{p}}} \approx 4.96 \times 10^{-18} \, \frac{V_{\mathrm{Te}}^3}{n_{\mathrm{p}}} \, , \; \mathrm{s} \, . \tag{8.74}$$

At typical temperatures of electrons in the corona $T_{\mathrm{e}} \approx 2 \times 10^6$ K, their thermal velocity (5.54) $V_{\mathrm{Te}} \approx 9.5 \times 10^8$ cm s^{-1}. Substituting this value in (8.74) and assuming $n_{\mathrm{p}} \approx n_{\mathrm{e}} \approx 2 \times 10^8$ cm^{-3}, we find that $\tau_{cl,\,\mathrm{ep}} \approx 22$ s.

According to (8.21) the characteristic time of distant collisions is $8 \ln \Lambda$ shorter than the close collision time (8.74). Hence, first, we have to find the value of the Coulomb logarithm (8.34):

$$\ln \Lambda = \ln \left[\left(\frac{3 k_{\mathrm{B}}^{3/2}}{2 \pi^{1/2} \, e^3} \right) \left(\frac{T_{\mathrm{e}}^3}{n_{\mathrm{e}}} \right)^{1/2} \right] \approx \ln \left[1.25 \times 10^4 \left(\frac{T_{\mathrm{e}}^3}{n_{\mathrm{e}}} \right)^{1/2} \right] . \tag{8.75}$$

At typical coronal temperature and density, formula (8.75) gives

$$\ln \Lambda \approx 22 \, .$$

With this value of $\ln \Lambda$ formula (8.21) gives

$$\tau_{\perp,\,\mathrm{ep}} = \frac{m_{\mathrm{e}}^2}{\pi e^4} \frac{1}{8 \, \ln \Lambda} \frac{V_{\mathrm{Te}}^3}{n_{\mathrm{p}}} \approx 2.87 \times 10^{-20} \, \frac{V_{\mathrm{Te}}^3}{n_{\mathrm{p}}} \, , \; \mathrm{s} \, . \tag{8.76}$$

In the solar corona $\tau_{\perp,\,\mathrm{ep}} \approx 0.1$ s. Therefore the *distant* collisions of thermal electrons with thermal protons in the corona are really much more frequent in comparison with close collisions.

Exercise 8.2 [Section 8.2] Evaluate the Debye radius and the plasma frequency in the solar corona.

Answer. From (8.31) it follows that for electron-proton plasma with $T_{\mathrm{e}} = T_{\mathrm{p}} = T$ and $n_{\mathrm{e}} = n_{\mathrm{p}} = n$ the Debye radius

$$r_{\mathrm{D}} = \left(\frac{k_{\mathrm{B}} T}{8 \pi e^2 \, n} \right)^{1/2} \approx 4.9 \left(\frac{T}{n} \right)^{1/2} , \; \mathrm{cm} \, . \tag{8.77}$$

Under conditions in the solar corona $r_{\mathrm{D}} \approx 0.5$ cm.

The electron plasma frequency (8.35)

$$\omega_{pl}^{(\mathrm{e})} = \left(\frac{4 \pi e^2 \, n_{\mathrm{e}}}{m_{\mathrm{e}}} \right)^{1/2} \approx 5.64 \times 10^4 \, \sqrt{n_{\mathrm{e}}} \, , \; \mathrm{rad \; s}^{-1} \, , \tag{8.78}$$

or

$$\nu_{pl}^{(\mathrm{e})} = \omega_{pl}^{(\mathrm{e})} / 2\pi \approx 10^4 \, \sqrt{n_{\mathrm{e}}} \, , \; \mathrm{Hz} \, . \tag{8.79}$$

In the solar corona $\omega_{pl}^{(\mathrm{e})} \sim 10^9$ rad s^{-1}.

Exercise 8.3 [Section 8.3] Under conditions of Exercise 8.1 evaluate the *exact* (determined by formulae (8.38) and (8.39)) *collisional times* between thermal electrons and between thermal protons, respectively. Compare these times with the characteristic time of energy exchange between electrons and protons in the coronal plasma.

Answer. By substituting $\ln \Lambda$ in (8.38), we have the following expression for the thermal electron collisional time

$$\tau_{ee} = \frac{m_e^2}{0.714\, e^4\, 8\pi\, \ln \Lambda}\, \frac{V_{Te}^3}{n_e} \approx 4.04 \times 10^{-20}\, \frac{V_{Te}^3}{n_e}\,,\ \text{s}\,. \tag{8.80}$$

In the solar corona $\tau_{ee} \approx 0.2$ s. For thermal protons formula (8.39) gives

$$\tau_{pp} = \frac{m_p^2}{0.714\, e^4\, 8\pi\, \ln \Lambda}\, \frac{V_{Tp}^3}{n_p} \approx 1.36 \times 10^{-13}\, \frac{V_{Tp}^3}{n_p}\,,\ \text{s}\,. \tag{8.81}$$

Assuming $T_p = T_e$ and $n_p = n_e$, we find the proton collisional time in the solar corona $\tau_{pp} \approx 7$ s; this is in a good agreement with formula (8.40), of course.

Let us find the time of energy exchange between electrons and protons. By using formula (8.44), we have

$$\tau_{ep}(\mathcal{E}) \approx 22\, \tau_{pp} \approx 164\ \text{s}\,. \tag{8.82}$$

So the energy exchange between electron and proton components in the coronal plasma is the slowest process determined by Coulomb collisions.

Exercise 8.4 [Section 8.4] Evaluate and compare Dreicer's electric fields in the solar corona and in the chromosphere.

Answer. From (8.70) it follows that

$$E_{Dr} = \frac{4\pi e^3 \ln \Lambda}{k_B}\, \frac{n_e}{T_e} \approx 7.54 \times 10^{-8}\, \frac{n_e\,(\text{cm}^{-3})}{T_e\,(\text{K})}\,,\ \text{V cm}^{-1}\,. \tag{8.83}$$

Here it was taken $\ln \Lambda \approx 21.6$ according to Exercise 8.1.

At typical temperature and number density of electrons in the solar corona $T_e \approx 2 \times 10^6$ K and $n_e \approx 2 \times 10^8$ cm^{-3}, we find that the Dreicer electric field $E_{Dr} \approx 7 \times 10^{-6}\,\text{V cm}^{-1} \sim 10^{-5}\,\text{V cm}^{-1}$. The same value follows, of course, from formula (8.71) with $r_D \approx 0.5$ cm (see Exercise 8.3).

In the solar chromosphere $n_e > 2 \times 10^{10}$ cm^{-3} and $T_e < 10^4$ K. According to formula (8.83), the Dreicer electric field $E_{Dr} > 0.1\,\text{V cm}^{-1}$ in the chromosphere is, at least, 10^4 times stronger than the coronal one.

Exercise 8.5. Define the dynamic friction by gravitational force as momentum loss by a massive moving object, for example a star in a galaxy, due to its gravitational interaction with its own gravitationally induced wake. Discuss two possibilities: (a) the background medium consists of collisionless matter (other stars in the galaxy), (b) the medium is entirely gaseous (e.g., Ostriker, 1999). The first case, the **gravitational drag** in collisionless systems (Chandrasekhar, 1943b), has widespread theoretical application in modern astrophysics.

Hint. At first, let us qualitatively understand why a friction force should arise in a collisionless gravitational system. Suppose a star has moved from a point A to a point B as shown in Figure 8.8.

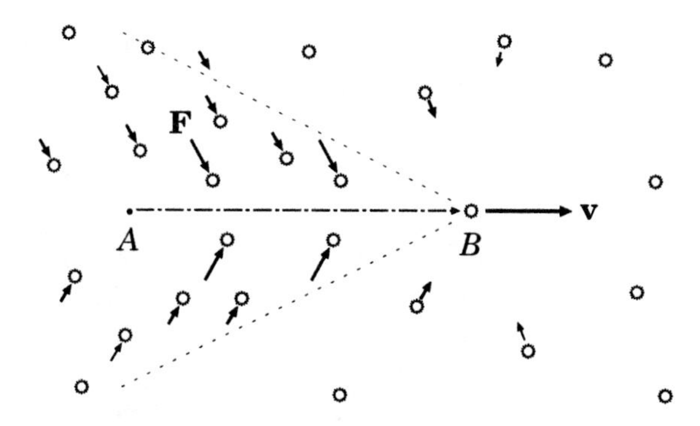

Figure 8.8: An illustration of the origin of dynamic friction in a collisionless gravitational system.

While passing from A to B, the star attracted the surrounding stars towards itself. Hence the number density of stars around AB should be slightly larger than that ahead of B. Therefore the star at the point B experiences a net gravitational attraction in the backward direction, i.e. in the direction opposite to the direction of the star velocity vector $\mathbf{v}$.

The variety of consequences of the gravitational drag force in collisionless astronomical systems includes the mass segregation in star clusters, sinking satellites in dark matter galaxy halo, orbital decay of binary supermassive black holes after galaxy mergers, etc. (Binney and Tremain, 1987).

Exercise 8.6. Discuss why the rate of escape of stars from a galactic claster, evaluated ignoring dynamic friction, is too rapid to be compatible with a life for the cluster (Chandrasekhar, 1943c). Show that the escape rate is drastically reduced when dynamic friction is allowed for.

Chapter 9

Macroscopic Description of Astrophysical Plasma

In this Chapter we are not concerned with individual particles but we will treat individual kinds of particles as continuous media interacting between themselves and with an electromagnetic field. This approach gives us the multi-fluid models of plasma, which are useful to consider many properties of astrophysical plasma.

9.1 Summary of microscopic description

The averaged Liouville equation or kinetic equation gives us a *microscopic* (though averaged in a statistical sense) description of the plasma state's evolution. Let us consider the way of transition to a less comprehensive *macroscopic* description of a plasma. We start from the kinetic equation for particles of kind k, in the form derived in Section 2.2:

$$\frac{\partial f_k(X,t)}{\partial t} + v_\alpha \frac{\partial f_k(X,t)}{\partial r_\alpha} + \frac{F_{k,\alpha}(X,t)}{m_k} \frac{\partial f_k(X,t)}{\partial v_\alpha} = \left(\frac{\partial \hat{f}_k}{\partial t} \right)_c . \tag{9.1}$$

Here the statistically averaged force is

$$F_{k,\alpha}(X,t) = \sum_l \int_{X_1} F_{kl,\alpha}(X, X_1)\, f_l(X_1, t)\, dX_1 \tag{9.2}$$

and the collisional integral

$$\left(\frac{\partial \hat{f}_k}{\partial t} \right)_c = - \frac{\partial}{\partial v_\alpha} J_{k,\alpha}(X,t)\,, \tag{9.3}$$

where the flux of particles of kind k

$$J_{k,\alpha}(X,t) = \sum_l \int_{X_1} \frac{1}{m_k} F_{kl,\alpha}(X,X_1)\, f_{kl}(X,X_1,t)\, dX_1 \qquad (9.4)$$

in the six-dimensional phase space $X = \{\mathbf{r}, \mathbf{v}\}$.

9.2 Transition to macroscopic description

Before turning our attention to the deduction of equations for the macroscopic quantities or macroscopic *transfer* equations, let us define the following *moments* of the distribution function.

(a) The zeroth moment (without multiplying the distribution function f_k by the velocity)

$$\int_{\mathbf{v}} f_k(\mathbf{r},\mathbf{v},t)\, d^3\mathbf{v} = n_k(\mathbf{r},t) \qquad (9.5)$$

is obviously the number of particles of kind k in a unit volume, i.e. the *number density* of particles of kind k. It is related to the *mass density* in a natural way:

$$\rho_k(\mathbf{r},t) = m_k\, n_k(\mathbf{r},t)\,.$$

The plasma mass density is accordingly

$$\rho(\mathbf{r},t) = \sum_k \rho_k(\mathbf{r},t) = \sum_k m_k\, n_k(\mathbf{r},t)\,. \qquad (9.6)$$

(b) The first moment of the distribution function, i.e. the integral of the product of the velocity to the first power and the distribution function f_k,

$$\int_{\mathbf{v}} v_\alpha\, f_k(\mathbf{r},\mathbf{v},t)\, d^3\mathbf{v} = n_k\, u_{k,\alpha} \qquad (9.7)$$

is the product of the number density of particles of kind k by their *mean velocity*

$$u_{k,\alpha}(\mathbf{r},t) = \frac{1}{n_k} \int_{\mathbf{v}} v_\alpha\, f_k(\mathbf{r},\mathbf{v},t)\, d^3\mathbf{v}\,. \qquad (9.8)$$

Consequently, the *mean momentum* of particles of kind k in a unit volume is expressed in terms of the first moment of the distribution function as follows

$$m_k\, n_k\, u_{k,\alpha} = m_k \int_{\mathbf{v}} v_\alpha\, f_k(\mathbf{r},\mathbf{v},t)\, d^3\mathbf{v}\,. \qquad (9.9)$$

(c) The second moment of the distribution function is defined to be

$$\Pi_{\alpha\beta}^{(k)}(\mathbf{r},t) = m_k \int\limits_{\mathbf{v}} v_\alpha v_\beta \, f_k(\mathbf{r},\mathbf{v},t)\, d^3\mathbf{v} = m_k n_k \, u_{k,\alpha} u_{k,\beta} + p_{\alpha\beta}^{(k)} . \tag{9.10}$$

Here we have introduced

$$v'_\alpha = v_\alpha - u_{k,\alpha}$$

which is the deviation of the particle velocity from its mean value

$$u_{k,\alpha} = \langle \, v_{k,\alpha} \, \rangle_v$$

in the sense of the definition (9.8), so that $\langle \, v'_\alpha \, \rangle = 0$; and

$$p_{\alpha\beta}^{(k)} = m_k \int\limits_{\mathbf{v}} v'_\alpha v'_\beta \, f_k(\mathbf{r},\mathbf{v},t)\, d^3\mathbf{v} , \tag{9.11}$$

is termed the *pressure tensor*.

$\Pi_{\alpha\beta}^{(k)}$ is the *tensor of momentum flux density* for particles of kind k. Its component $\Pi_{\alpha\beta}^{(k)}$ is the αth component of the momentum transported by the particles of kind k, in a unit time, across the unit area perpendicular to the axis r_β.

Once we know the distribution function $f_k(\mathbf{r},\mathbf{v},t)$, which contains all the statistically averaged information on the system of the particles of kind k at the *microscopic* level, we can derive all *macroscopic* quantities related to these particles. So, higher moments of the distribution function will be introduced as needed.

9.3 Macroscopic transfer equations

Note that the deduction of macroscopic equations is nothing but just the derivation of the equations for the distribution function moments.

9.3.1 Equation for the zeroth moment

Let us calculate the *zeroth* moment of the kinetic Equation (9.1):

$$\int\limits_{\mathbf{v}} \frac{\partial f_k}{\partial t}\, d^3\mathbf{v} + \int\limits_{\mathbf{v}} v_\alpha \frac{\partial f_k}{\partial r_\alpha}\, d^3\mathbf{v} + \int\limits_{\mathbf{v}} \frac{F_{k,\alpha}}{m_k} \frac{\partial f_k}{\partial v_\alpha}\, d^3\mathbf{v} = \int\limits_{\mathbf{v}} \left(\frac{\partial \hat{f}_k}{\partial t} \right)_{\mathrm{c}} d^3\mathbf{v} . \tag{9.12}$$

We interchange the order of integration over velocities and the differentiation with respect to time t in the first term and with respect to coordinates r_α in the second one. Under the second integral

$$v_\alpha \frac{\partial f_k}{\partial r_\alpha} = \frac{\partial}{\partial r_\alpha}(v_\alpha f_k) - f_k \frac{\partial v_\alpha}{\partial r_\alpha} = \frac{\partial}{\partial r_\alpha}(v_\alpha f_k) - 0 ,$$

since $\mathbf{r}$ and $\mathbf{v}$ are independent variables in phase space X.

Taking into account that the distribution function quickly approaches zero as $v \to \infty$, the integral of the third term is taken by parts and is equal to zero (Exercise 9.1).

Finally, the integral of the right-hand side of (9.12) describes the change in the number of particles of kind k in a unit volume, in a unit time, as a result of collisions with particles of other kinds. If the processes of transformation, during which the particle kind can be changed (such as ionization, recombination, charge exchange, dissociation etc., see Exercise 9.2), are not allowed for, then the last integral is zero as well:

$$\int_{\mathbf{v}} \left(\frac{\partial \hat{f}_k}{\partial t} \right)_c d^3\mathbf{v} = 0 \,. \tag{9.13}$$

Thus, by integration of (9.12), the following equation is found to result from (9.1)

$$\boxed{\frac{\partial n_k}{\partial t} + \frac{\partial}{\partial r_\alpha} \, n_k \, u_{k,\alpha} = 0 \,.} \tag{9.14}$$

This is the usual *continuity equation* expressing the conservation of particles of kind k or (that is the same, of course) conservation of their mass:

$$\frac{\partial \rho_k}{\partial t} + \frac{\partial}{\partial r_\alpha} \, \rho_k \, u_{k,\alpha} = 0 \,. \tag{9.15}$$

Here

$$\rho_k(\mathbf{r}, t) = m_k \, n_k(\mathbf{r}, t)$$

is the mass density of particles of kind k.

Equation (9.14) for the zeroth moment n_k depends on the unknown first moment $u_{k,\alpha}$. This is illustrated by Figure 9.1.

9.3.2 The momentum conservation law

Now let us calculate the *first* moment of the kinetic Equation (9.1) multiplied by the mass m_k:

$$m_k \int_{\mathbf{v}} \frac{\partial f_k}{\partial t} \, v_\alpha \, d^3\mathbf{v} + m_k \int_{\mathbf{v}} v_\alpha v_\beta \, \frac{\partial f_k}{\partial r_\beta} \, d^3\mathbf{v} + \int_{\mathbf{v}} v_\alpha F_{k,\beta} \, \frac{\partial f_k}{\partial v_\beta} \, d^3\mathbf{v} =$$

$$= m_k \int_{\mathbf{v}} v_\alpha \left(\frac{\partial \hat{f}_k}{\partial t} \right)_c d^3\mathbf{v} \,. \tag{9.16}$$

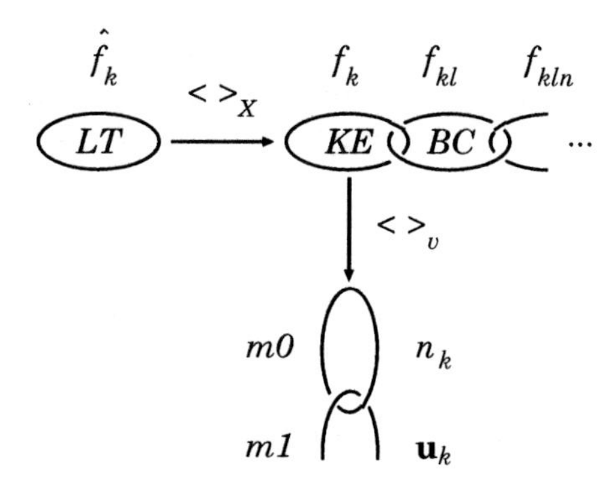

Figure 9.1: From the microscopic to the macroscopic view of a plasma. LT is the Liouville theorem (1.11) for an exact distribution function $\hat{f}_k$. KE and BC are the kinetic Equation (2.36) and the equation for the binary correlation function. $m0$ is the equation for the zeroth moment of the distribution function f_k, the number density n_k of the particles of kind k. This equation is unclosed.

With allowance made for the definitions (9.7) and (9.10), we obtain the *momentum conservation law*

$$\frac{\partial}{\partial t}(m_k n_k\, u_{k,\alpha}) + \frac{\partial}{\partial r_\beta}\left(m_k n_k\, u_{k,\alpha} u_{k,\beta} + p_{\alpha\beta}^{(k)}\right) -$$

$$- \langle\, F_{k,\alpha}(\mathbf{r},t)\,\rangle_v = \langle\, F_{k,\alpha}^{(c)}(\mathbf{r},t)\,\rangle_v \,. \tag{9.17}$$

Here $p_{\alpha\beta}^{(k)}$ is the pressure tensor (9.11).

The *mean force* acting on the particles of kind k in a unit volume (the mean force *per unit volume*) is (see Exercise 9.3):

$$\langle\, F_{k,\alpha}(\mathbf{r},t)\,\rangle_v = \int_{\mathbf{v}} F_{k,\alpha}(\mathbf{r},\mathbf{v},t)\, f_k(\mathbf{r},\mathbf{v},t)\, d^3\mathbf{v}\,. \tag{9.18}$$

This should not be confused with the *statistical mean* force acting on a single particle (see definition (9.2)). The statistically averaged force (9.2) is under the integral in formula (9.18).

In the particular case of the Lorentz force, we rewrite the mean force per unit volume as follows:

$$\langle\, F_{k,\alpha}(\mathbf{r},t)\,\rangle_v = n_k e_k \left[\, E_\alpha + \frac{1}{c}\,(\,\mathbf{u}_k \times \mathbf{B}\,)_\alpha \right]$$

or

$$\boxed{\langle\, F_{k,\alpha}(\mathbf{r},t)\,\rangle_v = \rho_k^{\mathrm{q}}\, E_\alpha + \frac{1}{c}\,(\,\mathbf{j}_k^{\mathrm{q}} \times \mathbf{B}\,)_\alpha \,.} \tag{9.19}$$

Here ρ_k^q and $\mathbf{j}_k^q$ are the mean densities of electric charge and current, produced by the particles of kind k. However note that

> the mean electromagnetic force couples all the charged components of cosmic plasma together

because the electric and magnetic fields, $\mathbf{E}$ and $\mathbf{B}$, act on all charged components and, at the same time, all charged components contribute to the electric and magnetic fields according to Maxwell's equations.

The right-hand side of Equation (9.17) contains the mean force resulting from collisions, i.e. the *mean collisional force* (see Exercise 9.4):

$$\langle F_{k,\alpha}^{(c)}(\mathbf{r},t)\rangle_v = m_k \int\limits_{\mathbf{v}} v_\alpha \left(\frac{\partial \hat{f}_k}{\partial t} \right)_{\rm c} d^3\mathbf{v} \,. \tag{9.20}$$

Substituting (9.3) in definition (9.20) gives us the following formula

$$\langle F_{k,\alpha}^{(c)}(\mathbf{r},t)\rangle_v = -\, m_k \int\limits_{\mathbf{v}} v_\alpha \frac{\partial}{\partial v_\beta} J_{k,\beta} \, d^3\mathbf{v} \,. \tag{9.21}$$

Let us integrate (9.21) by parts. For this purpose, at first, we find the derivative

$$\frac{\partial}{\partial v_\beta}(v_\alpha J_{k,\beta}) = J_{k,\beta} \frac{\partial v_\alpha}{\partial v_\beta} + v_\alpha \frac{\partial}{\partial v_\beta} J_{k,\beta} \,.$$

From this it follows that

$$v_\alpha \frac{\partial}{\partial v_\beta} J_{k,\beta} = -\, J_{k,\beta}\, \delta_{\alpha\beta} + \frac{\partial}{\partial v_\beta}(v_\alpha J_{k,\beta}) =$$

$$= -\, J_{k,\alpha} + \frac{\partial}{\partial v_\beta}(v_\alpha J_{k,\beta}) \,. \tag{9.22}$$

On substituting (9.22) and (9.4) in (9.20) and integrating, we obtain the most general formula for the mean collisional force

$$\langle F_{k,\alpha}^{(c)}(\mathbf{r},t)\rangle_v = m_k \int\limits_{\mathbf{v}} J_{k,\alpha}(\mathbf{r},\mathbf{v},t)\, d^3\mathbf{v} = \tag{9.23}$$

$$= \sum_{l \neq k} \int\limits_{\mathbf{v}} \int\limits_{\mathbf{v}_1} \int\limits_{\mathbf{r}_1} F_{kl,\alpha}(\mathbf{r},\mathbf{v},\mathbf{r}_1,\mathbf{v}_1)\, f_{kl}(\mathbf{r},\mathbf{v},\mathbf{r}_1,\mathbf{v}_1,t)\, d^3\mathbf{r}_1\, d^3\mathbf{v}_1\, d^3\mathbf{v} \,.$$

Note that

> for the particles of the same kind, the elastic collisions cannot change the total particle momentum per unit volume.

That is why $l \neq k$ in the sum (9.23).

Formula (9.23) contains the unknown binary correlation function f_{kl}. The last should be found from the correlation function Equation (2.46) indicated as the second link BC in Figure 9.1. Thus the equation for the first moment of the distribution function is as much unclosed as the initial kinetic Equation (9.1), which is the first equation of the chain for correlation functions (see KE in Figure 9.1).

If there are several kinds of particles, and if each of them is in the state of thermodynamic equilibrium, then the mean collisional force can conventionally be expressed in terms of the *mean momentum loss* during the collisions of a particle of kind k with the particles of other kinds:

$$\langle F_{k,\alpha}^{(c)} (\mathbf{r}, t) \rangle_v = - \sum_{l \neq k} \frac{m_k n_k (u_{k,\alpha} - u_{l,\alpha})}{\tau_{kl}} \,. \tag{9.24}$$

Here $\tau_{kl}^{-1} = \nu_{kl}$ is the mean frequency of collisions between the particles of kinds k and l. This force is zero, once the particles of all kinds have identical velocities. The mean collisional force, as well as the mean electromagnetic force, tends to make astrophysical plasma be a single hydrodynamic medium (see Section 12.1).

If $u_{l,\alpha} < u_{k,\alpha}$ then the mean collisional force is negative:

> the fastly moving particles of kind k slow down by dint of collisions with the slowly moving particles of other kinds.

Formula (9.24) has the status of a good approximation in plasma astrophysics.

9.3.3 The energy conservation law

The second moment (9.10) of a distribution function f_k is the tensor of momentum flux density $\Pi_{\alpha\beta}^{(k)}$. In general, in order to find an equation for this tensor, we should multiply the kinetic Equation (9.1) by the factor $m_k \, v_\alpha v_\beta$ and integrate over velocity space $\mathbf{v}$. In this way, we could arrive to a *matrix equation* in partial derivatives. If we take the trace of this equation we could obtain the partial differential *scalar equation* for energy density of the particles under consideration (e.g., Shkarofsky et al., 1966; § 9.2). This is the correct self-consistent way which is the basis of the moment method. For our aims, a more simple direct procedure is sufficient and correct.

In order to derive the *energy conservation law*, we multiply Equation (9.1) by the particle's kinetic energy $m_k v_\alpha^2/2$ and integrate over velocities, taking into account that

$$v_\alpha = u_{k,\alpha} + v_\alpha'$$

and

$$v_\alpha^2 = u_{k,\alpha}^2 + (v_\alpha')^2 + 2\, u_{k,\alpha}\, v_\alpha' \,.$$

A straightforward integration yields

$$\frac{\partial}{\partial t}\left(\frac{\rho_k u_k^2}{2} + \rho_k\,\varepsilon_k\right) + \frac{\partial}{\partial r_\alpha}\left[\rho_k u_{k,\alpha}\left(\frac{u_k^2}{2} + \varepsilon_k\right) + p_{\alpha\beta}^{(k)}\,u_{k,\beta} + q_{\,k,\alpha}\right] =$$

$$= \rho_k^{\mathrm{q}}\,(\mathbf{E}\cdot\mathbf{u}_k) + \left(\mathbf{F}_k^{(c)}\cdot\mathbf{u}_k\right) + Q_k^{(c)}\,(\mathbf{r},t)\,. \qquad (9.25)$$

Here

$$m_k\,\varepsilon_k(\mathbf{r},t) = \frac{1}{n_k}\int\limits_{\mathbf{v}} \frac{m_k\,(v_\alpha')^2}{2}\,f_k\,(\mathbf{r},\mathbf{v},t)\,d^3\mathbf{v} =$$

$$= \frac{m_k}{2n_k}\int\limits_{\mathbf{v}} (v_\alpha')^2\;f_k\,(\mathbf{r},\mathbf{v},t)\,d^3\mathbf{v} \qquad (9.26)$$

is the *mean kinetic energy* of chaotic (non-directed) motion per single particle of kind k. Thus the first term on the left-hand side of Equation (9.25) represents the time derivative of the energy of the particles of kink k in a unit volume, which is the sum of kinetic energy of a regular motion with the mean velocity $\mathbf{u}_k$ and the so-called *internal* energy.

The pressure tensor can be written as

$$p_{\alpha\beta}^{(k)} = p_k\,\delta_{\alpha\beta} + \pi_{\alpha\beta}^{(k)}\,. \qquad (9.27)$$

Thus, on rearrangement, we obtain the following general equation

$$\frac{\partial}{\partial t}\left(\frac{\rho_k u_k^2}{2} + \rho_k\,\varepsilon_k\right) + \frac{\partial}{\partial r_\alpha}\left[\rho_k u_{k,\alpha}\left(\frac{u_k^2}{2} + w_k\right) + \;\pi_{\alpha\beta}^{(k)}\,u_{k,\beta} + q_{\,k,\alpha}\right] =$$

$$= \rho_k^{\mathrm{q}}\,(\mathbf{E}\cdot\mathbf{u}_k) + \left(\mathbf{F}_k^{(c)}\cdot\mathbf{u}_k\right) + Q_k^{(c)}\,(\mathbf{r},t)\,. \qquad (9.28)$$

Here

$$w_k = \varepsilon_k + \frac{p_k}{\rho_k} \qquad (9.29)$$

is the *heat function* per unit mass. Therefore the second term on the left-hand side contains the energy flux

$$\rho_k u_{k,\alpha}\left(\frac{u_k^2}{2} + w_k\right),$$

which can be called the '*advective*' flux of kinetic energy.

Let us mention the well known astrophysical application of this term. *The advective cooling of ions* heated by viscosity might dominate the cooling by the electron-ion collisions, for example, in a low-density high-temperature plasma flow near a rotating black hole. In such an advection-dominated accretion flow (ADAF), the heat generated via viscosity is transferred inward rather than radiated away locally like in a standard accretion disk model (see Sections 8.3.4 and 13.2).

On the other hand, discussing the ADAF model as a solution for the important astrophysical problem should be treated with reasonable cautions. Looking at Equations (9.25) for electrons and ions separately, we see how many assumptions have to be made to arrive to the ADAF approximation. For example, this is not realistic to assume that plasma electrons are heated only due to Coulomb collisions with ions and, for this reason, the electrons are much cooler than the ions. The suggestions underlying the ADAF approximation ignore several physical effects including *reconnection* and dissipation of magnetic fields (regular and random) in astrophysical plasma. This makes a physical basis of the model very uncertain.

* * *

In order to clarify the physical meaning of the definitions given above, let us, for a while, come back to the general principles of plasma physics. If the particles of the kth kind are in the *thermodynamic equilibrium* state, then f_k is the Maxwellian function with the *temperature* T_k:

$$f_k^{(0)}(\mathbf{r}, \mathbf{v}) = n_k(\mathbf{r}) \left[\frac{m_k}{2\pi\, k_{\mathrm{B}} T_k(\mathbf{r})} \right]^{3/2} \exp \left\{ - \frac{m_k \,|\, \mathbf{v} - \mathbf{u}_k(\mathbf{r}) \,|^2}{2\, k_{\mathrm{B}} T_k(\mathbf{r})} \right\}, \qquad (9.30)$$

see Section 9.5. In this case, according to formula (9.26), the mean kinetic energy of chaotic motion per single particle of kind k

$$m_k\, \varepsilon_k = \frac{3}{2}\, k_{\mathrm{B}} T_k\,. \qquad (9.31)$$

The pressure tensor (9.11) is isotropic:

$$p_{\alpha\beta}^{(k)} = p_k\, \delta_{\alpha\beta}\,, \qquad (9.32)$$

where

$$p_k = n_k\, k_{\mathrm{B}} T_k \qquad (9.33)$$

is the gas pressure of the particles of kind k. This is also the equation of state for the *ideal* gas. Thus we have found that the pressure tensor is diagonal. This implies the absence of *viscosity* for the ideal gas, as we shall see below.

The heat function per unit mass or, more exactly, the *specific enthalpy* is

$$w_k = \varepsilon_k + \frac{p_k}{\rho_k} = \frac{5}{2}\, \frac{k_{\mathrm{B}} T_k}{m_k}\,. \qquad (9.34)$$

This is a particular case of the thermodynamic equilibrium state; it will be discussed in Section 9.5.

* * *

In general, we do not expect that the system of the particles of kind k has reached thermodynamic equilibrium. Nevertheless we use the mean kinetic energy (9.26) to define the *effective temperature* T_k according to definition (9.31).

Such a *kinetic* temperature is just a measure for the spread of the particle distribution in velocity space. The kinetic temperatures of different components in astrophysical plasma may differ from each other. Moreover, in an anisotropic plasma, the kinetic temperatures parallel and perpendicular to the magnetic field are different.

Without supposing thermodynamic equilibrium, in an *anisotropic* plasma, the part associated with the deviation of the distribution function from the isotropic one (which does not need to be a Maxwellian function in general) is distinguished in the pressure tensor:

$$p_{\alpha\beta}^{(k)} - p_k\, \delta_{\alpha\beta} = \pi_{\alpha\beta}^{(k)} . \tag{9.35}$$

Here $\pi_{\alpha\beta}^{(k)}$ is called the *viscous stress tensor*. So the term $\pi_{\alpha\beta}^{(k)}\, u_{k,\beta}$ in the energy-conservation Equation (9.25) represents the flux of energy released by the *viscous force* in the particles of kind k.

The vector

$$q_{\,k,\alpha} = \int\limits_{\mathbf{v}} \frac{m_k\,(v')^2}{2}\; v'_\alpha\, f_k\,(\mathbf{r}, \mathbf{v}, t)\; d^3\mathbf{v} \tag{9.36}$$

is the *heat flux density* due to the particles of kind k in a system of coordinates, in which the gas of these particles is immovable at a given point of space. Formula (9.36) shows that a third order term appears in the second order moment of the kinetic equation.

The right-hand side of the energy conservation law (9.25) contains the following three terms:

(a) The first term

$$\rho_k^{\mathrm{q}}\,(\mathbf{E} \cdot \mathbf{u}_k) = n_k e_k\, E_\alpha\, u_{k,\alpha} \tag{9.37}$$

is the work done by the Lorentz force (without the magnetic field, of course) in unit time on unit volume.

(b) The second term

$$\left(\mathbf{F}_k^{(\mathrm{c})} \cdot \mathbf{u}_k\right) = u_{k,\alpha} \int\limits_{\mathbf{v}} m_k\, v'_\alpha \left(\frac{\partial \hat{f}_k}{\partial t}\right)_{\mathrm{c}} d^3\mathbf{v} \tag{9.38}$$

is the work done by the collisional force of friction of the particles of kind k with all other particles in unit time on unit volume. This means that

> the work of friction force results from the mean momentum change of particles of kind k (*moving with the mean velocity* $\mathbf{u}_k$) owing to collisions with all other particles.

(c) The last term

$$Q_k^{(\mathrm{c})}\,(\mathbf{r}, t) = \int\limits_{\mathbf{v}} \frac{m_k\,(v')^2}{2} \left(\frac{\partial \hat{f}_k}{\partial t}\right)_{\mathrm{c}} d^3\mathbf{v} \tag{9.39}$$

is the rate of thermal energy release (heating or cooling) in a gas of the particles of kind k due to collisions with other particles. Recall that the collisional integral depends on the binary correlation function f_{kl}.

9.4 General properties of transfer equations

9.4.1 Divergent and hydrodynamic forms

Equations (9.14), (9.17), and (9.25) are referred to as the equations of particle, momentum and energy *transfer*, respectively; and the approximation in which they have been obtained is called the model of *mutually penetrating* charged gases. These gases are not assumed to be in the thermodynamic equilibrium. However the definition of the temperature (9.31) may be generally considered as formally coinciding with the corresponding definition pertaining to the gas of particles of kind k in thermodynamic equilibrium.

The equations of mass, momentum and energy transfer are written in the 'divergent' form. This essentially states the conservation laws and turns out to be convenient in numerical work, to constract the conservative schemes for computations. Sometimes, other forms are more convenient. For instance, the equation of momentum transfer or simply the equation of motion (9.17) can be brought into the frequently used form (with the aid of the continuity Equation (9.14) to remove the derivative $\partial \rho_k / \partial t$):

$$\rho_k \left(\frac{\partial\, u_{k,\alpha}}{\partial t} + u_{k,\beta}\, \frac{\partial\, u_{k,\alpha}}{\partial r_\beta} \right) = - \frac{\partial}{\partial r_\beta}\, p_{\alpha\beta}^{(k)} +$$

$$+ \langle\, F_{k,\alpha}\, (\mathbf{r}, t)\, \rangle_v + \langle\, F_{k,\alpha}^{(c)}\, (\mathbf{r}, t)\, \rangle_v \,. \tag{9.40}$$

The so-called *substantial derivative* appears on the left-hand side of this equation:

$$\boxed{\frac{d^{(k)}}{dt} = \frac{\partial}{\partial t} + u_{k,\beta}\, \frac{\partial}{\partial r_\beta} = \frac{\partial}{\partial t} + \mathbf{u}_k \cdot \nabla_{\mathbf{r}} \,.} \tag{9.41}$$

This substantial or *advective* derivative – the total time derivative following a *fluid element* of kind k – is typical of hydrodynamic-type equations, to which the equation of motion (9.40) belongs. The total time derivative with respect to the mean velocity $\mathbf{u}_k$ of the particles of kind k is different for each kind k. In Chapter 12 on the one-fluid MHD theory, we shall introduce the substantial derivative with respect to the average velocity of the plasma as a whole.

For the case of the Lorentz force (9.19), the equation of motion of the particles of kind k can be rewritten as follows:

$$\rho_k\, \frac{d^{(k)}\, u_{k,\alpha}}{dt} = - \frac{\partial}{\partial r_\beta}\, p_{\alpha\beta}^{(k)} + \rho_k^{\mathrm{q}}\, E_\alpha + \frac{1}{c}\, (\, \mathbf{j}_k^{\mathrm{q}} \times \mathbf{B}\,)_\alpha +$$

$$+ \langle F^{(c)}_{k,\alpha}(\mathbf{r},t)\rangle_v , \tag{9.42}$$

where the last term is the mean collisional force (9.20) or, more specifically, (9.24).

9.4.2 Status of conservation laws

As we saw in Section 9.3, when we treat a plasma as several continuous media (the mutually penetrating charged gases), for each of them,

> the main three average properties (density, velocity, and a quantity like temperature or pressure) are governed by the **basic conservation laws** for mass, momentum, and energy in the media.

These conservation equations are useful, of course, except they contain more unknowns than the number of equations. The transfer equations for local macroscopic quantities are as much unclosed as the initial kinetic Equation (9.1) which is the first equation of the chain for correlation functions (see KE in Figure 9.2). For example, formula (9.23) for the mean collisional force contains the unknown binary correlation function f_{kl}. The last should be found from the correlation function Equation (2.46) indicated as the second link BC in Figure 9.2. The terms (9.38) and (9.39) in the energy conservation Equation (9.25) also depend on the unknown binary correlation function f_{kl}.

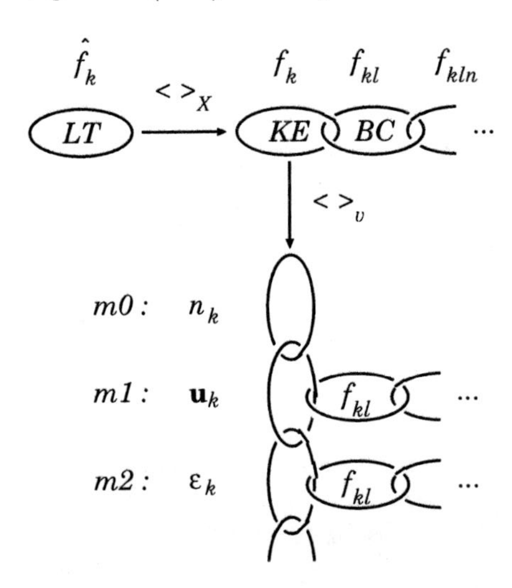

Figure 9.2: LT is the Liouville theorem for an exact distribution function. KE and BC are the kinetic equation and the equation for the binary correlation function. $m0$, $m1$ etc. are the chain of the equation for the moments of the distribution function f_k.

It is also important that the transfer equations are unclosed in 'orthogonal' direction: the Equation (9.14) for the zeroth moment (see $m0$ in Figure 9.2), density n_k, depends on the unknown first moment, the mean velocity $\mathbf{u}_k$, and so on. This process of generating equations for the higher moments could be extended indefinitely depending solely on how many primary variables (n_k, $\mathbf{u}_k$, ε_k, ...) one is prepared to introduce. However, if at any level the distribution function is known, or can be approximated to, in terms of the primary variables for which the equations have already been generated, then this set of equation should be closed. We will come back to this critical point in the next Section.

> Three basic conservation laws for mass, momentum, and energy in the components of astrophysical plasma represent the main transfer equations that are the first three links in the chain of the equations for the distribution function moments.

It certainly would not be possible to arrive to this fundamental conclusion and would be difficult to derive the conservation laws in the form of the transfer Equations (9.14), (9.17), and (9.25) in the way which is typical for the majority of textbooks: from simple specific knowledges to more general ones. Such generalization means that we could go from well-known things to more complicated ones, for example, from the Newton equation of motion of a particle to the ordinary hydrodynamic equation of fluid motion. Though this way makes a text easier to read, it does not give the reader complete knowledge of a subject. That is why we selected the opposite way: from general to specific knowledges.

The consecutive consideration of physical principles, starting from the most general ones, and of simplifying assumptions, which give us a simpler description of plasma under astrophysical conditions, allows us to find the answers to **two key questions**:

(1) what approximation is the best one (the simplest but sufficient) for description of a phenomenon in astrophysical plasma;

(2) how to build an adequate model for the phenomenon, for example, a solar flare or a flare in the corona of an accretion disk.

From a mathematical point of view, an elegant treatment of particle transfer in plasma can be based on the use of non-canonical conjugate variables (for example, $\mathbf{r}$ and $\mathbf{p}$ are not canonically conjugate for a system of particles moving under the Lorentz force) and the associated Lie algebra (see Balescu, 1988).

9.5 Equation of state and transfer coefficients

The transfer equations for a plasma component k would be closed with respect to the three unknown terms ρ_k, $\mathbf{u}_k$, and ε_k, if it were possible to express the other *unknown* quantities p_k, $\pi_{\alpha\beta}^{(k)}$, $q_{\alpha}^{(k)}$, etc. in terms of these three variables, or the variables ρ_k, $\mathbf{u}_k$ and the formally defined temperature T_k. For this purpose, we have to know the *equation of state* and the so-called *transfer coefficients*. How can we find them?

Formally, we should write equations for higher (than second) moments of the distribution function. However these equations will not be closed either. How shall we proceed?

According to the general principles of statistical physics,

> any distribution function tends, by virtue of collisions, to assume the Maxwellian form.

In this case the equation of state is that of the ideal gas.

The Maxwellian distribution is the kinetic equation solution for a stationary homogeneous plasma in the absence of any mean force in the thermal equilibrium state, i.e. for a plasma in **thermodynamic equilibrium**. Then spatial gradients and derivatives with respect to time are zero. In fact they are always nonzero. For this reason, the assumption of full thermodynamic equilibrium is replaced with the *local* thermodynamic equilibrium (LTE). Morover

> if the gradients and derivatives are *small*, then the real distribution function differs *little* from the local Maxwellian one, the difference being *proportional* to the small gradients or derivatives.

Thus if we are interested in the processes occurring in a time t, which is much greater than the characteristic collision time τ, and at a distance L, which is much larger than the particle mean free path λ,

$$t \gg \tau \,, \qquad L \gg \lambda \,, \tag{9.43}$$

then the particle distribution function $f_k(\mathbf{r}, \mathbf{v}, t)$ can be thought of as a sum of the local Maxwellian distribution

$$f_k^{(0)}(\mathbf{r}, \mathbf{v}, t) = n_k(\mathbf{r}, t) \left[\frac{m_k}{2\pi\, k_{\rm B} T_k(\mathbf{r}, t)} \right]^{3/2} \times$$

$$\times \exp \left\{ - \frac{m_k \,|\, \mathbf{v} - \mathbf{u}_k(\mathbf{r}, t) \,|^2}{2\, k_{\rm B} T_k(\mathbf{r}, t)} \right\} \tag{9.44}$$

and some **small additional term** $f_k^{(1)}(\mathbf{r}, \mathbf{v}, t)$. Therefore

$$f_k(\mathbf{r}, \mathbf{v}, t) = f_k^{(0)}(\mathbf{r}, \mathbf{v}, t) + f_k^{(1)}(\mathbf{r}, \mathbf{v}, t) \,, \qquad \left| f_k^{(1)} \right| < f_k^{(0)} . \tag{9.45}$$

According to (9.44), the function $f_k^{(0)}$ depends on t and $\mathbf{r}$ through $n_k(\mathbf{r}, t)$, $T_k(\mathbf{r}, t)$ and $\mathbf{u}_k(\mathbf{r}, t)$. Therefore we have derivatives $\partial f_k^{(0)} / \partial t$ and $\partial f_k^{(0)} / \partial r_\alpha$.

Now we substitute (9.44) in the kinetic Equation (9.1) and linearly approximate the collisional integral (9.3) by using one or another of the models introduced in Chapter 3; alternatively, see Exercise 9.5 as a specific example. Then we seek the additional term $f_k^{(1)}$ in the *linear* approximation with respect to the factors disturbing the Maxwellian distribution, such as gradients of physical parameters, electric fields etc. The quantities $q_\alpha^{(k)}$, $\pi_{\alpha\beta}^{(k)}$ etc., which in their turn are proportional to the same factors, can be expressed in terms of $f_k^{(1)}$. The proportionality coefficients are the sought-after *transfer coefficients*.

For example, in the case of the heat flux q_α, both the additional term $f_k^{(1)}$ and the flux q_α are chosen to be proportional to the temperature gradient. Thus, in a fully ionized plasma in the limit of a vanishing magnetic field, we find the heat flux in the electron component of plasma:

$$q_e = -\kappa_e \nabla T_e \,, \tag{9.46}$$

where

$$\kappa_e \approx \frac{1.84 \times 10^{-5}}{\ln\Lambda} \, T_e^{5/2} \tag{9.47}$$

is the coefficient of electron *thermal conductivity* (Spitzer, 1962).

In the presence of strong magnetic field, all the transport coefficients become highly anisotropic. Since the Maxwellian function (9.44) and its derivatives are uniquely determined by the parameters n_k, $\mathbf{u}_k$, and T_k, the transfer coefficients are expressed in terms of the same quantities and magnetic field B, of course.

This procedure makes it possible to close the set of transfer equations for astrophysical plasma

under the conditions (9.43). The first step is to calculate the departure $f_k^{(1)}$ from the Maxwellian distribution function by using some method of handling collisions. Several models have been suggested on different grounds to account for collisions in plasma (Shkarofsky et al., 1966; Krall and Trivelpiece, 1973).

The first three moment equations have been extensively used in astrophysics, for example, in the investigations of the solar wind. They have led to a significant understanding of phenomena such as escape, acceleration, and cooling. However, as more detailed solar wind observations become available, it appeared that the simplified, collisionally dominated models are not adequate for most of the interplanetary range and for most of the times, i.e. most physical states of the solar wind.

A higher order, closed set of equations for the six moments have been derived for multi-fluid, *moderately non-Maxwellian plasma* of the solar wind (Cuperman and Dryer, 1985). On the basis of these equations, for example, the generalized expression for heat flux relates the flux to the temperature gradients, relative streaming velocity, thermal anisotropies, temperature differences of the components.

Recommended Reading: Braginskii (1965), Hollweg (1986).

9.6 Gravitational systems

There is a big difference between astrophysical plasmas and astrophysical gravitational systems (Section 3.3). The gravitational attraction cannot be screened. A large-scale gravitational field always exists over a system. This follows from the formula (3.17) which shows that the averaged gravitational force cannot be equal to zero because the neutrality condition (3.18) cannot be satisfied if all the particles have the same charge sign.

The large-scale gravitational field makes an overall thermodynamic equilibrium impossible. On the contrary, the electric force in a plasma is screened beyond the Debye radius and does not come in the way of the plasma having a proper thermodynamic equilibrium. Therefore, as one might have anticipated,

> those results of plasma astrophysics which explicitly depend upon the plasma being in thermodynamic equilibrium do not hold for gravitational systems.

For gravitational systems, like the stars in a galaxy, we may hope that the final distribution function reflects something about the initial conditions rather than just reflecting the relaxation mechanism. The random motions of the stars may be not only non-Maxwellian but even direction dependent within the system. So galaxies may be providing us with clues on how they were formed (Palmer, 1994; Bertin, 1999; Peacock, 1999).

If we assume that the stars form a collisionless system (see, however, Section 3.3), they do not exert pressure. Such a pressureless gravitating system is unstable (Jean's instability). Presumably a real galaxy should possess something akin to pressure to withstand the collapsing action of its gravity. This 'pressure' is associated with the random motion of stars. So the role of sound speed is assumed to be played by the root mean speed of the stars.

Another justification for treating a galaxy in the hydrodynamic approximation is that we consider processes on a spatial scale which is large enough to contain a large number of stars – one of the two requirements of the continuum mechanics. Anyway, several aspects of the structure of a galaxy can be understood by assuming that it is made up of a continuum medium. More often than not,

> hydrodynamics provides a first level description of an astrophysical phenomenon governed predominantly by the gravitational force.

Magnetic fields are usually included later on in order to address additional issues. For example, the early stages of star formation during which an interstellar cloud of low density collapses under the action of its own gravity can be modeled in the hydrodynamic approximation. However, when we want to explain the difference between the angular momentum of the cloud and that of the born star, we have to include the effect of a magnetic field.

9.7 Practice: Exercises and Answers

Exercise 9.1 [Section 9.3] Show that the third integral in Equation (9.12) equals zero.

Answer. Let us find the derivative

$$\frac{\partial}{\partial v_\alpha}\left(\frac{F_{k,\alpha}}{m_k}\,f_k\right)=\frac{F_{k,\alpha}}{m_k}\,\frac{\partial f_k}{\partial v_\alpha}+\frac{f_k}{m_k}\,\frac{\partial F_{k,\alpha}}{\partial v_\alpha}=\frac{F_{k,\alpha}}{m_k}\,\frac{\partial f_k}{\partial v_\alpha}\,.$$

The condition (1.7) has been used on the right-hand side as the condition

$$\frac{\partial F_{k,\alpha}}{\partial v_\alpha}=0\,. \tag{9.48}$$

Hence

$$\int\limits_v \frac{F_{k,\alpha}}{m_k}\,\frac{\partial f_k}{\partial v_\alpha}\,d^3\mathbf{v}=\frac{F_k}{m_k}\,f_k(\mathbf{r},\mathbf{v},t)\Big|_{\mathbf{v}\to-\infty}^{\mathbf{v}\to+\infty}=0\,,$$

if the distribution function f_k quickly approaches zero as $v\to\infty$; q.e.d.

Exercise 9.2 [Section 9.3] Write the continuity equation with account of ionization and recombination.

Answer. The continuity equation including the source/sink terms related to ionization/recombination or charge exchange reads

$$\frac{\partial n_k}{\partial t}+\frac{\partial}{\partial r_\alpha}\,n_k\,u_{k,\alpha}=\sum_l\,(\,\gamma_{lk}\,n_l-\gamma_{kl}\,n_k\,)\,. \tag{9.49}$$

Here n_k denotes the particle density of species k, either neutral or ionized. The right-hand side of the equation is the change of n_k due to collisions. The coefficients γ_{kl} and γ_{lk} denote the rate of transformation of species k into species l and vice versa. These rates must obey the relation

$$\sum_k\sum_l\,(\,\gamma_{lk}\,n_l-\gamma_{kl}\,n_k\,)=0\,, \tag{9.50}$$

which ensures the total particle number density conservation.

Exercise 9.3 [Section 9.3] Consider the third integral in the first moment Equation (9.16).

Answer. Let us find the derivative

$$\frac{\partial}{\partial v_\beta}\,(v_\alpha F_{k,\beta}\,f_k)=v_\alpha F_{k,\beta}\,\frac{\partial f_k}{\partial v_\beta}+v_\alpha\,\frac{\partial F_{k,\beta}}{\partial v_\beta}\,f_k+F_{k,\beta}\,f_k\,\frac{\partial v_\alpha}{\partial v_\beta}=$$

$$=v_\alpha F_{k,\beta}\,\frac{\partial f_k}{\partial v_\beta}+0+F_{k,\beta}\,f_k\,\delta_{\alpha\beta}\,. \tag{9.51}$$

The condition (1.7) has been used on the right-hand side as the condition

$$\frac{\partial F_{k,\beta}}{\partial v_\beta}=0\,. \tag{9.52}$$

It follows from (9.51) that

$$v_\alpha F_{k,\beta} \frac{\partial f_k}{\partial v_\beta} = \frac{\partial}{\partial v_\beta} (v_\alpha F_{k,\beta} f_k) - F_{k,\alpha} f_k .$$

Thus

$$\int_{\mathbf{v}} v_\alpha F_{k,\beta} \frac{\partial f_k}{\partial v_\beta} d^3\mathbf{v} = v_\alpha F_k f_k \Bigg|_{\mathbf{v}\to-\infty}^{\mathbf{v}\to+\infty} - \int_{\mathbf{v}} F_{k,\alpha} f_k d^3\mathbf{v} . \tag{9.53}$$

The first term on the right-hand side equals zero, if the distribution function f_k quickly approaches zero as $v \to \infty$. Therefore, for the mean force acting on the particles of kind k in a unit volume, formula (9.18) has finally arrived.

Exercise 9.4 [Section 9.3] Find a condition under which the mean collisional force (9.20) is determined only by random motions of the particles of kind k.

Answer. In definition (9.20), let us take into account that

$$v_\alpha = u_{k,\alpha} + v'_\alpha .$$

Thus we obtain

$$\langle F^{(c)}_{k,\alpha} (\mathbf{r}, t) \rangle_v = m_k u_{k,\alpha} \int_{\mathbf{v}} \left(\frac{\partial \hat{f}_k}{\partial t} \right)_c d^3\mathbf{v} + m_k \int_{\mathbf{v}} v'_\alpha \left(\frac{\partial \hat{f}_k}{\partial t} \right)_c d^3\mathbf{v} . \tag{9.54}$$

The first integral on the right-hand side equals zero if condition (9.13) is satisfied. The remaining part

$$\langle F^{(c)}_{k,\alpha} (\mathbf{r}, t) \rangle_v = m_k \int_{\mathbf{v}} v'_\alpha \left(\frac{\partial \hat{f}_k}{\partial t} \right)_c d^3\mathbf{v} . \tag{9.55}$$

Thus the average transfer of momentum from the particles of kind k to the particles of other kinds is solely due to the random motions of the particles of kind k if the processes of transformation, during which the particle kind can be changed, are not allowed for.

Exercise 9.5 [Section 9.5] Let us approximate the collisional integral (9.3) by the following simple form (Bhatnagar et al., 1954):

$$\left(\frac{\partial \hat{f}_k}{\partial t} \right)_c = - \frac{f_k(\mathbf{r}, \mathbf{v}, t) - f_k^{(0)}(\mathbf{r}, \mathbf{v}, t)}{\tau_c} , \tag{9.56}$$

where an arbitrary distribution function $f_k(\mathbf{r}, \mathbf{v}, t)$ relaxes to the Maxwellian distribution function $f_k^{(0)}(\mathbf{r}, \mathbf{v}, t)$, as discussed in Section 9.5, in a collisional time τ_c. Discuss why this simple approximation illuminates much of the basic physics of transport phenomena in a relatively less-painful way for neutral gases but is not very reliable for plasmas, especially in the presence of magnetic fields.

Comment. The departure of the distribution function from the pure Maxwellian one, the function

$$f_k^{(1)}(\mathbf{r}, \mathbf{v}, t) = f_k(\mathbf{r}, \mathbf{v}, t) - f_k^{(0)}(\mathbf{r}, \mathbf{v}, t) \tag{9.57}$$

satisfies the following equation:

$$\frac{\partial f_k}{\partial t} + v_\alpha \frac{\partial f_k}{\partial r_\alpha} + \frac{F_{k,\alpha}}{m_k} \frac{\partial f_k}{\partial v_\alpha} = -\frac{f_k^{(1)}}{\tau_c}, \tag{9.58}$$

which is called the BGK (Bhatnagar, Gross and Krook) equation.

If a gradient in space, $\partial / \partial r_\alpha$, gives rise to the departure from the Maxwellian distribution, then in order to have a rough estimate of the effect, we may balance the second term on the left-hand side of Equation (9.58) with its right-hand side:

$$\frac{|\, v_\alpha \,|\, f_k^{(0)}}{L} \approx \frac{\left| f_k^{(1)} \right|}{\tau_c} . \tag{9.59}$$

Here $|\, v_\alpha \,|$ is the typical velocity of the particles of kind k, L is the typical length scale over which properties of the system change appreciably. From (9.59) it follows that

$$\frac{\left| f_k^{(1)} \right|}{f_k^{(0)}} \approx \frac{\lambda_c}{L} . \tag{9.60}$$

Thus the departure from the Maxwellian distribution will be small if the mean free path λ_c is small compared to the typical length scale. This is consistent with the second condition of (9.43).

Chapter 10

Multi-Fluid Models of Astrophysical Plasma

The multi-fluid models of plasma in electric and magnetic fields allow us to consider many important properties of astrophysical plasma, in particular the Langmuir and electromagnetic waves, as well as many other interesting applications.

10.1 Multi-fluid models in astrophysics

The transfer Equations (9.14), (9.17), and (9.25) give us the hydrodynamic-type description of multi-component astrophysical plasma in electric and magnetic fields. The problem is that, if we would like to solve the equations for one of the plasma components, we could not escape solving the transfer equations for all of the components since they depend on each other and on the electric and magnetic fields. For this reason, we should minimize the number of plasma components under consideration.

The 'two-fluid' hydrodynamic-type equations are often used to describe the flow of the electrons and protons of a fully-ionized astrophysical plasma under the action of an electric and magnetic fields. Such treatment yields, for example, the generalized Ohm's law in astrophysical plasma (Chapter 11) as well as a dynamical friction force which maximizes when the relative drift velocity is equal to the sum of the most probable random speeds of the electrons and ions. For relative drift velocities in excess of this value, the friction force decreases rapidly. The electron and ion currents flowing parallel to the existing magnetic fields increase steadily in time, i.e. runaway (Dreicer, 1959; see also Section 8.4).

The 'multi-fluid' models are useful, for example, to explore properties of the solar wind (e.g., Bodmer and Bochsler, 2000). The electrons, protons, and alpha particles in the solar wind constitute the main three components, while the less abundant elements and isotopes are treated as test species. To model the main gases, we have to study solutions for the conservation-law equations of the three components. The behaviour of minor ions depends in a complicated manner on their mass and on their charge, structured by the interplay of acceleration, gravity, pressure gradient, electromagnetic fields, Coulomb friction force, and thermal diffusion. Such models allow one to explore the efficiency of isotope fractionation processes in the solar corona.

10.2 Langmuir waves

Because a plasma consists of at least two components (electrons and ions), the number of possible waves is larger than in a normal fluid or gas, where sound or acoustic waves are the only possible waves. In this Section we shall discuss the *simplest* waves in plasma, whose properties can be deduced from the hydrodynamic-type equations for two mutually penetrating charged gases (Section 9.4).

Although astrophysical plasma is almost always magnetized, we can quite often neglect the magnetic field in discussing small-amplitude plasma waves; the condition will become clear later. The reduced complexity of the governing equations can be further simplified by approximations.

10.2.1 Langmuir waves in a cold plasma

Let us assume that the ions do not move at all (they are infinitely massive) and they are uniformly distributed in space. So the ions have a fixed number density n_0. This is a cold ion approximation.

Let us also neglect all magnetic fields. We shall assume that any variations of electron density n_e, electron velocity u_e, and related electric field $\mathbf{E}$ occur only in one dimension – the x axis. Then we are left with a set of three equations:

(a) the continuity equation (9.14) for electrons

$$\frac{\partial n_e}{\partial t} + \frac{\partial}{\partial x} n_e u_e = 0 \,, \tag{10.1}$$

(b) the motion equation (9.40)

$$m_e n_e \left(\frac{\partial u_e}{\partial t} + u_e \frac{\partial u_e}{\partial x} \right) = - \frac{\partial p_e}{\partial x} - e n_e E_x \,, \tag{10.2}$$

(c) the electric field equation

$$\frac{\partial E_x}{\partial x} = 4\pi e \, (n_0 - n_e) \,. \tag{10.3}$$

In general, we cannot solve these nonlinear equations exactly, except for very special cases. One of them is trivial:

$$n_e = n_0 \,, \quad u_e = 0 \,, \quad p_e = \text{const} \,, \quad E_x = 0 \,. \tag{10.4}$$

This solution corresponds to a *stationary* electron gas of uniform density.

Let us linearize Equations (10.1)–(10.3) with respect to the state (10.4). This yields the following set of *linear* equations:

$$\frac{\partial n_1}{\partial t} + n_0 \frac{\partial u_1}{\partial x} = 0 \,, \tag{10.5}$$

$$m_e n_0 \frac{\partial u_1}{\partial t} = - \frac{\partial p_1}{\partial x} - e n_e E_1 \,, \tag{10.6}$$

$$\frac{\partial E_1}{\partial x} = 4 \pi e \, n_1 \,. \tag{10.7}$$

Let us consider the special case of *cold* electrons:

$$p_e = 0 \,. \tag{10.8}$$

Now we eliminate u_1 and E_1 from the set of equation by taking the time derivative of Equation (10.5) to obtain the oscillator equation

$$\frac{\partial^2 n_1}{\partial t^2} + \left(\frac{4 \pi e^2 n_0}{m_e} \right) n_1 = 0 \,. \tag{10.9}$$

If we displace some electrons to produce an initial perturbation, we create a positive-charge density at the position where they started. This positive-charge perturbation attracts the electrons, which will tend to move back to their original position, but will overshoot it. They come back again, overshoot it, and so on. Without any damping, the energy put into the plasma to create the perturbation will remain in the plasma. So the oscillation will continue forever with the frequency

$$\boxed{\omega_{pl}^{(e)} = \pm \left(\frac{4 \pi e^2 n_e}{m_e} \right)^{1/2}} \tag{10.10}$$

called the *electron plasma frequency*.

Therefore, in a two-component cold plasma, there exist the **oscillations of charge density** – *Langmuir waves* which frequency is independent of the wave vector $\mathbf{k}$; so the group velocity, $\mathbf{V}_{gr} = d\omega / d\mathbf{k}$, is zero. Thus

> in a cold plasma, Langmuir waves are spatially localized oscillations of electric charge density which do not propagate at all.

Note that there is no equivalent to these oscillations in gasdynamics or gravitational dynamics, for which there is no charge separation and related electric-type force.

10.2.2 Langmuir waves in a warm plasma

What happens with behaviour of a Langmuir wave, if the electron temperature is not equal to zero? – Let us drop the assumption (10.8) of zero pressure in the linear equations (10.5)–(10.7). We then must include the perturbation of electron pressure

$$\frac{\partial p_1}{\partial x} = n_0 k_{\mathrm{B}} \frac{\partial T_1}{\partial x} + k_{\mathrm{B}} T_0 \frac{\partial n_1}{\partial x} \tag{10.11}$$

in Equation (10.6).

Now we must relate n_1 to T_1 and vice versa. For example, we could argue that for long-wavelength waves the compression is the one-dimensional ($N = 1$) adiabatic process with the index $\gamma = (N+2)/N = 3$. In this case, the perturbation of electron pressure becomes

$$\frac{\partial p_1}{\partial x} = 3 k_{\mathrm{B}} T_0 \frac{\partial n_1}{\partial x} \, . \tag{10.12}$$

Naturally we expect now an initial perturbation to propagate through the plasma as a wave. Thus a plane-wave solution of the form

$$f_1(x,t) = \tilde{f}_1 \exp\left[-\mathrm{i}\,(\omega t - kx)\right] \tag{10.13}$$

should satisfy the linear differential equations. The quantities with tildes are the complex amplitudes. They obey three linear algebraic equations:

$$-\mathrm{i}\,\omega\,\tilde{n}_1 + \mathrm{i}\,k\,n_0\,\tilde{u}_1 = 0\,,$$

$$-\mathrm{i}\,\omega\, m_e n_0\,\tilde{u}_1 + \mathrm{i}\,k\,3k_{\mathrm{B}} T_0\,\tilde{n}_1 + e n_0\,\tilde{E}_1 = 0\,,$$

$$\mathrm{i}\,k\,\tilde{E}_1 + 4\pi e\,\tilde{n}_1 = 0\,.$$

To have a nontrivial solution, the determinant must be zero. Its solution is

$$\boxed{\omega = \pm\,\omega_{pl}^{(e)} \left(1 + 3\, r_{\mathrm{D}}^2\, k^2\right)^{1/2},} \tag{10.14}$$

where

$$r_{\mathrm{D}} = \frac{1}{\sqrt{3}}\,\frac{V_{\mathrm{Te}}}{\omega_{pl}^{(e)}}\,, \tag{10.15}$$

is the electron Debye radius; V_{Te} is the mean thermal velocity (8.15) of electrons in a plasma.

The dispersion equation (10.14) can be also derived from the Vlasov equation, of course (see formula (49) in Vlasov, 1938). It is similar to the well-known relation for the propagation of transverse electromagnetic waves in a vacuum, except that the role of the light velocity c is here played by the thermal velocity V_{Te}. This dispersion relation is shown in Figure 10.1.

Therefore the frequency ω of Langmuir waves in a plasma with warm electrons depends on the wave vector $\mathbf{k}$ which is parallel to the x-axis. So

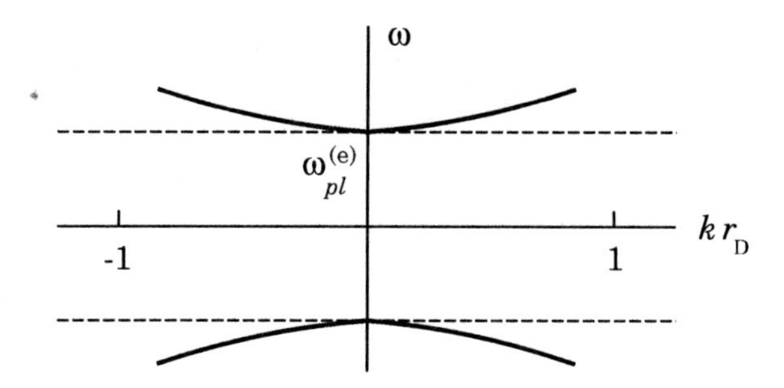

Figure 10.1: A dispersion diagram (solid curves) for Langmuir waves in a warm plasma. The ions do not move at all. Dashed straight lines are drawn for Langmuir waves in a cold plasma.

the group velocity, $\partial \omega / \partial k$, of Langmuir waves in a warm plasma without magnetic field is not equal to zero.

They oscillate at the electron plasma frequency $\omega_{pl}^{(e)}$ and propagate in a warm plasma. It follows from (10.14) and (10.15) that the group velocity is

$$V_{\text{gr}} = \frac{\partial \omega}{\partial k} = V_{\text{Te}}^2 \frac{k}{\omega} = \frac{3k_{\text{B}} T}{m_{\text{e}}} \frac{k}{\omega} . \tag{10.16}$$

Therefore the plasma waves are propagating as long as the electron temperature is non-zero. Moreover, due to the small mass of the electrons, the group velocity (10.16) is always relatively large.

10.2.3 Ion effects in Langmuir waves

Let us show that, when the ions are allowed to move, ion contributions are important only for slow variations or *low-frequency* waves because the ions cannot react quickly enough.

We are still dealing with linear waves which involve only the first-order electric field $\mathbf{E}^{(1)}$ directed along the wave vector $\mathbf{k}$ which is parallel to the x-axis. Linearizing the continuity equations for electrons and ions, the motion equations for electrons and ions, as well as the electric field equation, let us assume that the electrons and ions both obey the adiabatic Equation (10.12). Then we again use the wave solution (10.13) to reduce the linearized differential equations to algebraic ones and to obtain the determinant. Because $m_{\text{i}}/m_{\text{e}} \gg 1$, we neglect the term $m_{\text{e}}\,\omega^2$ in this determinant as compared with the term $m_{\text{i}}\,\omega^2$. By so doing, we obtain the relation

$$\omega^2 = k^2 \left(\frac{\gamma_{\text{i}}\, k_{\text{B}} T_{\text{i}}}{m_{\text{i}}} + \frac{\gamma_{\text{e}}\, k_{\text{B}} T_{\text{e}}}{m_{\text{i}}} \, \frac{1}{1 + \gamma_{\text{e}}\, k^2 r_{\text{D}}^2} \right) . \tag{10.17}$$

This dispersion relation is shown in Figure 10.2.

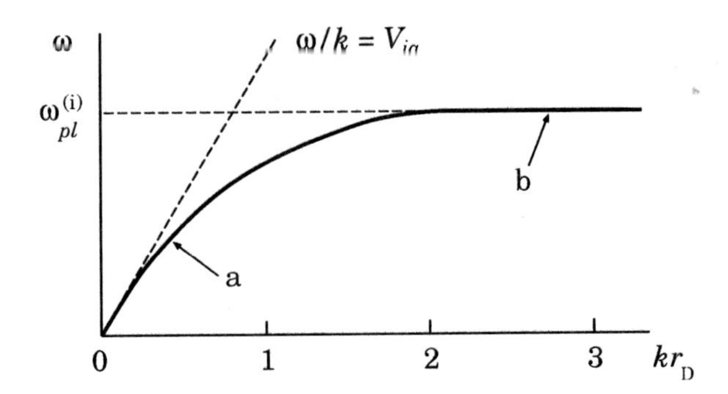

Figure 10.2: A dispersion diagram for ion-acoustic waves (part *a*) and for ion plasma waves (part *b*) in a warm plasma without magnetic field.

In the limit of small $kr_{\rm D}$

$$\omega^2 = k^2 \left(\frac{\gamma_{\rm i}\, k_{\rm B} T_{\rm i}}{m_{\rm i}} + \frac{\gamma_{\rm e}\, k_{\rm B} T_{\rm e}}{m_{\rm i}} \right) = k^2 V_{ia}^2 \,. \tag{10.18}$$

This is the so-called *ion-acoustic waves*. They are shown by a curve part (*a*) in Figure 10.2. The group velocity of the wave is independent of k:

$$V_{gr} = \frac{\partial\, \omega}{\partial\, k} = V_{ia} = \left(\frac{\gamma_{\rm i}\, k_{\rm B} T_{\rm i} + \gamma_{\rm e}\, k_{\rm B} T_{\rm e}}{m_{\rm i}} \right)^{1/2} . \tag{10.19}$$

An opposite limit is obtained for cold ions. If ion temperature $T_{\rm i} \to 0$, then $kr_{\rm D} \gg 1$, i.e., short wavelengths are under consideration. In this case, shown by the curve part (*b*) in Figure 10.2, the cold ions oscillate with a frequency

$$\omega_{pl}^{(\rm i)} = \pm \left(\frac{4\pi e^2 n_{\rm e}}{m_{\rm i}} \right)^{1/2} \tag{10.20}$$

called the *ion plasma frequency*.

Ion-acoustic waves are observed in many cases. They were registred, for example, by the spacecraft *Voyager 1* in the upstream side of the Jovian bow shock. Ion-acoustic waves presumably play an important role in solar flares, for example, in super-hot turbulent-current layers (see vol. 2, Section 6.3).

10.3 Electromagnetic waves in plasma

In this Section we still assume that the unperturbed plasma has no magnetic field: $\mathbf{B}^{(0)} = 0$. However we shall discuss waves that carry not only an electric field $\mathbf{E}^{(1)}$ but also a magnetic field $\mathbf{B}^{(1)}$.

Let us consider transversal waves, so that $\mathbf{k} \cdot \mathbf{E}^{(1)} = 0$ and $\mathbf{k} \cdot \mathbf{B}^{(1)} = 0$. The last equality is imposed by Equation (1.26) and is always true. We do not need Equation (1.27) in this case either. We shall neglect the ion motion, which is justified for high-frequency waves. So the remaining equations in their linearized form are

$$\frac{\partial \mathbf{u}_e^{(1)}}{\partial t} = -\nabla p_e^{(1)} - e n_e^{(0)} \mathbf{E}^{(1)}, \tag{10.21}$$

$$\frac{\partial n_e^{(1)}}{\partial t} + n_e^{(0)} \operatorname{div} \mathbf{u}_e^{(1)} = 0, \tag{10.22}$$

$$\operatorname{curl} \mathbf{B}^{(1)} = \frac{4\pi}{c} \mathbf{j}^{(1)} + \frac{1}{c} \frac{\partial \mathbf{E}^{(1)}}{\partial t}, \tag{10.23}$$

$$\operatorname{curl} \mathbf{E}^{(1)} = -\frac{1}{c} \frac{\partial \mathbf{B}^{(1)}}{\partial t}, \tag{10.24}$$

$$\mathbf{j}^{(1)} = e n_e^{(0)} \mathbf{u}_e^{(1)}. \tag{10.25}$$

The Lorentz force does not appear in the electron-motion Equation (10.21) because it is of the second-order small value proportional to $\mathbf{u}_e^{(1)} \times \mathbf{B}^{(1)}$. Furthermore vectors $\mathbf{E}^{(1)}$ and $\mathbf{u}_e^{(1)}$ are perpendicular to the wave vector $\mathbf{k}$, and thus $n_e^{(1)} = 0$ and $p_e^{(1)} = 0$. After assuming the exponential plane-wave form (10.13) and using usual algebra, we find the dispersion equation for electromagnetic waves:

$$\omega^2 = \omega_{pl}^{(e)\,2} + k^2 c^2 . \tag{10.26}$$

Here c is the speed of light in a vacuum. This dispersion relation is shown in Figure 10.3.

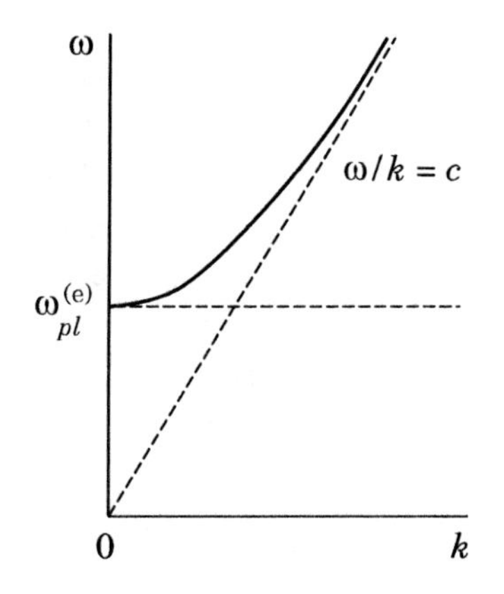

Figure 10.3: The dispersion diagram for electromagnetic waves in a cold plasma without magnetic field. For large values of k (short wavelengths), the group velocity (the slope of the solid curve) and phase velocity approach the speed of the light (dashed line). For small values of k (long wavelengths), the group velocity goes to zero.

If the wave frequency ω is much larger the electron plasma frequency $\omega_{pl}^{(e)}$, the wave becomes a free-space light wave with $\omega = kc$.

If $\omega \to \omega_{pl}^{(e)}$, a wave would decay in space and not propagate. In this case, the index of refraction

$$n_r = \frac{c}{V_{ph}} = \frac{ck}{\omega} = \left(1 - \frac{\omega_{pl}^{(e)\,2}}{\omega^2}\right)^{1/2} \tag{10.27}$$

goes to zero. If $\omega < \omega_{pl}^{(e)}$, the refraction index becomes imaginary.

Moving through astrophysical plasma of changing $\omega_{pl}^{(e)}$, a wave is reflected when $\omega = \omega_{pl}^{(e)}$ and, therefore, $n_r = 0$. This allows one to measure remotely the plasma density, for example, in the Earth ionosphere.

Another application is in ionospheric heating. At the height where $\omega_{pl}^{(e)}$ is equal to the wave frequency, the group velocity

$$V_{gr} = \frac{\partial \omega}{\partial k} = \frac{kc^2}{\omega} = n_r c \tag{10.28}$$

also goes to zero.

The wave amplitude becames large there because its flux of energy cannot propagate.

The large electric field of the wave can accelerate electrons and drive currents in the ionospheric plasma. In this way, the wave can heat and modify the plasma. If the power from a transmitter on the ground emitting a radiation at a frequency ω is large enough, the heating is quite significant.

10.4 What do we miss?

We have considered two basic types of waves in a two-fluid plasma. The Langmuir wave or plasma wave (as well as the ion plasma wave) does not have a wave magnetic field. The electromagnetic wave does have a magnetic field but can propagate only if its frequency is above the plasma frequency. We should see that, when there is a stationary magnetic field in the plasma, the wave properties become more complex and more interesting (e.g., Stix, 1992; Zheleznyakov, 1996).

In particular, we could find that the electromagnetic wave with its frequency below the plasma frequency can propagate through a magnetized plasma. For low-frequency waves this effect will be demonstrated in the magnetohydrodynamic (MHD) approximation in Chapter 15. What else has been lost in the above consideration?

The advantage of the hydrodynamic approach used in this Section to study the basic properties of waves in a two-fluid plasma is the relative simplicity. The hydrodynamic-type equations have three spatial dimensions and time, rather than the seven-dimensional phase space of the Vlasov kinetic theory (Section 3.1.2).

The obvious disadvantage is that some subtle **fine effects**, such as Landau damping (Section 7.1.2) which is caused by a resonance with particles moving at the phase velocity of a wave, cannot be obtained from the hydrodynamic-type equations. We have to use a kinetic treatment to specify how a distribution of particles responds to a wave. In this case we use the Vlasov equation to specify how the distribution functions of electrons and ions are affected by the wave fields (e.g., Chapter 7).

To calculate the collisional damping of plasma waves simply, the simplest hydrodynamic model is useful (Exercise 10.3).

The hydrodynamic-type models work only when a finite number of the low-order moments are sufficient to provide all the essential information about the system.

> If the distribution function has some unusual features, then a few low-order moments may not carry all the necassary information,

and we may lose important physics by restricting ourselves to the quasi-hydrodynamic description of cosmic plasma.

10.5 Practice: Exercises and Answers

Exercise 10.1. Show that in the solar corona a *dynamic viscosity* coefficient can be given by a simple formula (Hollweg, 1986):

$$\eta \approx 10^{-16}\, T_{\rm p}^{5/2}\,, \;\; {\rm g\ cm^{-1}\, s^{-1}}\,, \tag{10.29}$$

where $T_{\rm p}$ is the proton temperature, and the Coulomb logarithm has been taken to be 22. So, with $T_{\rm p} \approx 2 \times 10^6$ K, the viscosity coefficient in the corona

$$\eta \sim 1 \ {\rm g\ cm^{-1}\, s^{-1}}.$$

Why does the viscosity grow with the proton temperature? Why is it so large and does it grow with temperature so quickly?

Hint. Consider a fully-ionized hydrogen plasma in a magnetic field. Let $\tau_{\rm pp}$ represent the typical Coulomb collisional time (8.39) for thermal protons. Let $\omega_{\rm B}^{(\rm p)}$ denote the proton cyclotron frequency (5.52).

Write the viscous stress tensor (9.35) for the protons. This tensor involves five coefficients of viscosity, denoted η_0, η_1, ... η_4 by Braginskii (1965). Show that the coefficient η_0 is by far the largest one (10.29). The coefficients η_3 and η_4 are smaller by factors $\sim \left(\omega_{\rm B}^{(\rm p)} \tau_{\rm pp}\right)^{-1}$, while η_1 and η_2 are smaller than η_0 by factors $\sim \left(\omega_{\rm B}^{(\rm p)} \tau_{\rm pp}\right)^{-2}$. Thus the parts of the viscous stress tensor involving the off-diagonal terms can often be neglected. The part involving η_0 can be dynamically and thermodynamically important.

Exercise 10.2. Discuss a famous puzzle of plasma astrophysics. Solar flares generate electron beams that move through the solar corona and the interplanetary space at velocities $\sim 0.3\ c$ (see Exercise 5.3). These fast beams should lose their energy quickly to plasma waves. In fact, they do generate waves called solar type III radio bursts. However the solar fast electrons are still seen far beyond the orbit of the Earth. Why?

Hint. The link between the electron beams and the waves observed in space near the Earth or even on the ground is a little more complex than it seems. It involves the transformation of the electrostatic plasma oscillations with frequency near $\omega_{pl}^{(e)}$ into electromagnetic waves at the same frequency. In any realistic situation, the electrons in the beam are not cold but have a thermal spread. They cause a plasma wave to grow. But as the electric field in the wave grows, the electrons are heated.

The spreading and slowing of the beam in the velocity space cannot be described by fluid equations. This process is often referred to as *quasi-linear diffusion*. We can expect that the electron beam has slowed and spread in the velocity space to such a degree that waves do not grow anymore. A stable situation can occur, and a warm electron beam can propagate through the plasma without lossing energy.

Exercise 10.3. Show that Coulomb collisions damp the Langmuir plasma waves with the rate

$$\text{Im}\,\omega = -2\,\nu_{\text{ei}}\,. \tag{10.30}$$

Hint. Following formula (9.24), add to the right-hand side of the electron motion Equation (10.2) the collisional friction term

$$+\,m_{\text{e}}n_{\text{e}}\,\nu_{\text{ei}}\,(u_{\text{i},\alpha} - u_{\text{e},\alpha})\,.$$

Chapter 11

The Generalized Ohm's Law in Plasma

The multi-fluid models of the astrophysical plasma in magnetic field allow us to derive the generalized Ohm's law and to consider important physical approximations as well as many interesting applications.

11.1 The classic Ohm's law

The classic Ohm's law, $\mathbf{j} = \sigma \mathbf{E}$, relates the current $\mathbf{j}$ to the electric field $\mathbf{E}$ in a conductor in rest. The coefficient σ is electric conductivity. As we know, the electric field in plasma determines the electron and ion acceleration, rather than their velocity. That is why, generally, such a simple relation as the classic Ohm's law does not exist. Moreover, while considering astrophysical plasmas, it is necessary to take into account the presence of a magnetic field and the motion of a plasma as a whole, and as a medium consisting of several moving components.

Recall the way of deriving the usual classic Ohm's law in plasma without magnetic field. The electric current is determined by the relative motion of electrons and ions. Considering the processes in which all quantities vary only slightly in a time between the electron-ion collisions, electron inertia can be neglected. An equilibrium is set up between the electric field action and electrons-on-ions friction (see point A in Figure 8.7). Let us assume that the ions do not move. Then the condition for this equilibrium with respect to the electron gas

$$0 = -e n_{\mathrm{e}} E_{\alpha} + m_{\mathrm{e}} n_{\mathrm{e}} \, \nu_{\mathrm{ei}} \, (0 - u_{\mathrm{e},\alpha})$$

results in Ohm's law

$$j_{\alpha} = -e n_{\mathrm{e}} u_{\mathrm{e},\alpha} = \frac{e^2 n_{\mathrm{e}}}{m_{\mathrm{e}} \nu_{\mathrm{ei}}} \, E_{\alpha} = \sigma E_{\alpha} \, , \tag{11.1}$$

where

$$\sigma = \frac{e^2 n_e}{m_e \nu_{ei}} \tag{11.2}$$

is the *electric conductivity*.

In order to deduce the generalized Ohm's law for the plasma with magnetic field, we have to consider at least two equations of motion – for the electron and ion components. A crude theory of conductivity in a fully-ionized plasma can be given in terms of a two-fluid approximation. The more general case, with the motion of neutrals taken into account, has been considered by Schlüter (1951), Alfvén and Fälthammar (1963); see also different applications of the generalized Ohm's law in the *three-component* astrophysical plasma (Schabansky, 1971; Kunkel, 1984; Hénoux and Somov, 1991 and 1997; Murata, 1991).

11.2 Derivation of basic equations

Let us write the momentum-transfer Equations (9.17) for the electrons and ions, taking proper account of the Lorentz force (9.19) and the friction force (9.24). We have two following equations:

$$m_e \frac{\partial}{\partial t} (n_e u_{e,\alpha}) = - \frac{\partial \Pi^{(e)}_{\alpha\beta}}{\partial r_\beta} - e n_e \left[\mathbf{E} + \frac{1}{c} (\mathbf{u}_e \times \mathbf{B}) \right]_\alpha +$$

$$+ m_e n_e \nu_{ei} (u_{i,\alpha} - u_{e,\alpha}) , \tag{11.3}$$

$$m_i \frac{\partial}{\partial t} (n_i u_{i,\alpha}) = - \frac{\partial \Pi^{(i)}_{\alpha\beta}}{\partial r_\beta} + Z_i e n_i \left[\mathbf{E} + \frac{1}{c} (\mathbf{u}_i \times \mathbf{B}) \right]_\alpha +$$

$$+ m_e n_i \nu_{ei} (u_{e,\alpha} - u_{i,\alpha}) . \tag{11.4}$$

The last term in (11.3) represents the mean momentum transferred, because of collisions (formula (9.24)), from ions to electrons. It is equal, with opposite sign, to the last term in (11.4). It is assumed that there are just two kinds of particles, their total momentum remaining constant under the action of elastic collisions.

Suppose that the ions are protons ($Z_i = 1$) and electrical *neutrality* is observed:

$$n_i = n_e = n .$$

Let us multiply Equation (11.3) by $-e/m_e$ and add it to Equation (11.4) multiplied by e/m_i. The result is

$$\frac{\partial}{\partial t} [en (u_{i,\alpha} - u_{e,\alpha})] = \left[\frac{e}{m_i} F_{i,\alpha} - \frac{e}{m_e} F_{e,\alpha} \right] +$$

$$+ e^2 n \left(\frac{1}{m_e} + \frac{1}{m_i} \right) E_\alpha + \frac{e^2 n}{c} \left[\left(\frac{\mathbf{u}_e}{m_e} \times \mathbf{B} \right)_\alpha + \left(\frac{\mathbf{u}_i}{m_i} \times \mathbf{B} \right)_\alpha \right] -$$

$$- \nu_{\rm ei}\, en \left[(u_{{\rm i},\alpha} - u_{{\rm e},\alpha}) + \frac{m_{\rm e}}{m_{\rm i}} (u_{{\rm i},\alpha} - u_{{\rm e},\alpha}) \right] . \tag{11.5}$$

Here

$$F_{{\rm e},\alpha} = - \frac{\partial \Pi^{({\rm e})}_{\alpha\beta}}{\partial r_\beta} \quad \text{and} \quad F_{{\rm i},\alpha} = \frac{\partial \Pi^{({\rm i})}_{\alpha\beta}}{\partial r_\beta} . \tag{11.6}$$

Let us introduce the velocity of the centre-of-mass system

$$\mathbf{u} = \frac{m_{\rm i}\, \mathbf{u}_{\rm i} + m_{\rm e}\, \mathbf{u}_{\rm e}}{m_{\rm i} + m_{\rm e}} . \tag{11.7}$$

Since $m_{\rm i} \gg m_{\rm e}$,

$$\mathbf{u} = \mathbf{u}_{\rm i} + \frac{m_{\rm e}}{m_{\rm i}}\, \mathbf{u}_{\rm e} \approx \mathbf{u}_{\rm i} . \tag{11.8}$$

On treating Equation (11.5), we neglect the small terms of the order of the ratio $m_{\rm e}/m_{\rm i}$. On rearrangement, we obtain the equation for the current

$$\mathbf{j} = en\, (\mathbf{u}_{\rm i} - \mathbf{u}_{\rm e}) \tag{11.9}$$

in the system of coordinates (11.8). This equation is

$$\frac{\partial \mathbf{j}\,'}{\partial t} = \frac{e^2 n}{m_{\rm e}} \left[\mathbf{E} + \frac{1}{c} \left(\mathbf{u} \times \mathbf{B} \right) \right] - \frac{e}{m_{\rm e} c} \left(\mathbf{j}\,' \times \mathbf{B} \right) -$$

$$- \nu_{\rm ei}\, \mathbf{j}\,' + \frac{e}{m_{\rm i}}\, \mathbf{F}_{\rm i} - \frac{e}{m_{\rm e}}\, \mathbf{F}_{\rm e} . \tag{11.10}$$

The prime designates the electric current in the system of moving plasma, i.e. in the rest-frame of the plasma. Let $\mathbf{E}_{\rm u}$ denote the electric field in this frame of reference, i.e.

$$\mathbf{E}_{\rm u} = \mathbf{E} + \frac{1}{c}\, \mathbf{u} \times \mathbf{B} . \tag{11.11}$$

Now we divide Equation (11.10) by $\nu_{\rm ei}$ and represent it in the form

$$\mathbf{j}\,' = \frac{e^2 n}{m_{\rm e} \nu_{\rm ei}}\, \mathbf{E}_{\rm u} - \frac{\omega^{({\rm e})}_{\rm B}}{\nu_{\rm ei}}\, \mathbf{j}\,' \times \mathbf{n} - \frac{1}{\nu_{\rm ei}} \frac{\partial \mathbf{j}\,'}{\partial t} + \frac{1}{\nu_{\rm ei}} \left(\frac{e}{m_{\rm i}}\, \mathbf{F}_{\rm i} - \frac{e}{m_{\rm e}}\, \mathbf{F}_{\rm e} \right) , \tag{11.12}$$

where $\mathbf{n} = \mathbf{B}/B$ and $\omega^{({\rm e})}_{\rm B} = eB/mc$ is the electron gyro-frequency.

Thus we have derived a differential equation for the current $\mathbf{j}\,'$.

The third and the fourth terms on the right do not depend of magnetic field. Let us replace them by some *effective* electric field such that

$$\sigma \mathbf{E}_{\rm eff} = - \frac{1}{\nu_{\rm ei}} \frac{\partial \mathbf{j}\,'}{\partial t} + \frac{e}{\nu_{\rm ei}} \left(\frac{1}{m_{\rm i}}\, \mathbf{F}_{\rm i} - \frac{1}{m_{\rm e}}\, \mathbf{F}_{\rm e} \right) , \tag{11.13}$$

where

$$\boxed{\sigma = \frac{e^2 n}{m_{\rm e}\, \nu_{\rm ei}}} \tag{11.14}$$

is the *plasma conductivity* in the absence of magnetic field. Combine the fields (11.11) and (11.13),

$$\mathbf{E}' = \mathbf{E}_{\mathrm{u}} + \mathbf{E}_{\mathrm{eff}} \,,$$

in order to rewrite (11.12) in the form

$$\mathbf{j}' = \sigma \mathbf{E}' - \frac{\omega_{\mathrm{B}}^{(e)}}{\nu_{\mathrm{ei}}} \, \mathbf{j}' \times \mathbf{n} \,. \tag{11.15}$$

We will consider (11.15) as an *algebraic* equation in $\mathbf{j}'$, neglecting the $\partial \mathbf{j}'/\partial t$ dependence of the field (11.13). Note, however, that

> the term $\partial \mathbf{j}'/\partial t$ is by no means small in the problem of the particle acceleration by a strong electric field in astrophysical plasma.

Collisionless reconnection is an example in which **particle inertia** (usually combined with anomalous resistivity, see vol. 2, Section 6.3) of the current replaces classical resistivity in allowing fast reconnection to occur (e.g., Drake and Kleva, 1991; Horiuchi and Sato, 1994).

11.3 The general solution

Let us find the solution to (11.15) as a sum of three currents

$$\boxed{\mathbf{j}' = \sigma_{\parallel} \, \mathbf{E}'_{\parallel} + \sigma_{\perp} \, \mathbf{E}'_{\perp} + \sigma_{\mathrm{H}} \, \mathbf{n} \times \mathbf{E}'_{\perp} \,.} \tag{11.16}$$

Substituting formula (11.16) in Equation (11.15) gives

$$\sigma_{\parallel} = \sigma = \frac{e^2 n}{m_e \nu_{\mathrm{ei}}} \,, \tag{11.17}$$

$$\sigma_{\perp} = \sigma \, \frac{1}{1 + \left(\omega_{\mathrm{B}}^{(e)} \tau_{\mathrm{ei}} \right)^2} \,, \qquad \tau_{\mathrm{ei}} = \frac{1}{\nu_{\mathrm{ei}}} \,; \tag{11.18}$$

$$\sigma_{\mathrm{H}} = \sigma_{\perp} \left(\omega_{\mathrm{B}}^{(e)} \tau_{\mathrm{ei}} \right) = \sigma \, \frac{\omega_{\mathrm{B}}^{(e)} \tau_{\mathrm{ei}}}{1 + \left(\omega_{\mathrm{B}}^{(e)} \tau_{\mathrm{ei}} \right)^2} \,. \tag{11.19}$$

Formula (11.16) is called the *generalized* Ohm's law. It shows that the presence of a magnetic field in a plasma not only changes the magnitude of the conductivity, but the form of Ohm's law as well: generally, the electric field and the resulting current are not parallel, since $\sigma_{\perp} \neq \sigma_{\parallel}$. Therefore the electric conductivity of a plasma in a magnetic field is *anisotropic*. Moreover the current component $\mathbf{j}'_{\mathrm{H}}$ which is perpendicular to both the magnetic and electric fields, appears in the plasma. This component is the so-called Hall current (Figure 11.1).

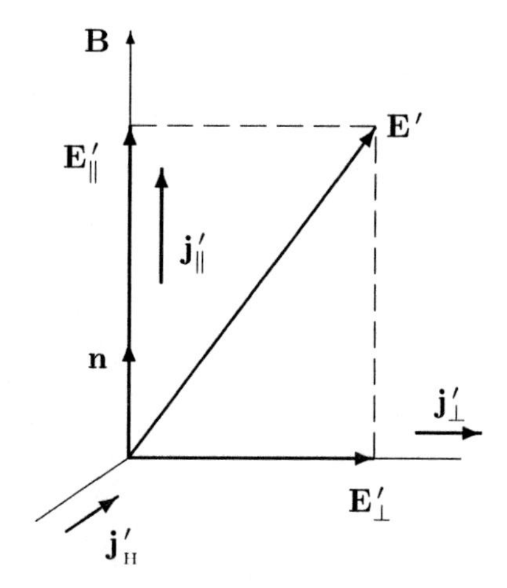

Figure 11.1: The generalized Ohm's law in a magnetized plasma: the direct ($\mathbf{j}'_{\parallel}$ and $\mathbf{j}'_{\perp}$) and Hall's ($\mathbf{j}'_{\rm H}$) currents in a plasma with electric ($\mathbf{E}'$) and magnetic ($\mathbf{B}$) fields.

11.4 The conductivity of magnetized plasma

11.4.1 Two limiting cases

The magnetic-field influence on the conductivity $\sigma_{\perp}$ of the 'direct' current $\mathbf{j}'_{\perp}$ across the magnetic field $\mathbf{B}$ and on the Hall-current conductivity $\sigma_{\rm H}$ is determined by the parameter $\omega_{\rm B}^{(e)} \tau_{\rm ei}$ which is the *turning angle* of an electron on the Larmor circle in the intercollisional time. Let us consider **two limiting cases**.

(a) Let the turning angle be small:)

$$\omega_{\rm B}^{(e)} \tau_{\rm ei} \ll 1 \, . \tag{11.20}$$

Obviously this inequality corresponds to the *weak* magnetic field or *dense cool* plasma, so that the electric current is scarcely affected by the magnetic field:

$$\sigma_{\perp} \approx \sigma_{\parallel} = \sigma \, , \qquad \frac{\sigma_{\rm H}}{\sigma} \approx \omega_{\rm B}^{(e)} \tau_{\rm ei} \ll 1 \, . \tag{11.21}$$

Thus in a frame of reference associated with the plasma, the usual Ohm's law with *isotropic* conductivity holds.

(b) The opposite case, when the electrons **spiral freely** between rare collisions:

$$\omega_{\rm B}^{(e)} \tau_{\rm ei} \gg 1 \, , \tag{11.22}$$

corresponds to the *strong* magnetic field and hot rarefied plasma. This plasma is termed the *magnetized* one. It is frequently encountered under astrophysical conditions. In this case

$$\sigma_{\parallel} = \sigma \, , \qquad \sigma_{\perp} \approx \sigma \left(\omega_{\rm B}^{(e)} \tau_{\rm ei} \right)^{-2} , \qquad \sigma_{\rm H} \approx \sigma \left(\omega_{\rm B}^{(e)} \tau_{\rm ei} \right)^{-1} , \tag{11.23}$$

or

$$\sigma_{\parallel} \approx \left(\omega_{\rm B}^{(e)} \tau_{\rm ei} \right) \sigma_{\rm H} \approx \left(\omega_{\rm B}^{(e)} \tau_{\rm ei} \right)^{2} \sigma_{\perp} \, . \tag{11.24}$$

Hence in the magnetized plasma, for example in the solar corona (see Exercises 11.1 and 11.2),

$$\boxed{\sigma_{\parallel} \gg \sigma_{\text{H}} \gg \sigma_{\perp} \,.} \tag{11.25}$$

In other words, the impact of the magnetic field on the direct current is especially strong for the component resulting from the electric field $\mathbf{E}'_{\perp}$. The current in the $\mathbf{E}'_{\perp}$ direction is considerably weaker than it would be in the absence of a magnetic field. Why is this so?

11.4.2 The physical interpretation

The physical mechanism of the perpendicular current $\mathbf{j}'_{\perp}$ is illustrated by Figure 11.2.

> The primary effect of the electric field $\mathbf{E}'_{\perp}$ in the presence of the magnetic field $\mathbf{B}$ is not the current in the direction $\mathbf{E}'_{\perp}$, but rather the electric *drift* in the direction perpendicular to both $\mathbf{B}$ and $\mathbf{E}'_{\perp}$.

The electric drift velocity (5.22) is independent of the particle's mass and charge. The electric drift of electrons and ions generates the motion of the plasma as a whole with the velocity $\mathbf{v} = \mathbf{v}_{\text{d}}$. This would be the case if there were no collisions at all (Figure 5.3).

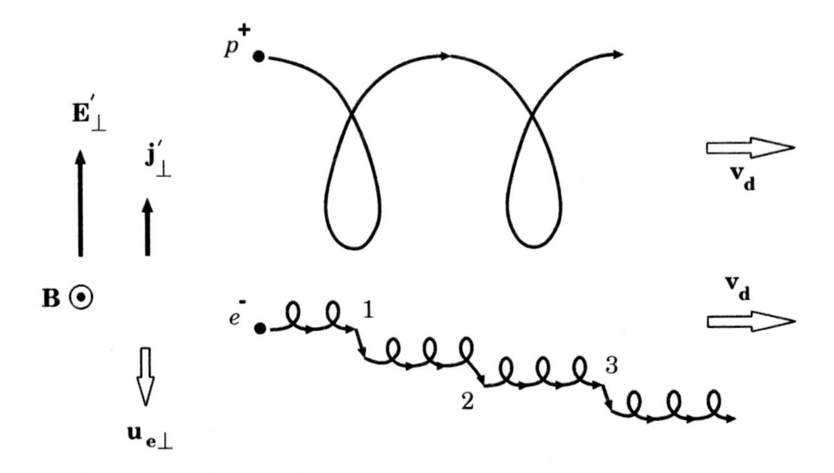

Figure 11.2: Initiation of the current in the direction of the perpendicular field $\mathbf{E}'_{\perp}$ as the result of rare collisions (1, 2, 3, ...) against a background of the electric drift. Only collisions of electrons are shown.

Collisions, even infrequent ones, result in a disturbance of the particle's Larmor motion, leading to a displacement of the ions (not shown in Figure 11.2) along $\mathbf{E}'_{\perp}$, and the electrons in the opposite direction as shown in

Figure 11.2. The small electric current $\mathbf{j}'_\perp$ (a factor of $\omega_{\rm B}^{(e)}\tau_{\rm ei}$ smaller than the drift one) appears in the direction $\mathbf{E}'_\perp$.

To ensure the current across the magnetic field, the so-called Hall electric field is necessary, that is the electric field component perpendicular to both the current $\mathbf{j}'_\perp$ and the field $\mathbf{B}$ (Braginskii, 1965; Sivukhin, 1996, Chapter 7, § 98). This is the secondary effect but it is not small in a strong magnetic field.

> The Hall electric field balances the Lorentz force acting on the carriers of the perpendicular electric current in plasma due to the presence of a magnetic field,

i.e. the force

$$\mathbf{F}\,(\,\mathbf{j}'_\perp) = \frac{en}{c}\,\mathbf{u}_{\,{\rm i}\perp} \times \mathbf{B} - \frac{en}{c}\,\mathbf{u}_{\,{\rm e}\perp} \times \mathbf{B} = \\ = \frac{1}{c}\,en\,(\mathbf{u}_{\,{\rm i}\perp} - \mathbf{u}_{\,{\rm e}\perp}) \times \mathbf{B} \tag{11.26}$$

Hence the magnitude of the Hall electric field is

$$\mathbf{E}'_{\rm H} = \frac{1}{enc}\,\mathbf{j}'_\perp \times \mathbf{B}\,. \tag{11.27}$$

The Hall electric field in plasma is frequently set up automatically, as a consequence of small charge separation within the limits of quasi-neutrality. In this case the 'external' field, which has to be applied to the plasma, is determined by the expressions

$$E'_\parallel = j'_\parallel / \sigma_\parallel \quad \text{and} \quad E'_\perp = j'_\perp / \sigma_\perp\,. \tag{11.28}$$

We shall not discuss here the dissipation process under the conditions of anisotropic conductivity. In general, the symmetric highest component of the *conductivity tensor* can play the most important role (see Landau et al., 1984, Chapter 3) in this process of fundamental significance for the flare energy release problem. In the particular case of a fully-ionized plasma, the tendency for a particle to spiral round the magnetic field lines insures the great reduction in the transversal conductivity (11.18). However, since the dissipation of the energy of the electric current into Joule heat is due solely to collisions between particles, the reduced conductivity does not lead to increased dissipation (Exercise 11.3).

On the other hand, the Hall electric field and Hall electric current can significantly modify conditions of magnetic reconnection (e.g., Bhattacharjee, 2004).

11.5 Currents and charges in plasma

11.5.1 Collisional and collisionless plasmas

Let us point out another property of the generalized Ohm's law in astrophysical plasma. Under laboratory conditions, as a rule, one cannot neglect the gradient forces (11.6). On the contrary, these forces usually play no part in astrophysical plasma. We shall ignore them. This simplification may be not well justified however in such important applications as reconnecting current layers (RCLs), shock waves and other discontinuities.

Moreover let us also restrict our consideration to very *slow* (say *hydrodynamic*) motions of plasma. These motions are supposed to be so slow that the following three conditions are fulfilled.

First, it is supposed that

$$\omega = \frac{1}{\tau} \ll \nu_{\rm ei} \quad \text{or} \quad \nu_{\rm ei}\, \tau \gg 1\,, \tag{11.29}$$

where τ is a characteristic time of the plasma motions. Thus

> departures of actual distribution functions for electrons and ions from the Maxwellian distribution are small.

This allows us to handle the transport phenomena in linear approximation.

Moreover, if a single-fluid model is to make physical sence, the electrons and ions could have comparable temperatures, ideally, the same temperature T which is the temperature of the plasma as a whole:

$$T_{\rm e} = T_{\rm p} = T\,.$$

Second, we neglect the electron inertia in comparison with that of the ions and make use of (11.8). This condition is usually written in the form

$$\omega \ll \omega_{_{\rm B}}^{(\rm i)} = \frac{eB}{m_{\rm i}c}\,. \tag{11.30}$$

Thus

> the plasma motions have to be so slow that their frequency is smaller than the lowest gyro-frequency of the particles.

Recall that the gyro-frequency of ions $\omega_{_{\rm B}}^{(\rm i)} \ll \omega_{_{\rm B}}^{(\rm e)}$.

The third condition

$$\nu_{\rm ei} \gg \omega_{_{\rm B}}^{(\rm e)} \quad \text{or} \quad \omega_{_{\rm B}}^{(\rm e)} \tau_{\rm ei} \ll 1\,. \tag{11.31}$$

Hence the hydrodynamic approximation can be used, the conductivity σ being isotropic. The generalized Ohm's law assumes the following form which is specific to *magnetohydrodynamics* (MHD):

$$\mathbf{j}' = \sigma \left(\mathbf{E} + \frac{1}{c}\, \mathbf{u} \times \mathbf{B} \right). \tag{11.32}$$

The MHD approximation is the subject of the next chapter. Numerous applications of MHD to various phenomena in astrophysical plasma will be considered in many places in the remainder of the book.

In the opposite case (11.22), when the parameter $\omega_{B}^{(e)}\tau_{ei}$ is large, charged particles revolve around magnetic field lines, and a typical particle may spend a considerable time in a region of a size of the order of the gyroradius (5.14). Hence, if the length scale of the plasma is much larger than the gyroradius, we may expect the hydrodynamic-type models to work.

It appears that, even when the parameter $\omega_{B}^{(e)}\tau_{ei}$ tends to infinity (like in the solar corona, see Exercise 11.2) and collisions are negligible, the *quasi-hydrodynamic* description of plasma, the Chew-Goldberger-Low (CGL) approximation (Chew et al., 1956) is possible (especially if the actual electric field **E** in a collisionless plasma is perpendicular to a sufficiently strong magnetic field **B**) and quite useful. This is because

> the strong magnetic field makes the plasma, even a non-collisional one, more 'interconnected', so to speak, more hydrodynamic in the directions perpendicular to the magnetic field.

That allows one to write down a well-justified set of two-dimensional MHD equations for the collisionless plasma in a magnetic field (see Volkov, 1966, Equations (42)–(45)). As for the motion of collisionless particles along the magnetic field, some important kinetic features and physical restrictions still are significant (Klimontovich and Silin, 1961; Shkarofsky et al., 1969, Chapter 10). Chew et al. (1956) emphasized that "a strictly hydrodynamic approach to the problem is appropriate only when some special circumstance suppresses the effects of pressure transport along the magnetic lines".

There is ample experimental evidence that strong magnetic fields do make astrophysical plasmas behave like hydrodynamic charged fluids. This does not mean, of course, that there are no pure kinetic phenomena in such plasmas. There are many of them indeed.

11.5.2 Volume charge and quasi-neutrality

One more remark concerning the generalized Ohm's law is important for the following. While deriving the law in Section 11.2, the *exact* charge neutrality of plasma or the exact electric neutrality was assumed:

$$Z_{i}n_{i} = n_{e} = n\,, \tag{11.33}$$

i.e. the absolute absence of the *volume charge* in plasma: $\rho^{q} = 0$. The same assumption was also used in Sections 8.2 and 3.2. However there is no need for such a strong restriction. It is sufficient to require *quasi-neutrality*, i.e. the numbers of ions (with account of their charge taken) and electrons per unit volume are very nearly equal:

$$\frac{Z_{i}n_{i} - n_{e}}{n_{e}} \ll 1\,. \tag{11.34}$$

So

the volume charge density has to be small in comparison to the plasma density.

Once the volume charge density

$$\rho^{\mathrm{q}} \neq 0\,, \tag{11.35}$$

yet another term must be taken into account in the generalized Ohm's law:

$$\mathbf{j}_{\mathrm{u}}^{\mathrm{q}} = \rho^{\mathrm{q}}\,\mathbf{u}\,. \tag{11.36}$$

This is the so-called *convective* current. It is caused by the volume charge transfer and must be added to the *conductive* current (11.16).

The volume charge, the associated electric force $\rho^{\mathrm{q}}\,\mathbf{E}$ and the convective current $\rho^{\mathrm{q}}\,\mathbf{u}$ are of great importance in electrodynamics of relativistic objects such as black holes (Novikov and Frolov, 1989) and pulsars (Michel, 1991). **Charge-separated plasmas** originate in magnetospheres of rotating black holes, for example, a super-massive black hole in active galactic nuclei. The shortage of charge leads to the emergence of a strong electric field along the magnetic field lines. The parallel electric field accelerates migratory electrons and/or positrons to ultrarelativistic energies (e.g., Hirotani and Okamoto, 1998).

Charge density oscillations in a plasma, the Langmuir waves, are considered in Section 10.2.

* * *

The volume charge density can be evaluated in the following manner. On the one hand, from Maxwell's equation $\operatorname{div}\mathbf{E} = 4\pi\rho^{\mathrm{q}}$ we estimate

$$\rho^{\mathrm{q}} \approx \frac{E}{4\pi L}\,. \tag{11.37}$$

On the other hand, the non-relativistic equation of plasma motion yields

$$e n_{\mathrm{e}} E \approx \frac{p}{L} \approx \frac{n_{\mathrm{e}} k_{\mathrm{B}} T}{L}\,,$$

so that

$$E \approx \frac{k_{\mathrm{B}} T}{eL}\,. \tag{11.38}$$

On substituting (11.38) in (11.37), we find the following estimate

$$\frac{\rho^{\mathrm{q}}}{e n_{\mathrm{e}}} \approx \frac{k_{\mathrm{B}} T}{eL}\,\frac{1}{4\pi L}\,\frac{1}{e n_{\mathrm{e}}} = \frac{1}{L^2}\left(\frac{k_{\mathrm{B}} T}{4\pi e^2\, n_{\mathrm{e}}}\right)$$

or

$$\boxed{\frac{\rho^{\mathrm{q}}}{e n_{\mathrm{e}}} \approx \frac{r_{\mathrm{D}}^{\,2}}{L^2}\,.} \tag{11.39}$$

Since the usual *concept of plasma* implies that the Debye radius

$$r_{\rm D} \ll L \,, \tag{11.40}$$

the volume charge density is small in comparison with the plasma density.

When we consider phenomena with a length scale L much larger than the Debye radius $r_{\rm D}$ and a time scale τ much larger than the inverse the plasma frequency, the charge separation in the plasma can be neglected.

11.6 Practice: Exercises and Answers

Exercise 11.1 [Section 11.4] Evaluate the characteristic value of the parallel conductivity (11.17) in the solar corona.

Answer. It follows from formula (11.17) that

$$\sigma_{\parallel} = \frac{e^2 n}{m_{\rm e}}\, \tau_{\rm ei} = 2.53 \times 10^8 \, n \, \tau_{\rm ei} \sim 10^{16} - 10^{17} \,, \ \ {\rm s}^{-1} \,, \tag{11.41}$$

if we take $\tau_{\rm ep} \sim 0.2 - 2.0$ s (Exercise 8.1).

Exercise 11.2 [Section 11.4] Estimate the parameter $\omega_{\rm B}^{(e)} \tau_{\rm ei}$ in the solar corona above a sunspot.

Answer. Just above a large sunspot the field strength can be as high as $B \approx 3000\,{\rm G}$. Here the electron Larmor frequency $\omega_{\rm B}^{(e)} \approx 5 \times 10^{10}$ rad s^{-1} (Exercise 5.1). Characteristic time of *close* electron-proton collisions $\tau_{\rm ep}(cl) \approx$ 22 s (see Exercise 8.1). Therefore $\omega_{\rm B}^{(e)} \tau_{\rm ei}(cl) \sim 10^{12}$ rad $\gg 1$.

Distant collisions are much more frequent (Exercise 8.1). However, even with $\tau_{\rm ep} \approx 0.1$ s, we obtain

$$\omega_{\rm B}^{(e)} \tau_{\rm ei} \sim 10^{10} \,{\rm rad} \gg 1 \,.$$

So, for anisoptropic conductivity in the corona, the approximate formulae (11.23) can be well used.

Exercise 11.3 [Section 11.4.2] Consider the generalized Ohm's law in the case when the electric field is perpendicular to the magnetic field $\mathbf{B} = B\,\mathbf{n}$. So

$$\mathbf{j}' = \sigma_{\perp}\, \mathbf{E}'_{\perp} + \sigma_{\rm H}\, \mathbf{n} \times \mathbf{E}'_{\perp} \,, \tag{11.42}$$

where

$$\sigma_{\perp} = \sigma \, \frac{1}{1 + \left(\omega_{\rm B}^{(e)} \tau_{\rm ei} \right)^2} \quad \text{and} \quad \sigma_{\rm H} = \sigma \, \frac{\omega_{\rm B}^{(e)} \tau_{\rm ei}}{1 + \left(\omega_{\rm B}^{(e)} \tau_{\rm ei} \right)^2} \,. \tag{11.43}$$

This indicates that the current $\mathbf{j}'_{\perp}$ in the direction of $\mathbf{E}'_{\perp}$ is reduced in the ratio

$$1 \Big/ \left(1 + \left(\omega_{\rm B}^{(e)} \tau_{\rm ei} \right)^2 \right) \approx \left(\omega_{\rm B}^{(e)} \tau_{\rm ei} \right)^{-2} , \quad \text{if} \quad \omega_{\rm B}^{(e)} \tau_{\rm ei} \gg 1 \,,$$

by the magnetic field. In addition, the other current $\left(\omega_{B}^{(e)}\tau_{ei}\right)^{2}$ times as large flows in the direction perpendicular to both **B** and $\mathbf{E}'_{\perp}$; this is the Hall current $\mathbf{j}'_{H}$.

Show that the reduction in the 'perpendicular' conductivity (Figure 11.1) does not increase the rate of dissipation of current energy (see Cowling, 1976, § 6.2).

Chapter 12

Single-Fluid Models for Astrophysical Plasma

Single-fluid models are the simplest but sufficient approximation to describe many large-scale low-frequency phenomena in astrophysical plasma: regular and turbulent dynamo, plasma motions driven by strong magnetic fields, accreation disks, and relativistic jets.

12.1 Derivation of the single-fluid equations

12.1.1 The continuity equation

In order to consider cosmic plasma as a *single* hydrodynamic medium, we have to sum each of the three transfer equations over all kinds of particles. Let us start from the continuity Equation (9.14). With allowance for the definition of the plasma mass density (9.6), we have

$$\frac{\partial \rho}{\partial t} + \operatorname{div} \left(\sum_{k} \rho_k \mathbf{u}_k \right) = 0 \, . \tag{12.1}$$

The mean velocities of motion for all kinds of particles are supposed to be equal to the plasma hydrodynamic velocity:

$$\mathbf{u}_1 (\mathbf{r}, t) = \mathbf{u}_2 (\mathbf{r}, t) = \cdots = \mathbf{u} (\mathbf{r}, t) \, , \tag{12.2}$$

as a result of action of the mean collisional force (9.24). However this is not a general case.

In general, the mean velocities are not the same, but a frame of reference can be chosen in which

$$\rho \, \mathbf{u} = \sum_{k} \rho_k \mathbf{u}_k \, . \tag{12.3}$$

Then from (12.1) and (12.3) we obtain the usual *continuity equation*

$$\frac{\partial \rho}{\partial t} + \operatorname{div} \rho \, \mathbf{u} = 0 \, . \tag{12.4}$$

12.1.2 The momentum conservation law in plasma

In much the same way as in previous Section, we handle the momentum transfer Equation (9.42). On summing over all kinds of particles, we obtain the following equation:

$$\rho \, \frac{d u_{\alpha}}{dt} = - \frac{\partial}{\partial r_{\beta}} \, p_{\alpha\beta} + \rho^{\mathrm{q}} E_{\alpha} + \frac{1}{c} \, (\, \mathbf{j} \times \mathbf{B} \,)_{\alpha} + \sum_{k} \langle \, F_{k,\alpha}^{(\mathrm{c})} \, (\mathbf{r}, t) \, \rangle_{v} \, . \tag{12.5}$$

Here the *volume charge* density in plasma is

$$\rho^{\mathrm{q}} = \sum_{k} n_k e_k = \frac{1}{4\pi} \operatorname{div} \mathbf{E} \, , \tag{12.6}$$

and the electric current density is

$$\mathbf{j} = \sum_{k} n_k e_k \, \mathbf{u}_k = \frac{c}{4\pi} \operatorname{curl} \mathbf{B} - \frac{1}{4\pi} \frac{\partial \mathbf{E}}{\partial t} \, . \tag{12.7}$$

The electric and magnetic fields, $\mathbf{E}$ and $\mathbf{B}$, involved in this description are averaged fields associated with the total electric charge density ρ^{q} and the total current density $\mathbf{j}$. They satiesfy the macroscopic Maxwell equations. In cosmic plasma, the magnetic *permeability* and the electric *permittivity* can almost always be replaced by their vacuum values. For this reason, the macroscopic Maxwell equations have the same structure as Equations (1.27) and (1.24) that have been used on the right-hand side of formulae (12.6) and (12.7).

Since elastic collisions do not change the total momentum, we have

$$\sum_{k} \langle \, F_{k,\alpha}^{(\mathrm{c})} \, (\mathbf{r}, t) \, \rangle_{v} = 0 \, . \tag{12.8}$$

On substituting (12.6)–(12.8) in Equation (12.5), the latter can be rearranged to give the *momentum conservation law* for plasma

$$\rho \, \frac{d u_{\alpha}}{dt} = - \frac{\partial}{\partial r_{\beta}} \, p_{\alpha\beta} + F_{\alpha}(\mathbf{E}, \mathbf{B}) \, . \tag{12.9}$$

Here the electromagnetic force is written in terms of the electric and magnetic field vectors:

$$F_\alpha(\mathbf{E}, \mathbf{B}) = -\frac{\partial}{\partial t} \frac{(\mathbf{E} \times \mathbf{B})_\alpha}{4\pi c} - \frac{\partial}{\partial r_\beta} \mathrm{M}_{\alpha\beta} \,. \tag{12.10}$$

The tensor

$$\mathrm{M}_{\alpha\beta} = \frac{1}{4\pi} \left[-E_\alpha E_\beta - B_\alpha B_\beta + \frac{1}{2} \delta_{\alpha\beta} \left(E^2 + B^2 \right) \right] \tag{12.11}$$

is called the *Maxwellian tensor* of stresses.

The divergent form of the **momentum conservation law** is

$$\boxed{\frac{\partial}{\partial t} \left[\rho u_\alpha + \frac{(\mathbf{E} \times \mathbf{B})_\alpha}{4\pi c} \right] + \frac{\partial}{\partial r_\beta} \left(\Pi_{\alpha\beta} + \mathrm{M}_{\alpha\beta} \right) = 0 \,.} \tag{12.12}$$

The operator $\partial/\partial t$ acts on two terms that correspond to momentum density: $\rho\,\mathbf{u}$ is the momentum of the motion of the plasma as a whole in a unit volume, $\mathbf{E} \times \mathbf{B}/4\pi c$ is the momentum density of the electromagnetic field. The divergency operator $\partial/\partial r_\alpha$ acts on

$$\Pi_{\alpha\beta} = p_{\alpha\beta} + \rho\, u_\alpha u_\beta \tag{12.13}$$

which is the *momentum flux density* tensor

$$\Pi_{\alpha\beta} = \sum_k \Pi_{\alpha\beta}^{(k)} \,, \tag{12.14}$$

see definition (9.10). Therefore the *pressure tensor*

$$p_{\alpha\beta} = p\, \delta_{\alpha\beta} + \pi_{\alpha\beta} \,, \tag{12.15}$$

where

$$p = \sum_k p_k \tag{12.16}$$

is the total plasma pressure, the sum of *partial pressures*, and

$$\pi_{\alpha\beta} = \sum_k \pi_{\alpha\beta}^{(k)} \tag{12.17}$$

is the *viscous stress* tensor (see definition (9.35)), which allows for the transport of momentum from one layer of the plasma flow to the other layers so that relative motions inside the plasma are damped out. If we accept condition (12.2) then the random velocities are now defined with respect to the macroscopic velocity $\mathbf{u}$ of the plasma as a whole.

The momentum conservation law in the form (12.9) or (12.12) is applied for a wide range of conditions in cosmic plasmas like **fluid relativistic flows**, for example, astrophysical jets (Section 13.3). The assumption that the astrophysical plasma behaves as a continuum medium, which is essential if these forms of the momentum conservation law are to be applied, is excellent in the cases in which we are often interested:

> the Debye length and the particle Larmor radii are much smaller than the plasma flow scales.

On the other hand, going from the multi-fluid description to a single-fluid model is a seriuos damage because we loose an information not only on the small-scale dynamics of the electrons and ions but also on the high-frequency processes in plasma.

> The single-fluid equations describe the *low-frequency large-scale* behaviour of plasma in astrophysical conditions.

12.1.3 The energy conservation law

In a similar manner as above, the energy conservation law is derived. We sum the general Equation (9.25) over k and then substitute in the resulting equation the total electric charge density (12.6) and the total electric current density (12.7) expressed in terms of the electric field $\mathbf{E}$ and magnetic field $\mathbf{B}$. On rearrangement, the following divergent form of the energy conservation law (cf. the simplified Equation (1.54) for electromagnetic field energy and kinetic energy of charged particles) is obtained:

$$\frac{\partial}{\partial t}\left(\frac{\rho u^2}{2}+\rho\varepsilon+\frac{E^2+B^2}{8\pi}\right)+$$
$$+\frac{\partial}{\partial r_\alpha}\left[\rho\, u_\alpha\left(\frac{u^2}{2}+w\right)+\frac{c}{4\pi}\,(\mathbf{E}\times\mathbf{B})_\alpha+\pi_{\alpha\beta}\,u_\beta+q_\alpha\right]=$$
$$=\left(u_\alpha F_\alpha^{(\mathrm{c})}\right)_{ff}. \tag{12.18}$$

On the left-hand side of this equation, an additional term has appeared: the operatop $\partial/\partial t$ acts on the energy density of the electromagnetic field

$$W=\frac{E^2+B^2}{8\pi}\,. \tag{12.19}$$

The divergency operator $\partial/\partial r_\alpha$ acts on the *Pointing vector*, the electromagnetic energy flux through a unit surface in space:

$$\mathbf{G}=\frac{c}{4\pi}\,[\mathbf{E}\times\mathbf{B}]\,. \tag{12.20}$$

The right-hand side of Equation (12.18) contains the total work of *friction forces* (9.38) in unit time on unit volume

$$\left(u_\alpha F_\alpha^{(\mathrm{c})}\right)_{ff}=\sum_k\left(F_{k,\alpha}^{(\mathrm{c})}\,u_{k,\alpha}\right)=\sum_k u_{k,\alpha}\int_{\mathbf{v}} m_k\,v'_\alpha\left(\frac{\partial f_k}{\partial t}\right)_{\mathrm{c}}d^3\mathbf{v}\,. \tag{12.21}$$

This work related to the relative motion of the plasma components is not zero.

By contrast, the total heat release under elastic collisions between particles of different kinds (see definition (9.39)) is

$$\sum_k Q_k^{(c)}(\mathbf{r}, t) = \sum_k \int\limits_{\mathbf{v}} \frac{m_k (v')^2}{2} \left(\frac{\partial f_k}{\partial t} \right)_c d^3\mathbf{v} = 0 . \tag{12.22}$$

Elastic collisions in a plasma conserve both the total momentum (see Equation (12.8)) and the total energy (see Equation (12.22)).

If we accept condition (12.2) then, with account of formula (9.24), the collisional heating (12.21) by friction force is also equal to zero. In this limit, there is not any term which contains the collisional integral. Collisions have done a good job.

Note, in conclusion, that we do not have any equations for the anisotropic part of the pressure tensor, which is the viscous stress tensor $\pi_{\alpha\beta}$, and for the flux q_α of heat due to random motions of particles. This is not unexpected, of course, but inherent at the method of the moments as discussed in Section 9.4. We have to find these transfer coefficients by using the procedure described in Section 9.5.

12.2 Basic assumptions and the MHD equations

12.2.1 Old and new simplifying assumptions

As we saw in Chapter 9, the set of transfer equations for local macroscopic quantities determines the behaviour of different kinds of particles, such as electrons and ions in astrophysical plasma, once two main conditions are complied with:

(a) many collisions occur in a characteristic time τ of the process or phenomenon under consideration:

$$\tau \gg \tau_c , \tag{12.23}$$

(b) the particle's path between two collisions – the particle's free path – is significantly smaller than the distance L, over which macroscopic quantities change considerably:

$$L \gg \lambda_c . \tag{12.24}$$

Here τ_c and λ_c are the collisional time and the collisional mean free path, respectively. Once these conditions are satisfied, we can close the set of hydrodynamic *transfer* equations, as has been discussed in Section 9.5.

While considering the generalized Ohm's law in Chapter 11, three other assumptions have been made, that are complementary to the restriction (12.23) on the characteristic time τ of the process.

The first condition can be written in the form

$$\tau \gg \tau_{\rm ei} = \nu_{\rm ei}^{-1}\,, \tag{12.25}$$

where $\tau_{\rm ei}$ is the electron-ion collisional time, the longest collisional relaxation time. Thus departures from the Maxwellian distribution are small. Moreover the electrons and ions should have comparable temperatures, ideally, the same temperature T being the temperature of the plasma as a whole.

Second, we neglect the electron inertia in comparison with that of the ions. This condition is usually written as

$$\tau \gg \left(\omega_{\rm B}^{(\rm i)}\right)^{-1}, \quad \text{where} \quad \omega_{\rm B}^{(\rm i)} = \frac{eB}{m_{\rm i}c}\,. \tag{12.26}$$

Thus the plasma motions have to be so slow that their frequency $\omega = 1/\tau$ is smaller than the lowest gyro-frequency of the particles. Recall that the gyro-frequency of ions $\omega_{\rm B}^{(\rm i)} \ll \omega_{\rm B}^{(\rm e)}$.

The third condition,

$$\omega_{\rm B}^{(\rm e)}\tau_{\rm ei} \ll 1\,, \tag{12.27}$$

is necessary to write down Ohm's law in the form

$$\mathbf{j} = \sigma\left(\mathbf{E} + \frac{1}{c}\,\mathbf{v}\times\mathbf{B}\right) + \rho^{\rm q}\,\mathbf{v}\,. \tag{12.28}$$

Here $\mathbf{v}$ is the macroscopic velocity of plasma considered as a continuous medium, $\mathbf{E}$ and $\mathbf{B}$ are the electric and magnetic fields in the 'laboratory' system of coordinates, where we measure the velocity $\mathbf{v}$. Accordingly,

$$\mathbf{E}_{\rm v} = \mathbf{E} + \frac{1}{c}\,\mathbf{v}\times\mathbf{B} \tag{12.29}$$

is the electric field in a frame of reference related to the plasma. The isotropic conductivity is (formula (11.14)):

$$\sigma = \frac{e^2 n}{m_{\rm e}\,\nu_{\rm ei}}\,. \tag{12.30}$$

Complementary to the restriction (12.24) on the characteristic length L of the phenomenon, we have to add the condition

$$L \gg r_{\rm D}\,, \tag{12.31}$$

where $r_{\rm D}$ is the Debye radius. Then the volume charge density $\rho^{\rm q}$ is small in comparison with the plasma density ρ.

Under the conditions listed above, we use the general hydrodynamic-type equations which are the conservation laws for mass (12.4), momentum (12.5) and energy (12.18).

These equations have a much wider area of applicability than the equations of ordinary magnetohydrodynamics derived below.

The latter will be much simpler than the equations derived in Section 12.1. Therefore **new additional simplifying assumptions** are necessary. Let us introduce them. There are two.

* * *

First assumption: the plasma conductivity σ is assumed to be large, the electromagnetic processes being not very fast. Then, in Maxwell's equation (1.24)

$$\operatorname{curl}\mathbf{B} = \frac{4\pi}{c}\,\mathbf{j} + \frac{1}{c}\,\frac{\partial \mathbf{E}}{\partial t}\,,$$

we ignore the *displacement* current in comparison to the *conductive* one. The corresponding condition is found by evaluating the currents as follows

$$\frac{1}{c}\,\frac{E}{\tau} \ll \frac{4\pi}{c}\,j \quad \text{or} \quad \omega E \ll 4\pi\sigma E\,.$$

Thus we suppose that

$$\boxed{\omega \ll 4\pi\sigma\,.} \tag{12.32}$$

In the same order with reference to the small parameter ω/σ (or, more exactly, $\omega/4\pi\sigma$), we can neglect the *convective* current (see formula (11.36) and its discussion in Section 11.5.2) in comparison with the *conductive* current in Ohm's law (12.28). Actually,

$$\rho^{q}\,v \approx v\operatorname{div}\mathbf{E}\,\frac{1}{4\pi} \approx \frac{L}{\tau}\,\frac{E}{L}\,\frac{1}{4\pi} \approx \frac{\omega}{4\pi}\,E \ll \sigma E\,,$$

once the condition (12.32) is satisfied.

The conductivity of astrophysical plasma, which is often treated in the MHD approximation, is very high (e.g., Exercise 11.1). This is the reason why condition (12.32) is satisfied up to frequencies close to optical ones.

Neglecting the displacement current and the convective current, Maxwell's equations and Ohm's law result in the following relations:

$$\mathbf{j} = \frac{c}{4\pi}\operatorname{curl}\mathbf{B}\,, \tag{12.33}$$

$$\mathbf{E} = -\,\frac{1}{c}\,\mathbf{v}\times\mathbf{B} + \frac{c}{4\pi\sigma}\operatorname{curl}\mathbf{B}\,, \tag{12.34}$$

$$\rho^{q} = -\,\frac{1}{4\pi c}\operatorname{div}(\,\mathbf{v}\times\mathbf{B}\,)\,, \tag{12.35}$$

$$\operatorname{div}\mathbf{B} = 0\,, \tag{12.36}$$

$$\frac{\partial \mathbf{B}}{\partial t} = \text{curl}\,(\,\mathbf{v} \times \mathbf{B}\,) + \frac{c^2}{4\pi\sigma}\,\Delta\,\mathbf{B}\,. \tag{12.37}$$

Once two vectors, **B** and **v**, are given, the current density **j**, the electric field **E**, and the volume charge density $\rho^{\,q}$ are completely determined by formulae (12.33)—(12.35). Thus

> the problem is reduced to finding the interaction of the magnetic field **B** and the hydrodynamic velocity field **v**.

As a consequence, the approach under discussion has come to be known as *magnetohydrodynamics* (Alfvén, 1950; Syrovatskii, 1957).

The corresponding equation of plasma motion is obtained by substitution of formulae (12.33)–(12.35) in the equation of momentum transfer (12.5). With due regard for the manner in which viscous forces are usually written in hydrodynamics, we have

$$\rho\,\frac{d\mathbf{v}}{dt} = -\nabla p + \rho^{\,q}\,\mathbf{E} - \frac{1}{4\pi}\,\mathbf{B} \times \text{curl}\,\mathbf{B} +$$

$$+\,\eta\,\Delta\,\mathbf{v} + \left(\zeta + \frac{\eta}{3}\right)\nabla\,\text{div}\,\mathbf{v}\,. \tag{12.38}$$

Here η is the *first viscosity* coefficient, ζ is the *second viscosity* coefficient (see Landau and Lifshitz, *Fluid Mechanics*, 1959a, Chapter 2, § 15). Formulae for these coefficients as well as for the viscous forces should be derived from the moment equation for the pressure tensor, which we were not inclined to write down in Section 9.3 being busy in the way to the energy conservation law.

* * *

A second additional simplifying assumption has to be introduced now. Treating Equation (12.38), the electric force $\rho^{\,q}\,\mathbf{E}$ can be ignored in comparison to the magnetic one if

$$v^2 \ll c^2\,, \tag{12.39}$$

that is in the non-relativistic approximation. To make certain that this is true, evaluate the electric force using (12.35) and (12.34):

$$\rho^{\,q}\,E \approx \frac{1}{4\pi c}\,\frac{vB}{L}\,\frac{vB}{c} \approx \frac{B^2}{4\pi}\,\frac{1}{L}\,\frac{v^2}{c^2}\,, \tag{12.40}$$

the magnetic force being proportional to

$$\frac{1}{4\pi}\,|\,\mathbf{B} \times \text{curl}\,\mathbf{B}\,| \approx \frac{B^2}{4\pi}\,\frac{1}{L}\,. \tag{12.41}$$

Comparing (12.40) with (12.41), we see that the electric force is a factor of v^2/c^2 short of the magnetic one.

In a great number of astrophysical applications of MHD, the plasma velocities fall far short of the speed of light. The Sun is a good case in point. Here the largest velocities observed, for example, in coronal transients and coronal mass ejections (CMEs) do not exceed several thousands of km/s, i.e. $\lesssim 10^8$ cm/s. Under these conditions, **we neglect the electric force acting upon the volume charge** in comparison with the magnetic force.

However the relativistic objects such as accretion disks near black holes (see Chapter 7 in Novikov and Frolov, 1989), and pulsar magnetospheres are at the other extreme (Michel, 1991; Rose, 1998). The electric force acting on the volume charge plays a crucial role in the electrodynamics of relativistic objects.

12.2.2 Non-relativistic magnetohydrodynamics

With the assumptions made above, the considerable simplifications have been obtained; and now we write the following set of equations of non-relativistic MHD:

$$\frac{\partial}{\partial t}\,\rho\,v_{\alpha} = -\frac{\partial}{\partial r_{\beta}}\,\Pi^{*}_{\alpha\beta}\,, \tag{12.42}$$

$$\frac{\partial \mathbf{B}}{\partial t} = \operatorname{curl}\,(\,\mathbf{v}\times\mathbf{B}\,) + \nu_{\mathrm{m}}\,\Delta\,\mathbf{B}\,, \tag{12.43}$$

$$\operatorname{div}\mathbf{B} = 0\,, \tag{12.44}$$

$$\frac{\partial \rho}{\partial t} + \operatorname{div}\rho\,\mathbf{v} = 0\,, \tag{12.45}$$

$$\frac{\partial}{\partial t}\left(\frac{\rho v^2}{2} + \rho\varepsilon + \frac{B^2}{8\pi}\right) = -\operatorname{div}\mathbf{G}\,, \tag{12.46}$$

$$p = p\,(\rho, T)\,. \tag{12.47}$$

In contrast to Equation (12.12), the momentum of electromagnetic field does not appear on the left-hand side of the non-relativistic Equation (12.42). It is negligibly small in comparison to the plasma stream momentum $\rho\,v_{\alpha}$. This fact is a consequence of neglecting the displacement current in Maxwell's equations.

On the right-hand side of Equation (12.42), the asterisk refers to the total (unlike (12.13)) momentum flux density tensor $\Pi^{*}_{\alpha\beta}$, which is equal to

$$\Pi^{*}_{\alpha\beta} = p\,\delta_{\alpha\beta} + \rho\,v_{\alpha}v_{\beta} + \frac{1}{4\pi}\left(\frac{B^2}{2}\,\delta_{\alpha\beta} - B_{\alpha}B_{\beta}\right) - \sigma^{\mathrm{v}}_{\alpha\beta}\,. \tag{12.48}$$

In Equation (12.43)

$$\nu_{\mathrm{m}} = \frac{c^2}{4\pi\sigma} \tag{12.49}$$

is the *magnetic diffusivity* (or magnetic viscosity). It plays the same role in Equation (12.43) as the kinematic viscosity $\nu = \eta/\rho$ in the equation of plasma motion (12.42). The vector **G** is defined as the energy flux density (cf. Equation (12.18))

$$G_\alpha = \rho\, v_\alpha \left(\frac{v^2}{2} + w \right) + \frac{1}{4\pi} \left[\mathbf{B} \times (\mathbf{v} \times \mathbf{B}) \right]_\alpha -$$

$$- \frac{\nu_{\rm m}}{4\pi} \left(\mathbf{B} \times \operatorname{curl} \mathbf{B} \right)_\alpha - \sigma^{\rm v}_{\alpha\beta}\, v_\beta - \kappa\, \nabla_\alpha T, \qquad (12.50)$$

where the *specific enthalpy* is

$$w = \varepsilon + \frac{p}{\rho} \qquad (12.51)$$

(see definition (9.34)).

The Poynting vector appearing as a part in expression (12.50) is rewritten using formula (12.34):

$$\mathbf{G}_{\rm P} = \frac{c}{4\pi}\, \mathbf{E} \times \mathbf{B} = \frac{1}{4\pi}\, \mathbf{B} \times (\mathbf{v} \times \mathbf{B}) - \frac{\nu_{\rm m}}{4\pi}\, \mathbf{B} \times \operatorname{curl} \mathbf{B} \,. \qquad (12.52)$$

As usually in electrodynamics, the flux of electromagnetic energy disappeares when electric field **E** is parallel to magnetic field **B**.

The energy flux density due to friction processes is written as the contraction of the velocity vector **v** and the viscous stress tensor

$$\sigma^{\rm v}_{\alpha\beta} = \eta \left(\frac{\partial v_\alpha}{\partial r_\beta} + \frac{\partial v_\beta}{\partial r_\alpha} - \frac{2}{3}\, \delta_{\alpha\beta}\, \frac{\partial v_\gamma}{\partial r_\gamma} \right) + \zeta\, \delta_{\alpha\beta}\, \frac{\partial v_\gamma}{\partial r_\gamma} \qquad (12.53)$$

(see Landau and Lifshitz, *Fluid Mechanics*, 1959a, Chapter 2, § 15). How should we find formula (12.53) and formulae for coefficients η and ζ? – In order to find an equation for the second moment (9.10), we should multiply the kinetic Equation (9.1) by the factor $m_k\, v_\alpha v_\beta$ and integrate over velocity space **v**. In this way, we could derive the equations for the anisotropic part of the pressure tensor and for the flux of heat due to random motions of particles (Shkarofsky et al., 1966; § 9.2). We restrict ourself just by recalling the expressions for the viscous stress tensor (12.53) and heat flux density $-\kappa \nabla T$, where κ is the plasma thermal conductivity.

* * *

The equation of state (12.47) can be rewritten in other thermodynamic variables. In order to do this, we have to make use of Equations (12.42)–(12.45) and the thermodynamic identities

$$d\varepsilon = T\, ds + \frac{p}{\rho^2}\, d\rho \quad \text{and} \quad dw = T\, ds + \frac{1}{\rho}\, dp \,.$$

Here s is the entropy per unit mass.

At the same time, it is convenient to transform the energy conservation law (12.46) from the divergent form to the hydrodynamic one containing the substantial derivative (9.41). On rearrangement, Equation (12.46) results in the *heat transfer equation*

$$\rho T \frac{ds}{dt} = \frac{\nu_{\rm m}}{4\pi} (\operatorname{curl} \mathbf{B})^2 + \sigma^{\rm v}_{\alpha\beta} \frac{\partial v_\alpha}{\partial r_\beta} + \operatorname{div} \kappa \nabla T \, . \tag{12.54}$$

It shows that

> the heat abundance change $dQ = \rho T \, ds$ in a moving element of unit volume is a sum of the Joule and viscous heating and conductive heat redistribution to neighbour elements.

The momentum conservation law (12.42) can be also recast into the equation of plasma motion in the hydrodynamic form:

$$\frac{d\mathbf{v}}{dt} = -\frac{\nabla p}{\rho} - \frac{1}{4\pi\rho} \mathbf{B} \times \operatorname{curl} \mathbf{B} + \frac{\eta}{\rho} \Delta \mathbf{v} + \frac{1}{\rho} \left(\zeta + \frac{\eta}{3} \right) \nabla \operatorname{div} \mathbf{v} \, . \tag{12.55}$$

Once again, we see that the momentum of electromagnetic field does not appear in the non-relativistic equation of plasma motion.

12.2.3 Relativistic magnetohydrodynamics

Relativistic MHD models are of considerable interest in several areas of astrophysics. The theory of gravitational collapse and models of supernova explosions are based on relativistic hydrodynamic models for a star. In most models a key feature is the occurrence of a relativistic shock, for example, to expel the bulk of the star. The effects of deviations from spherical symmetry due to an initial angular momentum and magnetic field require the use of relativistic MHD models.

In the theories of galaxy formation, relativistic fluid models have been used, for example, in order to describe the evolution of perturbations of the baryon and radiation components of the cosmic medium. Theories of relativistic stars are also based on relativistic fluid model (Zel'dovich and Novikov, 1978; Rose, 1998).

When the medium interacts electromagnetically and is highly conducting, the simplest description is in terms of relativistic MHD. From the mathematical viewpoint, the relativistic MHD was mainly treated in the framework of *general relativity*. This means that the MHD equations were studied in conjunction with Einstein's equations. Lichnerowicz (1967) has made a thorough and deep investigation of the initial value problem. Gravitohydromagnetics describes one of the most fascinating phenomena in the outer space (e.g., Punsly, 2001).

In many applications, however, one neglects the gravitational field generated by the conducting medium in comparison with the background gravitational field as well as in many cases one simply uses *special relativity*. Mathematically this amounts to taking into account only the **conservation laws**

for matter and the electromagnetic field, neglecting Einstein's equations. Such relativistic MHD theory is much simpler than the full general relativistic theory. So more detailed results can be obtained (Anile, 1989; Novikov and Frolov, 1989; Koide et al., 1999).

12.3 Magnetic flux conservation. Ideal MHD

12.3.1 Integral and differential forms of the law

Equations (12.45), (12.42), and (12.46) are the conservation laws for mass, momentum, and energy, respectively. Let us show that, with the proviso that $\nu_{\rm m} = 0$, Equation (12.43) is the magnetic flux conservation law.

Let us consider the derivative of the vector $\mathbf{B}$ flux through a surface S moving with the plasma (Figure 12.1).

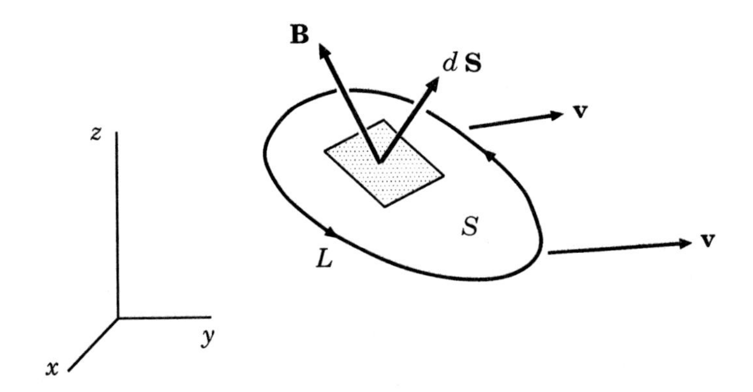

Figure 12.1: The magnetic field $\mathbf{B}$ flux through the surface S moving with a plasma with velocity $\mathbf{v}$.

According to the known formula of vector analysis (see Smirnov, 1965), we have

$$\frac{d}{dt} \int\limits_S \mathbf{B} \cdot d\mathbf{S} = \int\limits_S \left(\frac{\partial \mathbf{B}}{\partial t} + \mathbf{v} \, \mathrm{div} \, \mathbf{B} + \mathrm{curl} \, (\mathbf{B} \times \mathbf{v}) \right) \cdot d\mathbf{S} \, . \tag{12.56}$$

By virtue of Equation (12.44), formula (12.56) is rewritten as follows

$$\frac{d}{dt} \int\limits_S \mathbf{B} \cdot d\mathbf{S} = \int\limits_S \left(\frac{\partial \mathbf{B}}{\partial t} - \mathrm{curl} \, (\mathbf{v} \times \mathbf{B}) \right) \cdot d\mathbf{S} \, ,$$

or, making use of Equation (12.43),

$$\frac{d}{dt} \int_S \mathbf{B} \cdot d\mathbf{S} = \nu_{\mathrm{m}} \int_S \Delta \mathbf{B} \cdot d\mathbf{S} . \tag{12.57}$$

Thus, if we cannot neglect magnetic diffusivity ν_{m}, then

> the change rate of magnetic flux through a surface moving together with a conducting plasma is proportional to the magnetic diffusivity of the plasma.

The right-hand side of formula (12.57) can be rewritten with the help of the Stokes theorem:

$$\frac{d}{dt} \int_S \mathbf{B} \cdot d\mathbf{S} = -\nu_{\mathrm{m}} \oint_L \operatorname{curl} \mathbf{B} \cdot d\mathbf{l} . \tag{12.58}$$

Here L is the 'liquid' contour bounding the surface S. We have also used here that

$$\Delta \mathbf{B} = -\operatorname{curl} \operatorname{curl} \mathbf{B} .$$

By using Equation (12.33) we have

$$\frac{d}{dt} \int_S \mathbf{B} \cdot d\mathbf{S} = -\frac{c}{\sigma} \oint_L \mathbf{j} \cdot d\mathbf{l} . \tag{12.59}$$

The change rate of flux through a surface connected with the moving plasma is proportional to the electric resistivity σ^{-1} of the plasma.

Equation (12.59) is equivalent to the differential Equation (12.43) and presents an *integral* form of the magnetic flux conservation law.

> The magnetic flux through any surface moving with the plasma is conserved, once the electric resistivity σ^{-1} can be ignored.

Let us clarify the conditions when it is possible to neglect electric resistivity of plasma. The relative role of the dissipation processes in the *differential* Equation (12.43) can be evaluated by proceeding as follows. In a spirit similar to that of Section 5.2, we pass on to the dimensionless variables

$$\mathbf{r}^* = \frac{\mathbf{r}}{L} , \quad t^* = \frac{t}{\tau} , \quad \mathbf{v}^* = \frac{\mathbf{v}}{v} , \quad \mathbf{B}^* = \frac{\mathbf{B}}{B_0} . \tag{12.60}$$

On substituting definition (12.60) into Equation (12.43) we obtain

$$\frac{B_0}{\tau} \frac{\partial \mathbf{B}^*}{\partial t^*} = \frac{v B_0}{L} \operatorname{curl}^* (\mathbf{v}^* \times \mathbf{B}^*) + \nu_{\mathrm{m}} \frac{B_0}{L^2} \Delta^* \mathbf{B}^* .$$

Now we normalize this equation with respect to its left-hand side, i.e.

$$\frac{\partial \mathbf{B}^*}{\partial t^*} = \frac{v\tau}{L} \, \text{curl}^* \left(\mathbf{v}^* \times \mathbf{B}^* \right) + \frac{\nu_{\rm m}\tau}{L^2} \, \Delta^* \mathbf{B}^* \, . \tag{12.61}$$

The dimensionless Equation (12.61) contains two dimensionless parameters. The first one,

$$\delta = \frac{v\tau}{L} \, ,$$

will be discussed in the next Section. Here, for simplicity, we assume $\delta = 1$. The second parameter,

$$\boxed{\text{Re}_{\rm m} = \frac{L^2}{\nu_{\rm m}\,\tau} = \frac{vL}{\nu_{\rm m}} \, ,} \tag{12.62}$$

is termed the *magnetic* Reynolds number, by analogy with the *hydrodynamic* Reynolds number $\text{Re} = vL/\nu$. This parameter characterizes the ratio of the first term on the right-hand side of (12.61) to the second one. Omitting the asterisk, we write Equation (12.61) in the *dimensionless* form

$$\frac{\partial \mathbf{B}}{\partial t} = \text{curl} \left(\mathbf{v} \times \mathbf{B} \right) + \frac{1}{\text{Re}_{\rm m}} \, \Delta \mathbf{B} \, . \tag{12.63}$$

The larger the magnetic Reynolds number, the smaller the role played by magnetic diffusivity.

So the magnetic Reynolds number is the dimensionless measure of the relative importance of resistivity. If $\text{Re}_{\rm m} \gg 1$, we neglect the plasma resistivity and associated Joule heating and magnetic field dissipation, just as one neglects viscosity effects under large Reynolds numbers in ordinary hydrodynamics.

In laboratory experiments, for example in devices for studying the processes of current layer formation and rupture during magnetic reconnection (e.g., Altyntsev et al., 1977; Bogdanov et al., 1986, 2000), because of a small value L^2, the magnetic Reynolds number is usually not large: $\text{Re}_{\rm m} \sim 1 - 3$. In this case the electric resistivity has a dominant role, and Joule dissipation is important.

12.3.2 The equations of ideal MHD

Under astrophysical conditions, owing to the *low resistivity* and the enormously large length scales usually considered, the magnetic Reynolds number is also very large: $\text{Re}_{\rm m} > 10^{10}$ (for example, in the solar corona; see Exercise 12.1). Therefore, in a great number of problems of plasma astrophysics, it is sufficient to consider a medium with *infinite conductivity*:

$$\text{Re}_{\rm m} \gg 1 \, . \tag{12.64}$$

Furthermore the usual Reynolds number can be large as well (see, however, Exercise 12.2):

$$\mathrm{Re} \gg 1\,. \tag{12.65}$$

Let us also assume the heat exchange to be of minor importance. This assumption is not universally true either. Sometimes the thermal conductivity (due to thermal electrons or radiation) is so effective that the plasma behaviour must be considered as isothermal, rather than adiabatic. However, conventionally,

> while treating the 'ideal medium', all dissipative transfer coefficients as well as the thermal conductivity are set equal to zero

in the non-relativistic MHD equations (12.42)–(12.49):

$$\boxed{\nu_{\mathrm{m}} = 0\,, \quad \eta = \zeta = 0\,, \quad \kappa = 0\,.} \tag{12.66}$$

The complete set of the MHD equations for the ideal medium has two different (but equivalent) forms. The first one (with the energy Equation 12.54) is the form of *transfer* equations:

$$\frac{\partial \mathbf{v}}{\partial t} + (\mathbf{v}\cdot\nabla)\,\mathbf{v} = -\frac{\nabla p}{\rho} - \frac{1}{4\pi\rho}\,\mathbf{B}\times\mathrm{curl}\,\mathbf{B}\,,$$

$$\frac{\partial \mathbf{B}}{\partial t} = \mathrm{curl}\,(\mathbf{v}\times\mathbf{B})\,, \qquad \mathrm{div}\,\mathbf{B} = 0\,, \tag{12.67}$$

$$\frac{\partial \rho}{\partial t} + \mathrm{div}\,\rho\,\mathbf{v} = 0\,, \quad \frac{\partial s}{\partial t} + (\mathbf{v}\cdot\nabla)\,s = 0\,, \quad p = p\,(\rho, s)\,.$$

The other form of ideal MHD equations is the *divergent* form which also corresponds to the *conservation laws* for energy, momentum, mass and magnetic flux:

$$\frac{\partial}{\partial t}\left(\frac{\rho v^2}{2} + \rho\varepsilon + \frac{B^2}{8\pi}\right) = -\,\mathrm{div}\,\mathbf{G}\,, \tag{12.68}$$

$$\frac{\partial}{\partial t}\,\rho\,v_\alpha = -\,\frac{\partial}{\partial r_\beta}\,\Pi^*_{\alpha\beta}\,, \tag{12.69}$$

$$\frac{\partial \rho}{\partial t} = -\,\mathrm{div}\,\rho\,\mathbf{v}\,, \tag{12.70}$$

$$\frac{\partial \mathbf{B}}{\partial t} = \mathrm{curl}\,(\,\mathbf{v}\times\mathbf{B}\,)\,, \tag{12.71}$$

$$\mathrm{div}\,\mathbf{B} = 0\,, \tag{12.72}$$

$$p = p\,(\rho, s)\,. \tag{12.73}$$

Here the energy flux density and the momentum flux density tensor are, respectively, equal to (cf. (12.50) and (12.48))

$$\mathbf{G} = \rho \, \mathbf{v} \left(\frac{v^2}{2} + w \right) + \frac{1}{4\pi} \left(B^2 \, \mathbf{v} - (\mathbf{B} \cdot \mathbf{v}) \, \mathbf{B} \right), \tag{12.74}$$

$$\Pi^*_{\alpha\beta} = p \, \delta_{\alpha\beta} + \rho \, v_\alpha v_\beta + \frac{1}{4\pi} \left(\frac{B^2}{2} \, \delta_{\alpha\beta} - B_\alpha B_\beta \right). \tag{12.75}$$

The magnetic flux conservation law (12.71) written in the integral form

$$\frac{d}{dt} \int_S \mathbf{B} \cdot d\mathbf{S} = 0 \,, \tag{12.76}$$

where the integral is taken over an arbitrary surface moving with the plasma, is quite characteristic of ideal MHD. It allows us to clearly represent the magnetic field as a set of field lines attached to the medium, as if they were 'frozen into' it. For this reason, Equation (12.71) is frequently referred to as the 'freezing-in' equation.

The freezing-in property converts the notion of magnetic field line from the purely geometric to the material sphere.

> In the ideally conducting medium, the field lines move together with the plasma. The medium displacement conserves not only the magnetic flux but each of the field lines as well.

To convince ourselves that this is the case, we have to imagine a thin tube of magnetic field lines. There is no magnetic flux through any part of the surface formed by the collection of the boundary field lines that intersect the closed curve L. Let this flux tube evolve in time. Because of flux conservation, the plasma elements that are initially on the same magnetic flux tube must remain on the magnetic flux tube.

In ideal MHD flows, magnetic field lines inside the thin flux tube accompany the plasma. They are therefore materialized and are unbreakable because the flux tube links the same 'fluid particles' or the same 'fluid elements'. As a result its **topology cannot change**. Fluid particles which are not initially on a common field line cannot become linked by one later on. This general topological constraint restricts the ideal MHD motions, forbidding a lot of motions that would otherwise appear.

Conversely, the constraint that the thin flux tube follows the fluid particle motion, whatever its complexity, may create situations where the magnetic field structure becomes itself very complex (see vol. 2, Chapter 12).

In general, the field intensity $\mathbf{B}$ is a *local* quantity. However the magnetic field lines (even in vacuum) are *integral* characteristics of the field. Their analysis becomes more complicated. Nonetheless, a large number of actual fields have been studied because the general features of the morphology – an investigation of *non-local* structures – of magnetic fields are fairly important in plasma astrophysics.

The geometry of the field lines appears in different ways in the equilibrium criteria for astrophysical plasma. For example, much depends on whether the field lines are concave or convex, on the value of the gradient of the so-called *specific volume* of magnetic flux tubes (Chapter 19), on the presence of X-type points (Section 14.3) as well as on a number of other *topological* characteristics, e.g. magnetic *helicity* (see vol. 2, Chapter 12).

12.4 Practice: Exercises and Answers

Exercise 12.1 [Section 12.3.2] Estimate the magnetic diffusivity and the magnetic Reynolds number under typical conditions in the solar corona.

Answer. Let us take characteristic values of the parallel conductivity as they were estimated in Exercise 11.1:

$$\sigma_{\parallel} = \sigma \sim 10^{16} - 10^{17}\ \mathrm{s}^{-1}\,.$$

Substituting these values in formula (12.49) we obtain

$$\nu_{\mathrm{m}} = \frac{c^2}{4\pi\sigma} \approx 7.2 \times 10^{19}\,\frac{1}{\sigma} \sim 10^3 - 10^4\ \mathrm{cm}^2\,\mathrm{s}^{-1}. \qquad (12.77)$$

According to definition (12.62) the magnetic Reynolds number

$$\mathrm{Re}_{\mathrm{m}} = \frac{vL}{\nu_{\mathrm{m}}} \sim 10^{11} - 10^{12}\,, \qquad (12.78)$$

if the characteristic values of length and velocity, $L \sim 10^4$ km $\sim 10^9$ cm and $v \sim 10$ km s^{-1} $\sim 10^6$ cm s^{-1}, are taken for the corona. Thus the ideal MHD approximation can be well used to consider, for example, magnetic field diffusion in coronal linear scales.

Exercise 12.2 [Section 12.3.2] Show that

> in the solar corona, viscosity of plasma can be a much more important dissipative mechanism than its electric resistivity.

Answer. By using the formula (10.29) for viscosity, let us estimate the value of kinematic viscosity in the solar corona:

$$\nu = \frac{\eta}{\rho} \approx 3 \times 10^{15}\ \mathrm{cm}^2\,\mathrm{s}^{-1}. \qquad (12.79)$$

Here $T_{\mathrm{p}} \approx 2 \times 10^6$ K and $n_{\mathrm{p}} \approx n_{\mathrm{e}} \approx 2 \times 10^8\,\mathrm{cm}^{-3}$ have been taken as the typical proton temperature and density.

If the characteristic values of length and velocity, $L \sim 10^9$ cm and $v \sim 10^6$ cm s^{-1}, are taken (see Exercise 12.1), then the hydrodynamic Reynolds number

$$\mathrm{Re} = \frac{vL}{\nu} \sim 0.3\,. \tag{12.80}$$

The smallness of this number demonstrates the potential importance of viscosity in the solar corona. A comparison between (12.80) and (12.78) shows that $\mathrm{Re_m} \gg \mathrm{Re}$. Clearly, the viscous effects can dominate the effects of electric resistivity in coronal plasma.

Chapter 13

Magnetohydrodynamics in Astrophysics

Magnetohydrodynamics (MHD) is the simplest but sufficient approximation to describe many large-scale low-frequency phenomena in astrophysical plasma: regular and turbulent dynamo, plasma motions driven by strong magnetic fields, accreation disks, and relativistic jets.

13.1 The main approximations in ideal MHD

13.1.1 Dimensionless equations

The equations of MHD, even the ideal MHD, constitute a set of nonlinear differential equations in partial derivatives. The order of the set is rather high, while its structure is complicated. To formulate a problem in the context of MHD, we have to know the initial and boundary conditions admissible by this set of equations. To do this, in turn, we have to know the type of these equations, in the sense adopted in mathematical physics (see Vladimirov, 1971).

To formulate a problem, one usually uses one or another approximation, which makes it possible to isolate the main effect – the essence of the phenomenon. For instance, if the magnetic Reynolds number is small, then the plasma moves comparatively easily with respect to the magnetic field. This is the case in MHD generators and other laboratory and technical devices (Sutton and Sherman, 1965, § 1.3; Shercliff, 1965, § 6.5).

The opposite approximation is that of large magnetic Reynolds numbers, when magnetic field 'freezing in' takes place in the plasma (see Section 12.3.2). Obviously, the transversal (with respect to the magnetic field) plasma flows are implied. For any flow along the field, Equation (12.71) holds. This approximation is quite characteristic of the astrophysical plasma dynamics.

How can we isolate the main effect in a physical phenomenon and correctly formulate the problem? From the above examples concerning the magnetic Reynolds number, the following rule suggests itself:

> take the dimensional parameters characterizing the phenomenon at hand, combine them into dimensionless combinations and then, on calculating their numerical values, make use of the corresponding approximation in the set of *dimensionless* equations.

Such an approach is effective in hydrodynamics (Sedov, 1973, Vol. 1).

Let us start with the set of the ideal MHD Equations (12.67):

$$\frac{\partial \mathbf{v}}{\partial t} + (\mathbf{v} \cdot \nabla)\, \mathbf{v} = -\frac{\nabla p}{\rho} - \frac{1}{4\pi \rho}\, \mathbf{B} \times \operatorname{curl} \mathbf{B}\,, \tag{13.1}$$

$$\frac{\partial \mathbf{B}}{\partial t} = \operatorname{curl}\,(\mathbf{v} \times \mathbf{B})\,, \tag{13.2}$$

$$\frac{\partial \rho}{\partial t} + \operatorname{div} \rho\, \mathbf{v} = 0\,, \tag{13.3}$$

$$\frac{\partial s}{\partial t} + (\mathbf{v} \cdot \nabla)\, s = 0\,, \tag{13.4}$$

$$\operatorname{div} \mathbf{B} = 0\,, \tag{13.5}$$

$$p = p\,(\rho, s)\,. \tag{13.6}$$

Let the quantities L, τ, v, ρ_0, p_0, s_0, and B_0 be the characteristic values of length, time, velocity, density, pressure, entropy and field strength, respectively. Rewrite Equations (13.1)–(13.6) in the dimensionless variables

$$\mathbf{r}^* = \frac{\mathbf{r}}{L}\,, \quad t^* = \frac{t}{\tau}\,, \ \dots \quad \mathbf{B}^* = \frac{\mathbf{B}}{B_0}\,.$$

Omitting the asterisk, we obtain the equations in dimensionless variables (Somov and Syrovatskii, 1976b):

$$\varepsilon^2 \left\{ \frac{1}{\delta}\, \frac{\partial \mathbf{v}}{\partial t} + (\mathbf{v} \cdot \nabla)\, \mathbf{v} \right\} = -\gamma^2\, \frac{\nabla p}{\rho} - \frac{1}{\rho}\, \mathbf{B} \times \operatorname{curl} \mathbf{B}\,, \tag{13.7}$$

$$\frac{\partial \mathbf{B}}{\partial t} = \delta \operatorname{curl}\,(\mathbf{v} \times \mathbf{B})\,, \tag{13.8}$$

$$\frac{\partial \rho}{\partial t} + \delta \operatorname{div} \rho\, \mathbf{v} = 0\,, \tag{13.9}$$

$$\frac{\partial s}{\partial t} + \delta\, (\mathbf{v} \cdot \nabla)\, s = 0\,, \tag{13.10}$$

$$\operatorname{div} \mathbf{B} = 0\,, \tag{13.11}$$

$$p = p\,(\rho, s)\,. \tag{13.12}$$

Here

$$\boxed{\delta = \frac{v\tau}{L}\,, \quad \varepsilon^2 = \frac{v^2}{V_{\rm A}^2}\,, \quad \gamma^2 = \frac{p_0}{\rho_0 V_{\rm A}^2}} \tag{13.13}$$

are three dimensionless parameters characterizing the problem;

$$V_{\rm A} = \frac{B_0}{\sqrt{4\pi\rho_0}} \tag{13.14}$$

is the characteristic value of the Alfvén speed (see Exercise 13.1).

If the gravitational force were taken into account in (13.1), Equation (13.7) would contain another dimensionless parameter, $gL/V_{\rm A}^2$, where g is the gravitational acceleration. The analysis of these parameters allows us to gain an understanding of the approximations which are possible in the ideal MHD.

13.1.2 Weak magnetic fields in astrophysical plasma

We begin with the assumption that

$$\varepsilon^2 \gg 1 \quad \text{and} \quad \gamma^2 \gg 1\,. \tag{13.15}$$

As is seen from Equation (13.7), in the zero-order approximation relative to the small parameters ε^{-2} and γ^{-2}, we neglect the magnetic force as compared to the inertia force and the gas pressure gradient. In subsequent approximations, the magnetic effects are treated as a small correction to the hydrodynamic ones.

A lot of problems of plasma astrophysics are solved in this approximation, termed the *weak* magnetic field approximation. Among the simplest of them are the ones concerning the weak field's influence on hydrostatic equilibrium. An example is the problem of the influence of poloidal and toroidal magnetic fields on the equilibrium of a self-gravitating plasma ball (a star, the magnetoid of quasar's kernel etc., see examples in Section 19.1.3).

Some other problems are in fact analogous to the previously mentioned ones. They are called *kinematic* problems, since

> they treat the influence of a given plasma flow on the magnetic field; the reverse influence is considered to be negligible.

Such problems are reduced to finding the magnetic field distribution resulting from the known velocity field. An example is the problem of magnetic field amplification and support by stationary plasma flows (magnetic dynamo) or turbulent amplification. The simplest example is the problem of magnetic field amplification by plasma **differential rotation** (Elsasser, 1956; Moffat, 1978; Parker, 1979; Rüdiger and von Rekowski, 1998).

A leading candidate to explain the origin of large-scale magnetic fields in astrophysical plasma is the mean-field **turbulent magnetic dynamo** theory

(Moffat, 1978; Parker, 1979; Zel'dovich et al., 1983). The theory appeals to a combination of helical turbulence (leading to the so-called α effect), differential rotation (the Ω effect) and turbulent diffusion to exponentiate an initial seed mean magnetic field. The total magnetic field is split into a mean component and a fluctuating component, and the rate of growth of the mean field is sought.

The mean field grows on a length scale much larger than the outer scale of the turbulent velocity, with a growth time much larger than the eddy turnover time at the outer scale. A combination of kinetic and magnetic helicities provides a statistical correlation of small-scale loops favorable to exponential growth. Turbulent diffusion is needed to redistribute the amplified mean field rapidly to ensure a net mean flux gain inside the system of interest (a star or galaxy). Rapid growth of the fluctuating field necessarily accompanies the mean-field dynamo. Its impact upon the growth of the mean field, and the impact of the mean field itself on its own growth are controversial and depends crucially on the boundary conditions (e.g., Blackman and Field, 2000).

13.1.3 Strong magnetic fields in plasma

The opposite approximation – that of the *strong* magnetic field – has been less well studied. It reflects the specificity of MHD to a greater extent than the weak field approximation. The strong field approximation is valid when the **magnetic force**

$$\mathbf{F}_m = -\frac{1}{4\pi}\,\mathbf{B} \times \operatorname{curl}\mathbf{B} \tag{13.16}$$

dominates all the others (inertia force, gas pressure gradient, etc.). Within the framework of Equation (13.7), the magnetic field is referred to as a strong one if in some region under consideration

$$\varepsilon^2 \ll 1 \quad \text{and} \quad \gamma^2 \ll 1\,, \tag{13.17}$$

i.e. if the magnetic energy density greatly exceeds that of the kinetic and thermal energies:

$$\frac{B_0^2}{8\pi} \gg \frac{\rho_0 v^2}{2} \quad \text{and} \quad \frac{B_0^2}{8\pi} \gg 2 n_0 k_{\text{B}} T_0\,.$$

From Equation (13.7) it follows that, in the zeroth order with respect to the small parameters (13.17), the magnetic field is *force-free*, i.e. it obeys the equation

$$\mathbf{B} \times \operatorname{curl}\mathbf{B} = 0\,. \tag{13.18}$$

This conclusion is quite natural:

> if the magnetic force dominates all the others, then the magnetic field must balance itself in the region under consideration.

Condition (13.18) obviously means that electric currents flow parallel to magnetic field lines. If, in addition, electric currents are absent in some region (in the zeroth approximation relative to the small parameters ε^2 and γ^2), then the strong field is simply *potential* in this region:

$$\operatorname{curl} \mathbf{B} = 0 \,, \qquad \mathbf{B} = \nabla \Psi \,, \qquad \Delta \Psi = 0 \,. \tag{13.19}$$

In principle, the magnetic field can be force-free or even potential for another reason: due to the equilibrium of non-magnetic forces. However this does not happen frequently.

Let us consider the first order in the small parameters (13.17). If they are not equally significant, there are two possibilities.

(a) We suppose, at first, that

$$\varepsilon^2 \ll \gamma^2 \ll 1 \,. \tag{13.20}$$

Then we neglect the inertia force in Equation (13.7) as compared to the gas pressure gradient. Decomposing the magnetic force into a *magnetic tension* force and a *magnetic pressure* gradient force (see Exercises 13.2 and 13.3),

$$\mathbf{F}_m = -\frac{1}{4\pi}\, \mathbf{B} \times \operatorname{curl} \mathbf{B} = \frac{1}{4\pi}\, (\mathbf{B} \cdot \nabla)\, \mathbf{B} - \nabla \frac{B^2}{8\pi} \,, \tag{13.21}$$

we obtain the following dimensionless equation:

$$(\mathbf{B} \cdot \nabla)\, \mathbf{B} = \nabla \left(\frac{B^2}{2} + \gamma^2 p \right) . \tag{13.22}$$

Owing to the presence of the gas pressure gradient, the magnetic field differs from the force-free one at any moment of time:

> the magnetic tension force $(\mathbf{B} \cdot \nabla)\, \mathbf{B}/4\pi$ must balance not only the magnetic pressure gradient but that of the gas pressure as well.

Obviously the effect is proportional to the small parameter γ^2.

This approximation can be naturally called the *magnetostatic* one since $\mathbf{v} = 0$. It effectively works in regions of a strong magnetic field where the gas pressure gradients are large, for example, in coronal loops and reconnecting current layers (RCLs) in the solar corona (Exercise 13.4).

(b) The inertia force also causes the magnetic field to deviate from the force-free one:

$$\varepsilon^2 \left\{ \frac{1}{\delta} \frac{\partial \mathbf{v}}{\partial t} + (\mathbf{v} \cdot \nabla)\, \mathbf{v} \right\} = -\frac{1}{\rho}\, \mathbf{B} \times \operatorname{curl} \mathbf{B} \,. \tag{13.23}$$

Here we ignored (in the first order) the gas pressure gradient as compared with the inertia force. Thus it is not the relation (13.20) between the small parameters (13.17), but rather its converse, that should be obeyed, i.e.

$$\gamma^2 \ll \varepsilon^2 \ll 1 \,. \tag{13.24}$$

The problems on plasma flows in a strong magnetic field are of considerable interest in plasma astrophysics. To solve them, inequalities (13.24) can be assumed to hold. Then we can use (13.23) as the MHD equation of motion. The approximation corresponding inequalities (13.24) is naturally termed the approximation of *strong* field and *cold* plasma.

The main applications of the strong-field-cold-plasma approximation are concerned with the solar atmosphere (see vol. 2, Chapters 2 and 6) and the Earth's magnetosphere. Both astrophysical objects are well studied from the observational viewpoint. So we can proceed with confidence from qualitative interpretation to the construction of quantitative models. The presence of a sufficiently strong magnetic field and a comparatively rarefied plasma is common for both phenomena. This justifies the applicability of the approximation at hand.

> A sufficiently strong magnetic field easily moves a comparatively rarefied plasma in many non-stationary phenomena in space.

Analogous conditions are reproduced under laboratory modelling of these phenomena (e.g., Hoshino et al., 2001). Some other astrophysical applications of the strong-field-cold-plasma approximation will be discussed in the following two Sections.

* * *

In closing, let us consider the dimensionless parameter $\delta = v\tau/L$. As is seen from Equation (13.23), it characterizes the relative role of the local $\partial/\partial t$ and transport $(\mathbf{v} \cdot \nabla)$ terms in the substantial derivative d/dt.

If $\delta \gg 1$ then, in the zeroth approximation relative to the small parameter δ^{-1}, the plasma flow can be considered to be *stationary*

$$\varepsilon^2 \, (\mathbf{v} \cdot \nabla) \, \mathbf{v} = - \frac{1}{\rho} \, \mathbf{B} \times \operatorname{curl} \mathbf{B} \, . \tag{13.25}$$

If $\delta \ll 1$, i.e. plasma displacement is small during the magnetic field change, then the transport term $(\mathbf{v} \cdot \nabla)$ can be ignored in the substantial derivative and the equation of motion in the strong-field-cold-plasma approximation takes the form

$$\varepsilon^2 \, \frac{\partial \mathbf{v}}{\partial t} = - \frac{1}{\rho} \, \mathbf{B} \times \operatorname{curl} \mathbf{B} \, , \tag{13.26}$$

other equations becoming linear. This case corresponds to small plasma displacements from the equilibrial state, i.e. small perturbations. (If need be, the right-hand side of Equation (13.26) can be linearized in the usual way.)

Generally the parameter $\delta \approx 1$ and the set of MHD equations in the approximation of strong field and cold plasma for ideal medium assumes the following dimensionless form:

$$\varepsilon^2 \, \frac{d\mathbf{v}}{dt} = - \frac{1}{\rho} \, \mathbf{B} \times \operatorname{curl} \mathbf{B} \, , \tag{13.27}$$

$$\frac{\partial \mathbf{B}}{\partial t} = \text{curl}\,(\,\mathbf{v} \times \mathbf{B}\,)\,, \tag{13.28}$$

$$\frac{\partial \rho}{\partial t} + \text{div}\,\rho\,\mathbf{v} = 0\,. \tag{13.29}$$

In the next Chapter we shall consider some continuous plasma flows in a strong magnetic field, which are described by Equations (13.27)–(13.29).

13.2 Accretion disks of stars

13.2.1 Angular momentum transfer in binary stars

Magnetic fields were discussed as a possible means of angular transport in the development of *accretion disk* theory in the early seventies (Shakura and Sunyaev, 1973; Novikov and Thorne, 1973). Interest in the role of magnetic fields in binary stars steadily increased after the discovery of the nature of AM Herculis. It appeared that the optical counterpart of the soft X-ray source has linear and circular polarization in the V and I spectral bands, of a strength an order of magnitude larger than previously observed in any object. This suggested the presence of a very strong field, with $B \sim 10^8$ G, assuming the fundamental cyclotron frequency to be observed.

Similar systems were soon discovered. Evidence for strong magnetic fields was subsequently found in the X-ray binary pulsars and the intermediate polar binaries, both believed to include accretion disks. A magnetically channelled wind from the main sequence star has been invoked to explain the higher rates of mass transfer observed in binaries above the period gap, and in an explanation of the gap. The winds from accretion disks have been suggested as contributing to the inflow by removing angular momentum.

Magnetohydrodynamics in binary stars is now an area of central importance in stellar astrophysics (Campbell, 1997; Rose, 1998). Magnetic fields are believed to play a role even in apparently non-magnetic binaries. They provide the most viable means, through the so-called shear-type instabilities, of generating the MHD turbulence in an accretion disk necessary to drive the plasma inflow via the resulting magnetic and viscous stresses.

The fundamental problem is the role of magnetic fields in redistributing angular momentum in binary stars. The disk is fed by the plasma stream originated in the $L1$ region (Figure 13.1) of the secondary star. In a steady state,

> plasma is transported through the disk at the rate it is supplied by the stream and the angular momentum will be advected outwards.

Angular momentum avdection requires coupling between rings of rotating plasma; the ordinary hydrodynamic viscosity is too weak to provide this. Hence some form of **anomalous viscosity** must be invoked to explain the plasma flow through the disk.

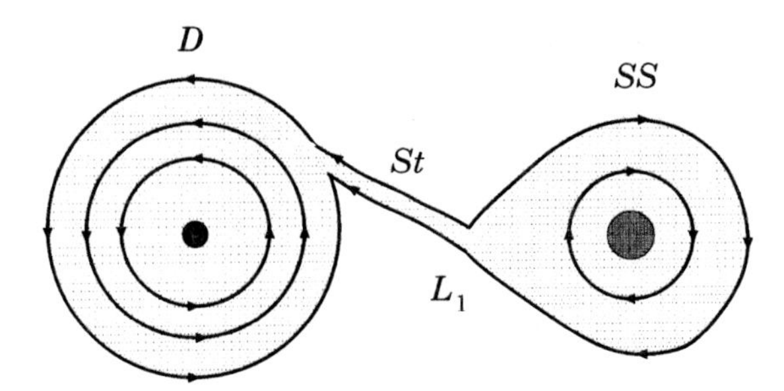

Figure 13.1: The standard model of a binary system viewed down the rotational pole. The tidally and rotationally distorted secondary star SS loses plasma from the unstable L_1 point. The resulting plasma stream St feeds an accretion disk D, centred on the primary star.

The key point is the recognition that a simple linear instability, which we refer to here as the standard *magnetorotational* instability (Hawley et al., 1995), generates MHD turbulence. This turbulence transports angular momentum outward through the disk, allowing accretion to proceed. Although turbulence seems like a natural and straightforward transport mechanism, it turns out that the **magnetic fields are essential**. Purely hydrodynamic turbulence is not self-sustaining and does not produce sustained outward transport of angular momentum (see Hawley and Balbus, 1999). MHD turbulence greatly enhances angular momentum transport associated with the so-called α-disks (Balbus and Papaloizou, 1999).

It is most probable that the accretion disks have turbulent motions generated by the shear instabilities. The turbulence and strong radial shear lead to the generation and maintenance of a large scale magnetic field.

> Viscous and magnetic stresses cause radial advection of the angular momentum via the azimutal forces.

Provided these forces oppose the large-scale azimutal motion, plasma will spiral in through the disk as angular momentum flows outwards. Presumably, the approximation of a weak field (Section 13.1.2) can be used inside the disk to model these effects. Most models to date involve a vertically averaged structure. The future aim is to find 3D solutions which self-consistently incorporate the magnetic shear instabilities and vertical structure.

The stellar spin dynamics and stability are also important, of course. For example, in spin evolution calculations, a compact white draft, or neutron star, is usually treated as a rigid body. This is valid provided the dynamic time-scale for adjustments in the stellar structure is short compared to the spin evolution time scale. In general, however, a strongly-magnetic primary star may experience significant distortions from spherical symmetry due to non-radial internal magnetic forces. This fact can be demonstrated by the tensor virial theorem in MHD (Section 19.1.3).

13.2.2 Magnetic accretion in cataclysmic variables

Cataclysmic variables (CVs) are interacting binary systems composed of a white dwarf (primary star) and a late-type, main-sequence companion (secondary star). The secondary star fills its Roche lobe, and plasma is transferred to the compact object through the inner Lagrangian point. The way this plasma falls towards the primary depends on the intensity of a magnetic field of the white dwarf.

If the magnetic field is weak, the mass transfer occurs through an optically thick accretion disk. Such CVs are classified as non-magnetic ones.

The strong magnetic field ($B \gtrsim 10^7$ G) may entirely dominate the geometry of the accretion flow. The magnetic field is strong enough to synchronize the white dwarf rotation (spin) with the orbital period. Synchronization occurs when the magnetic torque between primary and secondary overcomes the accretion torque, and no disk is formed. Instead, the field channels accretion towards its polar regions. Such synchronous systems are known as AM Herculis binaries or polars.

The intermediate ($B \sim 2 - 8 \times 10^6$ G) magnetic field primary stars harbor magnetically truncated accretion disks which can extend until magnetic pressure begins to dominate. A shock should appear when the plasma streams against the white dwarf's magnetosphere. The shock should occur close to the corotation radius (the distance from the primary at which the Keplerian and white dwarf angular velocities match), inside and above the disk plane. Presumably the plasma is finally accreted onto the magnetic poles of the white dwarf. The asynchronous systems are known as DQ Herculis binaries or Intermediate Polars (IPs).

General properties of plasma flows driven by a strong magnetic field will be discussed in Chapter 14.

The accretion geometry strongly influences the emission properties at all wavelengths and its variability. The knowledge of the behaviour in all energy domains can allow one to locate the different accreting regions (Bianchini et al., 1995). Reid et al. (2001) discovered the first magnetic white dwarf of the spectral type DZ, which shows lines of heavy elements like Ca, Mg, Na, and Fe. The cool white dwarf LHS 2534 offers the first empirical data in an astrophysical setting of the Zeeman effect on neutral Na, Mg, and both ionized and neutral Ca. The Na I splittings result in a mean surface field strength estimate of 1.92×10^6 G. In fact, there are direct laboratory measurements of the Na I D lines that overlap this field strength.

13.2.3 Accretion disks near black holes

In interacting binary stars there is an abundance of evidence for the presence of accretion disks: (a) double-peaked emission lines are observed; (b) eclipses of an extended light source centered on the primary occur, and (c) in some cases eclipses of the secondary star by the disk are also detected. The case for the presence of accretion disks in active galactic nuclei is less clear. Nonetheless the disk-fed accreation onto a super-massive black hole is the commonly accepted model for these astronomical objects. In fact, active galactic nuclei also exhibit the classical double-peaked, broad emission lines which are considered to be characteristic for a rotating disk.

As the plasma accretes in the gravitational field of the central mass, magnetic field lines are convected inwards, amplified and finally deposited on **the horizon of the black hole** (Section 8.3.4). As long as a magnetic field is confined by the disk, a differential rotation causes the field to wrap up tightly (see Section 20.1.5), becoming highly sheared and predominantly azimuthal in orientation. A dynamo in the disk may be responsible for the maintenance and amplification of the magnetic field.

In the standard model of an accreation disk (Shakura and Sunyaev, 1973; Novikov and Thorne, 1973), the gravitational energy is locally radiated from the optically thin disk, and the plasma keeps its Keplerian rotation. However **the expected power far exceeds the observed luminosity**.

There are two possible explanations for the low luminosities of nearby black holes: (a) the accretion occurs at extremely low rates, or (b) the accretion occurs at low radiative efficiency. *Advection* has come to be thought of as an important process and results in a structure different from the standard model. The advection process physically means that

> the energy generated via viscous dissipation is restored as entropy of the accreting plasma flow rather than being radiated.

The advection effect can be important if the radiation efficiency decreases under these circumstances (Section 8.3.4). An optically thin advection-dominated accretion flow (ADAF) seems to be a hydrodynamic model that can reproduce the observed hard spectra of black hole systems such as active galactic nuclei (AGN) and Galactic black hole candidates (e.g., Manmoto, 2000).

This situation is perhaps best illustrated by the case of nearby elliptical galaxy nuclei (Di Matteo et al., 2000). Assuming that the accretion occurs primarily from the hot, quasi-spherical interstellar medium (ISM), the Bondi (1952) theory can be used to estimate the accreation rates onto the supermassive black holes. Such estimates, however, require accurate measurements of both the density and the temperature of the ISM at the Bondi accreation radius, i.e., the radius at which the gravitational force of the black hole begins to dominate the dynamics of the hot plasma.

In order to determine unambiguously whether or not the low luminosities of nearby black holes are due to a low radiative efficiency in the accreting plasma, it is also necessary to measure the nuclear power. When combined with the estimated accretion rates, this gives us a direct measurement of the radiative efficiency η_r.

Thanks to its high spatial resolution and sensitivity, the *Chandra X-ray Observatory* is able, for the first time, to detect nuclear X-ray point sources in nearby galaxies and provide us with direct measurements of their luminosities. *Chandra* also allows us to measure the central temperatures and densities of the ISM close to accretion radii of the central black holes and therefore to determine the Bondi accretion rates in these systems to much greater accuracy than before.

Di Matteo et al. (2001) explored the implications of *Chandra* observations of the giant elliptical galaxy NGC 6166. They show that, if the central black hole of $\sim 10^9\ M_\odot$ is fed at the estimated Bondi rate, the inferred efficiency $\eta_r \lesssim 10^{-5}$. At the given accretion rate, ADAF models can explain the observed nuclear luminosity. However the presence of **fast outflows** in the accretion flow is also consistent with the present constraints. The power output from the jets in NGC 6166 is also important to the energetics of the system.

13.2.4 Flares in accretion disk coronae

Following the launch of several X-ray satellites, astrophysicists have tried to observe and analyze the violent variations of high energy flux from black hole candidates (e.g., Negoro et al., 1995; see also review in Di Matteo et al., 1999). So far, similar solar and astrophysical statistical studies have been done almost independently of each other. Ueno (1998) first compared X-ray light curves from the solar corona and from the accretion disk in Cyg X-1, a famous black hole candidate. He analyzed also the power spectral densities, the peak interval distributions (the interval of time between two consecutive flares), and the peak intensity distributions.

It has appeared that there are many relationships between flares in the solar corona and 'X-ray shots' in accretion disks. (Of course, there are many differences and unexplained features.) For example, the peak interval distribution of Cyg X-1 shows that the occurrence frequency of large X-ray shots is reduced. A second large shot does not occur soon after a previous large shot. This suggests the existence of energy-accumulation structures, such as magnetic fields in solar flares.

It is likely that accretion disks have a corona which interacts with a magnetic field generated inside a disk. Galeev et al. (1979) suggested that the corona is confined in strong magnetic loops which have buoyantly emerged from the disk. Buoyancy constitutes a mechanism able to channel a part of the energy released in the accretion process directly into the corona outside the disk.

> Magnetic reconnection of buoyant fields in the lower density surface regions may supply the energy source for a hot corona.

On the other hand, the coronal magnetic field can penetrate the disk and is stressed by its motions. The existence of a disk corona with a *strong* field

(Section 13.1.3) raises the possibility of a wind flow similar to the solar wind. In principle, this would result in angular momentum transport away from the disk, which could have some influence on the inflow. Another feature related to the accretion disk corona is the possibility of a flare energy release similar to solar flares (see vol. 2, Section 8.3).

When a plasma in the disk corona is optically thin and has a dominant magnetic pressure, the circumstances are likely to be similar to the solar corona. Therefore

> it is possible to imagine some similarity between the mechanisms of solar flares and X-ray shots in accretion disks.

Besides the effect of heating the the disk corona, reconnection is able to accelerate particles to high energies (Lesch and Pohl, 1992; Bednarek and Protheroe, 1999). Some geometrical and physical properties of the flares in accretion disk coronae can be inferred almost directly from soft- and hard X-ray observations of Galactic black hole candidates (Beloborodov, 1999; Di Matteo et al., 1999).

13.3 Astrophysical jets

13.3.1 Jets near black holes

Jet-like phenomena, including relativistic jets (Begelman et al., 1984; Birkinshaw, 1997), are observed on a wide range of scales in accretion disk systems. Active galactic nuclei (AGN) show extremely energetic outflows extending even to scales beyond the outer edge of a galaxy in the form of strongly collimated radio jets. The luminosities of the radio jets give an appreciable fraction of the luminosity of the underlying central object. There is substantial evidence that **magnetic forces are involved in the driving mechanism** and that the magnetic fields also provide the collimation of relativistic flows (see also Section 20.1.3). So numerucal simulations must incorporate relativistic MHD in a four-dimensional space-time (Nishikawa et al., 1999; Koide et al., 1999).

Rotating black holes are thought to be the prime-mover behind the activity detected in centers of galaxies. The gravitational field of rotating black holes is more complex than that of non-rotating ones. In addition to the ordinary gravitational force, $m\mathbf{g}$, the rotation generates the so-called *gravitomagnetic* force which is just an analogy of the Lorentz force. In fact, the full weak-gravity (far from the hole) low-velocity (replacing the relativistic unified space-time with an equivalent Galilean 'absolute-space-plus-universal-time') coordinate acceleration of uncharged particle (Macdonald et al., 1986; see also Chapter 4 in Novikov and Frolov, 1989)

$$\frac{d^2\mathbf{r}}{dt^2} = \mathbf{g} + \frac{d\mathbf{r}}{dt} \times \mathbf{H}_{gr} \tag{13.30}$$

looks like the Lorentz force with the electric field $\mathbf{E}$ replaced by $\mathbf{g}$, the magnetic field $\mathbf{B}$ replaced by the vector $\mathbf{H}_{gr} = \operatorname{curl} \mathbf{A}_{gr}$, and the electric charge e replaced by the particle mass m. These analogies lie behind the use of the words 'gravitoelectric' and 'gravitomagnetic' to describe the gravitational acceleration field $\mathbf{g}$ and to describe the 'shift function' $\mathbf{A}_{gr}$ and its derivatives (Exercise 13.6).

The analogy with electromagnetism remains strong so long as all velocities are small compared with that of light and gravity is weak enough to be linear. Thus, far from the horizon, the gravitational acceleration

$$\mathbf{g} = -\frac{M}{r^2}\, \mathbf{e}_r \tag{13.31}$$

is the radial Newtonian acceleration and the gravitomagnetic field

$$\mathbf{H}_{gr} = 2\, \frac{\mathbf{J} - 3\,(\mathbf{J} \cdot \mathbf{e}_r)\, \mathbf{e}_r}{r^3} \tag{13.32}$$

is a dipole field with the role of dipole moment played by the hole's angular momentum

$$\mathbf{J} = \int (\mathbf{r} \times \rho_m \mathbf{v})\; dV\,. \tag{13.33}$$

A physical manifestation of the gravitomagnetic field (13.32) is the precession that is induced in gyroscopes far from the hole. The electromagnetic analogy suggests that not only should the gravitomagnetic field exert a torque on a gyroscope outside a black hole, it should also exert a force. **The gravitomagnetic force drives an accretion disk into the hole's equatorial plane** and holds it there indefinitely regardless of how the disk's angular momentum may change (Figure 13.2).

Consequently, at radii where the bulk of the disk's gravitational energy is released and where the hole-disk interactions are strong, there is only one geometrically preferred direction along which a jet might emerge: the normal to the disk plane, which coincides with the rotation axis of the black hole. In some cases the jet might be produced by winds off the disk, in other cases by electrodynamic acceleration of the disk, and in others by currents in the hole's magnetosphere (see Begelman et al., 1984). However whatever the mechanism, the jet presumably is locked to the hole's rotation axis.

The black hole acts as a gyroscope to keep the jet aligned. The fact that it is very difficult to torque a black hole accounts for the constancy of the observed jet directions over length scales as great as millions of light years and thus over time scales of millions of years or longer.

A black hole by itself is powerless to produce the observed jets. It does so only with the aid of surrounding plasma and magnetic fields. A supermassive hole in a galactic nucleus can acquire surrounding matter either by gravitationally pulling interstellar gas into its vicinity, or by tidally disrupting passing stars and smearing their matter out around itself. In either case the

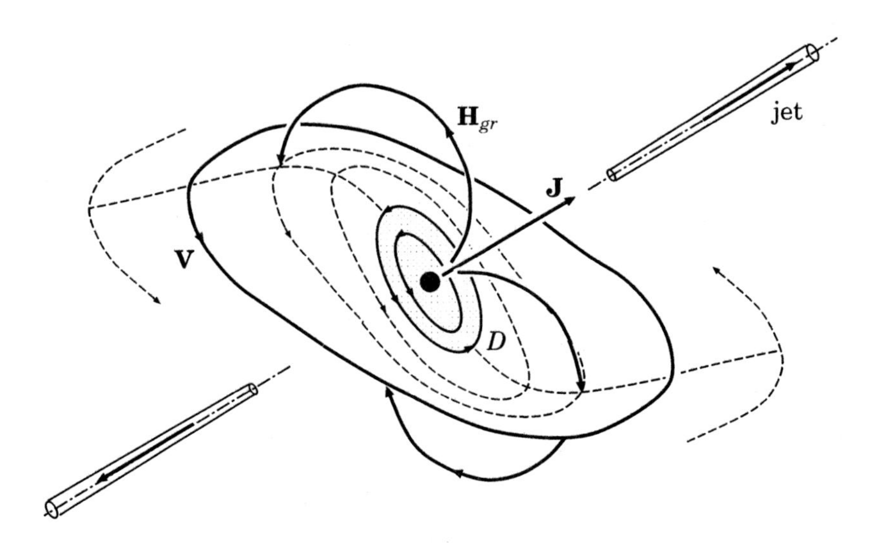

Figure 13.2: An accretion disk D around a rotating black hole is driven into the hole's equatorial plane at small radii by a combination of gravitomagnetic forces (action of the gravitomagnetic field $\mathbf{H}_{gr}$ on orbiting plasma) and viscous forces.

gas is likely to have so much angular momentum that, instead of being swallowed directly and radially into the hole, it forms an orbiting disk around the hole. The orientation of the disk at large radii is determined by the direction of the angular momentum of the recently acquired gas, see an external part of the accretion disk in Figure 13.2.

In the highly-conducting medium, the gravitomagnetic force couples with electromagnetic fields over Maxwell's equations. This effect has interesting consequences for the magnetic fields advected from the interstellar matter towards the black hole (Camenzind, 1990). It leads to a gravitomagnetic dynamo which amplifies any seed field near a rotating compact object. This process builds up the dipolar magnetic structures which may be behind the bipolar outflows seen as relativistic jets (for comparison with a non-relativistic process see Section 14.4).

Magnetic fields also influence the accretion towards the rotating black hole. For rapidly rotating holes, the accreation can carry negative angular momentum inwards, spinning down the black hole.

13.3.2 Relativistic jets from disk coronae

Relativistic jets are produced perpendicular to the accretion disk plane (see Figure 13.2) around a super-massive black hole in the central part of an AGN. The shock of the jets on intergalactic media, at a distance of several hundreds of kpc from the central engine, is considered as being able to accelerate particles up to the highest energies, say 10^{20} eV for cosmic rays. This hypothesis need, however, to be completed by some further and necessary ingradients since such powerful galaxies are rare objects.

Subramanian et al. (1999) consider the possibility that the relativistic jets observed in many active galactic nuclei may be powered by the Fermi acceleration of protons in a tenuous corona above a two-temperature accretion disk (Section 8.3.4). The acceleration arises, in this scenario, as a consequence of the shearing motion of the magnetic field lines in the corona, that are anchored in the underlying Keplerian disk. The protons in the corona have a power-law distribution because the density there is too low for proton-proton collisions (formula (8.39)) to thermalize the energy supplied via Fermi acceleration.

The same mechanism also operates in the disk itself. However there the density is high enough for thermalization to occur and consequently the disk protons have the Maxwellian distribution. Particle acceleration in the corona leads to the development of a pressure-driven wind that passes through a critical point and subsequently transforms into a relativistic jet at large distances from the black hole.

13.4 Practice: Exercises and Answers

Exercise 13.1 [Section 13.1.1] Evaluate the characteristic value of Alfvén speed in the solar corona above a large sunspot.

Answer. From definition (13.14) we find the following formula for Alfvén speed

$$V_{\rm A} \approx 2.18 \times 10^{11} \, \frac{B}{\sqrt{n}} \, , \quad {\rm cm\,s^{-1}} \, . \tag{13.34}$$

In this formula, in the coefficient, we have neglected a small contribution of the ions that are heavier than protons into the plasma density ρ. Another thing is much more important however.

Above a sunspot the field strength can be as high as $B \approx 3000$ G. Plasma density in the low corona $n \approx 2 \times 10^8 \, {\rm cm}^{-3}$. For these values formula (13.34) gives unacceptably high values of the Alfvén speed: $V_{\rm A} \approx 5 \times 10^{10} \, {\rm cm\,s}^{-1} > c$. This means that

> in a strong magnetic field and low density plasma, the Alfvén waves propagate with velocities approaching the light speed c.

So formula (13.34) has to be corrected by a *relativistic* factor which takes this fact into account.

Alfvén (1950) pointed out that the 'magnetohydrodynamic waves' are just an extreme case of electromagnetic waves (Section 15.2.2 and Exercise 15.3). Alfvén has shown that the transition between electromagnetic and Alfvén waves can be surveyed by the help of the following formula for the speed of propagation along the magnetic field:

$$V_{\text{A}}^{rel} = \frac{B}{\sqrt{4\pi\rho}} \frac{1}{\sqrt{1 + B^2/4\pi\rho c^2}} , \tag{13.35}$$

which agrees with (13.14) when $B^2 \ll 4\pi\rho c^2$. Therefore the relativistic Alfvén wave speed is always smaller than the light speed:

$$V_{\text{A}}^{rel} = \frac{c}{\sqrt{1 + 4\pi\rho c^2/B^2}} \leq c \, . \tag{13.36}$$

For values of the magnetic field and plasma density mentioned above, this formula gives $V_{\text{A}}^{rel} \approx 2 \times 10^{10}\,\text{cm}\,\text{s}^{-1} < c$.

Formula (13.36) shows that, in low desity cosmic plasmas, the Alfvén speed can easily approach the light speed c.

Exercise 13.2 [Section 13.1.3] Discuss properties of the Lorentz force (13.16) in terms of the Maxwellian stress tensor (12.11).

Answer. In non-relativistic MHD, the Maxwellian stress tensor has only magnetic field components (see formula (12.48))

$$\text{M}_{\alpha\beta} = \frac{1}{4\pi} \left(\frac{B^2}{2} \, \delta_{\alpha\beta} - B_\alpha B_\beta \right) . \tag{13.37}$$

Let us write down these components in the reference system which has the z coordinate in the direction of the magnetic field at a given point. In its neighbourhood, formula (13.37) implies

$$\text{M}_{\alpha\beta} = \left\| \begin{array}{ccc} B^2/8\pi & 0 & 0 \\ 0 & B^2/8\pi & 0 \\ 0 & 0 & -B^2/8\pi \end{array} \right\| . \tag{13.38}$$

According to definition (13.37) the zz component of the tensor has two parts:

$$\text{M}_{zz} = \frac{B^2}{8\pi} - \frac{B^2}{4\pi} \, . \tag{13.39}$$

The first part, $B^2/8\pi$, combines with the M_{xx} and M_{yy} components to give an isotropic pressure. The remaining part, $-B^2/4\pi$, corresponds to excess 'negative pressure' or *tension* in the z direction. Thus

> a magnetic field has a tension along the field lines in addition to having the isotropic pressure, $B^2/8\pi$.

The second term on the right-hand side of the Maxwellian stress tensor (13.37) describes the *magnetic tension* along field lines. Recall that the diagonal components of the pressure tensor (12.15), in exactly the same way, correspond to isotropic gas pressure and the off-diagonal components to viscous shear.

Exercise 13.3 [Section 13.1.3] Show that the *magnetic tension force* is directed to the local centre of curvature.

Answer. The Lorentz force is

$$\mathbf{F}_m = -\frac{1}{4\pi}\,\mathbf{B} \times \operatorname{curl} \mathbf{B} = \frac{1}{4\pi}\,(\mathbf{B} \cdot \nabla)\,\mathbf{B} - \nabla\,\frac{B^2}{8\pi}\,. \tag{13.40}$$

Here $(\mathbf{B} \cdot \nabla)$ is the directional derivative along a magnetic field line. Hence we can use formulae that are similar to (5.43) and (5.44) to rewrite the first term on the right-hand side of (13.40) as follows

$$\frac{1}{4\pi}\,(\mathbf{B} \cdot \nabla)\,\mathbf{B} = -\frac{B^2}{4\pi}\,\frac{\mathbf{e}_c}{\mathcal{R}_c} + \frac{\partial}{\partial l}\,\frac{B^2}{8\pi}\,\mathbf{n}\,. \tag{13.41}$$

Here $\mathbf{n} = \mathbf{B}/B$ is the unit vector along the magnetic field, l is the distance along the field line, $\mathcal{R}_c$ is a radius of *curvature* for the field line at a given point $\mathbf{R}$. At this point the unit vector $\mathbf{e}_c$ is directed from the curvature center 0_c as shown in Figure 5.8.

Let us decompose the second term on the right-hand side of (13.40) as

$$-\nabla\,\frac{B^2}{8\pi} = -\nabla_\perp\,\frac{B^2}{8\pi} - \frac{\partial}{\partial l}\,\frac{B^2}{8\pi}\,\mathbf{n}\,, \tag{13.42}$$

where the operator $\nabla_\perp$ operates in the planes normal to the magnetic field lines.

Now we combine formulae (13.41) and (13.42) to write the Lorentz force as

$$\boxed{\mathbf{F}_m = -\nabla_\perp\,\frac{B^2}{8\pi} - \frac{B^2}{4\pi}\,\frac{\mathbf{e}_c}{\mathcal{R}_c}\,.} \tag{13.43}$$

The first term in the Lorentz force is the magnetic pressure force which is isotropic in the planes normal to the magnetic field lines. It is directed from high magnetic pressure (strong magnetic field) to low magnetic pressure (low field strength) in the same way as the gas pressure. Therefore

> the magnetic pressure force acts when the strength of the magnetic field is not a constant in space.

The second term on the right-hand side of (13.43), the magnetic tension force, is directed to the local center of curvature (see point 0_c is Figure 5.8). It is inversely proportional to the curvature radius $\mathcal{R}_c$. Thus the more a field line is curved, the stronger the tension force is.

The magnetic tension force behaves in an identical way as the tension force in an elastic string.

It is present for magnetic fields with curved field lines and tendes to make curved field lines straight, for example, in an Alfvén wave (see Figure 15.1).

The sum of both terms, the Lorentz force, has no component along the magnetic field. We already knew this since the vector product $\mathbf{B} \times \operatorname{curl} \mathbf{B}$ is perpendicular to the vector $\mathbf{B}$.

Exercise 13.4 [Section 13.1.1] For the conditions in the low corona, used in Exercise 13.1, estimate the parameter γ^2.

Answer. Substitute $p_0 = 2n_0 k_{\rm B} T_0$ in definition (13.13):

$$\gamma^2 = \frac{n_0 k_{\rm B} T_0}{B_0^2/8\pi} \approx 3.47 \times 10^{-15} \, \frac{n_0 T_0}{B_0^2} \, . \tag{13.44}$$

Let us take as the characteristic values of temperature $T_0 \approx 2 \times 10^6$ K and magnetic field $B_0 \approx 3000$ G. For these values formula (13.44) gives the dimensionless parameter $\gamma^2 \sim 10^{-7}$. Hence, in the solar corona above sunspots, the conditions (13.24) of a strong field can be satisfied well for a wide range of plasma parameters.

Exercise 13.5 [Section 12.3.2] By using general formula (12.74) for the energy flux in ideal MHD, find the magnetic energy influx into a reconnecting current layer (RCL).

Answer. Let us consider a current layer as a neutral one (Figure 8.5). In this simplest approximation, near the layer, the magnetic field $\mathbf{B} \perp \mathbf{v}$. Therefore in formula (12.74) the scalar product $\mathbf{B} \cdot \mathbf{v} = 0$ and the energy flux density

$$\mathbf{G} = \rho \, \mathbf{v} \left(\frac{v^2}{2} + w \right) + \frac{B^2}{4\pi} \, \mathbf{v} \, . \tag{13.45}$$

If the approximation of a strong field is satisfied, the last term in (13.45) is dominating, and we find the magnetic energy flux density or the Poynting vector (cf. general definition (12.52)) directed into the current layer

$$\mathbf{G}_{\rm P} = \frac{B^2}{4\pi} \, \mathbf{v} \, . \tag{13.46}$$

For a quarter of the current layer assumed to be symmetrical and for a unit length along the current, the total flux of magnetic energy

$$\mathcal{E}^{\,in}_{\,mag} = \frac{B_0^2}{4\pi} \, v_0 b \, . \tag{13.47}$$

Here b is half-width of the layer (see vol. 2, Figure 6.1), B_0 is the field strength on the inflow sides of the current layer, v_0 is the inflow velocity.

Exercise 13.6 [Section 13.3] Consider a weakly gravitating, slowly rotating body such as the Earth or the Sun, with all nonlinear gravitational effects neglected. Compute the *gravitational* force and *gravitomagnetic* force (as in Section 13.3.1) from the linearized Einstein equations (see Landau and Lifshitz, *Classical Theory of Field*, 1975, Chapter 10, § 100). Show that, for a time-independent body, these equations are identical to the Maxwell equations (1.24)–(1.27):

$$\operatorname{curl}\mathbf{g} = 0\,, \qquad \operatorname{div}\mathbf{g} = -\,4\pi\, G\rho_m\,, \tag{13.48}$$

$$\operatorname{curl}\mathbf{H}_{gr} = -\,16\pi\, G\,\rho_m\mathbf{v}\,, \qquad \operatorname{div}\mathbf{H}_{gr} = 0\,. \tag{13.49}$$

Here the differences are: (a) two minus signs due to gravity being attractive rather than repulsive, (b) the factor 4 in the $\operatorname{curl}\mathbf{H}_{gr}$ equation, (c) the presence of the gravitational constant G, (d) the replacement of charge density ρ^q by mass density ρ_m, and (e) the replacement of electric current density $\mathbf{j}$ by the density of mass flow $\rho_m\mathbf{v}$ with $\mathbf{v}$ the velocity of the mass.

Chapter 14

Plasma Flows in a Strong Magnetic Field

A sufficiently strong magnetic field easily moves a comparatively rarified plasma in many non-stationary phenomena in space, for example in solar flares and coronal mass ejections which strongly influence the interplanetary and terrestrial space.

14.1 The general formulation of the problem

As was shown in Section 13.1.3, the set of MHD equations for an ideal medium in the approximation of strong field and cold plasma is characterized only by the small parameter $\varepsilon^2 = v^2/V_{\rm A}^2$:

$$\varepsilon^2 \frac{d\mathbf{v}}{dt} = -\frac{1}{\rho}\,\mathbf{B} \times \operatorname{curl}\mathbf{B}\,, \tag{14.1}$$

$$\frac{\partial \mathbf{B}}{\partial t} = \operatorname{curl}\,(\mathbf{v} \times \mathbf{B})\,, \tag{14.2}$$

$$\frac{\partial \rho}{\partial t} + \operatorname{div}\rho\mathbf{v} = 0\,. \tag{14.3}$$

Let us try to find the solution to this set as a power series in the parameter ε^2, i.e. representing all the unknown quantities in the form

$$f(\mathbf{r},t) = f^{\,(0)}(\mathbf{r},t) + \varepsilon^2 f^{\,(1)}(\mathbf{r},t) + \dots\,. \tag{14.4}$$

Then we try to find the solution in three consequent steps.

(a) To zeroth order with respect to ε^2, the magnetic field is determined by the equation

$$\mathbf{B}^{\,(0)} \times \operatorname{curl}\mathbf{B}^{\,(0)} = 0\,. \tag{14.5}$$

This must be supplemented with a boundary condition, which generally depends on time:

$$\mathbf{B}^{(0)}(\mathbf{r},t)\,\big|_{S} = \mathbf{f}_1(\mathbf{r},t)\,. \tag{14.6}$$

Here S is the boundary of the region G, in the interior of which the force-free-field Equation (14.5) applies.

> The strong force-free magnetic field, changing in time according to the boundary condition (14.6), sets the plasma in motion.

(b) The kinematics of this motion is uniquely determined by two conditions. The first one follows from the equation of motion and signifies the orthogonality of acceleration to the magnetic field lines

$$\mathbf{B}^{(0)} \cdot \frac{d\mathbf{v}^{(0)}}{dt} = 0\,. \tag{14.7}$$

This equation is obtained by taking the scalar product of Equation (14.1) and the vector $\mathbf{B}^{(0)}$.

The second condition is a consequence of the freezing-in Equation (14.2)

$$\frac{\partial \mathbf{B}^{(0)}}{\partial t} = \operatorname{curl}\left(\mathbf{v}^{(0)} \times \mathbf{B}^{(0)}\right). \tag{14.8}$$

Equations (14.7) and (14.8) determine the velocity field $\mathbf{v}^{(0)}(\mathbf{r},t)$, if the initial condition inside the region G is given:

$$\mathbf{v}_{\parallel}^{(0)}(\mathbf{r},0)\,\big|_{G} = \mathbf{f}_2(\mathbf{r})\,. \tag{14.9}$$

Here $\mathbf{v}_{\parallel}^{(0)}$ is the velocity component along the field lines. The velocity component across the field lines is uniquely defined, once the field $\mathbf{B}^{(0)}(\mathbf{r},t)$ is known, by the freezing-in Equation (14.8) at any moment, including the initial one.

(c) Since we know the velocity field $\mathbf{v}^{(0)}(\mathbf{r},t)$, the continuity equation

$$\frac{\partial \rho^{(0)}}{\partial t} + \operatorname{div} \rho^{(0)}\mathbf{v}^{(0)} = 0 \tag{14.10}$$

allows us to find the plasma density distribution $\rho^{(0)}(\mathbf{r},t)$, if we know its initial distribution

$$\rho^{(0)}(\mathbf{r},0)\,\big|_{G} = f_3(\mathbf{r})\,. \tag{14.11}$$

Therefore Equations (14.5), (14.7) and (14.8), together with the continuity equation (14.10), completely determine the unknown zero-order quantities $\mathbf{B}^{(0)}(\mathbf{r},t)$, $\mathbf{v}^{(0)}(\mathbf{r},t)$ and $\rho^{(0)}(\mathbf{r},t)$, once the boundary condition (14.6) at the boundary S is given, and the initial conditions (14.9) and (14.11) inside the region G are given (Somov and Syrovatskii, 1976b).

At any moment of time, the field $\mathbf{B}^{(0)}(\mathbf{r}, t)$ is found from Equation (14.5) and the boundary condition (14.6). Thereupon the velocity $\mathbf{v}^{(0)}(\mathbf{r}, t)$ is determined from Equations (14.7) and (14.8) and the initial condition (14.9). Finally the continuity Equation (14.10) and the initial condition (14.11) give the plasma density distribution $\rho^{(0)}(\mathbf{r}, t)$.

From here on we restrict our attention to the consideration of the zeroth order relative to the parameter ε^2, neglecting the magnetic field deviation from a force-free state. However the consecutive application of the expansion (14.4) to the set of Equations (14.1)–(14.3) allows us to obtain **a closed set of equations** for determination of MHD quantities **in any order** relative to the small parameter ε^2.

An important point, however, is that, during the solution of the problem in the zeroth order relative to ε^2, regions can appear, where the gas pressure gradient cannot be ignored. Here effects proportional to the small parameter γ^2 must be taken into account (Section 13.1.3). This fact usually imposes a limitation on the applicability of the strong-field-cold-plasma approximation.

The question of the existence of general solutions to the MHD equations in this approximation will be considered in Section 14.3, using two-dimensional problems as an example.

14.2 The formalism of two-dimensional problems

While being relatively simple from the mathematical viewpoint, two-dimensional MHD problems allow us to gain some knowledge concerning the flows of plasma with the frozen-in strong magnetic field. Moreover the two-dimensional problems are sometimes a close approximation of the real three-dimensional flows and can be used to compare the theory with experiments and observations, both qualitatively and quantitatively.

There are **two types of problems** (Somov, 1994a) treating the plane flows of plasma, i.e. the flows with the velocity field of the form

$$\mathbf{v} = \{ v_x(x, y, t),\ v_y(x, y, t),\ 0 \} . \tag{14.12}$$

All the quantities are dependent on the variables x, y and t.

14.2.1 The first type of problems

The first type incorporates the problems with a magnetic field which is everywhere parallel to the z axis of a Cartesian system of coordinates:

$$\mathbf{B} = \{ 0,\ 0,\ B(x, y, t) \} . \tag{14.13}$$

Thus the corresponding electric current is parallel to the (x, y) plane:

$$\mathbf{j} = \{ j_x(x, y, t),\ j_y(x, y, t),\ 0 \} . \tag{14.14}$$

As an example of a problem of the first type, let us consider the effect of a *longitudinal* magnetic field in a reconnecting current layer (RCL). Under real conditions, reconnection does not occur at the zeroth lines, but rather at the 'limiting lines' of the magnetic field or 'separators' (see vol. 2, Section 3.2). The latter differ from the zeroth lines only in that the separators contain the longitudinal component of the field as shown in Figure 14.1.

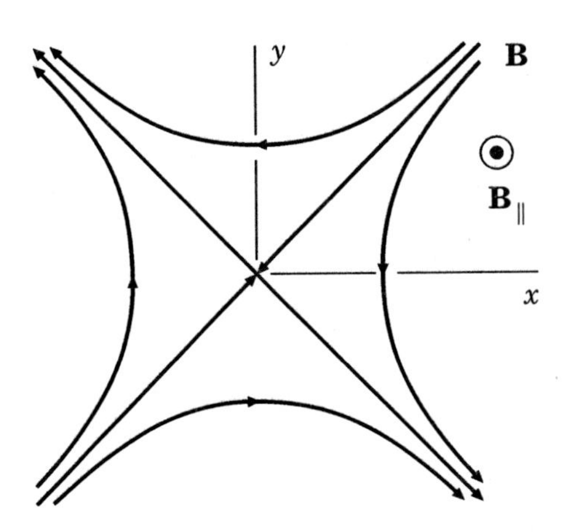

Figure 14.1: Structure of the magnetic field near a separator. A longitudinal field $\mathbf{B}_{\parallel}$ parallel to the z axis is superimposed on the two-dimensional hyperbolic field in the plane (x, y).

With the appearance of the longitudinal field, the force balance in the RCL that is formed at the separator is changed. The field and plasma pressure outside the layer must balance not only the gas pressure but also that of the longitudinal field inside the layer (Figure 14.2)

$$\mathbf{B}_{\parallel} = \left\{ 0, 0, B_{\parallel}(x, y, t) \right\}. \tag{14.15}$$

This effect is well known in the so-called theta-pinch. In axially symmetric geometry, in cylindrical coordinates r, θ, z, an azimuthal current density j_{θ} crossed with an axial field B_z can support a radial pressure gradient.

If the longitudinal field accumulated in the layer during reconnection, the field pressure $B_{\parallel}^2/8\pi$ would considerably limit the layer compression as well as the reconnection rate. However the solution of the problem of the first type with respect to $\mathbf{B}_{\parallel}$ (see vol. 2, Section 6.2.2) shows that another effect is of importance in the real plasma with finite conductivity.

The effect, in essence, is this: the **longitudinal field compression** in the RCL produces a gradient of this field and a corresponding electric current circulating in the transversal (relative to the main current j_z in the layer) plane (x, y). This current circulation is of the type (14.14); it is represented schematically in Figure 14.2.

The circulating current plays just the same role as the j_{θ}-current in the theta-pinch, a one-dimensional equilibrium in a cylindric geometry with an axial field $B_z(r)$. **Ohmic dissipation of the circulating current** under conditions of finite conductivity leads to longitudinal field diffusion outwards from the layer, thus limiting the longitudinal field accumulation in the RCL.

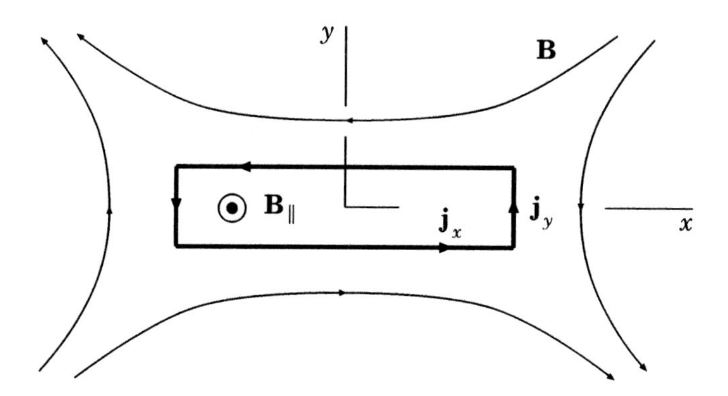

Figure 14.2: A model of a reconnecting current layer with a longitudinal component of a magnetic field $\mathbf{B}_{\parallel}$.

14.2.2 The second type of MHD problems

14.2.2 (a) Magnetic field and its vector potential

From this point on we shall be mainly interested in two-dimensional problems of the second type. They treat the plane plasma flows (14.12) associated with the plane magnetic field

$$\mathbf{B} = \{ B_x(x, y, t),\ B_y(x, y, t),\ 0 \}. \tag{14.16}$$

The electric currents corresponding to this field are parallel to the z axis

$$\mathbf{j} = \{ 0,\ 0,\ j\,(x, y, t) \}. \tag{14.17}$$

The vector-potential $\mathbf{A}$ of such a field has as its only non-zero component:

$$\mathbf{A} = \{ 0,\ 0,\ A\,(x, y, t) \}.$$

The magnetic field $\mathbf{B}$ is defined by the z-component of the vector-potential:

$$\mathbf{B} = \left\{ \frac{\partial A}{\partial y},\ -\frac{\partial A}{\partial x},\ 0 \right\}. \tag{14.18}$$

The scalar function $A\,(x, y, t)$ is often termed the *vector potential*. This function is quite useful, owing to its properties.

Property 1. Substitute (14.18) in the differential equations describing the magnetic field lines

$$\frac{dx}{B_x} = \frac{dy}{B_y} = \frac{dz}{B_z}. \tag{14.19}$$

Equations (14.19) imply parallelism of the vector $d\mathbf{l} = \{dx,\ dy,\ dz\}$ to the vector $\mathbf{B} = \{B_x, B_y, B_z\}$. In the case under study $B_z = 0,\ dz = 0$, and

$$\frac{dx}{\partial A/\partial y} = -\frac{dy}{\partial A/\partial x}$$

or

$$\frac{\partial A}{\partial x}\,dx + \frac{\partial A}{\partial y}\,dy = 0\,.$$

On integrating the last, we come to the conclusion that the relation

$$\boxed{A\,(x, y, t) = \text{const} \quad \text{for} \quad t = \text{const}} \tag{14.20}$$

is the equation for a family of magnetic field lines in the plane $z = \text{const}$ at the moment t.

Property 2. Let L be some curve in the plane (x, y) and $d\mathbf{l}$ an arc element along the curve in Figure 14.3.

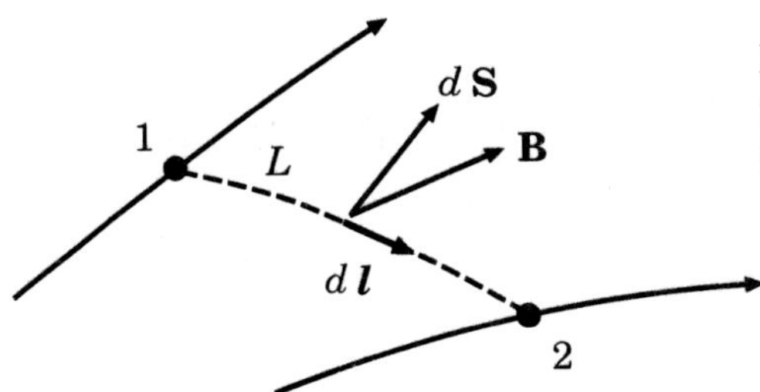

Figure 14.3: The curve L connects the points 1 and 2 situated in different field lines.

Let us calculate the magnetic flux $d\,\Phi$ through the arc element $d\mathbf{l}$. By definition,

$$d\,\Phi = \mathbf{B} \cdot d\,\mathbf{S} = \mathbf{B} \cdot (\mathbf{e}_z \times d\,\mathbf{l}) = \mathbf{B} \cdot \begin{vmatrix} \mathbf{e}_x & \mathbf{e}_y & \mathbf{e}_z \\ 0 & 0 & 1 \\ dx & dy & 0 \end{vmatrix} =$$

$$= \mathbf{B} \cdot \{\,(-dy)\,\mathbf{e}_x + dx\,\mathbf{e}_y\,\} = -B_x\,dy + B_y\,dx\,. \tag{14.21}$$

On substituting definition (14.18) in formula (14.21) we find that

$$d\,\Phi = -\frac{\partial A}{\partial y}\,dy - \frac{\partial A}{\partial x}\,dx = -\,dA\,. \tag{14.22}$$

On integrating (14.22) along the curve L from point 1 to point 2 we obtain the magnetic flux

$$\Phi = A_2 - A_1\,. \tag{14.23}$$

Thus the fixed value of the vector potential A is not only the field line 'tag' determined by formula (14.20);

> the difference of values of the vector potential A on two field lines is equal to the magnetic flux between them.

From this, in particular, the following simple rule holds: we have to plot the magnetic field lines corresponding to equidistant values of A.

Property 3. Let us substitute definition (14.18) in the freezing-in Equation (14.2). We obtain the following general equation

$$\operatorname{curl} \frac{d\mathbf{A}}{dt} = 0 \,. \tag{14.24}$$

Disregarding a gradient of an arbitrary function, which can be eliminated by a gauge transformation, and considering the second type of MHD problems, we have

$$\frac{dA}{dt} \equiv \frac{\partial A}{\partial t} + (\mathbf{v} \cdot \nabla) A = 0 \,. \tag{14.25}$$

This equation means that, in the plane (x, y), the lines

$$A\,(x, y, t) = \text{const} \tag{14.26}$$

are *Lagrangian* lines, i.e. they move together with the plasma. According to (14.20) they are composed of the field lines, hence Equation (14.25) expresses the magnetic field freezing in plasma.

Thus (formally it follows from (14.25) on passing to the Lagrangian variables) we have one of the integrals of motion

$$\boxed{A\,(x, y, t) = A\,(x_0, y_0, 0) \equiv A_0} \tag{14.27}$$

at an arbitrary t. Here x_0, y_0 are the coordinates of some 'fluid particle' at the initial moment of time; x, y are the coordinates of the same particle at a moment of time t or (by virtue of (14.27)) the coordinates of any other particle situated on the same field line A_0 at the moment t.

Property 4. Equation of motion (14.1) rewritten in terms of the vector potential $A(x, y, t)$ is of the form

$$\varepsilon^2 \frac{d\mathbf{v}}{dt} = -\frac{1}{\rho}\, \Delta A \, \nabla A \,. \tag{14.28}$$

In the zeroth order relative to ε^2, outside the zeroth points (where $\nabla A = 0$) and the magnetic field sources (where $\Delta A \neq 0$) we have:

$$\boxed{\Delta A = 0 \,.} \tag{14.29}$$

So the vector potential is a *harmonic* function of variables x and y. Hence, while considering the (x, y) plane as a complex plane $z = x + \mathrm{i}y$, it is convenient to relate an *analytic* function F to the vector potential A in the region under consideration:

$$F\,(z,t) = A\,(x,y,t) + \mathrm{i}\,A^{+}(x,y,t)\,. \tag{14.30}$$

Here $A^{+}(x,y,t)$ is a conjugate harmonic function connected with $A\,(x,y,t)$ by the Cauchy-Riemann condition

$$A^{+}(x,y,t) = \int \left(-\frac{\partial A}{\partial y}\,dx + \frac{\partial A}{\partial x}\,dy \right) + A^{+}(t) =$$

$$= -\int \mathbf{B}\cdot d\mathbf{l} + A^{+}(t)\,, \tag{14.31}$$

where $A^{+}(t)$ is a quantity independent of the coordinates x and y (see Lavrent'ev and Shabat, 1973, § 2).

The function $F(z,t)$ is termed the *complex potential.* The magnetic field vector, according to (14.18) and (14.30), is:

$$\mathbf{B} = B_x + \mathrm{i}\,B_y = -\,\mathrm{i}\left(\frac{dF}{dz}\right)^{*}, \tag{14.32}$$

the asterisk denoting the complex conjugation. After the introduction of the complex potential, we can widely apply the methods of the complex variable function theory, in particular the method of *conform mapping*, to determine the magnetic field in zeroth order in the small parameter ε^2 (e.g., Exercise 14.4).

This has been done successfully many times in order to determine the structure of the magnetic field: in vicinity of reconnecting current layer (RCL; Syrovatskii, 1971), in solar coronal streamers (Somov and Syrovatskii, 1972b) and the field of the Earth's magnetosphere (Oberz, 1973), the accretion disk magnetosphere (see vol. 2, Section 8.3). Markovskii and Somov (1989) suggested a generalization of the Syrovatskii model by attaching four shock MHD waves at the endpoints of the RCL. Under some simplifying assumptions, such model reduces exactly to the Riemann-Hilbert problem solved by Bezrodnykh and Vlasov (2002) in an analytical form on the basis of the Christoffel-Schwarz integral.

14.2.2 (b) Motion of the plasma and its density

In the strong field approximation, the plasma motion kinematics due to changes in a potential field is uniquely determined by two conditions:

(i) the freezing-in condition (14.25) or its solution (14.27) and

(ii) the acceleration orthogonality with respect to the field lines

$$\frac{d\mathbf{v}^{(0)}}{dt} \times \nabla A^{(0)} = 0 \tag{14.33}$$

(cf. Equation (14.7)). A point to be noted is that Equation (14.33) is a result of eliminating the unknown $\Delta A^{(1)}$, which has a first order in ε^2, from two components of the vector equation

$$\frac{d\mathbf{v}^{(0)}}{dt} = -\frac{1}{\rho^{(0)}}\,\Delta A^{(1)}\,\nabla A^{(0)}\,. \tag{14.34}$$

Once the kinematic part of the problem is solved, the trajectories of fluid particles are known:

$$x = x\,(x_0\,,\, y_0\,, t)\,, \qquad y = y\,(x_0\,,\, y_0\,, t)\,. \tag{14.35}$$

In this case the continuity Equation (14.3) solution presents no problem. In fact, the fluid particle density change on moving along the found trajectory is determined by the continuity Equation (14.3), rewritten in the Lagrangian form, and is equal to

$$\frac{\rho\,(x,\,y,\,t)}{\rho_0\,(x_0,\,y_0)} = \frac{dU_0}{dU} = \frac{\mathcal{D}(x_0,\,y_0)}{\mathcal{D}(x,\,y)}\,. \tag{14.36}$$

Here dU_0 is the initial volume of a particle, dU is the volume of the same particle at a moment of time t;

$$\frac{\mathcal{D}(x_0,\,y_0)}{\mathcal{D}(x,\,y)} = \frac{\partial x_0}{\partial x}\,\frac{\partial y_0}{\partial y} - \frac{\partial x_0}{\partial y}\,\frac{\partial y_0}{\partial x} \tag{14.37}$$

is the Jacobian of the transformation that is inverse to the transformation (14.35) of coordinates at a fixed value of time t.

The two-dimentional equations of the strong-field-cold-plasma approximation (Somov and Syrovatskii, 1976b) in the problem of the second type are relatively simple but rather useful for applications to space plasmas. In particular, they enable us to study the fast plasma flows in the solar atmosphere (Syrovatskii and Somov, 1980) and to understand some aspects of the reconnection process.

In spite of their numerous applications, the list of exact solutions to them is rather poor. Still, we can enrich it significantly,

relying on many astrophysical objects, for example in the accretion disk coronae (see vol. 2, Section 8.3), and some mathematical ideas.

Titov and Priest (1993) have shown that the equations of zeroth order can be reduced to a set of Cauchy-Riemann and ordinary differential equations, by using a conformal system of coordinates in which the positions of particles are fixed by magnitudes of two conjugate functions. These are the flux function and the potential of magnetic field. The set obtained has a special class of solutions. First, in such flows the conjugate potential is frozen into the moving medium as well as the vector potential $A(x, y, t)$. Second, each flow is realized as a contiuous sequence of conformal mappings. A linear diffusion-like equation describes such flows. The equation was solved analytically for examples describing the *magnetic collapse* (cf. vol. 2, Chapter 2) in the neighbourhood of the X-point.

14.3 On the existence of continuous flows

Thus, in the strong-field-cold-plasma approximation, the MHD equations for a plane two-dimensional flow of ideally conducting plasma (for second-type problems) are reduced, in the zeroth order in the small parameter ε^2, to the following set of equations:

$$\Delta A = 0 \,, \tag{14.38}$$

$$\frac{d\mathbf{v}}{dt} \times \nabla A = 0 \,, \tag{14.39}$$

$$\frac{dA}{dt} = 0 \,, \tag{14.40}$$

$$\frac{\partial \rho}{\partial t} + \operatorname{div} \rho \mathbf{v} = 0 \,. \tag{14.41}$$

Seemingly, the solution of this set is completely defined inside some region G (Figure 14.4) on the plane (x, y), once the boundary condition is given

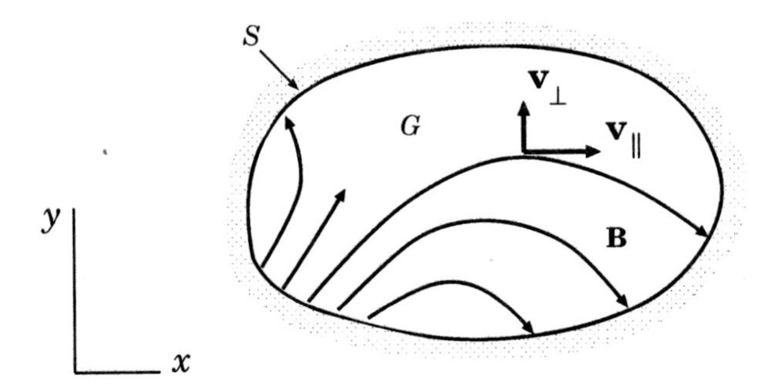

Figure 14.4: The boundary and initial conditions for the second-type MHD problems.

at the boundary S

$$A\,(x, y, t)\,\big|_{\,S} = f_1\,(x, y, t) \tag{14.42}$$

together with the initial conditions inside the region G

$$\mathbf{v}_\parallel\,(x, y, 0)\,\big|_{\,G} = \mathbf{f}_2\,(x, y)\,, \tag{14.43}$$

$$\rho\,(x, y, 0)\,\big|_{\,G} = f_3\,(x, y)\,. \tag{14.44}$$

Here $\mathbf{v}_\parallel$ is the velocity component along field lines. Once the potential $A\,(x, y, t)$ is known, the transversal velocity component is uniquely determined by the freezing-in Equation (14.40) and is equal, at any moment including the initial one, to

$$\mathbf{v}_\perp(x, y, t) = (\mathbf{v} \cdot \nabla A)\,\frac{\nabla A}{|\,\nabla A\,|^2} = -\,\frac{\partial A}{\partial t}\,\frac{\nabla A}{|\,\nabla A\,|^2}\,. \tag{14.45}$$

From Equation (14.38) and boundary condition (14.42) we find the vector potential $A(x, y, t)$ at any moment of time. Next, from Equations (14.39) and (14.40) and the initial condition (14.43), the velocity $\mathbf{v}(x, y, t)$ is determined; the density $\rho(x, y, t)$ is found from the continuity Equation (14.41) and the initial density distribution (14.44). The next Section is devoted to the consideration of an example which may have interesting applications.

14.4 Flows in a time-dependent dipole field

14.4.1 Plane magnetic dipole fields

Two straight parallel currents, equal in magnitude but opposite in direction, engender the magnetic field which far enough from the currents can be described by a complex potential

$$F(z) = \frac{\mathrm{i}\,\mathbf{m}}{z}\,, \qquad \mathbf{m} = m\,e^{\,\mathrm{i}\psi} \tag{14.46}$$

and is called the plane *dipole* field. The quantity $m = 2Il/c$ has the meaning of the *dipole moment*, I is the current magnitude, l is the distance between the currents. Formula (14.46) corresponds to the plane dipole situated at the origin of coordinates in the plane (x, y) and directed at an angle of ψ to the x axis. The currents are parallel to the z axis of the Cartesian system of coordinates.

Let us consider the plasma flow caused by the change with time of the strong magnetic field of the plane dipole. Let $\psi = \pi/2$ and $m = m(t)$, $m(0) = m_0$.

(a) Let us find the first integral of motion. According to (14.30) and (14.46), the complex potential

$$F(z, t) = \frac{\mathrm{i}\,m(t)\,e^{\,\mathrm{i}\pi/2}}{x + \mathrm{i}\,y} = \frac{-\,m(t)\,x + \mathrm{i}\,m(t)\,y}{x^2 + y^2}\,. \tag{14.47}$$

So, according to (14.20), the field lines constitute a family of circles

$$A(x, y, t) = -\,\frac{m(t)\,x}{x^2 + y^2} = \text{const} \quad \text{for} \quad t = \text{const}\,. \tag{14.48}$$

They have centres on the axis x and the common point $x = 0$, $y = 0$ in Figure 14.5.

Therefore the freezing-in condition (14.27) results in a first integral of motion

$$\frac{m\,x}{x^2 + y^2} = \frac{m_0\,x_0}{x_0^2 + y_0^2}\,. \tag{14.49}$$

Here x_0, y_0 are the coordinates of some fluid particle at the initial moment of time $t = 0$; Lagrangian variables x and y are the coordinates of the same particle at a moment t.

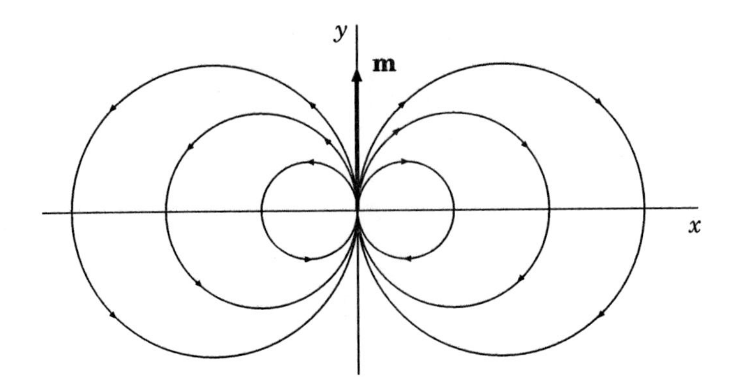

Figure 14.5: The field lines of a plane magnetic dipole.

(b) The second integral is easily found in the limit of small changes of the dipole moment $m(t)$ and respectively *small* plasma displacements. Assuming the parameter $\delta = v\tau/L$ to be small, Equation (13.26), which is *linear in velocity*, takes the place of (14.33). The integration over time (with zero initial values for the velocity) allows us to reduce Equation (13.26) to the form

$$\frac{\partial x}{\partial t} = K(x,y,t)\,\frac{\partial A}{\partial x}\,, \qquad \frac{\partial y}{\partial t} = K(x,y,t)\,\frac{\partial A}{\partial y}\,. \tag{14.50}$$

Here $K(x,y,t)$ is some function of coordinates and time. Eliminating it from two Equations (14.50), we arrive at

$$\frac{\partial y}{\partial x} = \frac{\partial A}{\partial y} \Big/ \frac{\partial A}{\partial x}\,. \tag{14.51}$$

Thus, in the approximation of small displacements, not only the acceleration but also the plasma displacements are normal to the field lines.

On substituting (14.48) in (14.51), we obtain an ordinary differential equation. Its integral

$$\frac{y}{x^2+y^2} = \text{const}$$

describes **a family of circles, orthogonal to the field lines**, and presents fluid particle trajectories. In particular, the trajectory of a particle, situated at a point (x_0, y_0) at the initial moment of time $t = 0$, is an arc of the circle

$$\frac{y}{x^2+y^2} = \frac{y_0}{x_0^2+y_0^2} \tag{14.52}$$

from the point (x_0, y_0) to the point (x, y) on the field line (14.49) as shown in Figure 14.6.

Thus the integrals of motion (14.49) and (14.52) completely determine the plasma flow in terms of the Lagrangian coordinates

$$x = x\,(x_0, y_0, t)\,, \qquad y = y\,(x_0, y_0, t)\,. \tag{14.53}$$

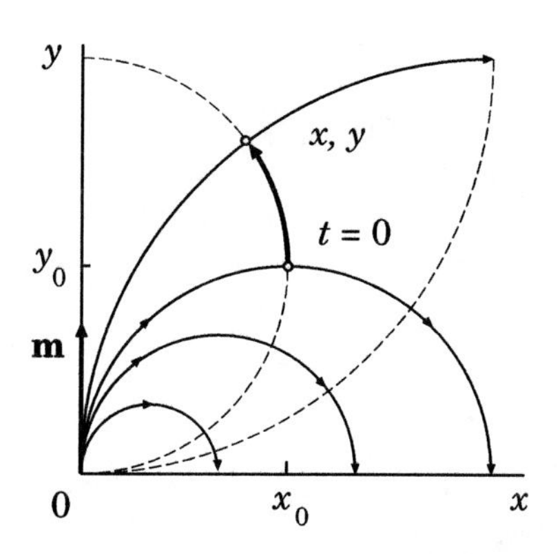

Figure 14.6: A trajectory of a fluid particle driven by a changing magnetic field of a plane dipole.

This flow has a simple form: the particles are connected with the magnetic field lines and move together with them in a transversal direction. Such simple kinematics is a result of considering small plasma displacements (from the state having zero initial velocity) under the action of the force perpendicular to the field lines.

The plasma density change is defined by Equation (14.36). On calculating the Jacobian for the transformation implicitly given by formulae (14.49) and (14.52), we obtain (for the case of a homogeneous initial density distribution ρ_0) the formula

$$\frac{\rho(x,y,t)}{\rho_0} = \left(\frac{m}{m_0}\right) \frac{m_0^4}{(m^2x^2 + m_0^2\, y^2)^4} \left\{ \left[m^2 x^4 + m_0^2\, y^4 + \right.\right.$$

$$\left.\left. + x^2 y^2 \left(3m^2 - m_0^2\right) \right]^2 - \left[2x^2 y^2 \left(m_0^2 - m^2\right) \right]^2 \right\} . \tag{14.54}$$

In particular, on the dipole axis ($x = 0$)

$$\boxed{\frac{\rho(0,y,t)}{\rho_0} = \frac{m}{m_0}\,,} \tag{14.55}$$

whereas in the 'equatorial plane' ($y = 0$)

$$\frac{\rho(x,0,t)}{\rho_0} = \left(\frac{m_0}{m}\right)^3 . \tag{14.56}$$

With increasing dipole moment m, the plasma density on the dipole axis grows proportionally to the moment,

whereas that at the equatorial plane falls in inverse proportion to the third power of the moment. The opposite process takes place as the moment decreases.

The result pertains to the case of small changes in the dipole moment and can demonstrate just the tendency of plasma behaviour in the strong magnetic field of a plane dipole. The exception is formula (14.55). It applies to any changes of the dipole moment. The reason is in the following. In the approximation of a strong field and cold plasma, the acceleration of plasma is perpendicular to the field lines and is zero at the dipole axis. Hence, if the plasma is motionless at the initial moment, arbitrary changes of the dipole moment do not cause a plasma motion on the dipole axis ($\mathbf{v} = 0$). Plasma displacements in the vicinity of the dipole axis always remain small ($\delta \ll 1$) and the solution obtained applies.

In the general case of arbitrarily large dipole moment changes,

> the inertial effects resulting in plasma flows along the magnetic field lines are of considerable importance

(Somov and Syrovatskii, 1972a). In this case, the solution of the problem requires the integration of Equation (14.33) or (14.34) together with the freezing-in Equation (14.25).

One can obtain exact analytical solutions for a linearly changing magnetic moment using the 'frozen-in coordinates' technique (Gorbachev and Kel'ner, 1988). These coordinates can be quite useful while solving nonstationary MHD problems. One introduces a set which is doubly Lagrangian: in the parameter s_1 along a stream line (along the velocity field $\mathbf{v}$) and in the parameter s_2 along a magnetic field line.

14.4.2 Axisymmetric dipole fields in plasma

Two-dimensional axisymmetric MHD problems can be better suited to astrophysical applications of the second-type problem considered. The MHD equations are written, using the approximation of a strong field and cold plasma, in spherical coordinates with due regard for axial symmetry. The role of the vector potential is fulfilled by the so-called *stream function*

$$\Phi\,(r, \theta, t) = r \sin\theta \, A_{\varphi}(r, \theta, t)\,. \tag{14.57}$$

Here A_{φ} is the only non-zero φ-component of the vector-potential $\mathbf{A}$.

In terms of the stream functions, the equations take the form

$$\frac{d\mathbf{v}}{dt} = \varepsilon^{-2} K(r, \theta, t)\, \nabla\Phi\,, \qquad \frac{d\,\Phi}{dt} = 0\,, \qquad \frac{d\rho}{dt} = -\rho \,\mathrm{div}\, \mathbf{v}\,, \tag{14.58}$$

where

$$K(r, \theta, t) = \frac{j_{\varphi}(r, \theta, t)}{\rho\, r \sin\theta} \tag{14.59}$$

(Somov and Syrovatskii, 1976b). The equations formally coincide with the corresponding Equations (14.28), (14.25) and (14.3) describing the plane flows in terms of the vector potential.

As a zeroth approximation in the small parameter ε^2, we may take, for example, the dipole field. In this case the stream function is of the form

$$\Phi^{(0)}\,(r,\theta,t) = m(t)\,\frac{\sin^2\theta}{r}\,, \tag{14.60}$$

where $m(t)$ is a time-varying moment.

Let us imagine a homogeneous magnetized ball of radius $R(t)$ with the frozen field $\mathbf{B}_{int}(t)$. The dipole moment of such a ball (a star or its envelope) is

$$m(t) = \frac{1}{2}\,B_{int}(t)\,R^3(t) = \frac{1}{2\pi}\,\left(B_0\,\pi R_0^2\right) R(t)\,, \tag{14.61}$$

where B_0 and R_0 are the values of $B_{int}(t)$ and $R(t)$ at the initial moment of time $t = 0$. The second equality takes account of the magnetic field freezing-in as conservation of the flux $B_{int}(t)\,R^2(t)$ through the ball. Formula (14.61) shows that the dipole moment of the ball is thereby proportional to its radius $R(t)$.

The solution to the problem (Somov and Syrovatskii, 1972a) shows that as the dipole moment grows (when the ball expands)

> the magnetic field rakes the plasma up to the dipole axis, compresses it and simultaneously accelerates it along the field lines.

A distinguishing characteristic of the solution is that the density at the axis grows in proportion to the dipole moment, just as in the two-dimensional plane case (formula (14.55)).

Envelopes of nova and supernova stars present a wide variety of different shapes. We can hardly find the ideally round envelopes, even among the ones of regular shape. It is more common to find either flattened or stretched envelopes. As a rule, their surface brightness is maximal at the ends of the main axes of an oval image. This phenomenon can sometimes be interpreted as a gaseous ring observed almost from an edge. However, if there is no luminous belt between the brightness maxima, which would be characteristic of the ring, then the remaining possibility is that single gaseous compressions – condensations – exist in the envelope.

At the early stages of the expansion during the explosion of a nova, the condensations reach such brightness that they give the impression that the nova 'bifurcates'. Consider one of the models in which a magnetic field plays a decisive role. Suppose that the star's magnetic field was a dipole one before the explosion. At the moment of the explosion a massive envelope with the frozen-in field separated from the star and began to expand. According to (14.61), the expansion results in the growth of the dipole moment. According to the solution of the problem considered above, the field will rake the interstellar plasma surrounding the envelope, as well as external layers of the envelope, up in the direction of the dipole axis.

The process of polar condensate formation can be conventionally divided into two stages (Somov and Syrovatskii, 1976b, Chapter 2). At the first one, the interstellar plasma is raked up by the magnetic field into the polar regions, a corresponding growth in density and pressure at the dipole axis taking place. At the second stage, the increased pressure hinders the growth of the density at the axis, thus stopping compression, but the plasma raking-up still continues. At the same time, the gas pressure gradient, arising ahead of the envelope, gives rise to the motion along the axis. As a result, by the time the magnetic force action stops, all the plasma is raked up into two compact condensates.

The plasma raking-up by the strong magnetic field seems to be capable of explaining some types of chromospheric ejections on the Sun (Somov and Syrovatskii, 1976b, Chapter 2, § 4).

If a magnetized ball compresses, plasma flows from the poles to the equatorial plane, thus forming a **dense disk or ring**. This case is the old problem of cosmic electrodynamics concerning the compression of a gravitating plasma cloud with the frozen-in field. The process of magnetic raking-up of plasma into dense disks or rings can effectively work in the atmospheres of collapsing stars.

14.5 Practice: Exercises and Answers

Exercise 14.1. Consider the properties of the vector-potential $\mathbf{A}$ which is determined in terms of two scalar functions α and β:

$$\mathbf{A} = \alpha \nabla \beta + \nabla \psi \, . \tag{14.62}$$

Here ψ is an arbitrary scalar function.

Answer. Formula (14.62) permits $\mathbf{B}$ to be written as

$$\mathbf{B} = \operatorname{curl} \mathbf{A} = \nabla \alpha \times \nabla \beta \, , \tag{14.63}$$

where the last step follows from the fact that the curl of a gradient vanishes.

This representation of $\mathbf{B}$ provides another way to obtain information about the magnetic field in three-dimensional problems. According to (14.63)

$$\mathbf{B} \cdot \nabla \alpha = 0 \quad \text{and} \quad \mathbf{B} \cdot \nabla \beta = 0 \, . \tag{14.64}$$

Thus $\nabla \alpha$ and $\nabla \beta$ are perpendicular to the vector $\mathbf{B}$, and functions α and β are constant along $\mathbf{B}$. The surfaces $\alpha = \text{const}$ and $\beta = \text{const}$ are orthogonal to their gradients and targent to $\mathbf{B}$. Hence

> a magnetic field line can be conveniently defined in terms of a pair of values: α and β.

A particular set of α and β labels a field line.

The functions α and β are referred to as Euler potentials or Clebsch variables. Depending on a problem to be examined, one form may have an advantage over another. The variables α and β, while in general not easily obtained, are available for some axisymmetric geometries.

Another advantage of these variables appears in the study of field line motions in the context of the *ideal* MHD theory (Section 5.7 in Parks, 2004). Since the time evolution of the magnetic field is governed by the induction Equation (14.2), the functions $\alpha(\mathbf{r}, t)$ and $\beta(\mathbf{r}, t)$ satisfy the equations:

$$\frac{\partial \alpha}{\partial t} + (\mathbf{v} \cdot \nabla)\, \alpha = 0 \quad \text{and} \quad \frac{\partial \beta}{\partial t} + (\mathbf{v} \cdot \nabla)\, \beta = 0\,. \tag{14.65}$$

That is, the functions $\alpha(\mathbf{r}, t)$ and $\beta(\mathbf{r}, t)$ take constant values for a point that moves with the plasma.

Exercise 14.2. Evaluate the typical value of the dipole moment for a neutron star.

Answer. Typical neutron stars have $B \sim 10^{12}$ G. With the star radius $R \sim 10$ km, it follows from formula (14.61) that $m \sim 10^{30}\,\text{G cm}^3$. Some of neutron stars, related to the so-called 'Soft Gamma-ray Repeaters' (SGRs), are the spinning super-magnetized neutron stars created by supernova explosions. The rotation of such stars called ***magnetars*** is slowing down so rapidly that a superstrong field of the unprecedented strength, $B \sim 10^{15}$ G, could provide so fast braking (see Section 19.1.3). For a magnetar the dipole moment $m \sim 10^{33}\,\text{G cm}^3$.

Exercise 14.3. Show that, prior to the onset of a solar flare, the magnetic energy density in the corona is of about three orders of magnitude greater than any of the other types. So the flares occur in a plasma environment well dominated by the magnetic field.

Hint. Take the coronal field of about 100 Gauss, and the coronal plasma velocity of order of 1 km s^{-1}.

Exercise 14.4. By using the method of conform mapping, determine the shape of a magnetic cavity created by a plane dipole inside a perfectly conducting uniform plasma with a gas pressure p_0. Determine the magnetic field inside the cavity.

Answer. The conditions to be satisfied along the boundary S of the magnetic cavity G are equality of magnetic and gas pressure,

$$\left.\frac{B^2}{8\pi}\right|_S = p_0 = \text{const}\,, \tag{14.66}$$

and tangency of the magnetic field,

$$\mathbf{B} \cdot \mathbf{n}\Big|_S = 0\,. \tag{14.67}$$

Condition (14.67) means that

$$\text{Re}\, F(z) = A\,(x, y) = \text{const}\,, \tag{14.68}$$

where a complex potential $F(z)$ is an analytic function (14.30) within the region G in the complex plane z except at the point $z = 0$ of the dipole.

Let us assume that a conform transformation $w = w(z)$ maps the region G onto the circle $|\, w \,| \leq 1$ in an auxiliary complex plane $w = u + iv$ so that the point $z = 0$ goes into the centre of the circle without rotation of the dipole as shown in Figure 14.7.

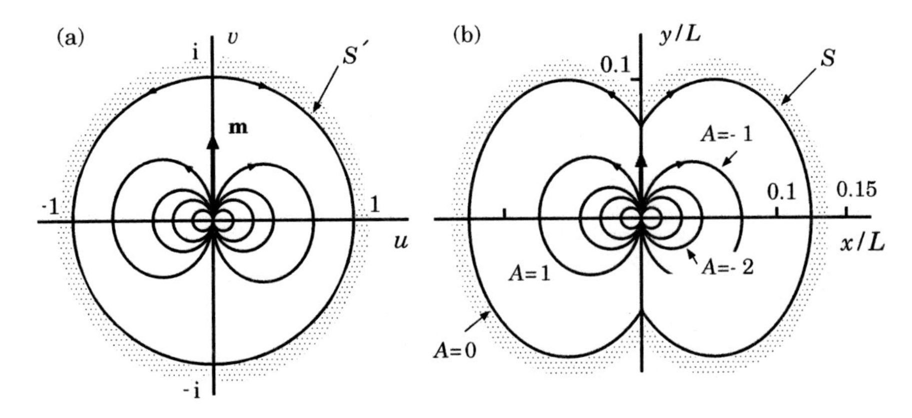

Figure 14.7: The field lines of a plane dipole **m** inside: (a) the unit circle in the plane w, (b) the cavity in a plasma of constant pressure.

The boundary $|\, w \,| = 1$ is the field line S' of the solution in the plane w, which we easily construct:

$$F(w) = \left(w - \frac{1}{w} \right) . \tag{14.69}$$

Note that we have used only the boundary condition (14.67).

The other boundary condition (14.66) will allow us to find an unknown conform transformation $w = w(z)$. With account of definition (14.32) taken, condition (14.66) gives us the following relation

$$\left| \frac{dz}{dw} \right|^2 = \frac{1}{8\pi\, p_0} \left| \frac{dF}{dw} \right|^2 . \tag{14.70}$$

At the boundary $|\, w \,| = 1$, this condition reduces to an ordinary differential equation relative to the real part, $x = x(u)$, of an unknown function $z = z(w)$:

$$\left(\frac{dx}{du} \right)^2 = M^2 u^4 , \quad \text{where} \quad M^2 = \frac{1}{2\pi} . \tag{14.71}$$

By integrating this equation we find

$$x = \pm M \frac{u^3}{3} + c_1 = \pm \frac{M}{3} \cos^3 \varphi + c_1 , \tag{14.72}$$

here φ is an argument of the complex number w, and c_1 is a constant of integration.

Since we know the real part $x = x(\varphi)$ on the circle boundary, we find the complex function $z = z(w)$ in the entire region $|\, w \,| \leq 1$, for example, by expanding the function $x = x(\varphi)$ in the Fourier series

$$x(\varphi) = c_1 + \frac{M}{4} \cos \varphi + \frac{M}{12} \cos 3\varphi \, . \tag{14.73}$$

So, inside the circle, the power series has only three terms:

$$x(r, \varphi) = c_1 + \frac{M}{4} r \cos \varphi + \frac{M}{12} r^3 \cos 3\varphi \, , \tag{14.74}$$

$$y(r, \varphi) = c_2 + \frac{M}{4} r \sin \varphi + \frac{M}{12} r^3 \sin 3\varphi \, . \tag{14.75}$$

Moreover $c_1 = c_2 = 0$ because $z(0) = 0$. Therefore

$$z(w) = \frac{M}{4} \left(w + \frac{w^3}{3} \right) . \tag{14.76}$$

The conform mapping (14.76) and the potential (14.69) determine the general solution of the problem, the complex potential (Oreshina and Somov, 1999):

$$F(z) = B_0 L^{2/3} \, \frac{K^4 - 3L^{2/3} K^2 + L^{4/3}}{K \left(K^2 - L^{2/3} \right)} \, . \tag{14.77}$$

Here $B_0 = p_0^{1/2}$ is the unit of magnetic field strength, the function

$$K(z) = \left(6\sqrt{2\pi} \cdot z + \sqrt{\, L^2 \, + 72\pi \cdot z^2} \right)^{1/3} , \tag{14.78}$$

and $L = m^{1/3} p_0^{-1/6}$ is the unit of length; it shows that, when the dipole moment m increases, the size of the magnetic cavity also increases. This is consistent with what we discussed in Section 14.4.

The field lines corresponding solution (14.77) are shown in Figure 14.7b. Therefore, in addition to the shape of the boundary (Cole and Huth, 1959), we have found an analytic solution for the magnetic field inside the static dipole cavity. This solution can be used in the zero-order approximation, described in Section 14.1, to analyse properties of plasma flows near collapsing or exploding astrophysical objects with strong magnetic fields.

Exercise 14.5. To estimate characteristic values of the large-scale magnetic field in the corona of an accretion disk (see vol. 2, Section 8.3.1), we have to find the structure of the field inside an open magnetosphere created by a dipole field of a star and a regular field generated by the disk.

Consider a simplified two-dimensional problem, demonstrated by Figure 14.8, on the shape of a magnetic cavity and the shape of the accretion

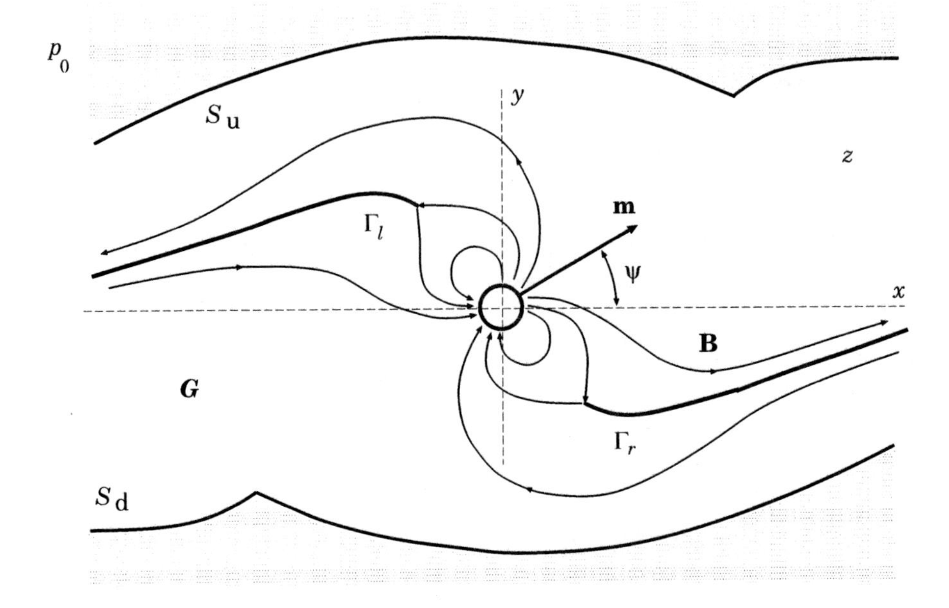

Figure 14.8: A two-dimensional model of the star magnetosphere with an accretion disk; Γ_l and Γ_r are the cross sections of the disk. The plane z corresponds to the complex variable $z = x + \mathrm{i}y$. S_u and S_d together with Γ_l and Γ_r constitute the boundary of the singly connected domain G in the plane z.

disk under assumption that this cavity, i.e. the magnetosphere, is surrounded by a perfectly conducting uniform plasma with a gas pressure p_0. Discuss a way to solve the problem by using the method of conform mapping (see vol. 2, Section 8.3.2).

Chapter 15

MHD Waves in Astrophysical Plasma

There are four different modes of magnetohydrodynamic waves in an ideal plasma with magnetic field. They can create turbulence, nonlinearly cascade in a wide range of wavenumbers, accelerate particles and produce a lot of interesting effects under astrophysical conditions.

15.1 The dispersion equation in ideal MHD

Small disturbances in a conducting medium with a magnetic field propagate as waves, their properties being different from those of the usual sound waves in a gas or electromagnetic waves in a vacuum. First, the conducting medium with a magnetic field has a characteristic anisotropy: the wave propagation velocity depends upon the direction of propagation relative to the magnetic field. Second, as a result of the interplay of electromagnetic and hydrodynamic phenomena, the waves in MHD are generally neither longitudinal nor transversal.

The study of the behaviour of small-amplitude waves, apart from being interesting in itself, has a direct bearing on the analysis of large-amplitude waves, in particular shock waves and other discontinuous flows in MHD.

Initially we shall study the possible types of small-amplitude waves, restricting ourselves to the *ideal* MHD Equations (12.67). Let us suppose a plasma in the initial stationary state is subjected to a small perturbation, so that velocity $\mathbf{v}_0$, magnetic field $\mathbf{B}_0$, density ρ_0, pressure p_0 and entropy s_0 acquire some small deviations $\mathbf{v}\,'$, $\mathbf{B}\,'$, $\rho\,'$, $p\,'$ and $s\,'$:

$$\begin{aligned} &\mathbf{v} = \mathbf{v}_0 + \mathbf{v}\,', \qquad \mathbf{B} = \mathbf{B}_0 + \mathbf{B}\,', \\ &\rho = \rho_0 + \rho\,', \qquad p = p_0 + p\,', \qquad s = s_0 + s\,'. \end{aligned} \tag{15.1}$$

The initial state is assumed to be a uniform flow of an homogeneous medium in a constant magnetic field:

$$\begin{array}{lll} \mathbf{v}_0 = \text{const}\,, & \mathbf{B}_0 = \text{const}\,, & \\ \rho_0 = \text{const}\,, & p_0 = \text{const}\,, & s_0 = \text{const}\,. \end{array} \tag{15.2}$$

Needless to say, the latter simplification can be ignored, i.e. we may study waves in inhomogeneous media, the coefficients in linearized equations being dependent upon the coordinates. For the sake of simplicity we restrict our consideration to the case (15.2).

It is convenient to introduce the following designations:

$$\mathbf{u} = \frac{\mathbf{B}_0}{\sqrt{4\pi\rho_0}}\,, \qquad \mathbf{u}' = \frac{\mathbf{B}'}{\sqrt{4\pi\rho_0}}\,. \tag{15.3}$$

Let us linearize the initial set of MHD equations for an ideal medium. We substitute definitions (15.1)–(15.3) in the set of Equations (12.67), neglecting the products of small quantities. Hereafter the subscript '0' for undisturbed quantities will be omitted. We shall get the following set of *linear differential* equations for the primed quantities characterizing small perturbations:

$$\begin{array}{l} \partial \mathbf{u}' / \partial t + (\mathbf{v} \cdot \nabla)\,\mathbf{u}' = (\mathbf{u} \cdot \nabla)\,\mathbf{v}' - \mathbf{u}\,\text{div}\,\mathbf{v}'\,, \qquad \text{div}\,\mathbf{u}' = 0\,, \\ \partial \mathbf{v}' / \partial t + (\mathbf{v} \cdot \nabla)\,\mathbf{v}' = -\,\rho^{-1}\,\nabla\,(p' + \rho\,\mathbf{u} \cdot \mathbf{u}') + (\mathbf{u} \cdot \nabla)\,\mathbf{u}'\,, \\ \partial \rho' / \partial t + (\mathbf{v} \cdot \nabla)\,\rho' = -\,\rho\,\text{div}\,\mathbf{v}'\,, \\ \partial s' / \partial t + (\mathbf{v} \cdot \nabla)\,s' = 0\,, \quad p' = (\partial p / \partial \rho)_s\,\rho' + (\partial p / \partial s)_\rho\,s'\,. \end{array} \tag{15.4}$$

The latter equation is the linearized equation of state. We rewrite it as follows:

$$p' = V_s^2 \rho' + b\,s'. \tag{15.5}$$

Here

$$V_s = (\partial p / \partial \rho)_s^{1/2} \tag{15.6}$$

is the velocity of *sound* in a medium without a magnetic field (Exercise 15.1), the coefficient $b = (\partial p/\partial s)_\rho$.

By virtue of (15.2), the set of Equations (15.4) is that of linear differential equations with *constant coefficients*. That is why we may seek a solution in the form of a superposition of plane waves with a dependence on coordinates and time of the type

$$f'(\mathbf{r}, t) \sim \exp\left[\,\mathrm{i}\,(\mathbf{k} \cdot \mathbf{r} - \omega t)\,\right], \tag{15.7}$$

where ω is the wave frequency and $\mathbf{k}$ is the wave vector. An arbitrary disturbance can be expanded into such waves by means of a Fourier transform. As this takes place, the set of Equations (15.4) is reduced to the following set of *linear algebraic* equations:

$$\begin{array}{l} (\omega - \mathbf{k} \cdot \mathbf{v})\,\mathbf{u}' + (\mathbf{k} \cdot \mathbf{u})\,\mathbf{v}' - \mathbf{u}\,(\mathbf{k} \cdot \mathbf{v}') = 0\,, \qquad \mathbf{k} \cdot \mathbf{u}' = 0\,, \\ (\omega - \mathbf{k} \cdot \mathbf{v})\,\mathbf{v}' + (\mathbf{k} \cdot \mathbf{u})\,\mathbf{u}' - \rho^{-1}\,(p' + \rho\,\mathbf{u} \cdot \mathbf{u}')\,\mathbf{k} = 0\,, \\ (\omega - \mathbf{k} \cdot \mathbf{v})\,\rho' - \rho\,(\mathbf{k} \cdot \mathbf{v}') = 0\,, \\ (\omega - \mathbf{k} \cdot \mathbf{v})\,s' = 0\,, \qquad p' - V_s^2\,\rho' - b\,s' = 0\,. \end{array} \tag{15.8}$$

The quantities $\mathbf{k}$ and ω appearing in this set are assumed to be known from the initial conditions. The unknown terms are the primed ones. With respect to these the set of Equations (15.8) is closed, linear and homogeneous (the right-hand sides equal zero). For this set to have nontrivial solutions, its determinant must be equal to zero.

The determinant can be conveniently calculated in a frame of reference with one of the axes along the wave vector $\mathbf{k}$. In addition, it is convenient to use the frequency

$$\omega_0 = \omega - \mathbf{k} \cdot \mathbf{v}\,, \tag{15.9}$$

i.e. the frequency in a frame of reference moving with the plasma.

Setting the determinant equal to zero, we get the following equation

$$\omega_0^2 \left[\omega_0^2 - (\mathbf{k} \cdot \mathbf{u})^2 \right] \times$$

$$\times \left[\omega_0^4 - k^2 \left(V_s^{\,2} + u^2 \right) \omega_0^2 + k^2 V_s^{\,2} \, (\mathbf{k} \cdot \mathbf{u})^2 \right] = 0\,. \tag{15.10}$$

This equation is called the *dispersion* equation. It defines four values of ω_0^2. Since they differ in absolute magnitude, **four different modes** of waves are defined, each of them having its own velocity of propagation with respect to the plasma

$$\mathbf{V}_{\mathrm{ph}} = \frac{\omega_0}{\mathbf{k}}\,. \tag{15.11}$$

Clearly this is the *phase* velocity of the wave. It should be distinguished from the *group* velocity

$$\mathbf{V}_{\mathrm{gr}} = \frac{d\,\omega_0}{d\,\mathbf{k}}\,. \tag{15.12}$$

Let us consider the properties of the waves defined by the dispersion Equation (15.10) in greater detail.

15.2 Small-amplitude waves in ideal MHD

15.2.1 Entropy waves

The first root of the dispersion Equation (15.10)

$$\omega_0 = \omega - \mathbf{k} \cdot \mathbf{v} = 0 \tag{15.13}$$

corresponds to the small perturbation which is immobile with respect to the medium:

$$\mathbf{V}_{\mathrm{ph}} = 0\,. \tag{15.14}$$

If the medium is moving, the disturbance is carried with it.

Substituting (15.13) in (15.8), we obtain the following equations:

$$(\mathbf{k}\cdot\mathbf{u})\,\mathbf{v}' - \mathbf{u}\,(\mathbf{k}\cdot\mathbf{v}') = 0\,, \tag{15.15}$$

$$\mathbf{k}\cdot\mathbf{u}' = 0\,, \tag{15.16}$$

$$(\mathbf{k}\cdot\mathbf{u})\,\mathbf{u}' - \rho^{-1}\,(p' + \rho\,\mathbf{u}\cdot\mathbf{u}')\,\mathbf{k} = 0\,, \tag{15.17}$$

$$\mathbf{k}\cdot\mathbf{v}' = 0\,, \tag{15.18}$$

$$p' - V_s^{\,2}\rho' - b\,s' = 0\,. \tag{15.19}$$

Let us make use of (15.18) in (15.15). Then we take the scalar product of Equation (15.17) with the vector **k** and make allowance for (15.16). We write

$$(\mathbf{k}\cdot\mathbf{u})\,\mathbf{v}' = 0\,, \tag{15.20}$$

$$\mathbf{k}\cdot\mathbf{u}' = 0\,, \tag{15.21}$$

$$p' + \rho\,\mathbf{u}\cdot\mathbf{u}' = 0\,, \tag{15.22}$$

$$\mathbf{k}\cdot\mathbf{v}' = 0\,, \tag{15.23}$$

$$p' - V_s^{\,2}\rho' - b\,s' = 0\,. \tag{15.24}$$

Substitution of (15.22) in (15.17) gives us the following set of equations:

$$(\mathbf{k}\cdot\mathbf{u})\,\mathbf{u}' = 0\,, \qquad (\mathbf{k}\cdot\mathbf{u})\,\mathbf{v}' = 0\,, \tag{15.25}$$

$$p' + \rho\,\mathbf{u}\cdot\mathbf{u}' = 0\,, \qquad p' - V_s^{\,2}\rho' - b\,s' = 0\,. \tag{15.26}$$

Since generally $\mathbf{k}\cdot\mathbf{u} \neq 0$, the velocity, magnetic field and gas pressure are undisturbed in the wave under discussion:

$$\mathbf{v}' = 0\,, \quad \mathbf{u}' = 0\,, \quad p' = 0\,. \tag{15.27}$$

The only disturbed quantities are the density and entropy related by the condition

$$\boxed{\rho' = -\frac{b}{V_s^{\,2}}\,s'\,.} \tag{15.28}$$

This is the reason why these disturbances are called the *entropy* waves. They are well known in hydrodynamics (Exercise 15.2). The meaning of an entropy wave is that regions containing hotter but more rarefied plasma can exist in a plasma flow.

The entropy waves are only arbitrarily termed *waves*, since their velocity of propagation with respect to the medium is zero. Nevertheless the entropy waves must be taken into account together with the real waves in such cases as the study of shock waves behaviour under small perturbations. Blokhintsev (1945) has considered the passage of small perturbations through a shock in ordinary hydrodynamics. He came to the conclusion that

> the entropy wave must be taken into account in order to match the linearized solutions at the shock front

(see Exercise 17.1). In MHD, the entropy waves are importnant in the problem of evolutionarity of the MHD discontinuities (Chapter 17) and reconnecting current layers (see vol. 2, Chapter 10). The entropy waves can be principally essential in astrophysical plasma where plasma motions are not slow, for example in *helioseismology* of the chromosphere and corona.

15.2.2 Alfvén waves

The second root of the dispersion Equation (15.10),

$$\omega_0^2 = (\mathbf{k} \cdot \mathbf{u})^2 \quad \text{or} \quad \omega_0 = \pm\, \mathbf{k} \cdot \mathbf{u}\,, \tag{15.29}$$

corresponds to waves with the phase velocity

$$\boxed{V_{\text{A}} = \pm \frac{B}{\sqrt{4\pi\rho}} \cos\theta\,.} \tag{15.30}$$

Here θ is the angle between the direction of wave propagation $\mathbf{k}/k$ and the ambient field vector $\mathbf{B}_0$ (Figure 15.1). In formula (15.30) the value $B = \mid \mathbf{B}_0 \mid$ and $\rho = \rho_0$. These are the *Alfvén* waves.

By substituting (15.29) in the algebraic Equations (15.8) we check that the thermodynamic characteristics of the medium remain unchanged

$$\rho' = 0\,, \quad p' = 0\,, \quad s' = 0\,, \tag{15.31}$$

while the perturbations of the velocity and magnetic field are subject to the conditions

$$\mathbf{v}' = \mp\, \mathbf{u}'\,, \quad \mathbf{u} \cdot \mathbf{u}' = 0\,, \quad \mathbf{k} \cdot \mathbf{u}' = 0\,. \tag{15.32}$$

Thus the Alfvén waves are the displacements of plasma together with the magnetic field frozen into it. They are transversal with respect to both the field direction and the wave vector as shown in Figure 15.1.

The Alfvén waves have no analogue in hydrodynamics. They are specific to MHD and were called the *magnetohydrodynamic* waves. This term emphasized that they do not change the density of a medium. The fact that the Alfvén waves are transversal signifies that

> a conducting plasma in a magnetic field has a characteristic elasticity resembling that of stretched strings under tension.

The *magnetic tension force* is one of the characteristics of MHD (see Exercise 13.3). According to (15.32), the perturbed quantities are related by an energy equipartition:

$$\frac{1}{2}\, \rho\, (v')^2 = \frac{1}{8\pi}\, (B')^2\,. \tag{15.33}$$

Let us note also that

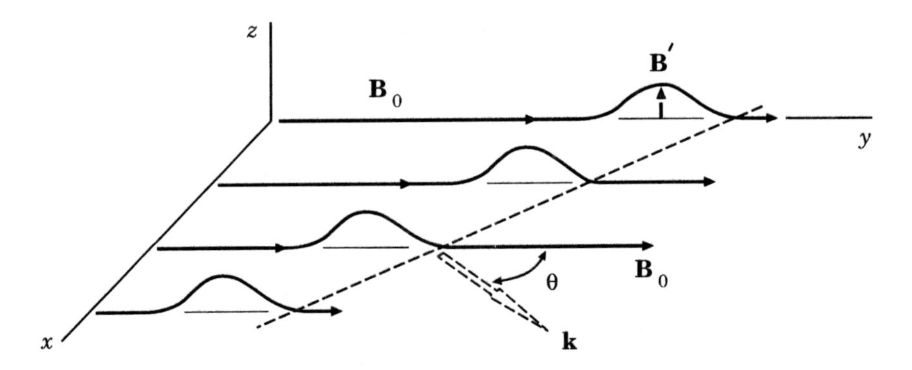

Figure 15.1: The transversal displacements of plasma and magnetic field in the Alfvén wave.

> the energy of Alfvén waves, much like the energy of oscillations in a stretched string, propagates along the field lines only.

Unlike the phase velocity, the group velocity of the Alfvén waves (15.12)

$$\mathbf{V}_{\rm gr} = \pm \frac{\mathbf{B}}{\sqrt{4\pi\rho}} \tag{15.34}$$

is directed strictly along the magnetic field; here $\mathbf{B} = \mathbf{B}_0$ of course.

In low density astrophysical plasmas with a strong field, like the solar corona, the Alfvén speed $V_{\rm A}$ can approach the light speed c (Exercise 15.3). The discovery of Alfvén waves was a major stage in the development of plasma astrophysics (Alfvén, 1950).

15.2.3 Magnetoacoustic waves

The dispersion Equation (15.10) has two other branches – two types of waves defined by a bi-square equation

$$\omega_0^4 - k^2 \left(u^2 + V_s^2\right) \omega_0^2 + k^2 V_s^2 \left(\mathbf{k} \cdot \mathbf{u}\right)^2 = 0\,. \tag{15.35}$$

Its solutions are two values of ω_0, which differ in absolute magnitude, corresponding to two different waves with the phase velocities V_+ and V_- which are equal to

$$V_\pm^2 = \frac{1}{2}\left[u^2 + V_s^2 \pm \sqrt{\left(u^2 + V_s^2\right)^2 - 4u^2 V_s^2 \cos^2\theta}\,\right]. \tag{15.36}$$

These waves are called the *fast* (+) and the *slow* (−) *magnetoacoustic* waves, respectively (van de Hulst, 1951). The point is that the entropy of the medium, as follows from Equations (15.8) under condition (15.35), does not change in such waves

$$s' = 0\,, \tag{15.37}$$

as is also the case in an usual sound wave. Perturbations of the other quantities can be expressed in terms of the density perturbation

$$p' = V_s^2 \rho' , \tag{15.38}$$

$$\mathbf{v}' = -\frac{\omega_0}{\rho k^2} \left(\frac{k^2 (\mathbf{k} \cdot \mathbf{u})\, \mathbf{u} - \omega_0^2 \, \mathbf{k}}{\omega_0^2 - (\mathbf{k} \cdot \mathbf{u})^2} \right) \rho' , \tag{15.39}$$

$$\mathbf{u}' = \frac{\omega_0^2}{\rho k^2} \left(\frac{k^2 \, \mathbf{u} - (\mathbf{k} \cdot \mathbf{u})\, \mathbf{k}}{\omega_0^2 - (\mathbf{k} \cdot \mathbf{u})^2} \right) \rho' . \tag{15.40}$$

Formulae (15.39) and (15.40) show that the magnetoacoustic waves are neither longitudinal nor transversal. Perturbations of the velocity and magnetic field intensity, $\mathbf{v}'$ and $\mathbf{u}'$, as differentiated from the Alfvén wave, lie in the $(\mathbf{k}, \mathbf{B}_0)$ plane in Figure 15.1. They have components both in the direction of the wave propagation $\mathbf{k}/k$ and in the perpendicular direction. That is why the magnetoacoustic waves generally have a linearly polarized electric field $\mathbf{E}'$ normal to both $\mathbf{B}_0$ and $\mathbf{k}$.

The perturbation of magnetic pressure $B^2/8\pi$ may be written in the form (see definition (15.3))

$$p'_{\mathrm{m}} = \rho \, \mathbf{u} \cdot \mathbf{u}' = \left(\frac{V_{\pm}^2}{V_s^2} - 1 \right) p' . \tag{15.41}$$

Therefore for the fast wave, by virtue of that $V_+^2 > V_s^2$, the perturbation of magnetic pressure p'_{m} is of the same sign as that of gas pressure p'.

The magnetic pressure and the gas pressure are added in the fast magnetoacoustic wave. The wave propagates faster, since the effective elasticity of the plasma is greater.

A different situation arises with the slow magnetoacoustic wave. In this case $V_-^2 < V_s^2$ and p'_{m} is *opposite* in sign to p'. Magnetic and gas pressure deviations partially compensate each other. That is why such a slow wave propagates slowly.

15.2.4 The phase velocity diagram

The dependence of the wave velocities on the angle θ between the undisturbed field $\mathbf{B}_0$ and the wave vector $\mathbf{k}$ is clearly demonstrated in a polar diagram – the phase velocity diagram. In Figure 15.2, the radius-vector length from the origin of the coordinates to a curve is proportional to the corresponding phase velocity (15.11). The horizontal axis corresponds to the direction of the undisturbed magnetic field.

As the angle $\theta \to 0$, the fast magnetoacoustic wave V_+ transforms to the usual sound one V_s if

$$V_s > V_{\mathrm{A}\,\|} = \frac{B}{\sqrt{4\pi\rho}} \equiv u_{\mathrm{A}} \tag{15.42}$$

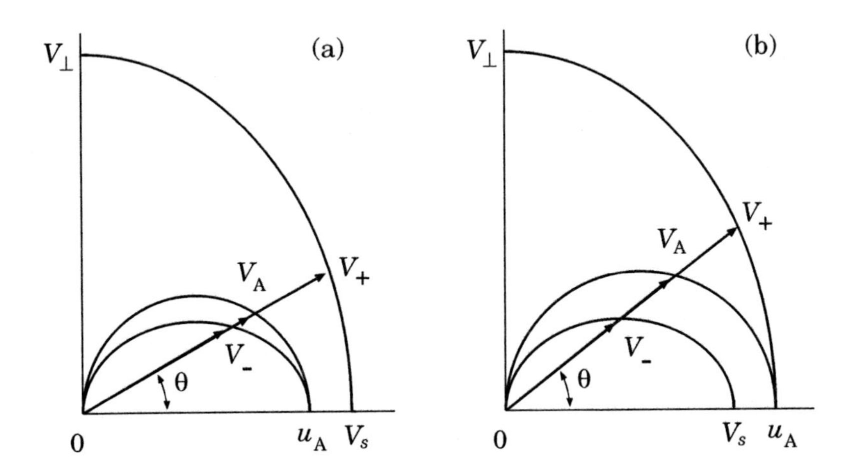

Figure 15.2: The phase velocities of MHD waves versus the angle θ for the two cases: (a) $u_{\text{A}} < V_s$ and (b) $u_{\text{A}} > V_s$.

in Figure 15.2a or to the Alfvén wave if $V_s < u_{\text{A}}$ in Figure 15.2b.

For the angle $\theta \to \pi/2$, the propagation velocities of the Alfvén and slow waves approach zero. As this takes place, both waves convert to the weak tangential discontinuity in which disturbances of velocity and magnetic field are parallel to the front plane. As $\theta \to \pi/2$, the fast magnetoacoustic wave velocity tends to

$$V_\perp = \sqrt{V_{\text{A}\,\|}^{\,2} + V_s^{\,2}} = \sqrt{u_{\text{A}}^2 + V_s^{\,2}}\,. \tag{15.43}$$

In the *strong* field limit ($V_{\text{A}\,\|}^{\,2} \gg V_s^{\,2}$) the diagram for the fast magnetoacoustic wave becomes practically isotropic as shown in Figure 15.3.

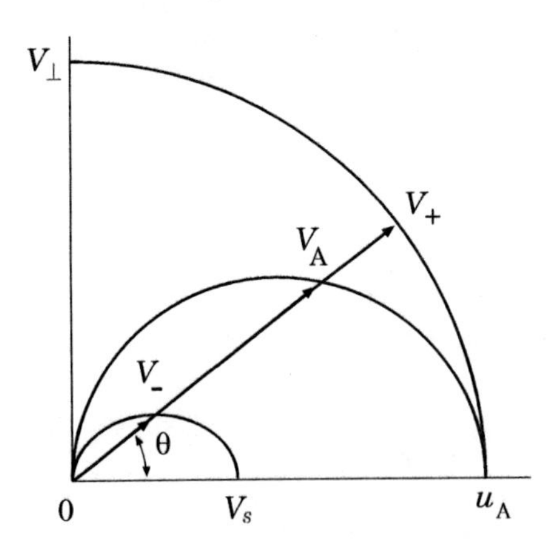

Figure 15.3: The phase velocity diagram for a plasma with a strong magnetic field.

Such a wave may be called the 'magnetic sound' wave since its phase velocity $V_+ \approx V_{\text{A}\,\|} \equiv u_{\text{A}}$ is almost independent of the angle θ.

Generally the sound speed is the minimum velocity of disturbance propagation in ordinary hydrodynamics. By contrast, there is *no minimum velocity* in magnetohydrodynamics.

This property is of fundamental importance for what follows in Chapters 16 and 17 – in study of the principal questions related to discontinuous flows of astrophysical plasma. The first of these questions is what kinds of discontinuities can really exist?

MHD waves produce a lot of effects in astrophysical plasma. The fast magnetoacoustic wave turbulence can presumably accelerate electrons in solar flares (see vol. 2, Section 12.3.1). The heavy ions observed in interplanetary space after impulsive flares can result from stochastic acceleration by the cascading Alfvén wave turbulence (vol. 2, Section 12.3.2).

15.3 Dissipative waves in MHD

15.3.1 Small damping of Alfvén waves

We shall start by treating a plane Alfvén wave propagating along a uniform field $\mathbf{B}_0$; so the angle $\theta = 0$ in Figure 15.1. Perturbations of the magnetic field and the velocity are small and parallel to the z axis:

$$\mathbf{B}' = \{\, 0,\, 0,\, b\,(t, y)\,\}\,, \quad \mathbf{v}' = \{\, 0,\, 0,\, v\,(t, y)\,\}\,. \tag{15.44}$$

In general, the damping effects for such a wave are determined by viscosity and conductivity. Let us consider, first, only the uniform finite conductivity σ. In this case we obtain the extended equation of the wave type with a dissipative term:

$$\frac{\partial^2 h}{\partial t^2} = u_{\rm A}^2\, \frac{\partial^2 h}{\partial y^2} + \nu_{\rm m}\, \frac{\partial^3 h}{\partial^2 y\, \partial t}\,. \tag{15.45}$$

Here $u_{\rm A} = V_{\rm A\,\parallel}$ and $\nu_{\rm m}$ is the magnetic diffusivity (12.49). In the case of infinite conductivity Equation (15.45) is reduced to the wave equation and represents an Alfvén wave with velocity $u_{\rm A}$.

Let us suppose that the conductivity is finite. We suppose further that the small perturbations are functions of t and y only:

$$b\,(t, y) = b_0 \exp\left(\,{\rm i}\,\omega t + \alpha y\right), \quad v\,(t, y) = v_0 \exp\left(\,{\rm i}\,\omega t + \alpha y\right). \tag{15.46}$$

Here ω, α, b_0, and v_0 are constants, all of which except ω may be complex numbers. Substituting (15.46) in (15.45) gives us the dispersion equation:

$$\omega^2 + \left(u_{\rm A}^2 + {\rm i}\,\nu_{\rm m}\,\omega\right)\alpha^2 = 0 \tag{15.47}$$

or

$$\alpha = \pm\,{\rm i}\,\frac{\omega}{u_{\rm A}}\left(1 + {\rm i}\,\frac{\nu_{\rm m}\,\omega}{u_{\rm A}^2}\right)^{-1/2}. \tag{15.48}$$

For small damping

$$\alpha = \pm \left(\mathrm{i}\,\frac{\omega}{u_{\mathrm{A}}} + \frac{\nu_{\mathrm{m}}\,\omega^2}{2u_{\mathrm{A}}^3} \right). \tag{15.49}$$

The distance y_0 in which the amplitude of the wave is reduced to $1/e$ is the inverse value of the real part of α. Thus we have

$$y_0 = \frac{2u_{\mathrm{A}}^3}{\nu_{\mathrm{m}}\,\omega^2} = \frac{8\pi\sigma u_{\mathrm{A}}^3}{\omega^2 c^2} = \frac{2\sigma u_{\mathrm{A}}}{\pi c^2}\,\lambda^2, \tag{15.50}$$

where $\lambda = 2\pi u_{\mathrm{A}}/\omega$ is the wave length. The **short waves suffer more damping** than do the long waves.

Since we treat the dissipative effects as small, the expression (15.50) is valid if $\lambda \ll y_0$. Thus we write

$$b\,(t, y) = b_0 \exp\left(-\frac{y}{y_0}\right) \exp\left[\mathrm{i}\,\omega\left(t - \frac{y}{u_{\mathrm{A}}}\right)\right], \tag{15.51}$$

$$v\,(t, y) = v_0 \exp\left(-\frac{y}{y_0}\right) \exp\left[\mathrm{i}\,\omega\left(t - \frac{y}{u_{\mathrm{A}}}\right)\right] \tag{15.52}$$

with

$$v_0 = u_{\mathrm{A}}\,\frac{b_0}{B_0}\left(1 - \mathrm{i}\,\frac{\nu_{\mathrm{m}}\,\omega}{2u_{\mathrm{A}}^2}\right). \tag{15.53}$$

The imaginary part indicates the phase shift of the velocity v in relation to the magnetic perturbation field b. Therefore

$$v\,(t, y) = u_{\mathrm{A}}\,\frac{b_0}{B_0} \exp\left(-\frac{y}{y_0}\right) \exp\left\{\mathrm{i}\left[\omega\left(t - \frac{y}{u_{\mathrm{A}}}\right) - \varphi\right]\right\}, \tag{15.54}$$

where

$$\varphi = \frac{\nu_{\mathrm{m}}\,\omega}{2u_{\mathrm{A}}^2} = \frac{\omega\,c^2}{8\pi\sigma u_{\mathrm{A}}^2} = \frac{\omega\,c^2\rho}{2\sigma B_0^2}. \tag{15.55}$$

So the existence of Alfvén waves requires an external field B_0 enclosed between two limits.

> The magnetic field should be strong enough to make the damping effects small but yet weak enough to keep the Alfvén speed well below the velocity of light,

because otherwise the wave becomes an ordinary electromagnetic wave (see Exercise 13.1). In optical and radio frequencies it is not possible to satisfy both conditions. However longer periods often observed in cosmic plasma leave a wide range between both limits so that Alfvén waves may easily exist.

One of favourable sites for excitation of MHD waves is the solar atmosphere. The chromosphere and corona are highly inhomogeneous media supporting a variety of filamentary structures in the form of arches and loops.

The foot points of these structures are anchored in the poles of the photospheric magnetic fields. They undergo a continuous twisting and turning due to convective motions in the subphotospheric layers. This twisting and turning excite MHD waves. The waves then dissipate and heat the corona (see vol. 2, Section 12.5). Presumably this energy is enough to explain coronal heating, but the unambiguous detection of the MHD waves heating the corona is still awaited.

15.3.2 Slightly damped MHD waves

The damping effects due to a finite conductivity σ and due to a kinematic viscosity $\nu = \eta/\rho$ (Section 12.2.2) can be included in a general treatment of MHD waves of small amplitudes (van de Hulst, 1951). Well developed waves are the waves that travel at least a few wave lengths before they lose a considerable fraction of their energy if the two dimensionless parameters

$$p_{\nu} = \frac{\omega\,\nu}{c^2} \quad \text{and} \quad p_{\nu_{\rm m}} = \frac{\omega\,\nu_{\rm m}}{c^2}\,, \tag{15.56}$$

that characterize two dissipative processes, are much smaller than the two small dimensionless parameters

$$p_s = \frac{V_s^2}{c^2} \quad \text{and} \quad p_{\rm A} = \frac{u_{\rm A}^2}{c^2}\,, \tag{15.57}$$

that characterize the propagation speeds of undamped waves.

Let us postulate the form

$$X \equiv c^2/\,V_{\rm ph}^2 = X_0\,(1 - {\rm i}\,q) \tag{15.58}$$

for a general solution of the linearized equations of dissipative MHD. Here

$$X_0 = c^2/\,V_{\rm ph,0}^2 \tag{15.59}$$

represents any solution for an undamped wave.

We shall not review all special cases here but shall mention only one, the same case as in previous Section. For Alfvén wave we find the following solution

$$X = X_{\rm m} \equiv c^2/u_{\rm A}^2\,, \quad q = (p_{\nu} + p_{\nu_{\rm m}})\,X_{\rm m}\,. \tag{15.60}$$

This shows that, if dissipative effects are small,

> the relative importance of resistivity and viscosity as damping effects in Alfvén wave is independent of frequency ω.

The damping length, i.e., the distance l_d, in which the amplitude of a wave decreases by a factor $1/e$, and the damping time τ_d, in which this distance is covered by the wave, can be found:

$$l_d = \frac{1}{kq} = \frac{u_{\rm A}}{q\,\omega} = \frac{u_{\rm A}^3}{\omega^2\,(\nu + \nu_{\rm m})}\,, \tag{15.61}$$

$$\tau_d = \frac{l_d}{u_{\rm A}} = \frac{1}{q\,\omega} = \frac{u_{\rm A}^2}{\omega^2\,(\nu + \nu_{\rm m})}\,. \qquad (15.62)$$

So the high frequency waves have a short damping length and time.

The magnetoacoustic waves (Section 15.2.3), being compressional, have an additional contribution to their damping rate from compressibility of the plasma. If dissipative effects are not small, they result in additional waves propagating in a homogeneous medium (see Section 17.3).

15.4 Practice: Exercises and Answers

Exercise 15.1. Evaluate the sound speed in the solar corona.

Answer. For an ideal gas with constant specific heats c_p and c_v, the sound speed (15.6) is

$$V_s = \left(\gamma_g \, \frac{p}{\rho} \right)^{1/2}, \qquad (15.63)$$

where $\gamma_g = c_p / c_v$. Let us consider the coronal plasma as a 'monatomic gas' ($\gamma_g = 5/3$) of electrons and protons with $T_{\rm e} = T_{\rm p} = T \approx 2 \times 10^6$ K and $n_{\rm e} = n_{\rm e} = n$. So $p = 2nk_{\rm B}T$ and $\rho = nm_{\rm p}$. Hence

$$V_s = \left(\frac{10}{3} \, \frac{k_{\rm B}}{m_{\rm p}} \right)_s^{1/2} \sqrt{T} = 1.66 \times 10^4 \, \sqrt{T({\rm K})}\,, \ {\rm cm \ s^{-1}}\,. \qquad (15.64)$$

In the solar corona $V_s \approx 230$ km s^{-1}.

Exercise 15.2. Consider entropy waves in ordinary hydrodynamics.

Answer. Let us take the linear algebraic Equations (15.25) and (15.26). In the absence of a magnetic field we put $\mathbf{u} = 0$ and $\mathbf{u}' = 0$. It follows from (15.25) that the perturbation of the velocity $\mathbf{v}'$ can be an arbitrary value except the gas pressure must be undisturbed. This follows from (15.26) and means that, instead of (15.27), we write

$$\mathbf{v}' \neq 0\,, \quad p' = 0\,. \qquad (15.65)$$

Perturbations of the density and entropy remain to be related by condition (15.28). So the velocity perturbation is independent of the entropy perturbation and, according to (15.13) and (15.23), satisfies the equation

$$\mathbf{k} \cdot \mathbf{v}' = \frac{\omega}{v} \, v_x' + k_y v_y' = 0\,. \qquad (15.66)$$

This is in the reference frame in which $v = v_x$.

Note that for such velocity perturbation (see Landau and Lifshitz, *Fluid Mechanics*, 1959a, Chapter 9):

$$\operatorname{curl} \mathbf{v}' \neq 0\,. \qquad (15.67)$$

That is why the wave is called the *entropy-vortex* wave.

In the presence of a magnetic field in plasma, it is impossible to create a vortex without a perturbation of the magnetic field.

For this reason, in a MHD entropy wave, the only disturbed quantities are the entropy and the density (see Equation (15.28)).

Exercise 15.3. Show that the inclusion of the displacement current modifies the dispersion relation for the Alfvén waves (15.29) to the following equation

$$\omega_0^2 = \frac{(\mathbf{k} \cdot \mathbf{u})^2}{1 + u^2/c^2} \quad \text{or} \quad \omega_0 = \pm \frac{\mathbf{k} \cdot \mathbf{u}}{\sqrt{1 + u^2/c^2}} \,. \tag{15.68}$$

So the phase velocity of the relativistic Alfvén waves

$$V_{\scriptscriptstyle A} = \pm \frac{B}{\sqrt{4\pi\rho}} \cos\theta \, \frac{1}{\sqrt{1 + B^2/4\pi\rho c^2}} \,, \tag{15.69}$$

which coinsides with the Alfvén formula (13.35).

Exercise 15.4. Discuss the following situation. A star of the mass M moves along a uniform magnetic field $\mathbf{B}_0$ at a constant velocity $\mathbf{v}_0$ which exceeds the phase velocity of a fast magnetoacoustic wave (Dokuchaev, 1964).

Hint. The moving star emits magnetoacoustic waves by the Cherenkov radiation (see Exercises 7.2–7.5).

Chapter 16

Discontinuous Flows in a MHD Medium

The phenomena related to shock waves and other dicontinuous flows in astrophysical plasma are so numerous that the study of MHD discontinuities on their own is of independent interest for space science.

16.1 Discontinuity surfaces in hydrodynamics

16.1.1 The origin of shocks in ordinary hydrodynamics

First of all, let us recall the way the shock waves are formed in ordinary hydrodynamic media without a magnetic field. Imagine a piston moving into a tube occupied by a gas. Let the piston velocity increase from zero by small jumps δv. As soon as the piston starts moving, it begins to rake the gas up and compress it. The front edge of the compression region thereby travels down the undisturbed gas inside the tube with the velocity of sound

$$V_s = \left(\frac{\partial p}{\partial \rho} \right)_s^{1/2} . \tag{16.1}$$

Each following impulse of compression $\delta \rho$ will propagate in a denser medium and hence with greater velocity. Actually, the derivative of the sound speed with respect to density

$$\frac{\partial V_s}{\partial \rho} = \frac{1}{2} \left(\frac{\partial^2 p}{\partial \rho^2} \right)_s \left(\frac{\partial p}{\partial \rho} \right)_s^{-1/2} \approx \sqrt{\gamma_g} \, (\gamma_g - 1) \, \rho^{\,(\gamma_g - 3)/2} > 0 \, ,$$

since for all real substances $\gamma_g > 1$ in the adiabatic process $p \sim \rho^{\gamma_g}$. Therefore $\delta V_s > 0$.

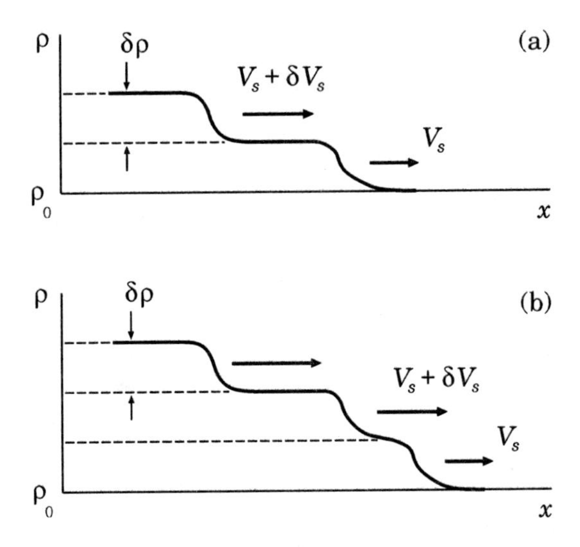

Figure 16.1: The behaviour of small perturbations in front of a piston.

As a consequence of this fact, successive compression impulses will catch up with each other as shown in Figure 16.1a. As a result, the compression region front steepens (Figure 16.1b). The gradients of the gas parameters become so large that the description of the gas as a hydrodynamic medium (Section 12.2) is no longer valid. The density, pressure and velocity of the gas change abruptly over a distance comparable to a particle's mean free path λ.

The physical processes inside such a jump, called a *shock wave*, are determined by the kinetic phenomena in the gas. As far as the hydrodynamic approximation is concerned,

> the surface, at which the continuity of the hydrodynamic parameters of a medium is violated, represents some discontinuity surface – a discontinuous solution of the hydrodynamic equations.

It stands to reason that some definite boundary conditions must hold at the discontinuity surface. What are they?

16.1.2 Boundary conditions and classification

Let us choose a frame of reference connected with a discontinuity surface. The frame is supposed to move with a constant velocity with respect to the medium. Generally, if the gas flow is non-stationary in the vicinity of the discontinuity, we could consider the discontinuity surface over a small period of time, so that the changes of velocity and other hydrodynamic quantities in time could be neglected.

In order to formulate the boundary conditions, let us consider an element of the discontinuity surface. Let the axis x be directed normally to it. **The**

flux of mass through such a surface element must conserve:

$$\rho_1 \, v_{x1} = \rho_2 \, v_{x2} \,. \tag{16.2}$$

Here the indices 1 and 2 refer to the two sides of the discontinuity surface.

In this chapter, the difference in a quantity across the discontinuity surface will be designated by curly brackets, e.g.

$$\{ \rho \, v_x \} = \rho_1 \, v_{x1} - \rho_2 \, v_{x2} \,.$$

Then Equation (16.2) is rewritten as

$$\{ \rho \, v_x \} = 0 \,. \tag{16.3}$$

The energy flux must also be continuous at the discontinuity surface. For a hydrodynamic medium without a magnetic field (cf. (12.74)) we obtain the following condition for the energy flux conservation:

$$\left\{ \rho \, v_x \left(\frac{v^2}{2} + w \right) \right\} = 0 \,. \tag{16.4}$$

Here w is the specific enthalpy (9.34).

The momentum flux must be also continuous (cf. (12.75)):

$$\Pi_{\alpha\beta} = p \, \delta_{\alpha\beta} + \rho \, v_\alpha v_\beta \,, \qquad \alpha = x \,.$$

The continuity of the x-component of the momentum flux means that

$$\{ p + \rho \, v_x^2 \} = 0 \,,$$

while the continuity of y- and z-components gives the two conditions

$$\{ \rho \, v_x v_y \} = 0 \,, \qquad \{ \rho \, v_x v_z \} = 0 \,.$$

Taking care of condition (16.3), let us rewrite the full set of boundary conditions at the discontinuity surface as follows:

$$\{ \rho \, v_x \} = 0 \,, \qquad \rho \, v_x \, \{ \mathbf{v}_\tau \} = 0 \,,$$

$$\rho \, v_x \left\{ \frac{v^2}{2} + w \right\} = 0 \,, \qquad \{ p + \rho \, v_x^2 \} = 0 \,. \tag{16.5}$$

Here the index τ identifies the tangential components of the velocity.

Obviously the set of Equations (16.5) falls into two *mutually exclusive* groups, depending on whether the matter flux across the discontinuity surface is zero or not. Consider these groups.

(a) If

$$v_x = 0$$

then the gas pressure is also continuous at the discontinuity surface,

$$\{ p \} = 0 \,, \tag{16.6}$$

while the tangential velocity component $\mathbf{v}_\tau$ as well as the density may experience an arbitrary jump:

$$\{ v_\tau \} \neq 0 \,, \quad \{ \rho \} \neq 0 \,, \quad \left\{ \frac{v^2}{2} + w \right\} \neq 0 \,.$$

Such discontinuities are called *tangential* (see Landau and Lifshitz, *Fluid Mechanics*, 1959a, Chapter 9, § 84).

(b) By contrast, if

$$v_x \neq 0$$

then

$$\{ \rho v_x \} = 0 \,, \quad \{ v_\tau \} = 0 \,, \quad \{ p + \rho v_x^2 \} = 0 \,, \quad \left\{ \frac{v^2}{2} + w \right\} = 0 \,. \tag{16.7}$$

Discontinuities of this type are termed *shock waves*. Their properties are also well known in hydrodynamics (Landau and Lifshitz, *Fluid Mechanics*, 1959a, Chapter 9, § 84).

Therefore

> the equations of ideal hydrodynamics in the conservation law form allow just two *mutually exclusive* types of discontinuities to exist: the shock wave and the tangential discontinuity.

16.1.3 Dissipative processes and entropy

The equations of ideal hydrodynamics, as a specific case ($\mathbf{B} = 0$) of the ideal MHD Equations (12.68)–(12.73), do not take into account either viscosity or thermal conductivity:

$$\eta = \zeta = 0 \,, \quad \kappa = 0 \,. \tag{16.8}$$

For this reason the ideal hydrodynamics equations describe three conservation laws: conservation of mass, momentum, and entropy. The last one,

$$\frac{\partial s}{\partial t} + (\mathbf{v} \cdot \nabla) \, s = 0 \,, \tag{16.9}$$

is the specific form of the energy conservation law (see Equation (12.54)) under assumption that the process under consideration is adiabatic. In Section 16.1.2 to obtain the boundary conditions at the discontinuity surface we used **conservation of mass, momentum, and energy**, but not entropy. The entropy increases across a shock (Exercise 16.6).

The increase in entropy indicates that *irreversible* dissipative processes (which can be traced to the presence of viscosity and heat conduction in a medium) occur in the shock wave. The model which does not take into account these processes (Section 16.1.2) admits the existence of discontinuities but is not capable of describing the continuous transition from the initial to the final state. The ideal hydrodynamics cannot describe either the mechanism of shock compression or the structure of the very thin but finite layer where the plasma undergoes a transition from the initial to the final state.

> The entropy increase across the shock is entirely independent of the dissipative mechanism and is defined exclusively by the conservation laws of mass, momentum, and energy

(see Exercise 16.6). Only the thickness of the discontinuity depends upon the rate of the irreversible heating of the plasma compressed by the shock. The following analogy in everyday life is interesting. A glass of hot water will invariably cool from a given temperature (the initial state) to a room temperature (the final state), independently of the mechanism of heat exchange with the surrounding air; the mechanism determines only the rate of cooling.

Recommended Reading: Zel'dovich and Raizer, *Physics of Shock Waves and High-Temperature Hydrodynamic Phenomena*, 1966, 2002, v. 1, Chapter 2.

16.2 Magnetohydrodynamic discontinuities

16.2.1 Boundary conditions at a discontinuity surface

Much like ordinary hydrodynamics, the equations of MHD for an ideal medium (Section 12.3) allow discontinuous solutions. De Hoffmann and Teller (1950) were the first to consider shock waves in MHD, based on the relativistic energy-momentum tensor for an ideal medium and the electromagnetic field.

Syrovatskii (1953) has given a more general formulation of the problem of the possible types of discontinuity surfaces in a conducting medium with a magnetic field. He has formulated a closed set of equations of ideal MHD and, using this, the *boundary conditions* at the discontinuity were written. We shall briefly reproduce the derivation of the boundary conditions.

We start from the equations of ideal MHD (12.68)–(12.73). Rewrite them (the Equation of state (12.73) is omitted for brevity) as follows:

$$\operatorname{div}\mathbf{B} = 0\,, \qquad \frac{\partial \mathbf{B}}{\partial t} = \operatorname{curl}\,(\,\mathbf{v}\times\mathbf{B}\,)\,, \qquad \frac{\partial \rho}{\partial t} = -\operatorname{div}\rho\mathbf{v}\,, \tag{16.10}$$

$$\frac{\partial}{\partial t}\left(\frac{\rho v^2}{2} + \rho\varepsilon + \frac{B^2}{8\pi}\right) = -\operatorname{div}\mathbf{G}\,, \qquad \frac{\partial}{\partial t}\,(\rho\, v_\alpha) = -\frac{\partial}{\partial r_\beta}\,\Pi^{*}_{\alpha\beta}\,.$$

In a frame of reference moving with the discontinuity surface, all the conditions are stationary ($\partial/\partial t = 0$). Hence

$$\operatorname{div}\mathbf{B} = 0\,, \tag{16.11}$$

$$\operatorname{curl}(\mathbf{v} \times \mathbf{B}) = 0, \tag{16.12}$$

$$\operatorname{div} \rho \mathbf{v} = 0, \qquad \operatorname{div} \mathbf{G} = 0, \qquad \frac{\partial}{\partial r_\beta} \Pi^*_{\alpha\beta} = 0. \tag{16.13}$$

Four of these conditions have the divergent form and are therefore reduced in the integral form to the conservation of fluxes of vectors appearing at the divergence. Thus the following quantities must conserve at the discontinuity: the perpendicular (to the surface S) component of the magnetic field vector B_{n}, the mass flux ρv_{n}, the energy flux G_{n}, and the momentum flux $\Pi^*_{\alpha\mathrm{n}}$.

The exception is condition (16.12). It is written as the curl of $\mathbf{v} \times \mathbf{B}$. Integration of (16.12) over the area enclosed by the contour shown in Figure 16.2 gives, by virtue of the Stokes theorem,

$$\int_S \operatorname{curl}(\mathbf{v} \times \mathbf{B}) \cdot d\mathbf{S} = \oint_L (\mathbf{v} \times \mathbf{B}) \cdot d\mathbf{l} = 0 .$$

Thus condition (16.12) demonstrates the continuity of the tangential component of the vector $(\mathbf{v} \times \mathbf{B})_\tau$, i.e. the electric field $\mathbf{E}_\tau$ in the discontinuity surface S.

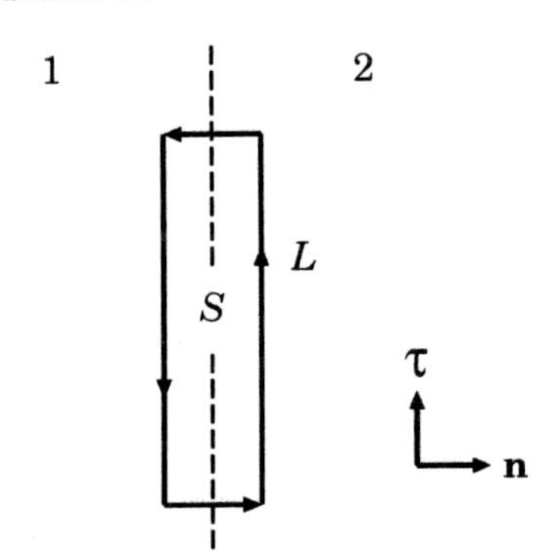

Figure 16.2: The contour L for the derivation of the boundary condition on electric field tangential component.

As in the previous section, the jump of a quantity on crossing the discontinuity surface is designated by curly brackets. The full system of boundary conditions at the surface is written as follows:

$$\{ B_{\mathrm{n}} \} = 0, \tag{16.14}$$

$$\{ (\mathbf{v} \times \mathbf{B})_\tau \} = 0, \tag{16.15}$$

$$\{ \rho v_{\mathrm{n}} \} = 0, \tag{16.16}$$

$$\{ G_{\mathrm{n}} \} = 0, \tag{16.17}$$

$$\{ \Pi^*_{\alpha\mathrm{n}} \} = 0. \tag{16.18}$$

The physical meaning of the boundary conditions obtained is obvious. The first two are the usual electrodynamic continuity conditions for the normal component of the magnetic field and the tangential component of the electric field. The last three equations represent the continuity of fluxes of mass, energy and momentum, respectively.

As distinct from that in ordinary hydrodynamics (see Equations (16.5)),

> the set of the MHD boundary conditions does not fall into mutually exclusive groups of equations.

This means that, with a few exceptions, any discontinuity, once accepted by these equations, can, generally speaking, transform to any other discontinuity under continuous change of the conditions of the motion (Syrovatskii, 1956).

Hence the classification of discontinuities in MHD seems to be a matter of convention. Any classification is based on the external properties of the flow near the surface, such as the absence or presence of normal components of the velocity $v_{\rm n}$ and magnetic field $B_{\rm n}$, continuity or jump in density. The classification given below is due to Syrovatskii (1953). It is quite convenient for investigating MHD discontinuities.

Before turning our attention to the discussion of the classification mentioned above, let us rewrite the boundary conditions obtained, using (12.74) and (12.75) for the densities of the energy and momentum fluxes and substituting (16.14) in (16.15) and (16.16) in (16.18). We get

$$\{ B_{\rm n} \} = 0 \,, \tag{16.19}$$

$$\{ v_{\rm n} \mathbf{B}_{\tau} \} = B_{\rm n} \{ \mathbf{v}_{\tau} \} \,, \tag{16.20}$$

$$\{ \rho \, v_{\rm n} \} = 0 \,, \tag{16.21}$$

$$\left\{ \rho \, v_{\rm n} \left(\frac{v^2}{2} + w \right) + \frac{1}{4\pi} \left(B^2 v_{\rm n} - (\mathbf{v} \cdot \mathbf{B}) \, B_{\rm n} \right) \right\} = 0 \,, \tag{16.22}$$

$$\left\{ p + \rho \, v_{\rm n}^2 + \frac{B^2}{8\pi} \right\} = 0 \,, \tag{16.23}$$

$$\rho \, v_{\rm n} \{ \mathbf{v}_{\tau} \} = \frac{B_{\rm n}}{4\pi} \{ \mathbf{B}_{\tau} \} \,. \tag{16.24}$$

For later use, we write down the boundary conditions in the Cartesian frame of reference, the x axis being perpendicular to the discontinuity surface:

$$\{ B_x \} = 0 \,, \tag{16.25}$$

$$\{ v_x B_y - v_y B_x \} = 0 \,, \tag{16.26}$$

$$\{ v_x B_z - v_z B_x \} = 0 \,, \tag{16.27}$$

$$\{ \rho \, v_x \} = 0 \,, \tag{16.28}$$

$$\left\{ \rho \, v_x \left(\frac{v^2}{2} + w \right) + \frac{1}{4\pi} \left(B^2 v_x - (\mathbf{v} \cdot \mathbf{B}) \, B_x \right) \right\} = 0 \,, \tag{16.29}$$

$$\left\{ p + \rho v_x^2 + \frac{B^2}{8\pi} \right\} = 0\,, \tag{16.30}$$

$$\left\{ \rho\, v_x v_y - \frac{1}{4\pi}\, B_x B_y \right\} = 0\,, \tag{16.31}$$

$$\left\{ \rho\, v_x v_z - \frac{1}{4\pi}\, B_x B_z \right\} = 0\,. \tag{16.32}$$

The set consists of eight boundary conditions. For $\mathbf{B} = 0$ it converts to the set of four Equations (16.5).

Let us consider the classification of discontinuity surfaces in MHD, which stems from the boundary conditions (16.19)–(16.24).

16.2.2 Discontinuities without plasma flows across them

Let us suppose the plasma flow through the discontinuity surface is absent

$$v_{\rm n} = 0\,. \tag{16.33}$$

The discontinuity type depends on whether the magnetic field penetrates through the surface or not. Consider both possibilities.

(a) If the perpendicular component of the magnetic field

$$B_{\rm n} \neq 0\,, \tag{16.34}$$

then the set of Equations (16.19)–(16.24) becomes

$$\{ B_{\rm n} \} = 0\,, \quad B_{\rm n}\, \{ \mathbf{v}_\tau \} = 0\,, \quad B_{\rm n}\, \{ \mathbf{B}_\tau \} = 0\,,$$

$$\left\{ p + \frac{B^2}{8\pi} \right\} = 0\,, \quad \{ \rho \} \neq 0\,. \tag{16.35}$$

The velocity, magnetic field strength and (by virtue of the fourth equation) gas pressure are continuous at the surface. The density jump does not have to be zero; otherwise, all values change continuously.

The discontinuity type considered is called the *contact* discontinuity and constitutes just **a boundary between two media**, which moves together with them. It is schematically depicted in Figure 16.3a.

(b) On the other hand, if

$$B_{\rm n} = 0 \tag{16.36}$$

then the velocity and magnetic field are parallel to the discontinuity surface (plane $x = 0$). In this case all the boundary conditions (16.19)–(16.24) are satisfied identically, with the exception of one. The remaining equation is

$$\left\{ p + \frac{B^2}{8\pi} \right\} = 0\,.$$

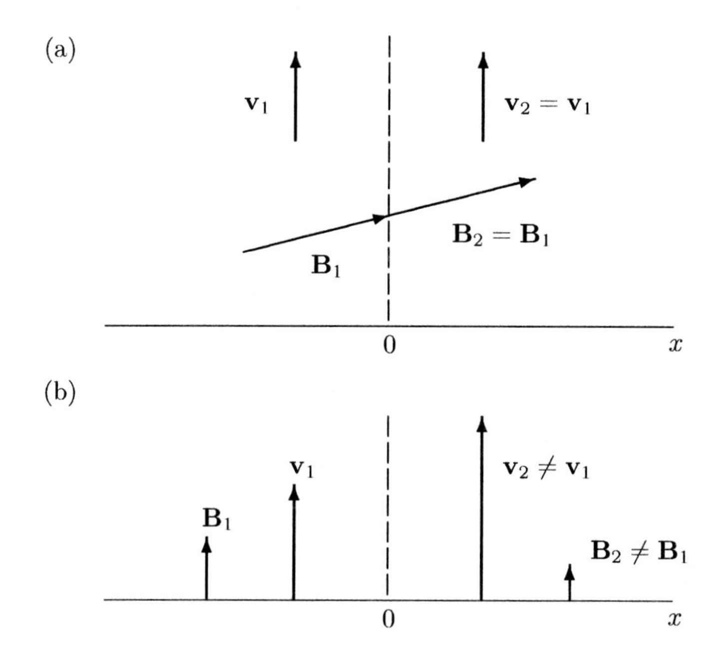

Figure 16.3: Discontinuity surfaces without a plasma flow across them: (a) contact discontinuity, (b) tangential discontinuity.

In other words, the velocity and magnetic field are parallel to the discontinuity surface and may experience arbitrary jumps in magnitude and direction, the only requirement being that the total pressure, that is the sum of the usual gas pressure and the magnetic one, remains continuous at the discontinuity surface:

$$p^* = p + \frac{B^2}{8\pi} \, . \tag{16.37}$$

Such a discontinuity is called a *tangential* discontinuity (Figure 16.3b). As treated in MHD, it has a remarkable property. The tangential discontinuity in ordinary hydrodynamics is always unstable (Syrovatskii, 1954; see also Landau and Lifshitz, *Fluid Mechanics*, Third Edition, Chapter 9, § 84, Problem 1). The velocity jump engenders vortices, thus resulting in a turbulence near the discontinuity. Another situation occurs in MHD.

Syrovatskii (1953) has shown that the magnetic field exerts a stabilizing influence on the tangential discontinuity. In particular, if the density ρ_0 and magnetic field $\mathbf{B}_0$ are continuous, the only discontinuous quantity being the tangential velocity component, $\mathbf{v}_2 - \mathbf{v}_1 = \mathbf{v}_0 \neq 0$, then the condition for the tangential discontinuity stability is especially simple:

$$\boxed{\frac{B_0^2}{8\pi} \geq \frac{1}{4}\,\frac{\rho_0 v_0^2}{2}\,.} \tag{16.38}$$

To put it another way, such a discontinuity (Figure 16.4a) becomes stable with respect to small perturbations (of the general rather than a particular type) once the magnetic energy density reaches one quarter of the kinetic energy density.

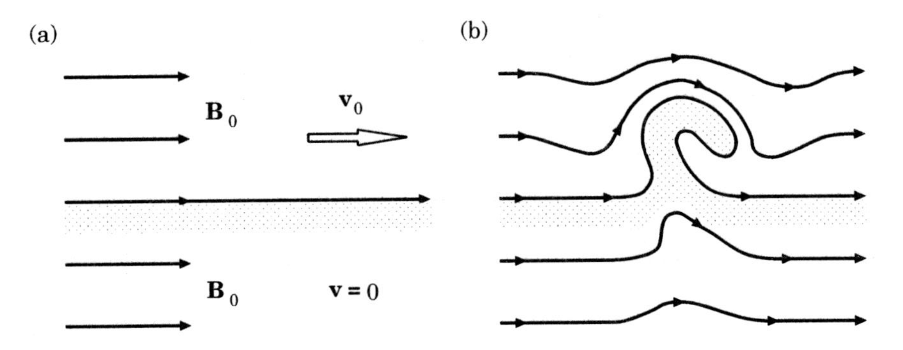

Figure 16.4: (a) The simplest type of the MHD tangential discontinuities. (b) Formation of a turbulent vortex gives rise to the magnetic field growth.

The general conclusion concerning the influence of the magnetic field on the stability of hydrodynamic motions of a conducting fluid is as follows:

> the magnetic field can only increase the stability of a given velocity distribution as compared to the stability of the same distribution in the absence of a magnetic field.

The point is that any flow instability and turbulence give rise, in view of the freezing-in of the field, to an increase of the magnetic energy (Figure 16.4b), which is always disadvantageous from the standpoint of the energetic principle of stability.

16.2.3 Perpendicular shock wave

Now let

$$v_{\rm n} \neq 0 \quad \text{and} \quad B_{\rm n} = 0\,, \tag{16.39}$$

i.e. a flow through the discontinuity surface is present whereas the magnetic field does not penetrate through the surface. Under these conditions, the following two statements result from Equations (16.19)–(16.24).

(a) From (16.24) the continuity of the tangential velocity component follows:

$$\{\, \mathbf{v}_\tau \} = 0 \, . \tag{16.40}$$

This makes it possible to transform to such a frame of reference in which the tangential velocity component is absent on either side of the discontinuity: $\mathbf{v}_{\tau 1} = \mathbf{v}_{\tau 2} = 0$.

(b) The tangential electric field continuity (16.20) results in

$$\{\, v_{\rm n}\, \mathbf{B}_\tau \} = 0 \, . \tag{16.41}$$

If the frame of reference is rotated with respect to the x axis in such a way that $B_z = 0$ on one side of the surface, then the same is true on the other side (for clarity see (16.27)). Thus a frame of reference exists in which, in view of (a),

$$\mathbf{v} = (\, v_{\rm n},\, 0,\, 0\,) = (\, v,\, 0,\, 0\,)$$

and in addition, by virtue of (b),

$$\mathbf{B} = (0,\, B_\tau,\, 0\,) = (\, 0,\, B,\, 0\,) \, .$$

In this frame of reference, the other boundary conditions take the form:

$$\{\, \rho\, v \,\} = 0 \, , \tag{16.42}$$

$$\{\, B/\rho \,\} = 0 \, , \tag{16.43}$$

$$\left\{\, \rho\, v^2 + p + \frac{B^2}{8\pi} \,\right\} = 0 \, , \tag{16.44}$$

$$\left\{\, \frac{v^2}{2} + w + \frac{B^2}{4\pi\rho} \,\right\} = 0 \, . \tag{16.45}$$

Such a discontinuity is called the *perpendicular* shock wave, since it constitutes the compression shock (see (16.7)) propagating perpendicular to the magnetic field as shown Figure 16.5.

Condition (16.43) reflects the fact of the field 'freezing-in' into the plasma. The role of pressure in such a wave is played by the total pressure

$$p^* = p + \frac{B^2}{8\pi} \, , \tag{16.46}$$

whereas the role of the specific enthalpy is fulfilled by

$$w^* = w + \frac{B^2}{4\pi\rho} \, . \tag{16.47}$$

Therefore the role of the internal energy density is played by the total internal energy

$$\varepsilon^* = w^* - \frac{p^*}{\rho} = \varepsilon + \frac{B^2}{8\pi\rho} \tag{16.48}$$

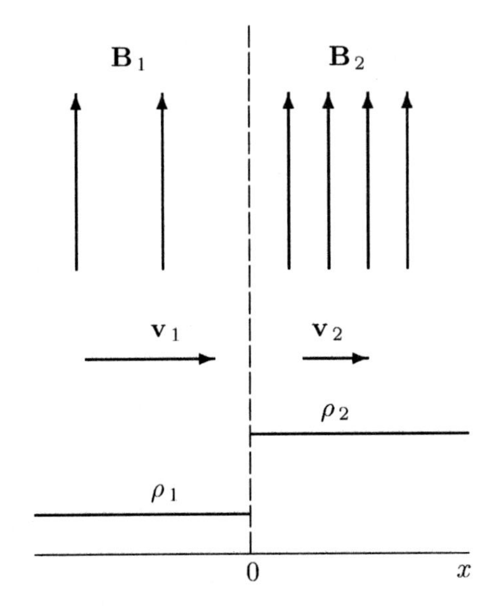

Figure 16.5: The character of the plasma motion and magnetic field compression ($B_2 > B_1$) in the perpendicular shock wave.

(cf. corresponding terms in Equations (12.68) and (12.69)).

For $\mathbf{B} = 0$, the perpendicular shock degenerates to the usual compression shock wave (Equations (16.7)).

For $\mathbf{B} \neq 0$, the propagation velocity of the perpendicular shock depends on the magnetic field strength.

> A magnetic field decreases the compressibility of plasma while increasing its elasticity.

This is seen from (16.46) and the freezing-in condition (16.43). Accordingly, the magnetic field increases the shock wave propagation velocity.

If the intensity of a perpendicular shock is diminished, it converts to a fast magnetoacoustic wave propagating across the magnetic field ($\theta = \pi/2$ in Figure 15.2) with the speed (15.43), i.e.

$$V_{\perp} = \sqrt{V_s^{\,2} + V_{\rm A}^{\,2}} \, . \tag{16.49}$$

16.2.4 Oblique shock waves

The types of discontinuity surfaces treated above are the limiting cases of a more general discontinuity type for which

$$v_{\rm n} \neq 0 \quad \text{and} \quad B_{\rm n} \neq 0 \, . \tag{16.50}$$

16.2.4 (a) The de Hoffmann-Teller frame of reference

In investigating the discontinuities (16.50), a frame of reference would be convenient in which $\mathbf{v}_1$ and $\mathbf{B}_1$ are parallel to each other. Such a frame does exist. It moves with respect to the laboratory one with the velocity

$$\mathbf{U} = \mathbf{v}_1 - \frac{v_{x1}}{B_{x1}}\,\mathbf{B}_1$$

parallel to the discontinuity surface. Actually, in this frame

$$\mathbf{v}_1\,(\mathbf{U}) = \mathbf{v}_1 - \mathbf{U} = \frac{v_{x1}}{B_{x1}}\,\mathbf{B}_1$$

and hence

$$\mathbf{v}_1 \times \mathbf{B}_1 = 0\,. \tag{16.51}$$

Then condition (16.20) in its coordinate form (16.26)–(16.27) can be used to obtain two equations valid to the right of the discontinuity, i.e. downstream of the shock:

$$v_{x2} B_{y2} - v_{y2} B_{x2} = 0\,, \qquad v_{x2} B_{z2} - v_{z2} B_{x2} = 0\,.$$

On rewriting these conditions as

$$\frac{v_{x2}}{v_{y2}} = \frac{B_{x2}}{B_{y2}} \quad \text{and} \quad \frac{v_{x2}}{v_{z2}} = \frac{B_{x2}}{B_{z2}}\,,$$

we ensure that the magnetic field is parallel to the velocity field (in the chosen reference frame) to the right of the discontinuity. In such frame of reference, called the de Hoffmann-Teller frame (de Hoffmann and Teller, 1950), the **electric field does not appear** according to (16.51).

This fact does not mean, of course, that the local cross-shock electric fields do not appear inside the shock transition layer, i.e. inside the discontinuity. The quasi-static electric and magnetic fields may determine the dynamics of particles in the shock front especially if Coulomb collisions play only a minor role. In collisionless shock waves, this dynamics depend on the particular mechanism of the energy redistribution among the perpendicular (with respect to the local magnetic field) and parallel degrees of freedom (see Section 16.4).

16.2.4 (b) Two types of shock waves

Thus $\mathbf{v}$ is parallel to $\mathbf{B}$ on either side of the discontinuity. As a consequence, of the eight boundary conditions initially considered (see (16.25)–(16.32)), there remain six equations:

$$\{\, B_x \,\} = 0\,, \tag{16.52}$$

$$\{\, \rho\, v_x \,\} = 0\,, \tag{16.53}$$

$$\left\{ \frac{v^2}{2} + w \right\} = 0\,, \tag{16.54}$$

$$\left\{ p + \rho\, v_x^2 + \frac{B^2}{8\pi} \right\} = 0\,, \tag{16.55}$$

$$\left\{ \rho\, v_x v_y - \frac{B_x B_y}{4\pi} \right\} = 0\,, \tag{16.56}$$

$$\left\{ \rho\, v_x v_z - \frac{B_x B_z}{4\pi} \right\} = 0\,. \tag{16.57}$$

Let us take account of the parallelism of **v** and **B** in the chosen reference frame:

$$\mathbf{v}_1 = q_1 \mathbf{B}_1\,, \qquad \mathbf{v}_2 = q_2 \mathbf{B}_2\,, \tag{16.58}$$

where q_1 and q_2 are some proportionality coefficients. On substituting (16.58) in (16.52)–(16.57) we obtain the following three conditions from (16.53), (16.56), and (16.57):

$$\{ \rho\, q \} = 0\,, \tag{16.59}$$

$$\left\{ \left(1 - \frac{1}{4\pi \rho\, q^2} \right) v_y \right\} = 0\,, \tag{16.60}$$

$$\left\{ \left(1 - \frac{1}{4\pi \rho\, q^2} \right) v_z \right\} = 0\,. \tag{16.61}$$

These equations admit two essentially different discontinuity types, depending on whether the density of the plasma is continuous or experiences a jump.

First we consider the discontinuity accompanied by a *density jump*:

$$\{ \rho \} \neq 0\,. \tag{16.62}$$

Discontinuities of this type are called *oblique* shock waves.

Rotate the reference frame with respect to the x axis in such a way that

$$v_{z1} = 0\,.$$

Then from (16.61) the following equation follows:

$$\left(1 - \frac{1}{4\pi \rho_2\, q_2^2} \right) v_{z2} = 0\,. \tag{16.63}$$

This suggests two possibilities: either

(**Case I**)

$$v_{z2} = 0\,, \tag{16.64}$$

i.e. the motion is planar (the velocity and magnetic field are in the plane (x, y) on either side of the discontinuity), or

(**Case II**)

$$v_{z2} \neq 0 \quad \text{but} \quad q_2^2 = \frac{1}{4\pi \rho_2}\,. \tag{16.65}$$

Note that in the latter case

$$q_1^2 \neq \frac{1}{4\pi \rho_1} \tag{16.66}$$

since concurrently valid equations

$$q_2^2 = \frac{1}{4\pi\rho_2} \quad \text{and} \quad q_1^2 = \frac{1}{4\pi\rho_1}$$

would imply that

$$\rho_2 q_2 = \frac{1}{4\pi q_2} \quad \text{and} \quad \rho_1 q_1 = \frac{1}{4\pi q_1}\,,$$

thus obviously contradicting (16.59) and (16.62). Therefore condition (16.66) must be valid.

Let us consider both cases indicated above.

16.2.4 (c) Fast and slow shock waves

Let us consider first the **Case I**. On the strength of (16.64), the boundary conditions (16.52)–(16.57) take the form

$$\{ B_x \} = 0\,, \qquad \{ \rho v_x \} = 0\,, \qquad \left\{ \frac{v^2}{2} + w \right\} = 0\,,$$

$$\left\{ p + \rho v_x^2 + \frac{B_y^2}{8\pi} \right\} = 0\,, \qquad \left\{ \rho v_x v_y - \frac{1}{4\pi} B_x B_y \right\} = 0\,. \tag{16.67}$$

The compression oblique shock wave interacts with the magnetic field in an intricate way. The relationship between the parameters determining the state of a plasma before and after the wave passage is the topic of a large body of research (see reviews: Syrovatskii, 1957; Polovin, 1961; monographs: Anderson, 1963, Chapter 5; Priest, 1982, Chapter 5).

Boundary conditions (16.67) can be rewritten in such a way as to represent the Rankine-Hugoniot relation (see Exercises 16.2 and 16.3 for an ordinary shock wave) for shocks in MHD (see Landau et al., 1984, Chapter 8). Moreover the Zemplen theorem on the **increase of density and pressure in a shock wave** can be proved in MHD (Iordanskii, 1958; Liubarskii and Polovin, 1958; Polovin and Liubarskii, 1958; see also Zank, 1991). The *fast* and the *slow* oblique shock waves are distinguished.

In the fast shock wave, the magnetic field increases across the shock and is bent towards the shock front surface $x = 0$ (Figure 16.6). So the magnetic pressure increases as well as the gas pressure:

$$\delta p_{\rm m} > 0\,, \qquad \delta p > 0\,. \tag{16.68}$$

In other words, and this seems to be a natural behaviour,

> compression of the plasma in a fast MHD shock wave is accompanied by compression of the magnetic field.

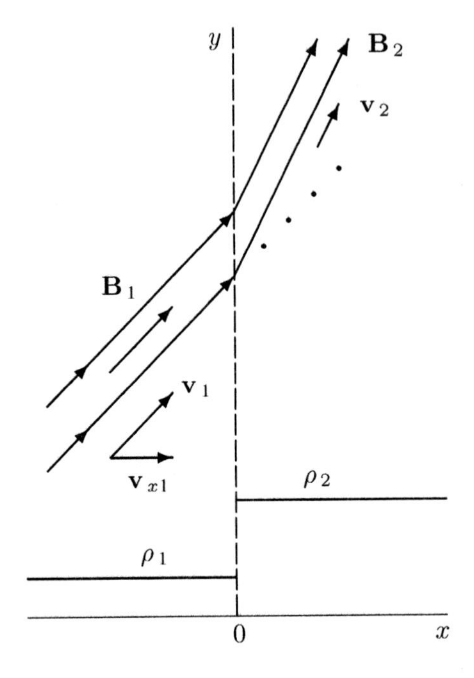

Figure 16.6: The magnetic field change ($B_2 > B_1$), velocity field and plasma density at the front of the fast shock wave.

In the limiting case of small intensity, the fast shock converts to the fast magnetoacoustic wave (see (16.46)). The speed of the fast shock wave with respect to the medium equals v_{x1}. It is greater than or equal to the speed of the fast magnetoacoustic wave:

$$v_{x1} \geq V_+ \, . \tag{16.69}$$

No small perturbation running in front of the shock can exist upstream of the fast shock wave.

In the slow shock wave, the magnetic field decreases across the shock and is bent towards the shock normal (Figure 16.7). Therefore

$$\delta p_{\rm m} < 0 \, , \qquad \delta p > 0 \, . \tag{16.70}$$

Compression of the plasma is accompanied by a *decrease* of the magnetic field strength in the slow MHD shock wave.

As the amplitude decreases, the slow shock wave will transform to the slow magnetoacoustic wave. The speed of the slow shock propagation is

$$V_- \leq v_{x1} \leq V_{\rm A} \, . \tag{16.71}$$

In the particular case

$$B_y = 0 \tag{16.72}$$

the set of boundary conditions (16.67) results in the set (16.5). This means that the oblique shock wave converts to the parallel (longitudinal) shock wave propagating along the magnetic field, mutual interaction being absent.

Figure 16.7: The magnetic field change ($B_2 < B_1$), velocity field and plasma density at the front of the slow shock wave.

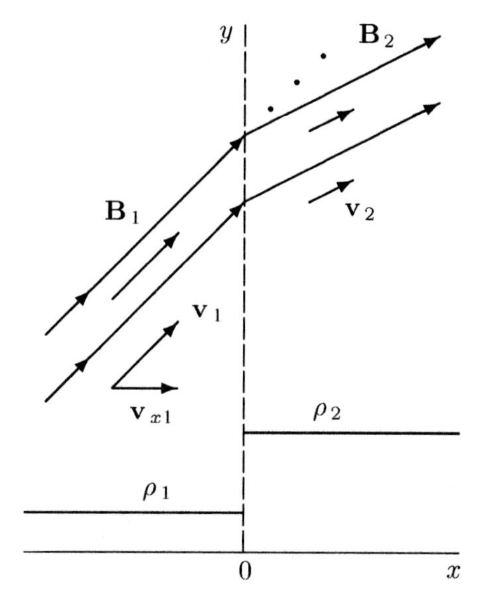

The set of boundary conditions (16.52)–(16.57) formally admits four other types of discontinuous solutions (Section 17.4.2), apart from those indicated above. These are the so called *intermediate* or *transalfvénic* shock waves (e.g., Shercliff, 1965, Chapter 7).

The peculiarity of these discontinuous solutions is that they have no counterpart among the small amplitude waves or simple waves.

This is the reason why the intermediate and transalfvénic shock waves are not included in the classification of discontinuities under consideration. What is more important is that the intermediate and transalfvénic shock waves are *non-evolutionary* (see Section 17.1).

The **Case II** shall be considered in the next Section.

16.2.5 Peculiar shock waves

We return to the consideration of the particular case (16.65) and (16.66):

$$v_{z2} \neq 0 \,, \quad q_1^2 \neq \frac{1}{4\pi \rho_1} \,, \quad q_2^2 = \frac{1}{4\pi \rho_2} \,. \tag{16.73}$$

On the strength of (16.60) and (16.61), the following conditions must be satisfied at such a discontinuity:

$$\left(1 - \frac{1}{4\pi \rho_1 q_1^2} \right) v_{y1} = 0 \,, \quad \left(1 - \frac{1}{4\pi \rho_1 q_1^2} \right) v_{z1} = 0 \,.$$

Because the expression in the parentheses is not zero, we get

$$v_{y1} = v_{z1} = 0 \,, \tag{16.74}$$

i.e. in front of such a discontinuity the tangential velocity component $\mathbf{v}_{\tau 1}$ is absent. The tangential field component $\mathbf{B}_{\tau 1}$ is also zero in front of the discontinuity, i.e. the motion follows the pattern seen in the parallel shock wave. However arbitrary tangential components of the velocity and magnetic field are permissible downstream of the shock, the only condition being that

$$\mathbf{v}_2 = \frac{\mathbf{B}_2}{\sqrt{4\pi\rho_2}} \, . \tag{16.75}$$

Such a discontinuity is called the *switch-on* shock. The character of motion of this wave is shown in Figure 16.8.

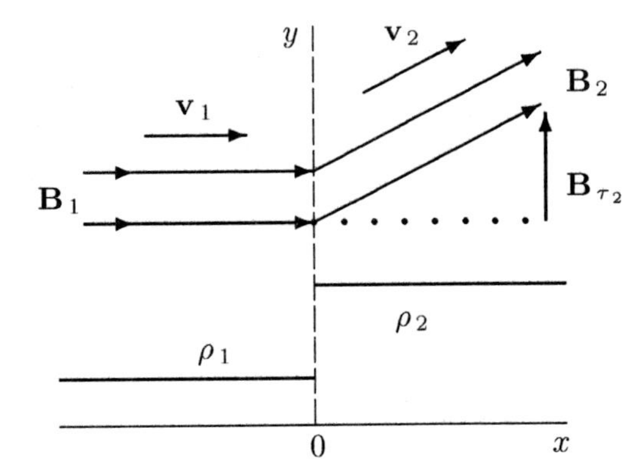

Figure 16.8: A switch-on wave: $\mathbf{B}_{\tau 1} = 0$, but $\mathbf{B}_{\tau 2} \neq 0$.

The switch-on shock exists in the interval

$$1 < \frac{v_{x1}^2}{V_{\mathrm{A}x1}^2} < \frac{4\, v_{x1}^2}{v_{x1}^2 + V_{\mathrm{s}1}^2}$$

(e.g., Liberman, 1978).

Assuming the tangential magnetic field component to be zero to the rear of the peculiar shock wave,

$$\mathbf{B}_{\tau 2} = 0 \, , \tag{16.76}$$

the fluid velocity in front of the discontinuity is the Alfvén one:

$$\mathbf{v}_1 = \frac{\mathbf{B}_1}{\sqrt{4\pi\rho_1}} \, . \tag{16.77}$$

Such a peculiar shock wave is called the *switch-off* shock (Figure 16.9).

16.2.6 The Alfvén discontinuity

Returning to the general set of Equations (16.52)–(16.57), consider the discontinuity at which the density is constant:

$$\{\, \rho \,\} = 0 \, . \tag{16.78}$$

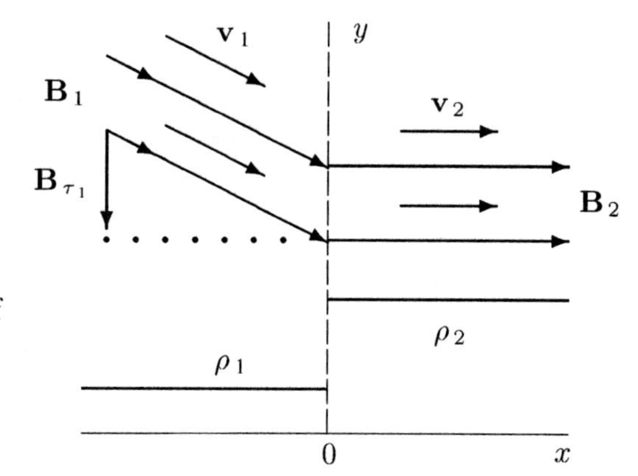

Figure 16.9: A switch-off wave: $\mathbf{B}_{\tau 2} = 0$ but $\mathbf{B}_{\tau 1} \neq 0$.

On substituting this condition in (16.53), we see that the normal component of the velocity is continuous at the discontinuity:

$$\{ v_x \} = 0 .$$

Furthermore, in view of Equation (16.59), the quantity q does not change at the discontinuity:

$$\{ q \} = 0 .$$

Hence the quantity

$$\left(1 - \frac{1}{4\pi \rho \, q^2} \right)$$

is also continuous and may be factored out in Equations (16.60) and (16.61). Rewrite them as follows:

$$\left(1 - \frac{1}{4\pi \rho \, q^2} \right) \{ \mathbf{v}_\tau \} = 0 . \tag{16.79}$$

If the expression in the parentheses is not zero then the tangential velocity component is continuous and all other quantities are easily checked to be continuous solutions. So, to consider the discontinuous solutions, we put

$$q = \pm \frac{1}{\sqrt{4\pi\rho}} .$$

Thus the velocity vector is connected with the magnetic field strength through the relations

$$\mathbf{v}_1 = \pm \frac{\mathbf{B}_1}{\sqrt{4\pi\rho}} , \qquad \mathbf{v}_2 = \pm \frac{\mathbf{B}_2}{\sqrt{4\pi\rho}} . \tag{16.80}$$

The following relations also hold at the discontinuity surface

$$\{ p \} = 0 , \quad \{ \mathbf{B}_\tau^2 \} = 0 . \tag{16.81}$$

Therefore the normal components and the absolute values of the tangential components of the magnetic field and velocity as well as all thermodynamical parameters conserve at the discontinuity. For given values of $\mathbf{B}_1$ and $\mathbf{v}_1$, possible values of $\mathbf{B}_2$ and $\mathbf{v}_2$ lie on a conical surface, the cone angle being equal to that between the normal to the discontinuity surface and the vector $\mathbf{B}_1$ (Figure 16.10). A discontinuity of this type is called *Alfvén* or *rotational*.

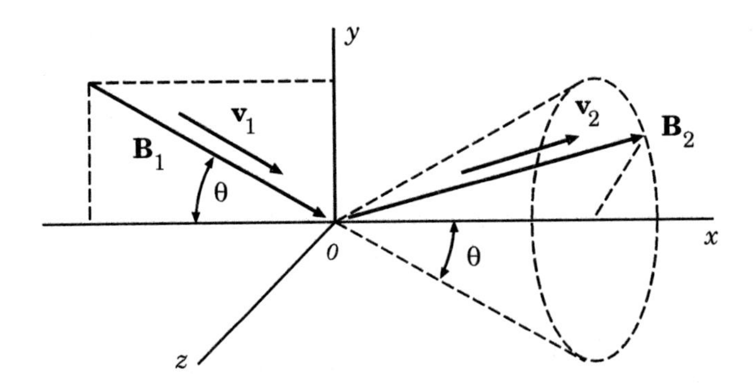

Figure 16.10: An Alfvén or rotational discontinuity.

Its peculiarity is reflected in the second name. On passing the discontinuity surface, a medium can acquire a directionally arbitrary tangential momentum, so that the flow is not generally planar.

The speed of the discontinuity propagation relative to the plasma

$$v_{x1} = \mp \frac{B_{x1}}{\sqrt{4\pi\rho}} \, . \tag{16.82}$$

In the limiting case of small intensity, the *Alfvén* or *rotational* discontinuity converts to the Alfvén wave (see (15.29)).

16.3 Transitions between discontinuities

As was shown by Syrovatskii (1956), *continuous* transitions occur between *discontinuous* MHD solutions of different types. This statement is easily verified on passing from the discontinuities (Section 16.2) to the limit of small-amplitude waves (Section 15.1). In this limit the fast and slow magnetoacoustic waves correspond to the oblique shocks, whereas the Alfvén wave corresponds to the Alfvén or rotational discontinuity.

The phase velocity diagrams for the small-amplitude waves are shown in Figure 15.2. Reasoning from it, the following scheme of continuous transitions between discontinuous solutions in ideal MHD can be suggested (Figure 16.11).

Let us recall that θ is the angle between the wave vector $\mathbf{k}$ and the magnetic field direction $\mathbf{B}_0/B_0$, i.e. axis x in Figure 15.2. If $\theta \to \pi/2$ then the fast

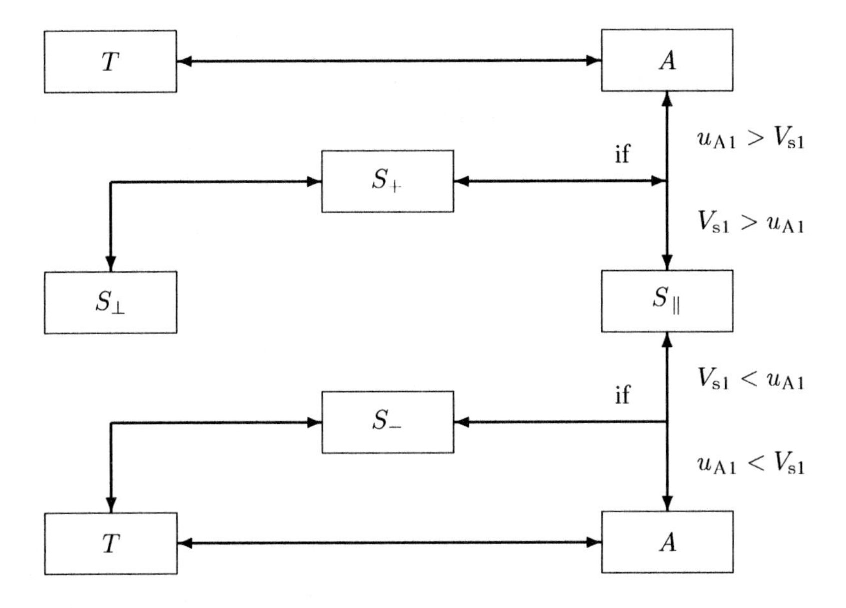

Figure 16.11: A scheme of the continuous transitions between discontinuous solutions in MHD, following from comparison of the properties of the discontinuities and small-amplitude waves on the phase velocity diagram.

magnetoacoustic wave (V_+) converts to the perpendicular wave propagating across the field with the velocity $V_\perp$ (15.43). In the limit of large-amplitude waves this corresponds to the transition from the fast shock (S_+) to the perpendicular one ($S_\perp$).

As $\theta \to 0$, the fast magnetoacoustic wave (V_+) converts to the usual sound one (V_s) if $V_s > V_A$ or to the Alfvén wave (V_A) if $V_A > V_s$. Therefore the fast shock (S_+) must convert, when $\theta \to 0$, either to the longitudinal shock ($S_\parallel$) if $V_{s1} > V_{A1}$ or to the Alfvén discontinuity (A) if $V_{A1} > V_{s1}$.

In much the same way, we conclude, reasoning from Figure 15.2, that the slow shock (S_-) converts either to the longitudinal shock ($S_\parallel$) for $V_{s1} < V_{A1}$ or to the Alfvén discontinuity (A) for $V_{s1} > V_{A1}$. This transition takes place as $\theta \to 0$. For $\theta \to \pi/2$, both the slow shock wave (S_-) and Alfvén discontinuity (A) transform to the tangential discontinuity (T) as demonstrated by the fact that the corresponding phase velocities of the slow magnetoacoustic (V_-) and Alfvén (V_A) waves tend to zero for $\theta \to \pi/2$.

* * *

How are such transitions realized? – They are effected through some discontinuities which may be called *transitional* since they conform to boundary conditions for both types of discontinuities and may be classified as either of the two. The existence of transitional discontinuities means that the discontinuity of one type can convert to the discontinuity of another type under a *continuous* change of parameters (Syrovatskii, 1956).

The absence of transitional discontinuities in MHD, manifested as the absence of transitions between small-amplitude waves in the phase velocity diagram (Figure 15.2), signifies the impossibility similar to that one in ordinary hydrodynamics because there exists a minimal velocity of shock propagation – the sound velocity V_s. That is why small perturbations in hydrodynamics cannot convert the shock wave (S) into the tangential discontinuity (T).

For the same reason the continuous transition between fast (S_+) and slow (S_-) shocks is impossible in MHD. This is shown in Figure 16.11 by the doubly crossed arrow. The fast shock (S_+) cannot continuously convert to the perpendicular one ($S_\perp$). These and other restrictions on continuous transitions between discontinuities in MHD will be explained in Chapter 17 from the viewpoint of evolutionarity conditions.

The classical theory of the MHD discontinous flows is of great utility in analysing the results of numerical calculations, for example time dependent numerical solutions of the dissipative MHD equations, in order to determine whether the numerical solutions are physically correct (e.g., Falle and Komissarov, 2001).

16.4 Shock waves in collisionless plasma

In ordinary collision-dominated gases or plasmas the density rise across a shock wave occurs in a distance of the order of a few collision mean free paths. The velocity distributions on both sides of the front are constrained by collisions to be Maxwellian and, if there is more than one kind of particles (for example, ions and electrons), the temperature of the various constituents of the plasma reach equality. Moreover, as we saw in Sections 16.1 and 16.2, the conditions (density, pressure, and flow velocity) on one side of the front are rigidly determined in terms of those on the other side by requirement that the flux of mass, momentum, and energy through the front be conserved. For weak shocks the front structure itself can be determined relatively simple, by taking into account collisional transfer coefficients representing viscosity, resistivity, and so on (Sirotina and Syrovatskii, 1960; Zel'dovich and Raizer, 1966, 2002; see also Section 17.4).

In a collisionless plasma the mechanisms by which the plasma state is changed by the passage of the shock front are more complex. Energy and momentum can be transferred from the plasma flow into electric and magnetic field oscillations for example by some **kinetic instabilities**. The energy of these collective motions must be taken into account when conservation laws are applied to relate the pre-shock state to the post-shock state. The ions and electrons are affected differently by instabilities. So there is no reason for their temperatures to remain equal. Since kinetic instabilities are seldom isotropic, it is unlikely that the temperatures will remain isotropic. These anisotropies further change the jump conditions.

The change in state derives from the collective interactions between particles and electric and magnetic fields. In general these fields are of two types. They can be: (a) **constant in time**, more exactly, quasi-static fields produced by charge separation, currents (e.g., Gedalin and Griv, 1999), or (b) **fluctuating in time**, produced by kinetic instabilities. The first situation is usually termed laminar, the second one turbulent. The fields often are turbulent. So the scattering of particles by turbulence can play the role of dissipation in the collisionless shock structure. This turbulence can be either a small-scale one generated by plasma instabilities inside a laminar shock front, or a large-scale turbulence associated with the dominant mode of the shock interaction itself (see Tidman and Krall, 1971).

Since we are discussing the kinetic processes which occur on a time scale much shorter than the time scale of Coulomb collisions, we may efficiently use the Vlasov equation (3.3) or the fluid-type descriptions derived from it (Chew et al., 1956; Klimontovich and Silin, 1961; Volkov, 1966) to study the properties of shock waves in collisionless plasma.

The high Much number collisionless shocks are well observed in some astrophysical objects, for example in young supernova remnants (SNRs). It has been suspected for many years that such shocks do not produce the electron-ion temperature equilibration. A clear hint for nonequilibration is the low electron temperature in young SNRs, which in no object seems to exceed 5 keV, whereas a typical shock velocity of 4000 km s^{-1} should give rise to a mean plasma temperature of about 20 keV. X-ray observations usually allow only the electron temperature to be determined.

The reflective grating spectrometer on board *XMM-Newton* allowed a direct measurement of an oxygen (O VII) temperature $T_i \approx 500$ keV in SN 1006 (Vink et al., 2003). Combined with the observed electron temperature $T_e \sim 1.5$ keV, this measurement confirms, with a high statistical confidence, that shock heating process resulted in only a small degree (~3 %) of electron-ion equilibration at the shock front and that the subsequent equilibration process is slow.

16.5 Practice: Exercises and Answers

Exercise 16.1. Relate the flow variables ρ, v, and p at the surface of an ordinary shock wave (Section 16.1.2).

Answer. From formula (16.7) with $v_\tau = 0$ and $v_x = v$, we find

$$\rho_1 v_1 = \rho_2 v_2 \,, \tag{16.83}$$

$$p_1 + \rho_1 v_1^2 = p_2 + \rho_2 v_2^2 \,, \tag{16.84}$$

$$\frac{v_1^2}{2} + w_1 = \frac{v_2^2}{2} + w_2 \,. \tag{16.85}$$

Here

$$w = \varepsilon + \frac{p}{\rho} \tag{16.86}$$

is the specific enthalpy; the thermodynamic relationship for the specific internal energy $\varepsilon(p, \rho)$ is assumed to be known.

Exercise 16.2. Assuming that the value of a parameter describing the strength of the shock in Exercise 16.1 is known (for example, the relative velocity $\delta v = v_1 - v_2$ which playes the role of the 'piston' velocity), find the general relationships that follow from the conservation laws (16.83)–(16.85).

Answer. Instead of the density let us introduce the specific volume $U = 1/\rho$. From (16.83) we obtain

$$\frac{U_2}{U_1} = \frac{v_2}{v_1} \, . \tag{16.87}$$

Eliminating the velocities v_1 and v_2 from Equations (16.84) and (16.85), we find

$$v_1^2 = U_1^2 \, \frac{p_2 - p_1}{U_1 - U_2} \, , \tag{16.88}$$

$$v_2^2 = U_2^2 \, \frac{p_2 - p_1}{U_1 - U_2} \, . \tag{16.89}$$

The velocity of the compressed plasma with respect to the undisturbed one

$$\delta v = v_1 - v_2 = \left[(p_2 - p_1)(U_1 - U_2) \right]^{1/2} . \tag{16.90}$$

Substituting (16.88) and (16.89) in the energy equation (16.85), we obtain

$$\delta w = w_2 - w_1 = \frac{1}{2} \, (p_2 - p_1)(U_1 + U_2) \, . \tag{16.91}$$

This is the most general form of the Rankine-Hugoniot relation.

Exercise 16.3. Consider the Rankine-Hugoniot relation for an ideal gas.

Answer. For an ideal gas with constant specific heats c_p and c_v, the specific enthalpy

$$w\,(p, U) = c_p \, T = \frac{\gamma_g}{\gamma_g - 1} \, p \, U \, , \tag{16.92}$$

where $\gamma_g = c_p / c_v$ is the specific heat ratio.

If we substitute (16.92) in (16.91), we obtain the Rankine-Hugoniot relation in the explicit form

$$\frac{p_2}{p_1} = \frac{(\gamma_g + 1)\, U_1 - (\gamma_g - 1)\, U_2}{(\gamma_g + 1)\, U_2 - (\gamma_g - 1)\, U_1} \, . \tag{16.93}$$

From here, the density ratio

$$r = \frac{\rho_2}{\rho_1} = \frac{U_1}{U_2} = \frac{(\gamma_g + 1)\, p_2 + (\gamma_g - 1)\, p_1}{(\gamma_g - 1)\, p_2 + (\gamma_g + 1)\, p_1} \, . \tag{16.94}$$

It is evident from (16.94) that **the density ratio** across a very strong shock, where the pressure p_2 behind the wave front is much higher than the initial pressure p_1, **does not increase infinitely** with increasing strength p_2/p_1, but approaches a certain finite value. This limiting density ratio is a fuction of the specific heat ratio γ_g only, and is equal to

$$r_\infty = \frac{\rho_2}{\rho_1} = \frac{\gamma_g + 1}{\gamma_g - 1} \,. \tag{16.95}$$

For a monatomic gas with $\gamma_g = 5/3$ the limiting compression ratio $r_\infty = 4$.

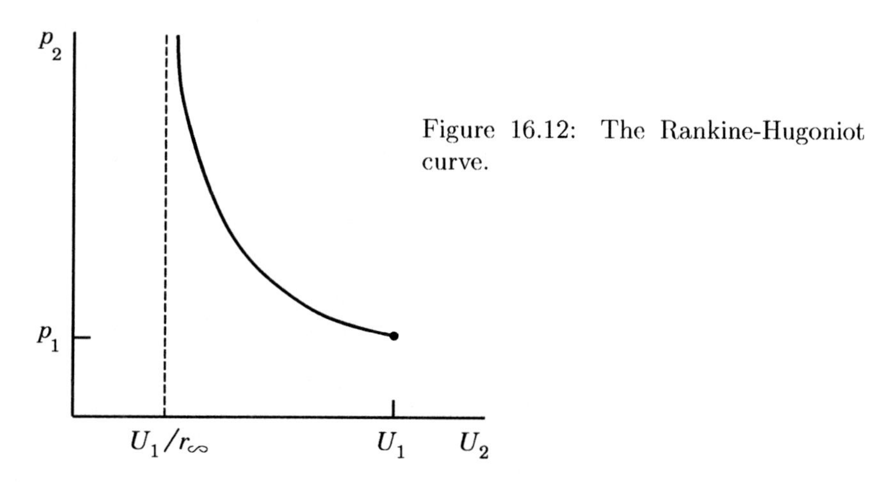

Figure 16.12: The Rankine-Hugoniot curve.

A curve on the diagram (p, U) passing through the initial state (p_1, U_1) according to (16.93) is called the Rankine-Hugoniot curve; it is shown in Figure 16.12.

Exercise 16.4. What is the value of the limiting density ratio r in relativistic shock waves?

Answer. Note that Equation (16.83) is valid only for nonrelativistic flows. In relativistic shock waves, the Lorentz factor (5.3) for the upstream and downstream flows must be included, and we have (de Hoffmann and Teller, 1950):

$$\gamma_{L,1}\, \rho_1\, v_1 = \gamma_{L,2}\, \rho_2\, v_2 \,. \tag{16.96}$$

The density ratio

$$r = \frac{\rho_2}{\rho_1} = \frac{v_1}{v_2}\, \frac{\gamma_{L,1}}{\gamma_{L,2}} \,. \tag{16.97}$$

In highly relativistic shock waves, the ratio v_1/v_2 remains finite, while the density ratio $r \to \infty$.

This is important fact for particle acceleration by shock waves (see Chapter 18).

Exercise 16.5. Write the density ratio r as a function of the upstream Mach number.

Answer. Let us use the definition of the sound speed (16.1) in an ideal gas with constant specific heats

$$V_s = \left(\gamma_g \, \frac{p}{\rho} \right)^{1/2} = \left(\gamma_g \, p \, U \right)^{1/2} . \tag{16.98}$$

The upstream Mach number (to the second power)

$$M_1^2 = \frac{v_1^2}{V_{s1}^2} = \frac{U_1}{\gamma_g \, p_1} \, \frac{p_2 - p_1}{V_1 - V_2} \, . \tag{16.99}$$

Here the solution (16.88) has been taken into account.

Substituting (16.99) in (16.94) gives us the compression ratio as a function of the upstream Mach number

$$r = \frac{(\gamma_g + 1) \, M_1^2}{(\gamma_g - 1) \, M_1^2 + 2} \, . \tag{16.100}$$

When $M_1 \to \infty$, the density ratio

$$r \to (\gamma_g + 1)/(\gamma_g - 1)$$

of course. This is the limiting case of a *strong* but nonrelativistic shock wave.

When $M_1 \to 1$, which is the limiting case of a *weak* shock wave, the density ratio $r \to 1$ too. By using formula (16.93), we see that the pressures on both sides of a weak shock wave are close to each other: $p_1 \approx p_2$ and $(p_2 - p_1)/p_1 \ll 1$. Thus a weak shock wave is practically the same as an acoustic compression wave.

For $M_1 < 1$ we could formally have an expansion shock wave with $r < 1$ and $p_2 < p_1$. However it can be shown (see the next Exercise) that such a transition would involve a *decrease* of entropy rather than an increase. So such transitions are ruled out by the second law of thermodynamics.

Exercise 16.6. Show that the entropy jump of a gas compressed by a shock increases with the strength of the shock wave but is entirely independent of the dissipative mechanism.

Answer. To within an arbitrary constant the entropy of an ideal gas with constant specific heats is given by formula (see Landau and Lifshitz, *Statistical Physics*, 1959b, Chapter 4):

$$S = c_v \ln p U^{\gamma_g} \, . \tag{16.101}$$

The difference between the entropy on each side of the shock front, as derived from (16.94), is

$$S_2 - S_1 = c_v \ln \left\{ \left(\frac{p_2}{p_1} \right) \left[\frac{(\gamma_g - 1)(p_2/p_1) + (\gamma_g + 1)}{(\gamma_g + 1)(p_2/p_1) + (\gamma_g - 1)} \right]^{\gamma_g} \right\} . \tag{16.102}$$

In the limiting case of a weak wave ($p_2 \approx p_1$) the expression in braces is close to unity. Therefore $S_2 \approx S_1$ and $S_2 > S_1$ if $p_2 > p_1$.

As the strength of the wave increases, that is, as the ratio p_2/p_1 increases beyond unity, the expression in braces increases monotonically and approaches infinity as $p_2/p_1 \to \infty$. Thus the entropy jump is positive and does increase with the strength of the shock wave.

> The increase in entropy indicates that irreversible dissipative processes occur in the shock front.

This can be traced to the presence of viscosity and heat conduction in the gas or plasma (see the discussion in Section 16.1.3).

Exercise 16.7. Consider a collisionless gravitational system described by the gravitational analog of the Vlasov equation (Exercise 3.9). Explain qualitatively why the Vlasov equation (3.44) does not predict the existence of a shock wave. In other words, unlike the case of gas or plasma, an evolution governed by the set of Equations (3.44)–(3.46) never leads to caustics or shocks.

Hint. By analogy with the discussion of the shock origin in ordinary hydrodynamics (Section 16.1.1), it is necessary to show that

> given sufficiently smooth initial data, the distribution function of a collisionless gravitational system will never diverge.

So the gravitational analog of the Vlasov equation manifests the so-called 'global existance' (Pfaffelmoser, 1992).

Chapter 17

Evolutionarity of MHD Discontinuities

A discontinuity cannot exist in astrophysical plasma with magnetic field if small perturbations disintegrate it into other discontinuities or transform it to a more general nonsteady flow.

17.1 Conditions for evolutionarity

17.1.1 The physical meaning and definition

Of concern to us is the issue of the stability of MHD discontinuities with respect to their decomposition into more than one discontinuity. To answer this question small perturbations must be imposed on the discontinuity surface. If they do not instantaneously lead to large changes of the discontinuity, then the discontinuity is termed *evolutionary*.

Obviously the property of evolutionarity does not coincide with stability in the ordinary sense. The usual instability means exponential ($e^{\gamma t}, \quad \gamma > 0$) growth of the disturbance, it remains small for some time ($t \leq \gamma^{-1}$). The discontinuity gradually evolves. By contrast,

> a disturbance **instantaneously** becomes large in a non-evolutionary discontinuity.

By way of illustration, the decomposition of a density jump $\rho(x)$ is shown in Figure 17.1. The disturbance $\delta\rho$ is not small, though it occupies an interval δx which is small for small t, when the two discontinuities have not become widely separated.

The problem of disintegration of discontinuities has a long history. Kotchine (1926) considered the disintegration of an arbitrary discontinuity into

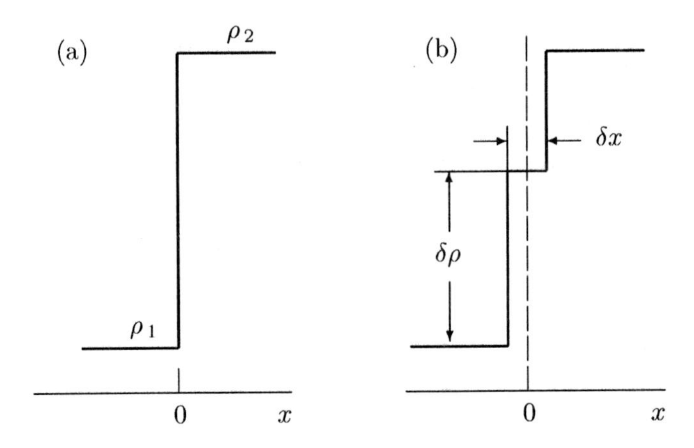

Figure 17.1: Disintegration of a density jump into two successive jumps.

a set of other discontinuities and rarefaction waves in the frame of hydrodynamics. Bethe (1942) studied the disintegration of a shock wave. The mathematical idea of evolutionarity was expressed for the first time in the context of the study of discontinuities in hydrodynamics (Courant and Friedrichs, 1985; see also Gel'fand, 1959).

> With respect to evolutionary discontinuities, the usual problem of linear stability can be formulated,

i.e. we find solutions to the linearized equations giving rise to small amplitudes which grow or decay in time.

The evolutionarity criterion may be obtained by counting the number of equations supplied by linearized boundary conditions at the discontinuity surface, and the number of independent parameters determining an arbitrary, initially small disturbance of the discontinuity. If the numbers are equal, then the boundary conditions uniquely define further development – evolution – of the disturbance which remains small for small $t > 0$. Such a discontinuity is *evolutionary*. By contrast, if the number of parameters is greater or less than the number of independent equations, then the problem of a small perturbation of the discontinuity has an infinitely large number of solutions or no solutions at all. Thus

> the initial assumption of the smallness of the disturbance for small t is incorrect, hence the discontinuity is *non-evolutionary*.

Such a discontinuity cannot exist as a stationary configuration because a small perturbation leads to a finite variation of the initial flow. This variation is the disintegration of the discontinuity into other discontinuities that move away from the place of their formation (Figure 17.1), or a transition to a more general nonsteady flow.

Let us count the number of equations which must be satisfied by an arbitrary small perturbation at the discontinuity. Let us take as the initial conditions the set of eight boundary conditions (16.25)–(16.32). It is to be linearized.

We consider perturbations of the discontinuity, which generate plane waves propagating along the x axis. Then the quantity B_x remains constant on either side of the discontinuity, and condition (16.25) (both exact and linearized) is satisfied identically. Hence, on either side of the discontinuity, seven quantities are perturbed: three velocity components (v_x, v_y, v_z), two field components (B_y, B_z), density ρ and pressure p. Small perturbations of these quantities,

$$\delta v_x \,,\ \delta v_y \,,\ \delta v_z \,,\ \delta B_y \,,\ \delta B_z \,,\ \delta \rho \,,\ \delta p,$$

on either side of the discontinuity surface are characterized by the coordinate and time dependence

$$\delta f\,(t,x) \sim \exp\left[\,\mathrm{i}\,(kx - \omega t)\,\right]$$

typical of the plane wave.

If the number of waves leaving the discontinuity is equal to the number of boundary conditions, then the problem of small perturbations has only one solution and the discontinuity is evolutionary. This form of evolutionarity conditions has been obtained for the first time by Lax (1957, 1973). The small perturbations must obey the linearized boundary conditions, i.e. linear algebraic equations following from (16.26)–(16.32). In addition to the seven quantities mentioned above, the velocity of propagation of the discontinuity surface is disturbed. It acquires a small increment δu_x relative to the chosen frame of reference in which the undisturbed discontinuity is at rest.

17.1.2 Linearized boundary conditions

Let us write down the linearized boundary conditions in a reference frame rotated with respect to the x axis in such a way that the undisturbed values $B_z = 0$ and $v_z = 0$. Thus we restrict our consideration to those discontinuity surfaces in which the undisturbed fields $\mathbf{B}_1$, $\mathbf{B}_2$ and the velocities $\mathbf{v}_1$, $\mathbf{v}_2$ lie in the plane (x, y).

From the boundary conditions (16.25)–(16.32) we find a set of linear equations which falls into two groups describing different perturbations:

(a) Alfvén perturbations $(\delta v_z,\ \delta B_z)$

$$\left\{ \rho\, v_x\, \delta v_z - \frac{1}{4\pi}\, B_x\, \delta B_z \right\} = 0\,, \tag{17.1}$$

$$\{\, v_x\, \delta B_z - B_x\, \delta v_z \,\} = 0\,; \tag{17.2}$$

(b) magnetoacoustic and entropy perturbations $(\delta v_x,\ \delta v_y,\ \delta B_y,\ \delta \rho,\ \delta p)$

$$\{\, \rho\, (\delta v_x - \delta u_x) + v_x\, \delta \rho \,\} = 0\,, \tag{17.3}$$

$$\left\{ \rho\, v_x\, \delta v_y + v_y \left[\, \rho\, (\delta v_x - \delta u_x) + v_x\, \delta \rho \,\right] - \frac{1}{4\pi}\, B_x\, \delta B_y \right\} = 0\,, \tag{17.4}$$

$$\left\{ \delta p + v_x^2\, \delta \rho + 2 \rho\, v_x\, (\delta v_x - \delta u_x) + \frac{1}{4\pi}\, B_y\, \delta B_y \right\} = 0\,, \tag{17.5}$$

$$\{\, B_x\, \delta v_y - B_y\, (\delta v_x - \delta u_x) - v_x\, \delta B_y \,\} = 0\,, \tag{17.6}$$

$$\begin{aligned} &\{\, \rho\, v_x \left[\, v_x\, (\delta v_x - \delta u_x) + v_y\, \delta v_y + \delta w \,\right] + \\ &+ \left(\frac{v_x^2 + v_y^2}{2} + w \right) \left[\, \rho\, (\delta v_x - \delta u_x) + v_x\, \delta \rho \,\right] + \\ &+ \frac{B_y}{4\pi} \left[\, B_y\, (\delta v_x - \delta u_x) + v_x\, \delta B_y - B_x\, \delta v_y \,\right] + \\ &\qquad + \frac{1}{4\pi}\, (v_x B_y - v_y B_x)\, \delta B_y \Big\} = 0\,. \end{aligned} \tag{17.7}$$

Condition (17.3) allows us to express the disturbance of the propagation velocity of the discontinuity surface δu_x in terms of perturbations of ρ and v_x:

$$\delta u_x\, \{\, \rho \,\} = \{\, \rho\, \delta v_x + v_x\, \delta \rho \,\}\ . \tag{17.8}$$

On substituting (17.8) in (17.4)–(17.7) there remain four independent equations in the second group of boundary conditions, since the disturbance of the velocity of the discontinuity surface δu_x can be eliminated from the set.

Therefore the MHD boundary conditions for perturbations of the discontinuity, which generate waves propagating perpendicular to the discontinuity surface, fall into two *isolated* groups. As this takes place,

> the conditions of evolutionarity (the number of waves leaving the MHD discontinuity is equal to the number of independent linearized boundary conditions) must hold not only for the variables in total but also for *each* isolated group

(Syrovatskii, 1959). The number of Alfvén waves leaving the discontinuity must be two, whereas there must be four magnetoacoustic and entropy waves. This make the evolutionary requirement more stringent.

Whether or not a discontinuity is evolutionary is clearly a *purely kinematic* problem. We have to count the number of small-amplitude waves leaving the discontinuity on either side. Concerning the boundary conditions the following comment should be made. As distinct from the unperturbed MHD equations, the perturbed ones are not stationary. Therefore the arguments used to derive Equations (16.19)–(16.24) from (16.10) are not always valid.

To derive boundary conditions at a disturbed discontinuity we have to transform to the reference frame connected with the surface. For example, for a perturbation (see Exercise 17.2)

$$\xi_x(y,t) = \xi_0 \exp\left[\,\mathrm{i}\,(k_y\, y - \omega\, t)\right] ,$$

where ξ_x is a displacement of the surface, this is equivalent to the following substitution in the linearized MHD equations

$$\frac{\partial}{\partial t}\,\delta \to -\,\mathrm{i}\,\omega\left(\delta - \xi_0\,\frac{\partial}{\partial y}\right), \qquad \frac{\partial}{\partial y}\,\delta \to \mathrm{i}\,k_y\left(\delta - \xi_0\,\frac{\partial}{\partial y}\right),$$

where $-\,\mathrm{i}\,\omega\,\xi_0 = \delta u_x$ is the amplitude of the time derivative of ξ. Consider, for example, the linearized continuity equation which after the integration over the discontinuity thickness takes the form

$$\mathrm{i}\int\limits_{-a}^{+a} (\omega - k_y v_y)\,\delta\rho\, dx - \mathrm{i}\,k_y \int\limits_{-a}^{+a} \rho\,\delta v_y\, dx =$$

$$= \{\, v_x\,\delta\rho + \rho\,[\,\delta v_x + \mathrm{i}\,(\omega - k_y v_y)\,\xi_0\,]\,\}\,. \tag{17.9}$$

If the integrals on the left-hand side of Equation (17.9) are equal to zero in the limit $a \to 0$ then, for $k_y = 0$, formula (17.9) transforms to (17.3). However this possibility is based on the supposition that $\delta\rho$ and δv_y inside the discontinuity do not increase in the limit $a \to 0$. We shall see in vol. 2, Chapter 10 that this supposition is not valid at least for more complicated, two-dimensional, configurations such as a reconnecting current layer.

17.1.3 The number of small-amplitude waves

If the discontinuity is immovable with respect to the plasma (no flow across the discontinuity), then *on either side* of the surface there exist three waves leaving it as shown in Figure 17.2:

$$-V_{+x1}\,,\;\; -V_{\mathrm{A}x1}\,,\;\; -V_{-x1}\,,\;\; V_{-x2}\,,\;\; V_{\mathrm{A}x2}\,,\;\; V_{+x2}\,. \tag{17.10}$$

Let the discontinuity move with a velocity v_{x1} relative to the plasma (Figure 17.3). The positive direction of the axis x is chosen to coincide with the direction of the plasma motion at the discontinuity surface. The index '1' refers to the region in front of the surface ($x < 0$) whereas the index '2' refers to the region behind the discontinuity ($x > 0$), i.e. downstream of the flow. Then there exist fourteen different phase velocities of propagation of small-amplitude waves:

$$v_{x1} \pm V_{+x1}\,,\;\; v_{x1} \pm V_{\mathrm{A}x1}\,,\;\; v_{x1} \pm V_{-x1}\,,\;\; v_{x1}\,,$$
$$v_{x2} \pm V_{-x2}\,,\;\; v_{x2} \pm V_{\mathrm{A}x2}\,,\;\; v_{x2} \pm V_{+x2}\,,\;\; v_{x2}\,.$$

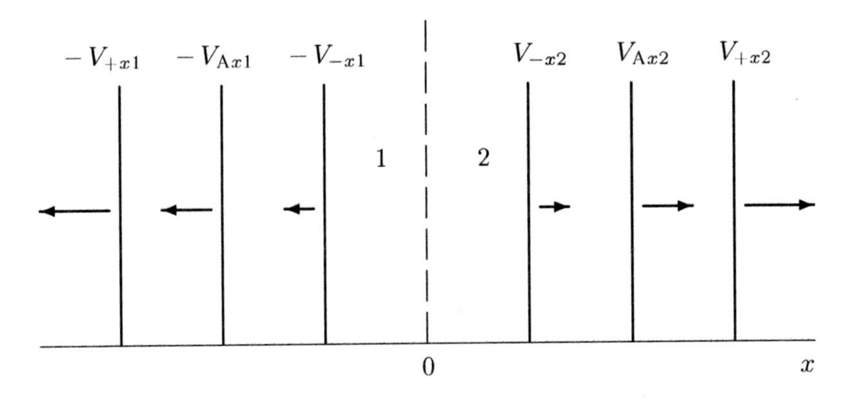

Figure 17.2: Six small-amplitude waves leaving an immovable discontinuity surface ($x = 0$) being perturbed.

Waves leaving the discontinuity have negative phase velocities in the region 1 and positive phase velocities in the region 2.

In the region 1, four velocities, corresponding to the waves moving toward the discontinuity surface, can be immediately discarded:

$$v_{x1} + V_{+x1}\,,\quad v_{x1} + V_{Ax1}\,,\quad v_{x1} + V_{-x1}\,,\quad v_{x1}\,.$$

The remaining three waves $(7 - 4)$ can leave the discontinuity or propagate toward it, depending on the plasma flow velocity towards the discontinuity v_{x1}.

In the region 2, four waves always have positive phase velocities:

$$v_{x2} + V_{+x2}\,,\quad v_{x2} + V_{Ax2}\,,\quad v_{x2} + V_{-x2}\,,\quad v_{x2}\,. \tag{17.11}$$

These waves leave the discontinuity. Other waves will be converging or diverging, depending on relations between the quantities

$$v_{x2}\,,\quad V_{+x2}\,,\quad V_{Ax2}\,,\quad V_{-x2}\,.$$

Let

$$0 < v_{x1} < V_{-x1}\,. \tag{17.12}$$

Then there are three waves leaving the discontinuity in the region 1:

$$v_{x1} - V_{-x1}\,,\ v_{x1} - V_{Ax1}\,,\ v_{x1} - V_{-x1}\,.$$

If

$$0 < v_{x2} < V_{-x2}\,, \tag{17.13}$$

then four waves (17.11) propagate downstream of the discontinuity since the waves

$$v_{x2} - V_{-x2}\,,\quad v_{x2} - V_{Ax2}\,,\quad v_{x2} - V_{+x2}$$

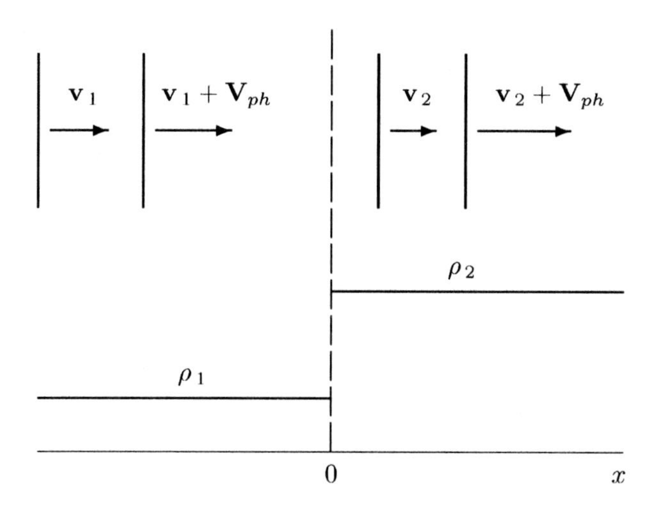

Figure 17.3: Small-amplitude waves in a plasma moving through the MHD discontinuity.

converge to the discontinuity.

We shall write down the number of diverging waves to the left (in front of) and to the right (behind) the discontinuity as their sum (e.g. 3 + 4 = 7 in the case considered) in the corresponding rectangle in the plane (v_{x1}, v_{x2}) presented in Figure 17.4. This rectangle is the lower left one. In the rectangle situated to the right of this one, two rather than three waves are diverging in the region 1:

$$v_{x1} - V_{\mathrm{A}x1}\,, \quad v_{x1} - V_{+x1}\,.$$

The wave $v_{x1} - V_{-x1}$ is carried by the flow to the discontinuity since

$$\boxed{V_{-x1} < v_{x1} < V_{\mathrm{A}x1}\,.} \tag{17.14}$$

Thus we write 2 + 4 = 6 in this rectangle. The whole table is filled up in a similar manner.

17.1.4 Domains of evolutionarity

If one considers the total number of boundary conditions (six), without allowance being made for their falling into two groups, then just three rectangles in Figure 17.4 should be inspected for possible evolutionarity. The boundaries of these rectangles are shown by solid lines.

However, as indicated above, the equality of the total number of independent boundary conditions to the number of diverging waves is insufficient for the existence and uniqueness of the solutions in the class of small perturbations (Syrovatskii, 1959). Take into account that

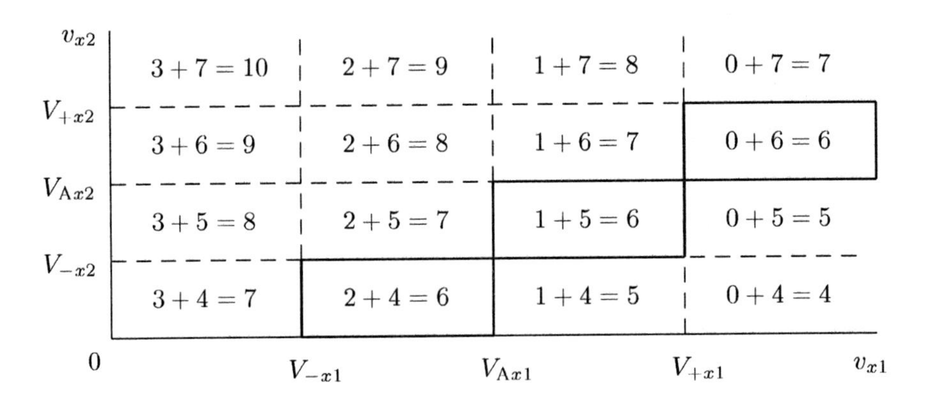

Figure 17.4: The number of small-amplitude waves leaving a discontinuity surface.

> the linearized boundary conditions fall into two groups, and hence the number of Alfvén waves must equal two and that of diverging magnetoacoustic and entropy waves must equal four.

Then one of the three rectangles becomes the point A in Figure 17.5.

The figure shows that there exist two domains of evolutionarity of shock waves:

(a) fast shock waves (S_+) for which

$$v_{x1} > V_{+x1}\,, \qquad V_{Ax2} < v_{x2} < V_{+x2}\,, \tag{17.15}$$

(b) slow shock waves (S_-) for which

$$V_{-x1} < v_{x1} < V_{Ax1}\,, \qquad v_{x2} < V_{-x2}\,. \tag{17.16}$$

Recall that our treatment of the Alfvén discontinuity was not quite satisfactory. It was treated as a flow in the plane (x, y). Generally this is not the case (Figure 16.10). The result of the above analysis is also not quite satisfactory: the evolutionarity of the Alfvén discontinuity, as well as the switch-on and switch-off shocks, is more complicated. While investigating the evolutionarity of these discontinuities, dissipative effects must be allowed for (Section 17.3).

Although *dissipative* waves quickly damp as they propagate away from the discontinuity surface, they play an important role in the system of small-amplitude waves leaving the discontinuity. Thus only one solution exists for the switch-off shock, i.e. it is evolutionary. By contrast,

> the switch-on shock wave, as well as the Alfvén or rotational discontinuity, are non-evolutionary

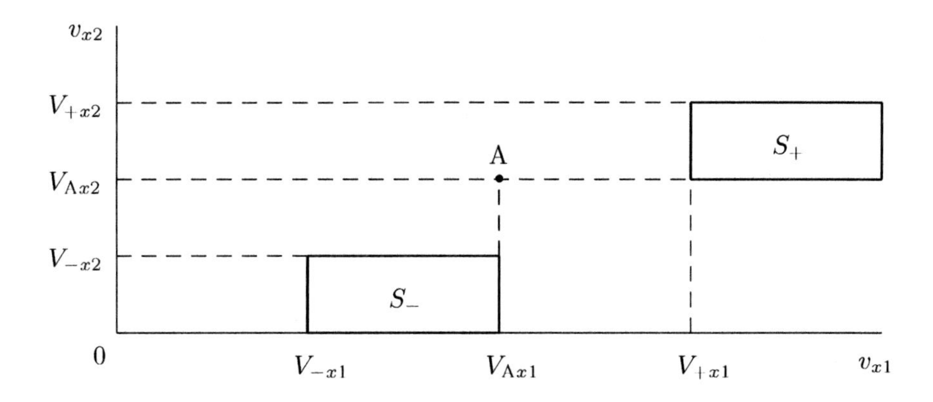

Figure 17.5: The evolutionarity domains for the fast (S_+) and slow (S_-) shocks and the Alfvén discontinuity.

in the linear approximation.

Roikhvarger and Syrovatskii (1974) have shown that attention to dissipation in the dispersion equation for magnetoacoustic and entropy waves leads to the appearance of dissipative waves and, as a consequence, to the non-evolutionarity of tangential and contact discontinuities (Section 17.3).

Recall that in an ideal medium the disintegration of a discontinuity is instantaneous in the sense that the secondary discontinuities become separated in the beginning of the disintegration process (Figure 17.1). In a dissipative medium the spatial profiles of the MHD discontinuities are continuous. Nevertheless the principal result remains the same. The steady flow is rearranged toward a nonsteady state, and after a large enough period of time the disintegration manifests itself (Section 17.4).

17.2 Consequences of evolutionarity conditions

17.2.1 The order of wave propagation

Some interesting inferences concerning the order of shock propagation result from the evolutionarity conditions (17.15) and (17.16).

If a shock wave follows another one of the same type (fast or slow), the back shock will catch up with the front one (Akhiezer et al., 1959). Let us consider, as an example, two slow shock waves, S_-^{A} and S_-^{B}, propagating in the direction of the x axis as shown in Figure 17.6.

In a reference frame connected with the front of the first shock S_-^{A}, we get, by virtue of the evolutionary condition (17.16),

$$V_{-x1}^{\mathrm{A}} < v_{x1}^{\mathrm{A}} < V_{\mathrm{A}x1}^{\mathrm{A}}\,, \quad v_{x2}^{\mathrm{A}} < V_{-x2}^{\mathrm{A}}\,. \tag{17.17}$$

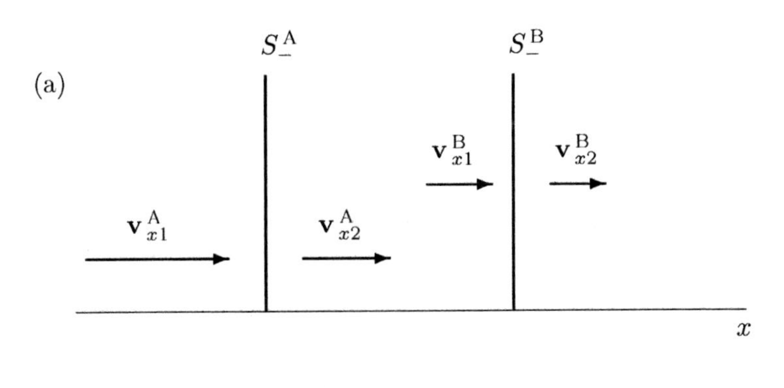

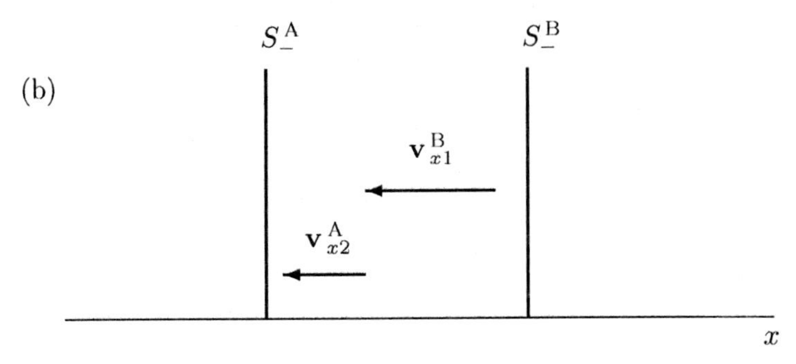

Figure 17.6: Plasma flow velocities relative to: (a) shock wave fronts, (b) the plasma between the shock waves.

In a reference frame connected with the front of the second shock S_-^B, analogous conditions hold:

$$V_{-x1}^B < v_{x1}^B < V_{Ax1}^B \,, \qquad v_{x2}^B < V_{-x2}^B \,. \tag{17.18}$$

Since the velocities of slow magnetoacoustic waves of small amplitude V_{-x2}^A and V_{-x1}^B refer to the same region (between the shocks), they are equal

$$V_{-x2}^A = V_{-x1}^B \,. \tag{17.19}$$

Substituting (17.19) in the second part of (17.17) and in the first part of (17.18) gives the inequality

$$v_{x2}^A < v_{x1}^B \,. \tag{17.20}$$

Hence, relative to the plasma between the shocks (Figure 17.6b), the shock S_-^B catches up with the shock S_-^A, which was to be proved.

As for different types of waves, the following inferences can be drawn: the Alfvén discontinuity will catch up with the slow shock, whereas the fast shock will catch up with all possible types of discontinuities. If shock waves are generated by a single source (for example, a flare in the solar atmosphere), then no more than three shocks can move in the same direction: the fast shock is followed by the Alfvén discontinuity, the slow shock being to the rear of the Alfvén discontinuity.

17.2.2 Continuous transitions between discontinuities

Reasoning from the polar diagram for phase velocities of small-amplitude waves, in Section 16.3 we have treated the possibility of continuous transitions between different types of discontinuous solutions in MHD. However the evolutionarity conditions have not been taken into account. They are known to impose limitations on possible continuous transitions between the discontinuities under changes of external parameters (magnetic field, flow velocity, etc.).

Continuous transition is impossible between the fast and slow shock waves. This stems from the fact that the evolutionarity domains for fast (S_+) and slow (S_-) shocks have no common points (Figure 17.5). Similarly, the lines of phase velocities V_+ and V_- in polar diagrams (Figures 15.2 and 15.3) are out of contact. That was the basis for banning transitions between the fast and slow shocks in Figure 16.11.

The fast shock (S_+) cannot continuously convert to the tangential discontinuity (T) since that would go against the evolutionarity condition $v_{x1} > V_{\mathrm{A}x1}$. The same ban stems from the consideration of the phase velocity diagram (Section 16.3). The perpendicular shock ($S_\perp$) is the limiting case of the fast shock. That is why the continuous transition of the perpendicular shock to the tangential discontinuity is forbidden, as shown in Figure 16.11.

As was indicated in the previous section, the issue of evolutionarity of the Alfvén discontinuity has no satisfactory solution in the framework of ideal MHD. The established viewpoint is that the continuous transition of shock waves (S_- and S_+) to the Alfvén discontinuity (A) is impossible, as is predicted by the phase velocity diagram with $\theta \to 0$. Transitions between the Alfvén (A) and tangential (T) discontinuities, between the tangential discontinuity and the slow shock (S_-), between the tangential and contact (C) discontinuities are assumed to be possible. These discontinuities convert to the tangential discontinuity in the limiting case $B_x \to 0$ (Polovin, 1961; Akhiezer et al., 1975).

We shall consider the evolutionarity conditions and their consequences for reconnecting current layers (RCLs) as a MHD discontinuity in vol. 2, Chapter 10.

17.3 Dissipative effects in evolutionarity

Roikhvarger and Syrovatskii (1974) have taken into account the effect of dissipation on the peculiar shocks. In this case the dispersion relation of the

Alfvén waves has the form:

$$k^2 V_{\mathrm{A}x}^2 - \left(\omega - kv_x - \mathrm{i}k^2 \nu_{\mathrm{m}}\right)\left(\omega - kv_x - \mathrm{i}k^2 \nu\right) = 0 . \tag{17.21}$$

Here $\mathbf{k}$ is directed along the x axis, ν_{m} is the magnetic diffusivity, and $\nu = \eta/\rho$ is the kinematic viscosity. After expansion of the solutions of this equation in powers of a small ω (the conditions under which ω is small will be discussed below) the expression for k reads as follows:
(a) for $v_x = V_{\mathrm{A}x}$

$$k^d = \pm \sqrt{\frac{\omega}{\nu_{\mathrm{m}} + \nu}}\,(1 - \mathrm{i})\,, \tag{17.22}$$

$$k^{\mathrm{A}} = \frac{\omega}{2v_x} - \mathrm{i}\,\frac{(\nu_{\mathrm{m}} + \nu)\,\omega^2}{16\, v_x^3}\,, \tag{17.23}$$

$$k^* = -\frac{\omega\left(\nu_{\mathrm{m}}^2 + \nu^2\right)}{v_x\left(\nu_{\mathrm{m}} + \nu\right)^2} + \mathrm{i}\,\frac{v_x\left(\nu_{\mathrm{m}} + \nu\right)}{\nu_{\mathrm{m}}\,\nu}\,; \tag{17.24}$$

(b) for $v_x \neq V_{\mathrm{A}x}$

$$k^{\mathrm{A}} = \frac{\omega}{v_x \pm V_{\mathrm{A}x}} - \mathrm{i}\,\frac{(\nu_{\mathrm{m}} + \nu)\,\omega^2}{2\left(v_x \pm V_{\mathrm{A}x}\right)^3}\,, \tag{17.25}$$

$$k^* = -\frac{\omega\left[\,(\nu_{\mathrm{m}} - \nu)^2\, v_x \pm (\nu_{\mathrm{m}} + \nu)\, K\,\right]}{4\, V_{\mathrm{A}x}^2\, \nu_{\mathrm{m}}\, \nu + v_x^2\,(\nu_{\mathrm{m}} - \nu)^2 \pm v_x\,(\nu_{\mathrm{m}} + \nu)\, K} + $$

$$+\,\mathrm{i}\,\frac{v_x\,(\nu_{\mathrm{m}} + \nu) \pm K}{2\,\nu_{\mathrm{m}}\,\nu}\,, \tag{17.26}$$

where

$$K = \left[\, v_x^2\,(\nu_{\mathrm{m}} - \nu)^2 + 4\, V_{\mathrm{A}x}^2\, \nu_{\mathrm{m}}\, \nu \,\right]^2 .$$

Thus

> the dissipative effects result in additional small-amplitude waves propagating in a homogeneous MHD medium.

The width of an MHD shock (at least of small amplitude) is proportional, in order of magnitude, to the dissipative transport coefficients and inversally proportional to the shock intensity (Sirotina and Syrovatskii, 1960). The intensity is determined by the difference $v_x - V_{\mathrm{A}x}$ on the side of the discontinuity on which it is not zero. Since the switch-off shock, as a slow one, has a finite intensity, and the switch-on shock exists in the interval (see Section 16.2.5)

$$1 < \frac{v_{x1}^2}{V_{\mathrm{A}x1}^2} < \frac{4v_{x1}^2}{v_{x1}^2 + V_{\mathrm{s}1}^2}\,,$$

the width of the peculiar shock can be estimated as

$$l \sim \frac{\nu_{\mathrm{m}} + \nu}{\mid v_x - V_{\mathrm{A}x} \mid} \tag{17.27}$$

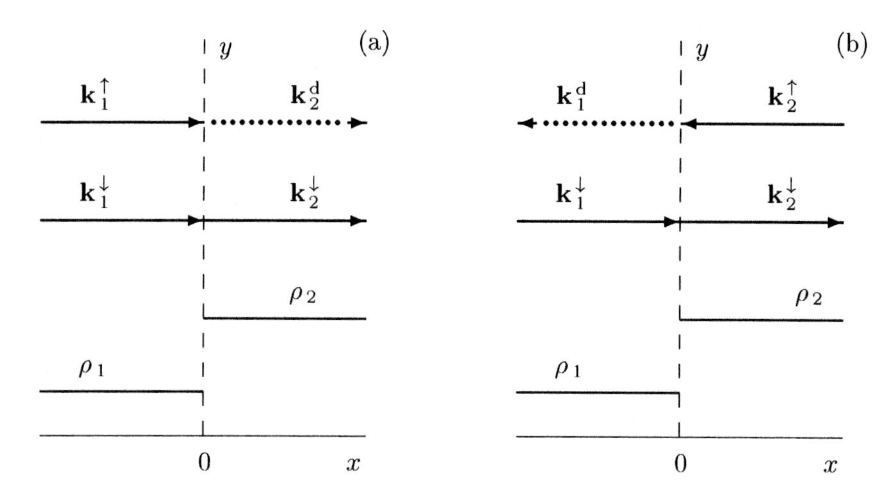

Figure 17.7: The direction of the wave propagation in the case of a switch-on shock (a) and a switch-off shock (b).

(Roikhvarger and Syrovatskii, 1974). It is just this distance within which the perturbations k^* from (17.24) and (17.26) damp considerably.

Therefore outside the shock front these waves are absent, and their amplitudes do not enter into the boundary conditions which relate perturbations outside the shock front.

The situation is different for the remaining perturbations, in particular, for the *purely dissipative* waves k^d from (17.22). For small enough ω their wave numbers are much larger than the thickness l of the shock. This is true under the condition

$$\omega \ll \frac{(v_x - V_{Ax})^2}{\nu_m + \nu}\,, \tag{17.28}$$

which coinsides with that used to derive (17.22)–(17.26). Since the characteristic length scale of such perturbations is much larger than the shock thickness l, their amplitudes satisfy the boundary conditions at the discontinuity surface (17.1) and (17.2) obtained for an ideal medium.

The classification of dissipative perturbations on incoming and outgoing waves should be made according to the sign of the imaginary part of the wave vector, because in a stable medium such waves damp in the direction of the propagation (Section 15.3). Consequently, there are two outgoing perturbations leaving the peculiar shock, one of them being the dissipative wave. Much like the case of non-peculiar shocks, both waves propagate downstream away from the (fast) switch-on shock, while there is one outgoing wave on each side of the (slow) switch-off shock (Figure 17.7).

With the precision adopted when deriving (17.22)–(17.26), the perturbations δv_z and δB_z in the dissipative wave k^* from (17.22) are related by the formula

$$\delta v_z^{\,d} = \left(1 \pm \frac{\nu_{\rm m} - \nu}{v_x} \sqrt{\frac{{\rm i}\,\omega}{2\,(\nu_{\rm m} + \nu)}} \right) \frac{\delta B_z^{\,d}}{\sqrt{4\pi\rho}} \, . \tag{17.29}$$

From here and, (17.1) and (17.2), it follows that if an Alfvén wave is incident onto the switch-off shock from upstream or downstream then the amplitude of the dissipative wave equals respectively

$$\delta B_{z1}^{\,d} = - \frac{2\, v_{x1}}{\nu_{\rm m} - \nu} \sqrt{\frac{2\,(\nu_{\rm m} + \nu)}{{\rm i}\,\omega}} \; \delta B_{z1}^{\downarrow} \, , \tag{17.30}$$

or

$$\delta B_{z1}^{\,d} = - \frac{2\, v_{x1}}{\nu_{\rm m} - \nu} \sqrt{\frac{2\,(\nu_{\rm m} + \nu)}{{\rm i}\,\omega}} \; \delta B_{z2}^{\uparrow} \, . \tag{17.31}$$

The amplitude $\delta B_{z2}^{\downarrow}$ of the travelling (non-dissipative) wave equals zero in the first case and $-\delta B_{z2}^{\uparrow}$ in the second case. Thus only one solution exists for the switch-off shock. Consequently, the switch-off shock is evolutionary.

On the contrary, the switch-on shock is non-evolutionary. Indeed, Equations (17.1) and (17.2), with regard for the relation at the switch-on shock

$$v_{x1}\, v_{x2} = V_{{\rm A}x1}^{2} \quad \text{and} \quad \frac{\rho_2}{\rho_1} = \frac{v_{x1}^2}{V_{{\rm A}x1}^2} \, , \tag{17.32}$$

can be rewritten as

$$v_{x1} \left(\delta v_{z2} - \frac{\delta B_{z2}}{\sqrt{4\pi\rho}} \right) = v_{x1}\, \delta v_{z1} - V_{{\rm A}x1} \frac{\delta B_{z1}}{\sqrt{4\pi\rho}} \, , \tag{17.33}$$

$$V_{{\rm A}x1} \left(\delta v_{z2} - \frac{\delta B_{z2}}{\sqrt{4\pi\rho}} \right) = V_{{\rm A}x1}\, \delta v_{z1} - v_{x1} \frac{\delta B_{z1}}{\sqrt{4\pi\rho}} \, . \tag{17.34}$$

The set of Equations (17.33) and (17.34) is incompatible with a non-zero amplitude of the incident wave, i.e. when δv_{z1} and δB_{z1} are not equal to zero. Note that if the incident wave is absent, this set has an infinite number of solutions. Hence the switch-on shock is non-evolutionary.

Finally it should be mentioned that the additional dissipative waves appear only for $v_x = V_{{\rm A}x}$. This means that

> the dissipative effects do not alter the evolutionarity conditions for non-peculiar (fast and slow) MHD shock waves.

At the same time the Alfvén discontinuity becomes non-evolutionary with respect to dissipative Alfvén waves. This is consistent with the fact that in the presence of dissipation it cannot have a stationary thickness and smooths out with time (see Landau et al., 1984).

It was also pointed out by Roikhvarger and Syrovatskii (1974) that the inclusion of dissipation into the dispersion relation for magnetoacoustic and entropy waves results in the appearence of dissipative waves, and, as a consequence, in non-evolutionarity of tangential, contact, and *weak* discontinuities (discontinuities of the derivatives of the MHD properties).

17.4 Discontinuity structure and evolutionarity

17.4.1 Perpendicular shock waves

It is natural to assume that

> the stationary problem of the structure of an evolutionary MHD discontinuity has a unique solution, while for the non-evolutionary one this problem does not have a solution.

To illustrate this assumption let us obtain the structure of the perpendicular shock. With this aim the one-dimensional dissipative MHD equations should be integrated over x. After that the conservation laws of mass, momentum, and energy, and Maxwell equations take the form (see Polovin and Demutskii, 1990):

$$\rho v = J \,, \tag{17.35}$$

$$Jv + p + \frac{B^2}{8\pi} - \mu \frac{\mathrm{d}v}{\mathrm{d}x} = S \,, \tag{17.36}$$

$$J \left[\frac{v^2}{2} + \frac{p}{\rho(\gamma_g - 1)} \right] + pv + \frac{vB^2}{4\pi} - \mu v \frac{\mathrm{d}v}{\mathrm{d}z} - \frac{\nu_{\mathrm{m}}}{4\pi} B \frac{\mathrm{d}B}{\mathrm{d}x} = Q \,, \tag{17.37}$$

$$vB - \nu_{\mathrm{m}} \frac{\mathrm{d}B}{\mathrm{d}x} = cE \,. \tag{17.38}$$

Here the thermal conductivity of the medium is assumed to be zero. J, S, and Q are constants of integration, γ_g is the adiabatic index, $\mu = (4/3)\,\eta + \zeta$, and ζ is a bulk viscosity (the indexes x and y at the quantities v_x and B_y are omitted).

From (17.35)–(17.38) we obtain the set of ordinary differential equations which describes the structure of the perpendicular shock:

$$\mu \frac{\mathrm{d}v}{\mathrm{d}x} = f(v, B) \,, \tag{17.39}$$

$$\nu_{\mathrm{m}} \frac{\mathrm{d}B}{\mathrm{d}x} = g(v, B) \,, \tag{17.40}$$

where

$$f(v, B) = \frac{\gamma_g + 1}{2} Jv - \gamma_g \left(S - \frac{B^2}{2\pi} \right) + \frac{\gamma_g - 1}{v} \left(Q - \frac{cEB}{4\pi} \right), \tag{17.41}$$

$$g(v, B) = vB - cE \,. \tag{17.42}$$

The curves $f(v, B) = 0$ and $g(v, B) = 0$ on the plane (v, B) are shown schematically in Figure 17.8. At the points 1 and 2 of intersection of these curves the derivatives $\mathrm{d}v/\mathrm{d}x$ and $\mathrm{d}B/\mathrm{d}x$ equal zero simultaneously.

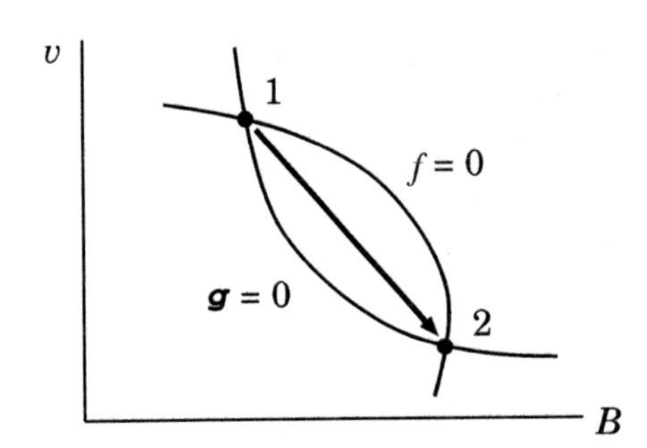

Figure 17.8: The structure of the perpendicular shock (bold arrow) connecting the states 1 and 2.

The points (B_1, v_1) and (B_2, v_2) correspond to the states ahead of the shock ($x \to -\infty$) and behind the shock ($x \to +\infty$). These are stationary points of the set of differential Equations (17.39) and (17.40). The structure of the shock

$$v = v(x), \quad B = B(x) \tag{17.43}$$

is a solution to the set (17.39), (17.40) which leaves the initial point 1 and enters into the final point 2.

To consider the behaviour of the integral curves in the vicinity of the stationary points 1 and 2 (Figure 17.8) the quantities J, S, and Q should be expressed in termes of the MHD properties v_i and B_i ahead of the shock ($i = 1$) and behind the shock ($i = 2$). Then, by virtue of the fact that the derivatives $\mathrm{d}v/\mathrm{d}x$ and $\mathrm{d}B/\mathrm{d}x$ tend to zero for $x \to \pm\infty$, Equations (17.35)–(17.37) yield

$$J = \rho_i \, v_i \,, \tag{17.44}$$

$$S = J v_i + p_i + \frac{B_i^2}{8\pi} \,, \tag{17.45}$$

$$Q = J \left(\frac{v_i^2}{2} + \frac{\gamma_g}{\gamma_g - 1} \frac{p_i}{\rho_i} \right) + \frac{v_i B_i^2}{4\pi} \,, \tag{17.46}$$

where $i = 1, 2$.

Let us now represent the quantities B and v in the form

$$B = B_i + \delta B_i \,, \qquad v = v_i + \delta v_i \,, \tag{17.47}$$

with δ being a small perturbation. Substituting this together with (17.44)–(17.46) in (17.41) and (17.42), and expanding the result in powers of δB_i and δv_i, we find to the first order

$$\mu \, \frac{\mathrm{d}\,\delta v_i}{\mathrm{d}x} = \frac{\rho_i}{v_i} \left(v_i^2 - V_{si}^{\,2} \right) \delta v_i + \frac{B_i}{4\pi} \, \delta B_i \,, \tag{17.48}$$

$$\nu_{\mathrm{m}} \, \frac{\mathrm{d}\,\delta B_i}{\mathrm{d}x} = B_i \, \delta v_i + v_i \, \delta B_i \,. \tag{17.49}$$

As is known (e.g., Fedoryuk, 1985), a stationary point $\delta v_i = 0$, $\delta B_i = 0$ of the set of autonomous differential equations

$$\frac{d\,\delta v_i}{dx} = a_{11}\,\delta v_i + a_{12}\,\delta B_i\,, \tag{17.50}$$

$$\frac{d\,\delta B_i}{dx} = a_{21}\,\delta v_i + a_{22}\,\delta B_i \tag{17.51}$$

is a saddle if the roots of characteristic equation

$$(a_{11} - \lambda)\,(a_{22} - \lambda) - a_{12}\,a_{21} = 0 \tag{17.52}$$

are real numbers and have opposite signs, i.e. if

$$(a_{11} - a_{22})^2 + 4a_{12}\,a_{21} > 0\,, \qquad a_{11}\,a_{22} - a_{12}\,a_{21} < 0\,. \tag{17.53}$$

In this case only two integral curves enter the stationary point $\delta v_i = 0$, $\delta B_i = 0$ from the opposite directions (Figure 17.9a). And in the orthogonal way only two curves leave the stationary point.

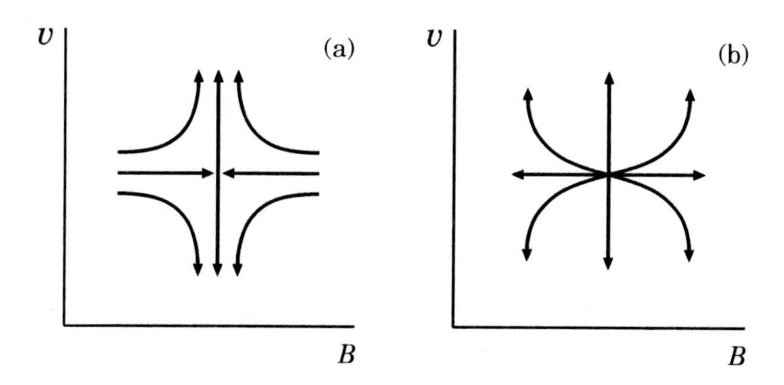

Figure 17.9: Stationary points of the set of autonomous differential equations. (a) Saddle. (b) Unstable node.

In the case when the roots of characteristic Equation (17.52) are real numbers and have the same sign, i.e. if

$$(a_{11} - a_{22})^2 + 4a_{12}\,a_{21} > 0\,, \qquad a_{11}\,a_{22} - a_{12}\,a_{21} > 0\,, \tag{17.54}$$

then the stationary point is a node. If in addition

$$a_{11} + a_{22} > 0 \tag{17.55}$$

then the node is unstable, and all the integral curves leave the stationary point (Figure 17.9b).

In a perpendicular MHD shock

$$a_{11} a_{22} - a_{12} a_{21} = \frac{\rho_i \left(v_i^2 - V_{\perp i}^2 \right)}{\mu \nu_m}, \tag{17.56}$$

as follows from Equations (17.48) and (17.49). Here

$$V_\perp = \sqrt{V_{A\,\|}^2 + V_s^2} = \sqrt{u_A^2 + V_s^2}. \tag{17.57}$$

(Section 15.2.4). So the second inequality (17.54) is always valid. As for the quantity $a_{11} + a_{22}$, it equals

$$a_{11} + a_{22} = \frac{\rho_i \left(v_i^2 - V_{si}^2 \right)}{\mu v_i} + \frac{v_i}{\nu_m}. \tag{17.58}$$

It follows from (17.56) and (17.58) that in the case of the perpendicular shock the stationary points of (17.48), (17.49) can be only of two types: either a saddle or an unstable node (recall that v_i is assumed to be positive).

Let us consider at first the case when

$$v_1 > V_{\perp 1}, \quad v_2 < V_{\perp 2}. \tag{17.59}$$

Then point 2 is a saddle, while point 1 is an unstable node. The only integral curve enters into point 2 in Figure 17.8 from the side of larger values of v. If the quantities v and B vary along this curve in the opposite direction, i.e. upstream of the shock, then they will inevitably reach the values (v_1, B_1), i.e. point 1, because all integral curves leave point 1 (unstable node in the case under consideration). This curve describes a unique structure of the perpendicular shock. The inequalities (17.59) coincide with the conditions of evolutionarity of the perpendicular shock (see (17.15)), because $V_+ = V_\perp$ for perpendicular propagation. Therefore

> the conditions that the perpendicular shock wave has the unique structure coincide with the conditions of its evolutionarity.

Now we consider the structure of a non-evolutionary perpendicular shock wave. If

$$v_2 > V_{\perp 2}, \tag{17.60}$$

then point 2 is an unstable node. Neither integral curve enters this point, i.e. the problem of structure of the shock does not have a solution.

If

$$v_1 < V_{\perp 1}, \quad v_2 < V_{\perp 2}, \tag{17.61}$$

then both stationary points 1 and 2 are saddles. In this case one of two integral curves, leaving point 1, may coinside with one of two curves entering point 2. However this takes place only for the definite exclusive values of the parameters ahead of the shock front. An infinitesimal perturbation of the state upstream of the shock destroys its structure. In other words, the integral curve cannot connect the states 1 and 2 in a general case.

17.4.2 Discontinuities with penetrating magnetic field

Let us turn to the discontinuity type for which

$$v_x \neq 0 \quad \text{and} \quad B_x \neq 0 \tag{17.62}$$

(Sections 16.2.4 and 16.2.5). Consider at first the discontinuity accompanied by a density jump:

$$\{ \rho \} \neq 0 \,. \tag{17.63}$$

(oblique shock waves). In this case the boundary conditions (16.67) can be rewritten in such a way as to represent the Rankine-Hugoniot relation for shock waves in MHD. Germain (1960) has shown that the boundary conditions allow four states (see also Shercliff, 1965):

$$\begin{aligned} \text{I} &: \quad v_x > V_+ \,, \\ \text{II} &: \quad V_+ > v_x > V_{\text{A}x} \,, \\ \text{III} &: \quad V_{\text{A}x} > v_x > V_- \,, \\ \text{IV} &: \quad V_- > v_x \,. \end{aligned} \tag{17.64}$$

The states are arranged in order of increasing entropy. The second law of thermodynamics requires that a shock transition is possible only from a lower state of entropy to an upper one. There are thus six transitions shown in Figure 17.10.

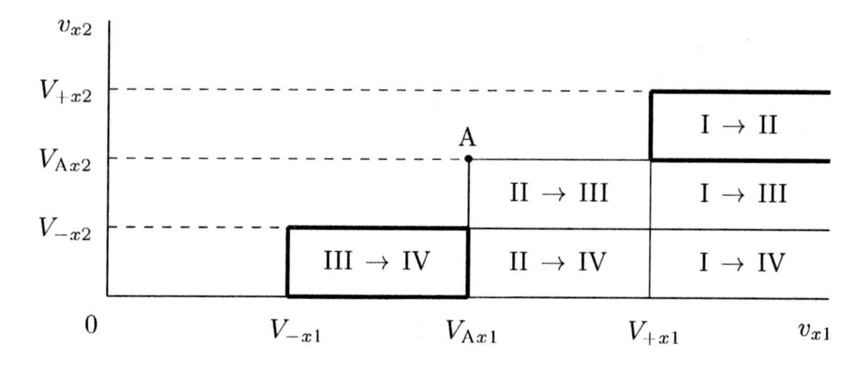

Figure 17.10: Transitions with increasing entropy. Evolutionarity domains (bold rectangles) for the fast (I → II) and slow (III → IV) shock waves.

The evolutionarity of an oblique shock wave is related to its structure in the following way (Germain, 1960; Kulikovskii and Lyubimov, 1961; Anderson, 1963). **The evolutionary fast and slow shocks always have a unique structure.** The shock transition II → III has a unique structure only for the definite relationship between the dissipative transport coefficients. If these coefficients fall into the certain intervals, the I → III and II → IV shocks may have a unique structure, while the I → IV transition may be connected by an infinite number of integral curves.

Besides, as shown by Liberman (1978) with the help of the method discussed in Section 17.4.1, the switch-on shock, which is not evolutionary with respect to dissipative waves, has a unique structure. The possible reason is that the peculiarity of the switch-on and switch-off shocks is related to the absence of B_τ on one side of the discontinuity surface. The small asymmetry, that is assumed when studying the stationary points, removes the degeneration, and thus makes the shock evolutionary.

17.5 Practice: Exercises and Answers

Exercise 17.1. Show that an ordinary shock wave is evalutionary.

Answer. From (16.7) it follows that there exist three boundary conditions at the surface of a shock wave in ordinary hydrodynamics:

$$\{ \rho v_x \} = 0 , \quad \{ p + \rho v_x^2 \} = 0 , \quad \left\{ \frac{v^2}{2} + w \right\} = 0 . \tag{17.65}$$

The boundary condition

$$\{ v_\tau \} = 0 \tag{17.66}$$

makes it possible to transform to such a frame of reference in which the tangential velocity component is absent on either side of the discontinuity: $\mathbf{v}_{\tau 1} = \mathbf{v}_{\tau 2} = 0$. So we obtain three linearized conditions for small perturbations. Since the disturbance of the velocity of the shock front surface δu_x can be eliminated from the set of boundary conditions, there remain two independent equations in the set.

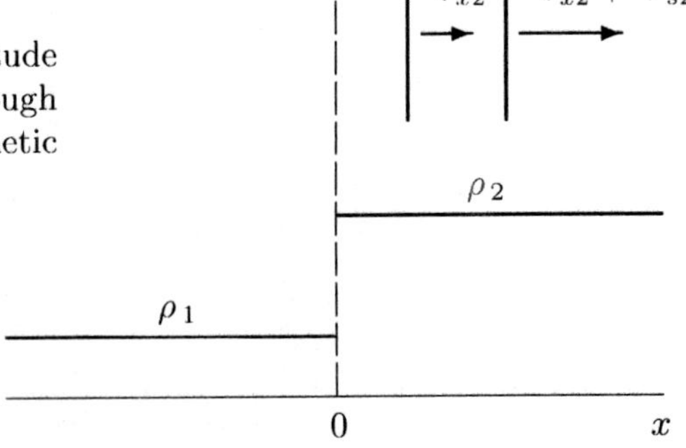

Figure 17.11: Small-amplitude waves in a plasma moving through a shock wave without a magnetic field.

Let us count the number of outgoing small-amplitude waves. There are no such waves upstream the shock because of the condition

$$v_{x1} > V_{s1} = 0 , \tag{17.67}$$

where V_{s1} is the upstream sound velocity. At the downstream side of the shock there are two waves: the sound wave propagating with velocity $v_{x2}+V_{s2}$ and the entropy-vortex wave (Exercise 15.2) propagating with velocity v_{x2} as shown in Figure 17.11. Therefore the number of waves leaving the shock is equal to the number of independent linearized boundary conditions; q.e.d.

Exercise 17.2. Since an ordinary shock wave is evolutionary, consider the linear problem of its stability in the ordinary sense of small perturbations.

Answer. Suppose that the surface of a shock is perturbed in the following way:

$$\xi = \xi_0 \exp \left[\mathrm{i} \left(k_y \, y - \omega \, t \right) \right] , \tag{17.68}$$

where ξ is a displacement of the surface. The shock front thus becomes corrugated. The corrugation causes a perturbation of the flow. An arbitrary hydrodynamic perturbation is represented as a sum of the entropy-vortex wave and the sound wave. Since the flow is stationary and homogeneous in the y direction, all perturbations have the same frequency ω and tangential component of the wave vector k_y.

Since the flow velocity ahead of the shock $v_1 > V_{s1}$, only the downstream flow is perturbed. The usual condition of compatibility of the linear equation set is that the determinant of the coefficients at unknown quantities is zero, which yields the dispersion equation

$$\frac{\omega v_2}{v_1} \left(k_y^2 + \frac{\omega^2}{v_2^2} \right) - \left(\frac{\omega^2}{v_1 v_2} + k_x^2 \right) (\omega - k_y v_2) \left[1 + J^2 \left(\frac{\partial U_2}{\partial p_2} \right)_{\mathrm{RH}} \right] = 0 \, . \tag{17.69}$$

Here $U = 1/\rho$ is a specific volume, $J = \rho_1 v_1 = \rho_2 v_2$. The subscript RH means that the derivative is taken along the Rankine–Hugoniot curve.

The shock front as a discontinuity is unstable if

$$\mathrm{Im}\, \omega > 0 \, , \qquad \mathrm{Im}\, k_x > 0 \, . \tag{17.70}$$

The second condition (17.70) means that the perturbation is excited by the shock itself, but not by some external source. As shown by D'yakov (1954), Equation (17.69) has solution which satisfies the condition (17.70), when

$$J^2 \left(\frac{\partial U_2}{\partial p_2} \right)_{\mathrm{RH}} < -1 \tag{17.71}$$

or

$$J^2 \left(\frac{\partial U_2}{\partial p_2} \right)_{\mathrm{RH}} > 1 + 2 \, \frac{v_2}{V_{s2}} \, . \tag{17.72}$$

If the parameters of the flow fall into the interval (17.71) or (17.72) then the small perturbation of the shock grows exponentially with time. This is the so-called *corrugational* instability of shock waves in ordinary hydrodynamics.

Along with this there is a possibility that Equation (17.69) has solutions with real ω and k_x which correspond to non-damping waves outgoing from the discontinuity (D'yakov, 1954). In this case

> the shock spontaneously radiates sound and entropy-vortex waves, with the energy being supplied from the whole moving medium.

Apparently this instability is the reason of the flow inhomogeneities observed, for example, in laboratory experiments when a strong shock propagates in a gas (see Markovskii and Somov, 1996).

Exercise 17.3. Show that an ordinary tangential discontinuity introduced in Section 16.1.2 is non-evalutionary.

Answer. From (16.6) it follows that there exists only one boundary condition at the surface of a tangential discontinuity in ordinary hydrodynamics. However two sound waves can propagate from the discontinuity at its both sides. Therefore the number of small-amplitude waves is greater than the number of linearized boundary equations.

Chapter 18

Particle Acceleration by Shock Waves

Sir Charles Darwin (1949) presumably thought that shock waves are responsible for accelerating cosmic rays. Nowadays shocks are widely recognized as a key to understanding high-energy particle acceleration in a variety of astrophysical environments.

18.1 Two basic mechanisms

Astrophysical plasma, being tenuous, differs from laboratory plasma in many ways. One of them is the following. In most environments where accelerated paricles are observed, typical sound speeds are considerably less than easily obtainable bulk flow velocities, and shock waves are expected to develop. In fact, shocks are associated with most energetic particle populations seen in space.

In the heliosphere, collisionless shocks are directly observable with spacecrafts and they have been well studied. In every case where direct observations have been made, shocks are seen to accelerate particles, often to power-law distributions. Investigations of heliospheric shocks, along with a great deal of theoretical work, also show that collective field-particle interactions control the shock dissipation and structure. The physics of shock dissipation and particle acceleration seem to be intimately related.

In this Chapter, we introduce only the most important aspects of the shock acceleration theory including two fundamental mechanisms of particle acceleration by a shock wave. Analytical models and numerical simulations (Jones and Ellison, 1991; Blandford, 1994; Giacalone and Ellison, 2000; Parks, 2004) illustrate the possible high efficiency of *diffusive* and *drift* accelerations to high energies.

18.2 Shock diffusive acceleration

18.2.1 The canonical model of diffusive mechanism

Axford et al. (1977) and Krymskii (1977) considered the idealized problem of the particle acceleration by a shock wave of plane geometry propagating in a medium containing small-scale inhomogeneities of a magnetic field which scatter fast particles. The origin of these scatterers will be discussed later on. This may be, for example, the case of parallel or nearly parallel MHD shocks. In shocks of this kind (see case (16.72)) the avarage magnetic field plays essentially no role since it is homogeneous, while fluctuations in the avarage field play a secondary role producing particle scattering. Assuming this, we consider a shock wave as an ordinary hydrodynamic shock with scatterers.

If the medium is homogeneous, and if the propagation of the shock is stationary, then the front of the shock separates the two half-spaces: $x < 0$ and $x > 0$, and the velocity of the medium is given by the following formula:

$$v(x) = \begin{cases} v_1 & \text{for} \quad x < 0\,, \\ v_2 = r^{-1} v_1 & \text{for} \quad x > 0\,. \end{cases} \tag{18.1}$$

Here

$$r = \frac{\rho_2}{\rho_1} = \frac{v_1}{v_2} \tag{18.2}$$

is the compression ratio. It follows from formula (16.94) that, in a very strong (but nonrelativistic) shock wave, the ratio

$$r \to r_\infty = \frac{\gamma_g + 1}{\gamma_g - 1}$$

and

$$v_2 = \frac{v_1}{r_\infty} = \frac{\gamma_g - 1}{\gamma_g + 1}\, v_1\,. \tag{18.3}$$

The adiabatic index γ_g is considered constant on both sides of the shock front $x = 0$.

Following Axford et al. (1977) and Krymskii (1977), let us assume that the distribution function in space and the scalar momentum of the accelerated particles, $f(\mathbf{r}, p)$, is isotropic (see generalization in Gieseler et al., 1999; Ruffolo, 1999). This means that $f(\mathbf{r}, p)$ is the same in all reference frames to first order in the small parameter v/v_p, where v_p and p are the individual particle velocity and momentum measured in the local plasma frame.

As long as scattering is strong enough to insure the isotropy assumption, the kinetic Equation (2.15) describing the transport of particles with $v_p \gg v$ in space and velocity can be written in the form of a *diffusion-convection* equation (see Krymskii (1977) and references therein):

$$\frac{\partial f}{\partial t} = \nabla_{\mathbf{r}} \left(D\, \nabla_{\mathbf{r}} f \right) - \nabla_{\mathbf{r}} \left(f \mathbf{v} \right) + \frac{1}{3}\, \frac{\partial \left(f p \right)}{\partial p}\, \operatorname{div} \mathbf{v}\,. \tag{18.4}$$

Here $D = D(\mathbf{r}, p)$ is the coefficient of diffusion of fast particles.

For our problem under consideration, with one-dimensional geometry, we have in the *stationary* case

$$\frac{\partial}{\partial x}\left[v f(x,p) - D(x,p)\,\frac{\partial f(x,p)}{\partial x} \right] = \frac{1}{3}\,\frac{\partial v}{\partial x}\,\frac{\partial}{\partial p}\left[p f(x,p) \right] . \qquad (18.5)$$

Let us integrate Equation (18.5) over x from $x = -\infty$ to $x = +\infty$. By employing the boundary conditions

$$f(x = -\infty, p) = f_1(p) \quad \text{and} \quad f(x = +\infty, p) = f_2(p) \, , \qquad (18.6)$$

where $f_2(p)$ is an unknown spectrum of accelerated particles, we obtain the following differential equation in p

$$v_2 f_2(p) - v_1 f_1(p) - 0 + 0 = \frac{1}{3}\,(v_2 - v_1)\,\frac{d}{dp}\left[p f_2(p) \right] . \qquad (18.7)$$

Using the definition of the compression ratio (18.2), we obtain an ordinary differential equation for the downstream distribution function $f_2(p)$ in the form

$$p\,\frac{d}{dp}\,f_2(p) + \frac{r+2}{r-1}\,f_2(p) = \frac{3r}{r-1}\,f_1(p) \, ; \qquad (18.8)$$

recall that $r > 1$.

The general solution of this equation is

$$f_2(p) = \frac{3}{r-1}\,p^{-\gamma_p} \int\limits_{p_0}^{p} f_1(p')\,(p')^{-\gamma_p}\,dp' + c_1\,p^{-\gamma_p} \, . \qquad (18.9)$$

Here

$$\boxed{\gamma_p = \frac{r+2}{r-1}} \qquad (18.10)$$

plays the role of the *spectral index* of the accelerated particles, c_1 is an arbitrary constant of integration which multiplies the homogenious term, the distribution function $f_1(p)$ is the far upstream spectrum of ambient particles that are accelerated by the shock, and p_0 is large enough so that the assumption $v_p \gg v$ holds.

So the solution of the diffusion-convection equation does show that a planar shock, propagating through a region in which fast particles are diffusing, produces a superthermal population of particles with the power-law momentum distribution

$$f_2(p) \sim p^{-\gamma_p} \, . \qquad (18.11)$$

The property which gave the diffusive acceleration process a wide appeal is the fact that, with the simplest assumptions made above, the spectral index (18.10) of the accelerated particles depends only on the compression ratio r of the shock wave.

Most astrophysical shocks, since they are strong, have compression ratios constrained to a rather narrow range of values near $r_\infty = 4$ assuming $\gamma_g = 5/3$. For a shock with Mach number M (see Exercise 16.5) greater than 3 say, as we see in Figure 18.1, the compression ratio $3 < r < 4$ and the spectral index $2 < \gamma_p < 2.5$.

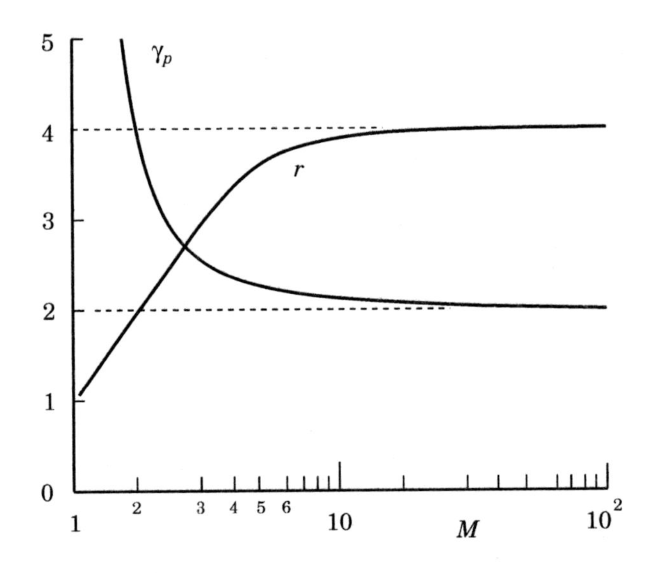

Figure 18.1: The compression ratio r and spectral index γ_p versus the Mach number M.

A spectral index of $\gamma_p \approx 2$ is characteristic of energy particle spectra observed in a wide range of astrophysical environments (Jones and Ellison, 1991; Blandford, 1994). For example, $\gamma_p \approx 2$ closely fits the inferred source spectrum of **Galactic cosmic rays** for high energies below approximately 10^{15} eV (e.g., Gombosi, 1999). In the cosmic rays observed at Earth, the spectrum of cosmic-ray ions is an unbroken power law from 10^9 to 10^{15} eV. The supernova shocks are one of the few mechanisms known to be capable of providing adequate energy to supply the pool of Galactic cosmic rays. Supernova remnants (SNRs) have long been suspected as the primary site of Galactic cosmic-ray acceleration.

The earliest evidence of non-thermal X-ray emission in a SNR came from featureless observed spectra interpreted as the extrapolation of a radio synchrotron spectrum. However early data were poor and the models were simplistic. New observations and theoretical results (Dyer et al., 2001) indicate that joint thermal and non-thermal fitting, using sophisticated models, will be required for analysis of most supernova-remnant X-ray data in the future to answer two questions: (a) Do SNRs accelerate ions? (b) Are they capable of accelerating particles to energies of 10^{15} eV?

In the **solar wind** the shock-associated low-energy-proton events seem to be well studied. The most intensive of them have a power-low energy spectrum, suggesting that protons are accelerated by the diffusive-shock acceleration mechanism (e.g., Rodriguez-Pacheco et al., 1998). Nevertheless the correlation between the spectral exponent γ with the solar wind velocity compression ratio is found to be linear. This result differs from that presented above. The discrepancy of the spectral-exponent dependence on the shock-wave parameters could lie on the event selection criterion or on the account of nonlinear effects (Section 18.2.3) or on another mechanism of acceleration.

18.2.2 Some properties of diffusive mechanism

As we saw above, the spectral index γ_p of energetic particles produced by diffusive shock acceleration does not depend on the diffusion coefficient D. However the diffusion coefficient D, together with the characteristic flow velocity $v \sim v_1$, determines the overall length scale of the acceleration region

$$l_{_D} \sim D(p)/v \tag{18.12}$$

and acceleration time

$$t_{_D} \sim D(p)/v^2 \,. \tag{18.13}$$

The first-order Fermi or diffusive shock acceleration is a statistical process in which particles undergo spatial diffusion and are accelerated as they scatter back and forth across the shock, thereby being compressed between scattering centers fixed in the converging upstream and downstream flows.

> Particle energies are derived just from the relative motion, the converging flow with velocity $v_1 - v_2$, between scatterers (waves) on either side of a shock front.

This is a main advantage of the diffusive mechanism. Its disadvantage is that particles can achieve very high energies by diffusion acceleration, but

> since particles spend most of their time random walking in the upstream or downstream plasma, the acceleration time can become excessively large

compared with, for example, the shock's life time.

Another disadvantage in applying it to some astrophysical phenomena, for example solar flares, consists of the lack of actual knowledge about the assumed scattering waves. However diffusion determines only the length scale (18.12) and characteristic time (18.13) of the acceleration process. In this context, let us recall once more (Section 16.1.3) the following analogy from everyday life. A glass of hot water with a temperature T_1 will invariably cool to a given room temperature T_2, independently of the mechanism of heat exchange with the surrounding medium, while the mechanism determines only the time of cooling.

In the presence of a magnetic field in plasma, the diffusive acceleration requires that the particles are able to traverse the shock front in both directions either along the field or by scattering across the field, in order that they may couple to the shock compression by pitch-angle scattering both upstream and downstream of the shock. At quasi-parallel shocks this condition on particle mobility is easily met. For sufficiently fast shocks, downstream shock-heated particles can be kinematically able to return to the shock along the downstream magnetic field to initiate the process of diffusive shock acceleration. At quasi-perpendicular shocks (Section 18.3.2), however, this condition is stringent. Although the diffusive mechanism is rapid since particles are confined closer to the shock front, there is a **high threshold speed**, significantly exceeding v_1, in order that diffusive acceleration can occur (Webb et al., 1995).

18.2.3 Nonlinear effects in diffusive acceleration

The test particle (i.e., *linear*) model demonstrated above yields the most important result: the power law (18.11) with the spectral index (18.10) is the natural product of the diffusive acceleration in shock waves. The equally important question of the actual efficiency of the process can only be adequately addressed to a fully *nonlinear* (and more complex) theory. Using observations of the Earth bow shock and interplanetary observations, numerical modeling of different shocks shows that the inherent efficiency of shock acceleration implies that

> the hydrodynamic feedback effects between the accelerated particles and the shock structure are important

and therefore essential to any complete description of the process. This has turned out to be a formidable task because of the wide range of spatial and energy scales that must be self-consistently included in numerical simulations.

On the one hand, the plasma microprocesses of the shock dissipation control injection from the thermal population. On the other hand, the highest energy particles (extending to $10^{14} - 10^{15}$ eV in the case of galactic cosmic rays) with extremely long diffusion lengths (18.12) are dynamically significant in strong shock waves and feed back on the shock structure. Ranges of interacting scales of many orders of magnitudes must be described self-consistently (for review see Parks, 2004).

18.3 Shock drift acceleration

The principal process whereby a particle gains energy upon crossing a shock wave with a magnetic field may be the so-called shock drift acceleration (Hudson, 1965). The drift mechanism, in contrast to the diffusive one, neglects any shock-front associated turbulence. So many not-well-justified assumptions concerning the physics of scatterers have not to be made in applying the drift acceleration model to an astrophysical phenomenon.

If the fast particle Larmor radius

$$r_{\rm L} = \frac{c p_{\perp}}{eB} \gg l_f \,, \tag{18.14}$$

where l_f is the front thickness, we can replace the shock by a simple discontinuity (the shock surface) and can approximate the particle motion as scatter-free on both sides of the shock. Let us begin by considering an interaction of individual particles with such a discontinuity. We shall consider very fast particles:

$$v_p \gg v_1 > v_2 \,. \tag{18.15}$$

These assumptions are basic for further considerations that we start from the simplest case – a perpendicular shock (Section 16.2.3).

18.3.1 Perpendicular shock waves

As shown in Figure 16.5, the magnetic fields $\mathbf{B}_1$ and $\mathbf{B}_2$ are parallel to the shock front $x = 0$; and plasma moves perpendicularly to the front. According to (16.41), there exists an identical electric field on both sides of the shock:

$$\mathbf{E} = -\frac{1}{c}\, \mathbf{v}_1 \times \mathbf{B}_1 = -\frac{1}{c}\, \mathbf{v}_2 \times \mathbf{B}_2 \,. \tag{18.16}$$

The fast particles rotate on the magnetic field lines and move together with the field lines with the plasma speed across the front as shown in Figure 18.2.

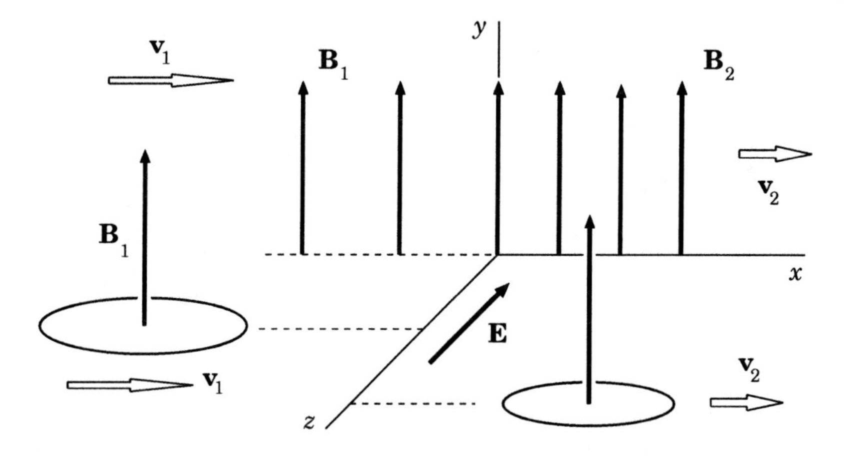

Figure 18.2: The Larmor ring moves together with the plasma and the magnetic field across the perpendicular shock front.

Nothing will happen before the Larmor ring touches the front; a particle simply drifts to the front. For what follows it is important that the particle will make many rotations (Figure 18.3) during the motion of the Larmor ring across the front because of the condition (18.15). A 'single encounter' consists of many individual penetrations by the particle through the shock surface as the particle follows its nearly helical trajectory. Because of the difference between the Larmor radius ahead of and behind the front, a drift parallel to the front will appear, accompanying the drift across the front.

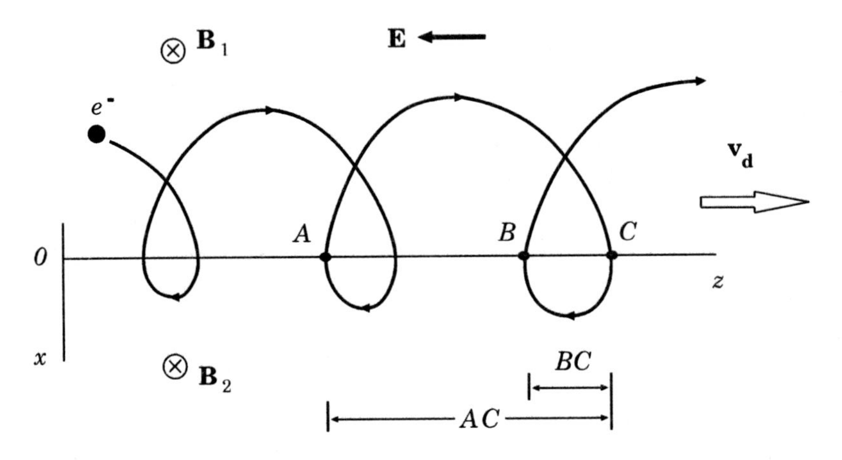

Figure 18.3: The trajectory of a negatively charged particle (an electron) multiply crossing the perpendicular shock front.

During each rotation, the electric field **E** accelerates a particle on the upstream side ($x < 0$) of the shock and decelerates it on the downstream side ($x > 0$). However the work of the field **E** on a larger circle exceeds the work on the smaller circle:

$$\delta A_1 = + eE \times AC > - \delta A_2 = eE \times BC\,, \tag{18.17}$$

since the length AC is larger than the length AB. Therefore, during each rotation, the particle is slightly accelerated. How much energy does the particle take during the motion of its Larmor ring across the shock front?

Since we consider the shock as a discontinuity, the adiabatic approximation is formally not suitable. However it appears that the *transversal* invariant (Section 6.2) conserves:

$$\frac{p_\perp^2}{B} = \text{const} \tag{18.18}$$

(Hudson, 1965; Alekseyev and Kropotkin, 1970). From (18.18) it follows that

$$p_{\perp 2}^2 = p_{\perp 1}^2 \times \frac{B_2}{B_1}\,.$$

Therefore the transversal kinetic energy of a nonrelativistic particle

$$\frac{\mathcal{K}_{\perp 2}}{\mathcal{K}_{\perp 1}} = \frac{p_{\perp 2}^{2}}{p_{\perp 1}^{2}} \propto \frac{B_2}{B_1} = r\,. \tag{18.19}$$

An increase of transversal energy (18.19) is relatively small when the Larmor ring of a particle crosses the front only once.

Multiple interactions of a particle with the shock is a necessary condition for a considerable increase of energy.

Drift acceleration typically involves several shock crossings and results from a net displacement δz of an ion (electron) guiding center parallel (antiparallel) to the convection electric field **E**. The energy gain is proportional to this displacement, which in general depends upon the plasma and shock parameters, the particle species and velocity, and the intensity of possible electromagnetic fluctuations in the vicinity of the shock as well as within the shock front itself. It is popular to discuss the displacement δz as the consequence of a gradient drift (see formula (5.14) in Jones and Ellison, 1991). Such a treatment is not reasonable when we consider the shock as a discontinuity; so formally $\nabla B \to \infty$. A wondeful thing is that the adiabatic approximation is not applicable for such a situation but the first adiabatic invariant (18.18) conserves.

18.3.2 Quasi-perpendicular shock waves

18.3.2 (a) Classical model of acceleration

The basic aspects of drift acceleration of fast particles by an almost perpendicular shock wave, as a discontinuity, emerge from a simple model which is valid for a certain range of incident pitch angles and which allows us to derive analytical expressions for the reflection and transmission coefficients, the energy and the angular distributions (Toptyghin, 1980; Decker, 1983).

By definition, in a quasi-perpendicular shock, the angle Ψ_1 (Figure 18.4) between the shock normal **n** and the upstream magnetic field vector $\mathbf{B}_1$ is greater than about 80°. Hence the field lines form small angles α_1 and α_2 with the shock front plane $x = 0$. Under this condition, as well as for the perpendicular shock case considered above, **the first adiabatic invariant is conserved** (Hudson, 1965; see also Section 4 in Wentzel, 1964). This enables analytical calculations of the energy increase on the front of a quasi-perpendicular shock as well as the reflection and transmission of fast particles (Sarris and Van Allen, 1974).

Since the particles conserve the first adiabatic invariant (Section 6.2.1), all particles with pitch angles

$$\theta > \theta_0 \tag{18.20}$$

will be reflected. To find the critical pitch angle θ_0, consider two frames of reference: S and S'.

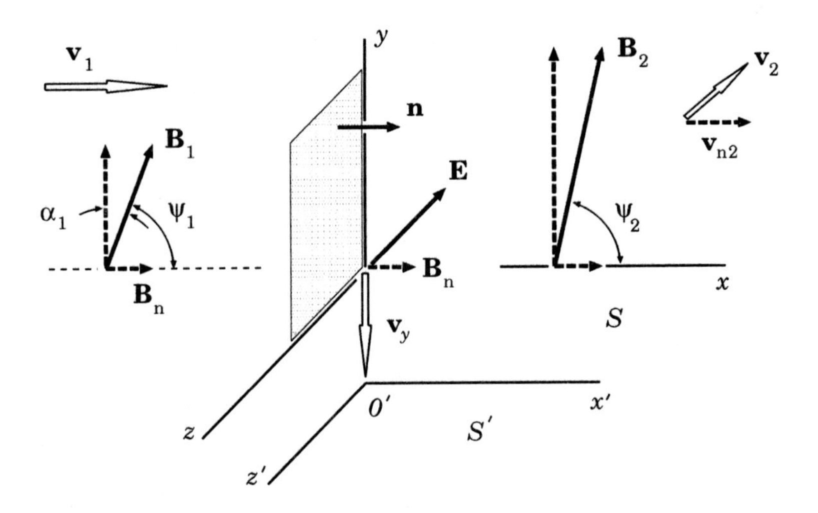

Figure 18.4: A quasi-perpendicular shock wave in the frame of reference S where $\mathbf{v}_1 \parallel \mathbf{n}$.

In the frame of reference S, where the shock front is in the plane (y, z) and the shock normal n is along the x axis, there is an electric field

$$\mathbf{E} = -\frac{1}{c}\,\mathbf{v}_1 \times \mathbf{B}_1 = -\frac{1}{c}\,\mathbf{v}_2 \times \mathbf{B}_2 \,. \tag{18.21}$$

In the frame of reference S', where $\mathbf{B}_1 \parallel \mathbf{v}_1$ and $\mathbf{B}_2 \parallel \mathbf{v}_2$ (Section 16.2.4), there is **no electric field**. The system S' moves along the y axis (perpendicular to the vector $\mathbf{E}$) with velocity

$$\mathbf{v}_y = c\,\frac{\mathbf{E} \times \mathbf{B}_n}{B_n^2}\,, \tag{18.22}$$

where $\mathbf{B}_n$ is the normal component of the magnetic field. Since $\mathbf{E}' = 0$, there is no change in the energy of a fast particle after reflection from the front: $\delta\mathcal{E}' = 0$.

We shall assume that B_n is very small but $v_y < c$. Using the relativistic Lorentz transformation for the energy-momentum 4D-vector with condition $\delta\mathcal{E}' = 0$, we obtain the relative energy increment of the reflected fast particles (see Exercise 18.1):

$$\frac{\delta\mathcal{K}}{\mathcal{K}} \approx \frac{4v_1^2}{v_p^2}\left[\frac{v_p \cos\theta}{v_1} + \mathrm{tg}\,\Psi_1\right]\mathrm{tg}\,\Psi_1\,. \tag{18.23}$$

Here $\mathcal{K} = mv_p^2/2$ is the kinetic energy of a particle in the shock wave frame of reference S, v_p is the particle velocity in the same frame, and θ is the pitch angle also in the frame S. The connection between θ' and θ is given by

$$\cos\theta' = \frac{v_p \cos\theta + v_1 \operatorname{tg}\Psi_1}{\left[v_p^2 + (v_1 \operatorname{tg}\Psi_1)^2 + 2v_p\,(v_1 \operatorname{tg}\Psi_1)\,\cos\theta \right]^{1/2}}\,. \tag{18.24}$$

In the S' frame of reference, where the electric field is zero, the first adiabatic invariant can be written as (see definition (6.11)):

$$\frac{\sin^2\theta'}{B} = \text{const}\,. \tag{18.25}$$

So the critical pitch angle θ_0' satisfies equation

$$\sin^2\theta_0' = \frac{B_1}{B_2}\,. \tag{18.26}$$

This allows us to calculate the critical pitch angle θ_0 in the shock-front frame S. For example, if a non-relativistic proton has an initial energy $\mathcal{K} =$ 0.3 MeV and if a shock wave has an upstream velocity $v_1 = 150$ km/s, the ratio $B_1/B_2 = 1/3$, and the angle $\Psi_1 = 88°$ and $89°$, then we find, correspondingly, $\theta_0 = 55°$ and $77°$. As the angle Ψ_1 increases toward $90°$, most of the particles are really transmitted into the downstream side. At $\Psi_1 = 90°$, which is the perpendicular shock case, there are no reflecting particles.

Formula (18.23) shows that

> the relative increment of kinetic energy of a fast particle increases when the angle Ψ_1 increases toward $90°$.

The model under consideration predicts **high field-aligned anisotropies** for a large Ψ_1 because of conservation of first adiabatic invariant and the large energy gains.

It is widely believed that the slow **thermal particles** inside the shock front can also be considered as adiabatic, at least, in thick collisionless shocks: the electron magnetic moment is conserved throughout the shock and $v_\perp^2/B = \text{const}$ (Feldman et al., 1982). In very thin collisionless shock (with a large cross-shock potential) the adiabaticity may break down, so that electrons become demagnetized. It means that the magnetic moment is no longer conserved, and a more substantial part of the energy may be transferred into the perpendicular degree of freedom (Balikhin et al., 1993; Gedalin and Griv, 1999).

18.3.2 (b) Some astrophysical applications

Observations of interplanetary shocks (e.g., Balogh and Erdös, 1991) show that the intensive acceleration of protons occurs when the upstream magnetic field is almost parallel to the shock front. Energetic particles entering the shock front stay with it, crossing it many times and being accelerated by the electric field of the front. After the direction of the interplanetary magnetic field changes again away from the parallel to the front, the intensive acceleration ceases.

Owing to interplanetary magnetic field fluctuations the upstream field vector $\mathbf{B}_1$, if it is found to be parallel to the shock front, stays as such for only a short time (a few minutes, in general). This time is enough for the low-energy protons ($\mathcal{K}_p < 1$ MeV) to be accelerated to about 2–3 times their original energy but not enough for the high-energy protons ($\mathcal{K}_p < 10$ MeV) to be noticeably affected by the shock wave.

Single scatter-free shock drift interactions at quasi-perpendicular shocks can accelerate particles to at most a few times the shock compression ratio. Weak scattering during single drift interactions can increase this upper limit for a small fraction of an incident particle distribution, but the energy spectra will be still rather steep. One anticipates large energy gains and flatter spectra that extend to high energies if some particles can return to the shock for many drift interactions.

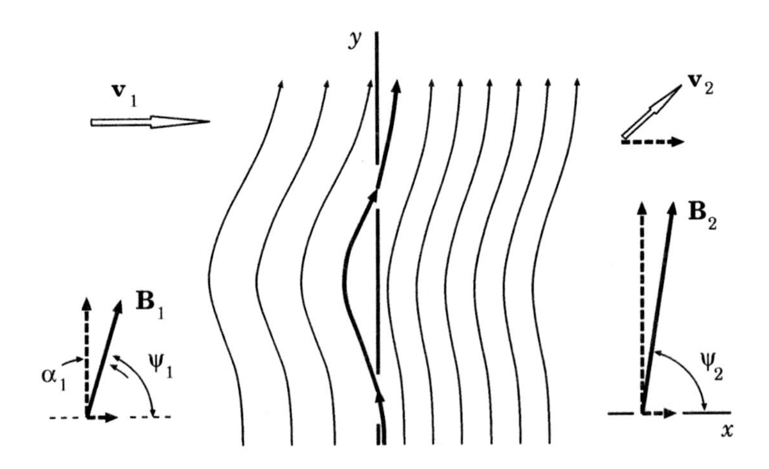

Figure 18.5: A collapsing magnetic trap on the upstream side of a quasi-perpendicular shock wave.

This is suggestive of the classical case of a **collapsing magnetic trap** (Section 6.2), and is the basis of the model of proton trapping and acceleration due to multiple drift interactions along magnetic loops that convect through a planar quasi-perpendicular shock (Wentzel, 1963, 1964; Gisler and Lemons, 1990). Figure 18.5 represents a quasi-perpendicular shock, with a small perturbation of the magnetic field superimposed on the unshocked homogeneous field $\mathbf{B}_1$. The heavy line displays a particular field line which intersects the shock front plane $x = 0$ two times, forming a magnetic loop in the upstream region.

> Upstream particles bounce back and forth along a loop and gain parallel energy at each reflection until they fall within the loss cone and transmit downstream.

Simple analytic models and detailed numerical study have shown that the collapse of the trap by the convection of the loop field lines through the shock is accompanied by a considerable increase of the accelerated proton flux, which may be responsible for the 'shock spike' events observed near fast mode interplanetary shock waves (Decker, 1993; Erdös and Balogh, 1994).

In general, if the magnetic field contains fluctuations with wavelengths that are considerably larger than the gyroradii of the fast particles, a fraction of particles is accelerated by a quasi-perpendicular shock to energy well above the thermal energy (Giacalone and Ellison, 2000).

18.3.3 Oblique shock waves

If values of the angle α between the magnetic field and the shock front plane are arbitrary, then the phase-averaged coefficients of reflection and transmission are complicated and can be found, in principle, by numerical calculations. When

$$\frac{v_1}{v_p} \leq \alpha_1 \leq \frac{\pi}{2} \tag{18.27}$$

and the pitch angle θ is arbitrary, the order of magnitude of the energy increase

$$\delta \mathcal{K} \approx \frac{p\, v_1}{\alpha_1} \ll \mathcal{K} = \mathcal{E} - mc^2 \tag{18.28}$$

is small in comparison with the initial kinetic energy. In a general case, the increase of particle energy is small when the Larmor ring of a particle crosses the front once. Multiple interactions of a particle with the shock front is the necessary condition for a considerable increase of energy.

One possibility for multiple interactions of a particle with the shock is a strong MHD turbulence. More exactly, it is assumed that in a sufficiently large region of space there exists an ensemble of MHD shocks which interact successively with the particles. The investigation of particle acceleration by a random shock wave ensemble is of certain interest in astrophysical applications but the conditions of such an acceleration mechanism are not totally clear yet.

Another possibity is the propagation of one shock in a turbulent medium or of an oblique collisionless shock when magnetic turbulence exists in the regions upstream and downstream of the shock (Decker and Vlahos, 1986). It is important, however, that the particle acceleration near the shock front in a turbulent medium, i.e. the diffusive mechanism (Section 18.2) will take place in the absence of a regular electric field. No terms should be added to the basic diffusion-convection equation (18.4) to take account of the drift mechanism in an oblique shock. The process is already included in the energy change which is proportional to the div $\mathbf{v}$ term. This, of course, assumes that there is sufficient scattering and that other assumptions used in deriving the diffusion-convection equations are also valid. That is not trivial.

The interesting possibility discussed in Section 18.3.2 is a combination of a magnetic trap with an oblique shock wave. In vol. 2, Section 7.3, this idea is applied to the particle acceleration in solar flares.

18.4 Practice: Exercises and Answers

Exercise 18.1. Derive formula (18.23) in Section 18.3.2.

Answer. According to the geometry shown in Figure 18.4, the frame of reference S' moves with respect to the shock wave frame of reference S with velocity (18.22):

$$\mathbf{v}_y = -\frac{cE}{B_n}\,\mathbf{e}_y\,. \tag{18.29}$$

In the frame S' there is no electric field; therefore there is no change in the energy of a particle reflecting at the shock front, that is $\delta\mathcal{E}' = 0$, where $\mathcal{E}'$ is the energy of the particle in S'. Let us transform this condition back to S by using the Lorentz transformation of the energy-momentum 4D-vector (Landau and Lifshitz, *Classical Theory of Field*, 1975, Chapter 2, § 9):

$$p_x = p_x', \qquad p_y = \gamma_{\rm L}\left(p_y' + \frac{v_y}{c^2}\,\mathcal{E}'\right), \qquad p_z = p_z',$$

$$\mathcal{E} = \gamma_{\rm L}\left(\mathcal{E}' + v_y\,p_y'\right). \tag{18.30}$$

Since $\delta\mathcal{E}' = 0$, it follows from (18.30) that

$$\delta\mathcal{E} = \gamma_{\rm L}\,v_y\,\delta p_y'. \tag{18.31}$$

The change in the y component of momentum of the reflected particle in the frame of reference S' is

$$\delta\,\mathbf{p}_y' = -2\,\mathbf{p}_y' = 2\gamma_{\rm L}\left(\mathbf{p}_y - \frac{v_y}{c^2}\,\mathcal{E}\,\mathbf{e}_y\right). \tag{18.32}$$

Note that vectors $\mathbf{p}_y$ and $\mathbf{v}_y$ point in opposite directions. Substituting (18.32) into (18.31) gives us

$$\delta\mathcal{E} = \frac{2v_y}{1 - v_y^2/c^2}\left[p_y + \frac{v_y}{c^2}\left(\mathcal{K} + mc^2\right)\right], \tag{18.33}$$

where $\mathcal{K} = mv_p^2/2$ is kinetic energy of a particle. Assuming $\mathcal{K} \ll mc^2$ and using (18.29), we obtain

$$\delta\mathcal{E} = \delta\mathcal{K} = \frac{2E}{B_n^2 - E^2}\left(\frac{v_{p,y}}{c}\,B_n + E\right)mc^2, \tag{18.34}$$

where $v_{p,y}$ is the y component of the particle velocity.

According to (18.21) the electric field

$$E = \frac{1}{c}\, v_1 B_{y1}\,, \tag{18.35}$$

where B_{y1} is the y component of the vector $\mathbf{B}_1$. So we rewrite formula (18.34) as follows

$$\delta\mathcal{K} = 2mv_1^2\, \frac{(v_{p,y}/v_1)(B_n/B_{y1}) + 1}{(B_n/B_{y1})^2 - (v_1/c)^2}\,. \tag{18.36}$$

The condition $v_y < c$ can equivalently be written as

$$\frac{B_n}{B_{y1}} > \frac{v_1}{c} \quad \text{or} \quad \mathrm{tg}\,\Psi_1 < \frac{v_1}{c}\,. \tag{18.37}$$

If we further assume that

$$\frac{B_n}{B_{y1}} \gg \frac{v_1}{c}\,, \tag{18.38}$$

we obtain from (18.36) the following formula

$$\delta\mathcal{K} = 2mv_1^2 \left(\frac{v_p \cos\theta}{v_1}\, \mathrm{tg}\,\Psi_1 + \mathrm{tg}^2\,\Psi_1 \right), \tag{18.39}$$

where θ is the pitch angle in the shock-front frame of reference S. Dividing (18.39) by $\mathcal{K}$, we obtain formula (18.23).

Chapter 19

Plasma Equilibrium in Magnetic Field

The concept of equilibrium is fundamental to any discussion of the energy contained in an astrophysical object or phenomenon. The MHD non-equilibrium is often related to the onset of dynamic phenomena in astrophysical plasma.

19.1 The virial theorem in MHD

19.1.1 A brief pre-history

An integral equality relating different kinds of energy (kinetic, thermal, gravitational, etc.) of some region with a volume V and a surface S, is commonly referred to as the *virial theorem*. It has been proved for mechanical systems for the first time by Clausius (1870). The derivation of the virial theorem for a mechanical system executing a motion in some finite region of space, velocities also being finite, can be found, for example, in Landau and Lifshitz (1976, *Mechanics*, Chapter 2, § 10). Its relativistic form is presented in Landau and Lifshitz (1975, *Classical Theory of Field*, Chapter 4, § 34).

The generalization of the virial theorem to include the magnetic energy in the context of MHD was achieved by Chandrasekhar and Fermi (1953) when addressing the question of the gravitational stability of infinitely conductive masses of cosmic dimensions in the presense of a magnetic field. Although "most students of physics will recognize the name of the virial theorem, few can state it correctly and even fewer appreciate its power" (Collins, 1978).

19.1.2 Deduction of the scalar virial theorem

The virial theorem is deduced from the momentum conservation law (see the ideal MHD motion Equation (12.69) or Equation (13.1)) rather than the energy conservation law. We have

$$\rho \frac{dv_\alpha}{dt} \equiv \rho \left(\frac{\partial v_\alpha}{\partial t} + v_\beta \frac{\partial v_\alpha}{\partial r_\beta} \right) = - \frac{\partial p}{\partial r_\alpha} - \frac{\partial \mathrm{M}_{\alpha\beta}}{\partial r_\beta} - \rho \frac{\partial \phi}{\partial r_\alpha} . \tag{19.1}$$

Here

$$\mathrm{M}_{\alpha\beta} = \frac{1}{4\pi} \left(\frac{B^2}{2} \delta_{\alpha\beta} - B_\alpha B_\beta \right) \tag{19.2}$$

is the Maxwellian stress tensor. So we consider an ideal MHD plasma distributed within a limited region V of space. The gravitational potential at a point $\mathbf{r}$ is

$$\phi(\mathbf{r}) = -G \int \frac{\rho(\mathbf{r}')}{|\mathbf{r} - \mathbf{r}'|} \, d^3\mathbf{r}', \tag{19.3}$$

where G is the gravitational constant (Appendix 3), $d^3\mathbf{r}' = dx'\,dy'\,dz'$.

The partial differential Equations (19.1) are often very difficult to solve. Moreover, in astrophysics, we may have such incomplete knowledge of a system that it may not be worthwhile to work out an elaborate solution. In many situations, it is possible to make important conclusions if we know some global relationships among the different forms of energy in the system.

Let us multiply the plasma motion Equation (19.1) by r_α and integrate it over the volume V. We observe in passing that multiplication of (19.1) by r_γ rather than r_α would result, on integrating, in the *tensor* virial theorem and not in the *scalar* one (Chandrasekhar, 1981; see also Strittmatter, 1966; Choudhuri, 1998).

First let us integrate the left-hand side of Equation (19.1) multiplied by r_α. We get

$$\int \rho r_\alpha \frac{dv_\alpha}{dt} \, dV = \int r_\alpha \frac{d^2 r_\alpha}{dt^2} \, \rho \, dV = \int r_\alpha \frac{d^2 r_\alpha}{dt^2} \, dm . \tag{19.4}$$

Here we have passed from the integration over volume to integration over mass: $dm = \rho\, dV$. We rearrange formula (19.4) as follows

$$r_\alpha \frac{d^2 r_\alpha}{dt^2} = \frac{d}{dt} \left(r_\alpha \frac{dr_\alpha}{dt} \right) - \left(\frac{dr_\alpha}{dt} \right)^2 =$$

$$= \frac{d}{dt} \left(\frac{1}{2} \frac{dr_\alpha^2}{dt} \right) - \left(\frac{dr_\alpha}{dt} \right)^2 = \frac{1}{2} \frac{d^2}{dt^2} r_\alpha^2 - v_\alpha^2 .$$

On substituting this into (19.4), we obtain

$$\int \rho r_\alpha \frac{dv_\alpha}{dt} \, dV = \frac{1}{2} \frac{d^2}{dt^2} \int r^2 \, dm - \int v^2 \, dm = \frac{1}{2} \frac{d^2 I}{dt^2} - 2T. \tag{19.5}$$

Here

$$I = \int r^2 \, dm \tag{19.6}$$

is the *moment of inertia* in the reference frame related to the mass center of the system. When the system expands, its moment of inertia I increases.

$$T = \int \frac{v^2}{2} \, dm \tag{19.7}$$

is *kinetic energy* or (to be more specific) the kinetic energy of macroscopic motions inside the system.

Let us multiply the first term on the right-hand side of Equation (19.1) by r_α and integrate it over volume:

$$- \int_V r_\alpha \frac{\partial p}{\partial r_\alpha} \, dV = - \oint_S p \, r_\alpha \, dS_\alpha + 3 \int_V p \, dV, \tag{19.8}$$

since

$$\frac{\partial}{\partial r_\alpha} (p r_\alpha) = r_\alpha \frac{\partial p}{\partial r_\alpha} + p \frac{\partial r_\alpha}{\partial r_\alpha} = r_\alpha \frac{\partial p}{\partial r_\alpha} + 3p \, .$$

The Gauss theorem was used to integrate the divergence over the volume in formula (19.8).

If U_{th} is the *thermal* energy of the plasma, γ_g is the ratio of specific heats at constant pressure and at constant volume, then

$$\int_V p \, dV = (\gamma_g - 1) \, U_{th} \, . \tag{19.9}$$

Therefore

$$- \int_V r_\alpha \frac{\partial p}{\partial r_\alpha} \, dV = - \oint_S p \, (\mathbf{r} \cdot d\mathbf{S} \,) + 3 \, (\gamma_g - 1) \, U_{th} \, . \tag{19.10}$$

Similarly we calculate the integral

$$- \int_V r_\alpha \frac{\partial \mathrm{M}_{\alpha\beta}}{\partial r_\beta} \, dV = - \int_S \mathrm{M}_{\alpha\beta} \, r_\alpha \, dS_\beta + \int_V \mathrm{M}_{\alpha\beta} \, \delta_{\alpha\beta} \, dV \tag{19.11}$$

since

$$\frac{\partial}{\partial r_\beta} (r_\alpha \, \mathrm{M}_{\alpha\beta}) = r_\alpha \frac{\partial \mathrm{M}_{\alpha\beta}}{\partial r_\beta} + \mathrm{M}_{\alpha\beta} \, \delta_{\alpha\beta} \, .$$

On rearranging, we find from (19.11) and (19.2)

$$- \int_V r_\alpha \frac{\partial \mathrm{M}_{\alpha\beta}}{\partial r_\beta} \, dV = \mathcal{M} - \int_S \left[\frac{B^2}{8\pi} \, (\mathbf{r} \cdot d\,\mathbf{S}) - \frac{1}{4\pi} \, (\mathbf{B} \cdot \mathbf{r}) \, (\mathbf{B} \cdot d\,\mathbf{S}) \right] , \tag{19.12}$$

where

$$\mathcal{M} = \int\limits_V \frac{B^2}{8\pi}\, dV \tag{19.13}$$

is the *magnetic* energy of the system.

The third term on the right-hand side of Equation (19.1) gives

$$- \int\limits_V r_\alpha \frac{\partial \phi}{\partial r_\alpha}\, \rho\, dV = \int\limits_V \rho\, r_\alpha \frac{\partial}{\partial r_\alpha} \int\limits_{V'} \frac{G \rho(\mathbf{r}')}{|\mathbf{r} - \mathbf{r}'|}\, dV'\, dV =$$

$$= G \int\limits_V \int\limits_{V'} \rho\, \rho'\, r_\alpha \frac{\partial}{\partial r_\alpha} \frac{1}{\sqrt{\left(r_\beta - r'_\beta\right)^2}}\, dV\, dV'. \tag{19.14}$$

We rewrite the expression as follows. Let the distance $R = \sqrt{\left(r_\beta - r'_\beta\right)^2}$. Then

$$r_\alpha \frac{\partial}{\partial r_\alpha} \frac{1}{R} = \frac{1}{2} \left(r_\alpha \frac{\partial}{\partial r_\alpha} \frac{1}{R} + r'_\alpha \frac{\partial}{\partial r'_\alpha} \frac{1}{R} \right) = - \frac{1}{R}$$

and

$$- \int\limits_V r_\alpha \frac{\partial \phi}{\partial r_\alpha}\, \rho\, dV = \Omega\,, \tag{19.15}$$

where

$$\Omega = - \frac{G}{2} \int\limits_V \int\limits_{V'} \frac{\rho\, \rho'}{R}\, dV\, dV', \tag{19.16}$$

is the *gravitational energy* of the system. Obviously, the energy is negative.

Combining (19.5), (19.10), (19.12), and (19.15) into a single equation, we finally obtain

$$\frac{1}{2} \frac{d^2 I}{dt^2} = 2T + 3\,(\gamma_g - 1)\, U_{th} + \mathcal{M} + \Omega - \oint\limits_S p\,(\mathbf{r} \cdot d\mathbf{S}) -$$

$$- \oint\limits_S \left[\frac{B^2}{8\pi} (\mathbf{r} \cdot d\mathbf{S}) - \frac{1}{4\pi} (\mathbf{B} \cdot \mathbf{r})(\mathbf{B} \cdot d\mathbf{S}) \right]. \tag{19.17}$$

Formula (19.17) is called the virial theorem. It has repeatedly been used in astrophysics when 'discussing the question of the stability' of equilibrium systems of various types. More exactly, this integral force balance relation is nothing more than a **necessary condition for equilibrium**. So it may be well used as a *non-existance theorem* for the equilibrium problem to find circumstances when non-equilibrium may occur.

19.1.3 Some astrophysical applications

The positive terms on the right-hand side of Equation (19.17) lead to an increase in the moment of inertia I of an astrophysical system under consideration. It is no wonder that the kinetic energy T or the thermal energy U_{th} tends to expand the system. The effect of magnetic field is more subtle. The magnetic field has tension along field lines and magnetic pressure. So we expect the overall average effect to be expansive. On the other hand, a negative term on the right-hand side, which is the gravitational energy Ω, tries to make the system more compact. Gravity is the only force which introduces a confining tendency in the system.

By way of illustration, let us consider some consequences of the virial theorem for the case of a *steady* system, i.e. when gravity balances the expansive forces so that

$$\frac{d^2 I}{dt^2} = 0\,. \tag{19.18}$$

Moreover let the kinetic energy of macroscopic motions be equal to zero

$$T = 0\,, \tag{19.19}$$

i.e. the system is in *static* equilibrium. Both assumptions must be justified carefully, if they are applied to astrophysical plasma.

Let us suppose also that the system is finite and the surface S, over which the integration in (19.10) and (19.12) is performed, can be moved sufficiently far away (formally speaking, to infinity), so that

$$\oint_S p\,(\mathbf{r}\cdot d\mathbf{S}) = 0 \tag{19.20}$$

and

$$\oint_S \left[\frac{B^2}{8\pi}\,(\mathbf{r}\cdot d\mathbf{S}) - \frac{1}{4\pi}\,(\mathbf{B}\cdot\mathbf{r})\,(\mathbf{B}\cdot d\mathbf{S}) \right] = 0\,. \tag{19.21}$$

Then from the virial theorem (19.17) it follows that

$$3\,(\gamma_g - 1)\,U_{th} + \mathcal{M} + \Omega = 0\,. \tag{19.22}$$

Introduce the 'total' (without what has been neglected) energy of the system

$$\mathcal{E} = U_{th} + \mathcal{M} + \Omega\,. \tag{19.23}$$

Eliminating the thermal energy U_{th} from Equations (19.22) and (19.23), the total energy is expressed as follows

$$\boxed{\mathcal{E} = -\frac{(3\gamma_g - 4)}{3\,(\gamma_g - 1)}\,(\,|\Omega| - \mathcal{M}\,)\,.} \tag{19.24}$$

In a sense, the equilibrium is stable if $\mathcal{E} < 0$, i.e.

$$\frac{(3\gamma_g - 4)}{3\,(\gamma_g - 1)}\,(\,|\,\Omega\,| - \mathcal{M}\,) > 0\,, \tag{19.25}$$

which is equivalent, once $\gamma_g > 4/3$, to

$$|\,\Omega\,| > \mathcal{M}\,. \tag{19.26}$$

It is self-evident that inequality (19.26) is just a *necessary* condition for the dynamical *global* stability of a system. The condition is by no means sufficient. It can be used to show a non-existence of equilibrium of the system.

Let us consider **two particular cases of astrophysical interest**.

(a) If $\mathcal{M} = 0$ then the system can be stable only for $\gamma_g > 4/3$. This condition is easy to understand. The pressure inside the system under adiabatic compression ($p \sim \rho^{\gamma_g}$) must grow faster than the gravitational pressure $p_g \sim \rho^{4/3}$. It is in this case that the system, for instance a star, can be sufficiently resilient to resist the gravitational collapse. That is why a star consisting of a monatomic gas (with $\gamma_g = 5/3$) can be dynamically stable.

(b) Let $\mathcal{M} > 0$. Generally, the necessary condition for stability (19.25) can be, in principle, violated. What this means is that the field diminishes the stability of a star. Given a sufficiently strong field, gravitational attraction forces cannot balance the magnetic repulsion of the constituents of the system. However, such a situation is difficult to conceive.

In actuality, **gravitational compression** cannot result in $\mathcal{M} > |\,\Omega\,|$ since, given the freezing-in condition and isotropic compression, $p_{\rm mag} \sim \rho^{4/3}$ in common with $p_g \sim \rho^{4/3}$. It is also impossible to obtain $\mathcal{M} > |\,\Omega\,|$ by dint of magnetic field amplification owing to **differential rotation**, since $|\,\Omega\,| > 2T$ in a gravitationally bound system. On the other hand, the energy of a magnetic field generated by differential rotation must remain less than the kinetic energy T of the rotation motion, i.e. $\mathcal{M} < T$. Hence $\mathcal{M} < |\,\Omega\,|$.

At most, the condition $\mathcal{M} \sim |\,\Omega\,|$ can be realized. This situation is probably realized in stars of the cold giant type with a large radius. Perhaps such stars are at the limit of stability, which reveals itself as non-steady behaviour.

Condition (19.26) allows us to evaluate the upper limit of the mean intensity of a magnetic field inside a star or other equilibrium configuration. Substitute the gravitational energy of a uniform ball,

$$\Omega = -\frac{3}{5}\,\frac{GM^2}{R}\,, \tag{19.27}$$

in (19.26). The result is (Syrovatskii, 1957)

$$B < B_{cr} = 2 \times 10^8 \left(\frac{M}{M_\odot}\right)\left(\frac{R}{R_\odot}\right)^{-2}. \tag{19.28}$$

For the Sun, magnetic field B must be less than 2×10^8 Gauss. For the most magnetic stars of the spectral class A, which are observed to have fields $\sim 10^4$

Gauss, the condition $B < 3 \times 10^7$ Gauss must hold. Hence these magnetic stars called the Ap stars, because they possess some peculiar properties (e.g., Hubrig et al., 2000), still are very far from the stability limit. As is seen from the Syrovatskii condition (19.28), the cold giants with large radii could be closer to such a limit.

Given a uniform field inside a star, on approaching the limit established by (19.28), the form of the star increasingly deviates from a sphere:

> the magnetic field resists gravitational compression of a collapsing star in the direction perpendicular to the field, whereas the plasma may freely flow along the field lines.

As a result, the equilibrium configuration is represented by a rotation ellipsoid compressed in the field direction. The virial theorem can be written (e.g., Nakano, 1998) for an axisymmetric oblate magnetic cloud of mass M and semimajor axes $a_{\perp}$ and $a_{\|}$, respectively, embedded in a medium of pressure p_s. This is typical for the problem of star formation in magnetic clouds.

The action of a magnetic field is analogous to rotation (Strittmatter, 1966). Furthermore, both the **strong field and fast rotation are typical of pulsars**, especially of the magnetars (see Exercise 14.2). So both these factors determine the real flattening of a neutron star. The flattening can be calculated using the tensor virial theorem. Note, however, that for a neutron star with $M \sim M_{\odot}$ and $R \sim 10$ km the critical magnetic field (19.28) is still unprecedentelly high: $B_{cr} \sim 10^{18}$ G. We call such fields *ultrastrong*.

Magnetars, or 'magnetically powered neutron stars', could form via a magnetic dynamo action in hot, nascent neutron stars if they are born spinning rapidly enough. Magnetism may be strong enough within these stars to evolve diffusively, driving internal heat dissipation that would keep the neutron stars hot and X-ray bright. Above a field strength of $\sim 10^{14}$ G, the evolving field inevitably induces stresses in the solid crust. Observations (e.g., Feroci et al., 2001) indicate that giant flares, involved a relativistic outflow of pairs and hard gamma rays, can plausibly be triggered by a large fracture in the crust of a neutron star with a field exceeding 10^{14} G. So the observed giant flares are presumably due to *local* magnetic instabilities in magnetars.

On the other hand, numerical studies (Bocquet et al., 1995) have confirmed that neutron stars with the ultrastrong internal magnetic fields are *globally* stable up to the order of 10^{18} G. They also have found that, for such values, the maximum mass of neutron stars increases by 13–29 % relative to the maximum mass of non-magnetized neutron stars.

If ultrastrong fields exist in the interior of neutron stars, such fields will primarily affect the behavior of the residual charged particle. Moreover, contributions from the anomalous magnetic moment of the particles in a magnetic field should also be significant (Broderic et al., 2000). In particular, in a ultrastrong field, complete spin polarization of the neutrons occurs as a result of the interaction of the neutron magnetic moment with the magnetic field. The presence of a sufficiently strong field changes the ratio of protons to neutrons as well as the neutron drip density (Suh and Mathews, 2001).

The virial theorem is sometimes applied in solar physics, for example, while studying active regions (Section 19.5). It allows us to evaluate the energy of equilibrium electric currents and show that the energy can be large enough to explain the flaring activity (Litvinenko and Somov, 1991a); see also discussion of the problem of the global MHD equilibria and filament eruptions in the solar corona (Litvinenko and Somov, 2001).

19.2 Force-free fields and Shafranov's theorem

19.2.1 The simplest examples of force-free fields

A particular case of equilibrium configurations of astrophysical plasma in a magnetic field is the *force-free field*, i.e. the field which does not require external forces. As was noted in Section 13.1, force-free fields naturally occur when the magnetic force dominates all the others, and hence the magnetic field must balance itself

$$\mathbf{B} \times \operatorname{curl} \mathbf{B} = 0 . \tag{19.29}$$

Let us consider several examples of such equilibrium configurations.

19.2.1 (a) The Syrovatskii force-free field

Let the magnetic field vector be situated in the plane parallel to the plane (x, y), but depend only on z

$$\mathbf{B} = \{ \, B_x(z), \; B_y(z), \; 0 \, \} \, . \tag{19.30}$$

Substitute (19.30) in Equation (19.29):

$$\operatorname{curl} \mathbf{B} = \left\{ -\frac{\partial B_y}{\partial z} \, , \; \frac{\partial B_x}{\partial z} \, , \; 0 \right\} , \tag{19.31}$$

$$\mathbf{B} \times \operatorname{curl} \mathbf{B} = \left\{ 0 \, , \; 0 \, , \; B_x \frac{\partial B_x}{\partial z} + B_y \frac{\partial B_y}{\partial z} \right\} = 0 \, . \tag{19.32}$$

The resulting equation is

$$\frac{\partial}{\partial z} \left(B_x^2 + B_y^2 \right) = 0 \, , \tag{19.33}$$

with the solution

$$B^{\,2} = B_x^{\,2} + B_y^{\,2} = \text{const} \, . \tag{19.34}$$

This is the simplest example of a force-free field. The magnitude of the field vector is independent of z. A one-dimensional force-free field of this type may be considered to be a *local* approximation of an arbitrary force-free field in a region of the magnetic 'shear' in the solar atmosphere. As a particular example, suitable for formal analysis, one may adopt the force-free field of the type

$$\mathbf{B} = \{ B_0 \cos kz \, , \, B_0 \sin kz \, , \, 0 \} \tag{19.35}$$

(Bobrova and Syrovatskii, 1979). The field lines, and hence the electric current, lie in the plane (x, y). The direction of the lines rotates with increasing z.

19.2.1 (b) The Lundquist force-free field

The magnetic field of a direct current flowing along the z axis tends to compress the plasma to the axis, owing to the tension of the field lines (see Section 13.1.3). By contrast,

> a bundle of parallel field lines tends to expand by the action of the magnetic pressure gradient.

Given the superposition of these fields for a certain relationship between them, the total magnetic force can be zero. Field lines for such a force-free field have the shape of spirals shown in Figure 19.1.

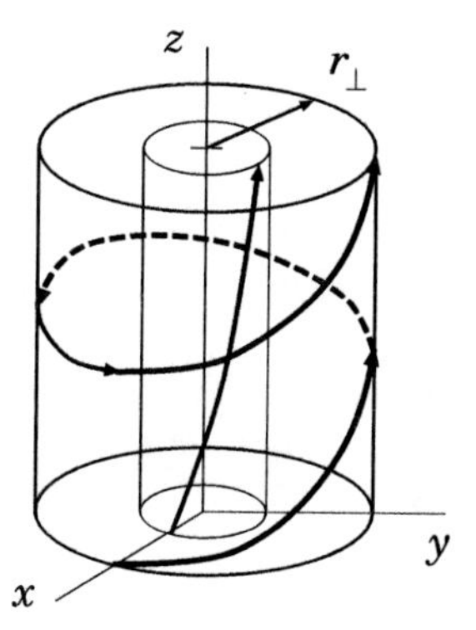

Figure 19.1: A helical magnetic field in the form of a spiral of constant slope on a cylindrical surface $r_\perp$ = const.

The corresponding *axially symmetric* solution to Equation (19.29) in cylindrical coordinates $r_\perp$, ϕ, z is of the form (Lundquist, 1951):

$$B_z = A \, J_0 \, (\alpha \, r_\perp) \, , \qquad B_\phi = A \, J_1 \, (\alpha \, r_\perp) \, , \qquad B_r = 0 \, . \tag{19.36}$$

Here J_0 and J_1 are the Bessel functions, A and α are constants.

A distinguishing feature of the field is that $B^2 \sim r_\perp^{-1}$ for large $r_\perp$ since Bessel functions $J_n \sim r_\perp^{-1/2}$ as $r_\perp \to \infty$ $(n = 0, 1)$. The magnetic energy

$$\mathcal{M} = \int \frac{B^2}{8\pi} \, dV \sim r_\perp^{-1} \, r_\perp^2 \sim r_\perp \tag{19.37}$$

diverges for large $r_\perp$. Such a **divergence of magnetic energy** is known to be typical of force-free fields and will be explained below.

19.2.2 The energy of a force-free field

Let us retain only magnetic terms in the virial theorem; we have

$$\mathcal{M} - \oint\limits_S \left[\frac{B^2}{8\pi} (\mathbf{r} \cdot d\mathbf{S}) - \frac{1}{4\pi} (\mathbf{B} \cdot \mathbf{r}) (\mathbf{B} \cdot d\mathbf{S}) \right] = 0 . \tag{19.38}$$

Provided the electric currents occupy a *finite* region, the value of the magnetic field is proportional to r^{-3} (or higher degrees of r^{-1}). Once the surface of integration S is expanded to infinity, the surface integral tends to zero. Equality (19.38) becomes impossible.

Therefore any finite magnetic field cannot contain itself. There must be *external* forces to balance the outwardly directed pressure due to the *total magnetic* energy $\mathcal{M}$.

The same statement may be formulated as follows. **The force-free field cannot be created in the whole space**. This is the so-called *Shafranov theorem* (Shafranov, 1966). While stresses may be eliminated in a given region V, they cannot be canceled everywhere. In general

> a force-free configuration requires the forces needed to balance the outward pressure of the magnetic field to be reduced in magnitude by spreading them out over the bounding surface S.

In this way, the virial theorem sets limits on the space volume V that can be force-free.

The Shafranov theorem is the counterpart of the known Irnshow theorem (see Sivukhin, 1996, Chapter 1, § 9) concerning the equilibrium configuration of a system of electric charges. Such a configuration also can be stable only in the case that some external forces, other than the electric ones, act in the system.

In fact, Shafranov (1966) has proved a stronger statement than the above theorem on the force-free field. He has taken into account not only the terms corresponding to the magnetic force in (19.17) but the gas pressure as well:

$$\int\limits_V \left(3p + \frac{B^2}{8\pi} \right) dV =$$

$$= \oint\limits_S \left[\left(p + \frac{B^2}{8\pi} \right) (\mathbf{r} \cdot d\mathbf{S}) - \frac{1}{4\pi} (\mathbf{B} \cdot \mathbf{r}) (\mathbf{B} \cdot d\mathbf{S}) \right] . \tag{19.39}$$

If the plasma occupies some finite volume V, the pressure outside of this volume being zero, and if electric currents occupy a finite region, then the surface integral tends to zero, once the surface of integration, S, is expanded to infinity. On the other hand, the expression under the integral sign on the left-hand side is always positive. Hence the integral is positive. Thus the equality (19.39) turns out to be impossible. Therefore any finite equilibrium configuration of a plasma with a magnetic field can exist only in the presence of external forces which, *apart from the gas pressure*, serve to fix the electric currents.

In a laboratory, fixed current conductors must be present. In this case the right-hand side of (19.39) is reduced to the integral over the surface of the conductors.

Under astrophysical conditions, the role of the external force is frequently played by the gravitational force or by an external magnetic field having its sources outside the volume under investigation. However these sources must be kept and driven by non-magnetic forces.

A typical example of such a situation is the magnetic field of an active region on the Sun. This is the sum of the proper field created by currents flowing inside the active region, and the external field with the sources situated (and fixed) below the photosphere (Litvinenko and Somov, 1991a). In this case the formula for the magnetic energy of the equilibrium system contains a term due to the interaction of internal currents (in particular current sheets in the regions of reconnection) with the external magnetic field.

19.3 Properties of equilibrium configurations

19.3.1 Magnetic surfaces

Let us consider the case of *magnetostatic* equilibrium

$$-\nabla p + \frac{1}{4\pi}\,\mathrm{curl}\,\mathbf{B}\times\mathbf{B} - \rho\,\nabla\phi = 0\,. \tag{19.40}$$

The gravitational force is supposed to be negligible

$$\rho\,\nabla\phi = 0\,, \tag{19.41}$$

On dropping the third term in Equation (19.40) and taking the scalar product with vector $\mathbf{B}$ we obtain

$$\boxed{\mathbf{B}\cdot\nabla p = 0\,,} \tag{19.42}$$

i.e. magnetic field lines in an equilibrium configuration are situated on the surface $p =$ const. Therefore

in order to contain a plasma by the magnetic field, the field lines are forbidden to leave the volume occupied by the plasma.

There is a common viewpoint that, by virtue of the condition

$$\mathrm{div}\,\mathbf{B} = 0\,, \tag{19.43}$$

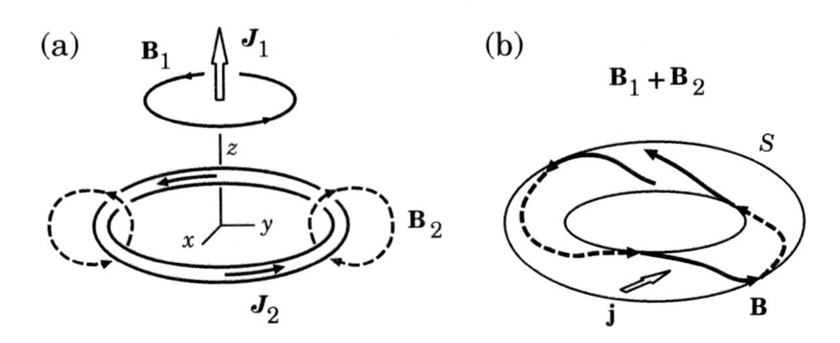

Figure 19.2: (a) A line current J_1 and a ring current J_2. (b) The field lines of the total field $\mathbf{B}_1 + \mathbf{B}_2$ form a toroidal surface S.

field lines may either close or go to infinity. However the other variant is possible, when a field line fills up an entire surface – *magnetic surface*.

Let us consider the field of two electric currents – a line current J_1 flowing along the vertical z axis (Figure 19.2a) and a plane current ring J_2 (see Tamm, 1989, Chapter 4, § 53). If there were only the current J_1, the field lines of this current $\mathbf{B}_1$ would constitute circumferences centred at the z axis. The field lines $\mathbf{B}_2$ of the ring current J_2 lie in meridional planes. The total field $\mathbf{B} = \mathbf{B}_1 + \mathbf{B}_2$ forms a helical line on a toroidal surface S. The course of this spiral depends on the ratio B_1/B_2. Once this is a rational number, the spiral will close. However, in general, it does not close but continuously fills up the entire toroidal surface S (Figure 19.2b).

By virtue of condition (19.42), the plasma pressure at such a surface (called the magnetic one) is constant. Such a magnetic field can serve as a trap for the plasma. This fact constitutes the basis for constructing laboratory devices for plasma containment in stellarators, suggested by Spitzer.

Take the scalar product of Equation (19.40), without the gravitational force, with the electric current vector

$$\mathbf{j} = \frac{c}{4\pi}\,\text{curl}\,\mathbf{B}\,. \tag{19.44}$$

The result is

$$\boxed{\mathbf{j}\cdot\nabla p = 0} \tag{19.45}$$

which, in combination with (19.42), signifies that, in an equilibrium configuration, the electric current flows on magnetic surfaces (Figure 19.2b).

In general, magnetic fields do not form magnetic surfaces. Such surfaces arise in magnetohydrostatic equilibria and for some highly symmetric field configurations. In the case of the latter, Equations (14.19) for the magnetic field lines admit an exact integral which is the equation for the magnetic surface.

19.3.2 The specific volume of a magnetic tube

Let us consider two closed magnetic surfaces: $p = \text{const}$ and $p + dp = \text{const}$. Construct a system of noncrossing partitions between them (Figure 19.3). Let $d\mathbf{l}_1$ be the line element directed normally to the surface $p = \text{const}$:

$$d\mathbf{l}_1 = \frac{\nabla p}{|\nabla p|^2}\, dp\,. \tag{19.46}$$

The vectors $d\mathbf{l}_2$ and $d\mathbf{l}_3$ are directed along the two independent contours l_2 and l_3 which may be drawn on a toroidal surface: for example the curve l_2 is directed along a large circle of the toroid while l_3 lies along the small one. The surface element of this partition is

$$d\mathbf{S}_3 = d\mathbf{l}_1 \times d\mathbf{l}_2\,. \tag{19.47}$$

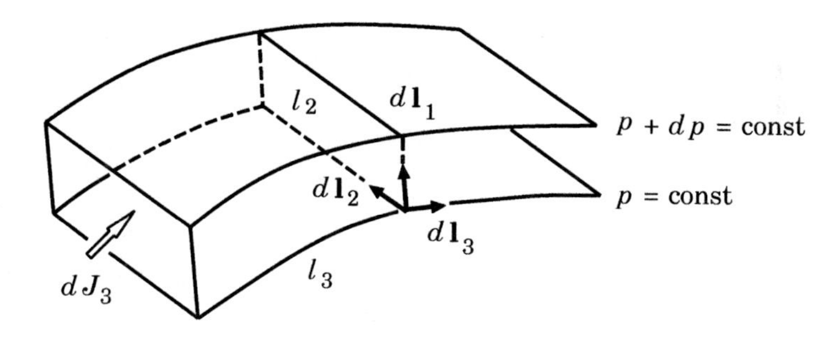

Figure 19.3: The calculation of the electric current between two magnetic surfaces.

The total current dJ_3 flowing through the partition situated on the contour l_2 is

$$dJ_3 = \oint\limits_{l_2} \mathbf{j} \cdot (d\mathbf{l}_1 \times d\mathbf{l}_2)\,. \tag{19.48}$$

According to Equation (19.45), the total current flowing through the system of noncrossing partitions between the two magnetic surfaces is constant. In other words, dJ_3 is independent of the choice of the integration contour. We are concerned with the physical consequences of this fact.

In order to find the expression for the current density $\mathbf{j}$ in an equilibrium configuration, take the vector product of Equation (19.40) with the magnetic field $\mathbf{B}$. The result is

$$\mathbf{B} \times \nabla p = \frac{1}{c}\, \mathbf{B} \times (\mathbf{j} \times \mathbf{B})\,,$$

which, on applying the formula for a double vector product to the right-hand side, becomes

$$c\mathbf{B} \times \nabla p = \mathbf{j} B^2 - \mathbf{B}\,(\mathbf{j} \times \mathbf{B})\,.$$

Thus we have

$$\mathbf{j} = c\,\frac{\mathbf{B}\times\nabla p}{B^2} + f\,\mathbf{B}\,, \tag{19.49}$$

where $f = f(\mathbf{r})$ is an arbitrary function. If need be, it can be found from the condition $\operatorname{div}\mathbf{j} = 0$.

Substitute (19.49) in the integral (19.48). The last takes the following form (see Exercise 19.4):

$$dJ_3 = -\,c\,dp \oint\limits_{l_2} \frac{\mathbf{B}\cdot d\mathbf{l}_2}{B^2} + \oint\limits_{l_2} f(\mathbf{r})\,\mathbf{B}\cdot(d\mathbf{l}_1\times d\mathbf{l}_2)\,. \tag{19.50}$$

Provided the contour l_2 coincides with a closed field line, the vector

$$d\mathbf{l}_2 = \frac{\mathbf{B}}{B}\,dl\,,$$

and, therefore, the second term on the right-hand side of Formula (19.50) vanishes.

Once a magnetic field line closes on making one circuit of the toroid, the expression

$$dJ_{\mathrm{n}} = -c\,dp \oint \frac{dl}{B} \tag{19.51}$$

defines the total current flowing between neighbouring magnetic surfaces normal (the subscript n) to the field line. Since the magnitude of this current is independent of the choice of contour, for each field line on a magnetic surface the integral

$$\boxed{U = \oint \frac{dl}{B}} \tag{19.52}$$

is constant. The condition of constancy of U can be generalized to include the surface with unclosed field lines (Shafranov, 1966). Thus (Kadomtsev, 1966),

> under the condition of magnetostatic equilibrium, the magnetic surface consists of the field lines with the same value of U.

Let us introduce the notion of the *specific volume* of a magnetic tube (Rosenbluth and Longmire, 1957) or simply the *specific magnetic volume* as the ratio of its geometric volume dV to the magnetic flux $d\Phi$ through the tube. If dS_{n} is the cross-sectional surface of the tube, its geometric volume is

$$dV = \oint dS_{\mathrm{n}}\,dl$$

whereas the magnetic flux

$$d\Phi = B\,dS_{\mathrm{n}}\,.$$

On the basis of the magnetic flux constancy inside the tube of field lines, i.e. $d\,\Phi = \text{const}$, we deduce that

$$\frac{dV}{d\,\Phi} = \oint \frac{dS_{\rm n}}{B\,dS_{\rm n}}\,dl = \oint \frac{dl}{B} = U. \tag{19.53}$$

The stability of an equilibrium MHD configuration can be judged by the condition (19.52). This property will be discussed in the next Section.

19.3.3 The flute or convective instability

Much like any gas with a finite temperature, the plasma in a magnetic field tends to expand. However, given a high conductivity, it cannot move independently of the magnetic field. The plasma moves together with the field lines in such a way that it travels to a region of the field characterized by a greater specific volume.

In order for an equilibrium configuration to be stable with respect to a given perturbation type – deformation of a tube of magnetic field lines – the following condition is necessary (Rosenbluth and Longmire, 1957):

$$\delta U = \delta \oint \frac{dl}{B} < 0\,. \tag{19.54}$$

To put it another way,

> the magnetostatic equilibrium is stable once the given type of deformation does not facilitate the plasma spreading,

i.e. increasing its specific volume.

As an example, let us consider the plasma in the magnetic field of a linear current J:

$$B_{\varphi} = \frac{2J}{cr}\,, \tag{19.55}$$

here r, z, φ are cylindrical coordinates. In such a field there exists an equilibrium plasma configuration in the form of an infinite hollow cylinder C as shown in Figure 19.4a.

Let us calculate the specific volume for such a configuration. The geometric volume of the tube of field lines is

$$dV = 2\pi r\,dr\,dz\,,$$

whereas the magnetic flux

$$d\,\Phi = B_{\varphi}\,dr\,dz = \frac{2J}{cr}\,dr\,dz\,.$$

Hence the specific volume

$$U = \frac{dV}{d\,\Phi} = \frac{\pi c}{J}\,r^2. \tag{19.56}$$

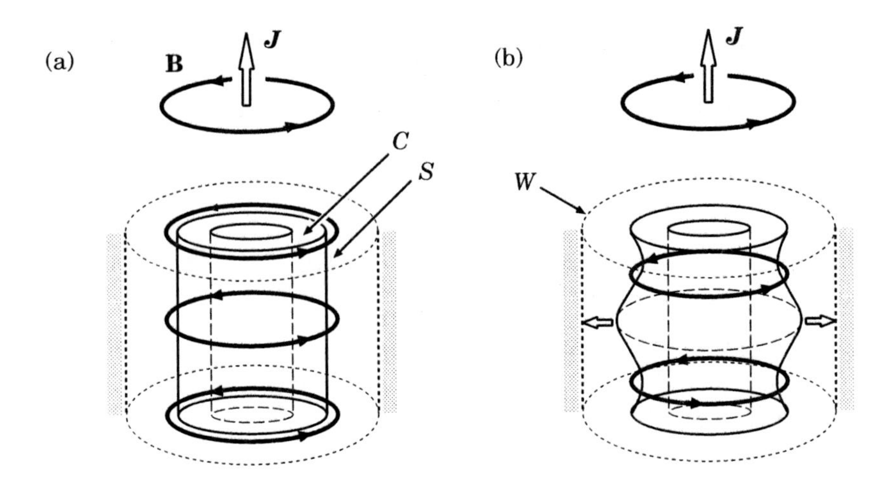

Figure 19.4: (a) An equilibrium plasma configuration. (b) Unstable perturbations of the outer boundary.

It is seen from (19.56) that the specific volume grows with the radius. In particular, for small perturbations δr of the external surface S of the plasma cylinder C

$$\delta U = \frac{2\pi c}{J}\, r\, \delta r > 0$$

once $\delta r > 0$. It is sufficient to have a small perturbation of the external boundary of the plasma to obtain ring flutes which will rapidly grow towards the wall W of the chamber as shown in Figure 19.4b.

19.3.4 Stability of an equilibrium configuration

The problems of plasma **equilibrium and stability** are of great value for plasma astrophysics as a whole (Zel'dovich and Novikov, 1971; Chandrasekhar, 1981), and especially for solar physics (Parker, 1979; Priest, 1982).

> The Sun seems to maintain stability of solar prominences and coronal loops with great ease

(Tandberg-Hanssen, 1995; Acton, 1996) in contrast to the immense difficulty of containing plasmas in a laboratory.

Therefore, sometimes, we need to explain how an equilibrium can remain stable for a very long time. This is, for example, the case of reconnecting current layers (RCLs) in the solar atmosphere and the geomagnetic tail (see vol. 2, Sections 8.2 and 11.6.3). At other times, we want to understand

> why magnetic structures on the Sun suddenly become unstable and produce dynamic events

of great beauty such as eruptive prominences and solar flares, coronal transients, and coronal mass ejections (CMEs).

The methods employed to investigate the stability of an equilibrium MHD system are natural generalizations of those for studying a particle in one-dimensional motion. One approach is to seek normal mode solutions as we did it in Chapter 15.

An alternative approach for tackling stability is to consider the change in potential energy due to a displacement from equilibrium. The main property of a stable equilibrium is that it is at the minimum of the potential energy. So any perturbations around the equilibrium ought to increase the total potential energy. Hence, in order to determine if an equilibrium is stable, one finds out if all types of perturbations increase the potential energy of the system (Bernstein et al., 1958).

Recommended Reading: Morozov and Solov'ev (1966a), Kadomtsev (1960, 1966), Shu (1992).

19.4 The Archimedean force in MHD

19.4.1 A general formulation of the problem

Now we return to the equation of magnetostatic equilibrium (19.40). Let us rewrite it as follows:

$$\nabla p = \rho \, \mathbf{g} + \mathbf{f} \, , \tag{19.57}$$

where

$$\mathbf{f} = \frac{1}{c} \, \mathbf{j} \times \mathbf{B} \tag{19.58}$$

is the Lorentz force, $\mathbf{g} = -g \, \mathbf{e}_z$ is the gravity acceleration.

We begin by considering an incompressible conducting fluid situated in a uniform magnetic field $\mathbf{B}_0$ and electric field $\mathbf{E}_0$ as illustrated by Figure 19.5. Provided the current $\mathbf{j}_0$ flowing in the fluid is *uniform*, the Lorentz force created is uniform as well:

$$\mathbf{f}_0 = \frac{1}{c} \, \mathbf{j}_0 \times \mathbf{B}_0 \, . \tag{19.59}$$

By virtue of Equation (19.57), **the Lorentz force makes the fluid heavier or lighter**. In both cases the uniform volume force is potential and, much like the gravity force, will be balanced by an additional pressure gradient appearing in the fluid. As will be shown later, that allows the creation of a regulated *expulsion* force (Figure 19.5) analogous to the Archimedean force in ordinary hydrodynamics.

A body plunged into the fluid is acted upon by the force

$$\mathbf{F} = \int\limits_V (\rho_1 \, \mathbf{g} + \mathbf{f}_1) \, dV + \oint\limits_S p_0 \, \mathbf{n} \, dS \, . \tag{19.60}$$

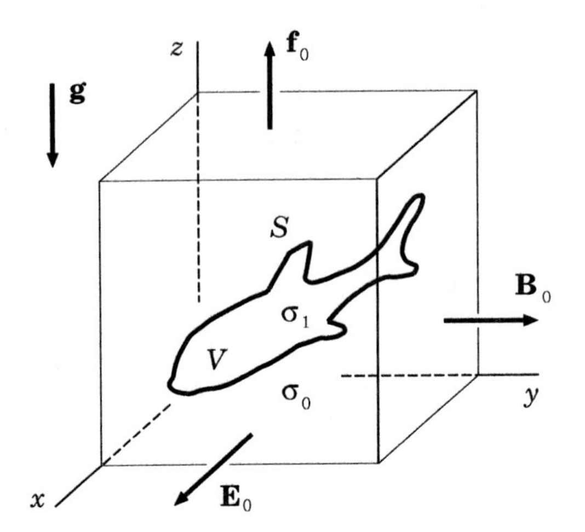

Figure 19.5: Formulation of the problem concerning the Archimedean force in magnetohydrodynamics (see Somov, 1994b).

Here ρ_1 is the density of the submerged body, which is generally not equal to that of the fluid ρ_0;

$$\mathbf{f}_1 = \frac{1}{c}\, \mathbf{j}_1 \times \mathbf{B}_0 \tag{19.61}$$

is the volume Lorentz force, $\mathbf{j}_1$ is the current inside the body, $\mathbf{n}$ is the inward normal to the surface S, and p_0 is the pressure on the body from the fluid, resulted from (19.57):

$$\nabla p_0 = \rho_0\, \mathbf{g} + \mathbf{f}_0\, . \tag{19.62}$$

19.4.2 A simplified consideration of the effect

If the current $\mathbf{j}_0$ was uniform, the right-hand side of Equation (19.62) would be a uniform force, and formula (19.60) could be rewritten as

$$\mathbf{F} = \int\limits_V (\, \rho_1\, \mathbf{g} + \mathbf{f}_1\,)\, dV - \int\limits_V \nabla p_0\, dV \tag{19.63}$$

or

$$\boxed{\mathbf{F} = \int\limits_V (\rho_1 - \rho_0)\, \mathbf{g}\, dV + \frac{1}{c} \int\limits_V (\mathbf{j}_1 - \mathbf{j}_0\,) \times \mathbf{B}_0\, dV.} \tag{19.64}$$

The first term in (19.64) corresponds to the usual Archimedean force in hydrodynamics. It equals zero once $\rho_1 = \rho_0$. When $\rho_1 > \rho_0$, the direction of this force coinsides with the gravitational acceleration $\mathbf{g}$. The second term describes the *magnetic expulsion* force. It vanishes once $\mathbf{j}_1 = \mathbf{j}_0$, i.e. $\sigma_1 = \sigma_0$.

The second term in formula (19.64) shows that the magnetic expulsion force, different from the known Parker's magnetic buoyancy force (see Chapter 8 in Parker, 1979) by its origin, appears provided $\sigma_1 \neq \sigma_0$. This fact has been used to construct, for example, MHD devices for the separation of mechanical mixtures. In what follows we shall call the second term in (19.64) the magnetic *σ-dependent* force:

$$\mathbf{F}_\sigma = \frac{1}{c} \int_V (\sigma_1 - \sigma_0)\, \mathbf{E}_0 \times \mathbf{B}_0 \, dV. \tag{19.65}$$

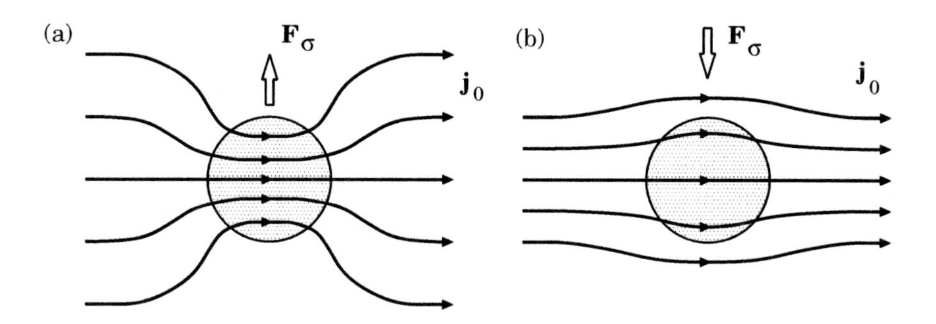

Figure 19.6: Opposite orientation of the σ-dependent force in two opposite cases: (a) $\sigma_1 > \sigma_0$ and (b) $\sigma_1 < \sigma_0$. Appearence of a non-uniform distribution of electric current is shown.

Note, however, that the simplest formula (19.64) is of purely illustrative value since **the electric field and current density are not uniform in the presence of a body** with conductivity σ_1 which is not equal to that of the fluid σ_0 (Figure 19.6). In this case, the appearing *σ-dependent* force is generally not potential. Hence it cannot be balanced by potential forces. That is the reason why

> the magnetic σ-dependent force generates MHD vortex flows of the conducting fluid.

The general analysis of the corresponding MHD problem was made by Andres et al. (1963). The stationary solutions for a ball and a cylinder were obtained by Syrovatskii and Chesalin (1963) for the specific case when both the magnetic and usual Reynolds numbers are small; similar stationary solutions for a cylinder see also in Marty and Alemany (1983), Gerbeth et al. (1990). The character of the MHD vortex flows and the forces acting on submerged bodies will be analyzed in Sections 20.3 and 20.4.

19.5 MHD equilibrium in the solar atmosphere

The magnetic configuration in an active region in the solar atmosphere is, in general, very complex and modelling of dynamical processes in these regions requires a high degree of idealization. First, as regards the most powerful and fascinating of these processes, the two-ribbon flare, the typical preflare magnetic field distribution seems to conform to a certain standard picture: a magnetic arcade including a more or less pronounced plage filament, prominence. Second, instead of dynamics, models deal with a static or steady-state equilibrium in order to understand the causes of a flare or another transient activity in the solar atmosphere as a result of some instability or lack of equilibrium.

So it is assumed that initially the configuration of prominence and overlying arcade is in equilibrium but later the eruption takes place.

> Either the MHD equilibrium of solar plasma has become unstable or the equilibrium has been lost.

One limiting possibility is that the magnetic field around the prominence evolves into an unstable or non-equilibrium configuration and then drives the overlying magnetic arcade. However observations imply that this is unlikely. An alternative is that the overlying arcade evolves until it is no longer in stable equilibrium and then its eruption stimulates the prominence to erupt by removing stabilising field lines. Presumably this is the case of a coronal loop transient and coronal mass ejection (CME).

The idealized models used in theoretical and numerical studies of this problem usually consider two-dimensional force-free arcade configurations with foot points anchored in the photosphere which are energized, for example, by photospheric shear flows in the direction along the arcade (see Biskamp and Welter, 1989). Some other models take into account the gas pressure gradient and the gravitational force (Webb, 1986).

However it is important to investigate more general circumstances when equilibrium and non-equilibrium may occur. The electromagnetic expulsion force – a MHD analogue of the usual Archimedean force – plays an important part in the dynamics of coronal plasma with a non-uniform distribution of temperature and, hence, electric conductivity. More exactly, the condensation mode of the radiatively-driven thermal instability in an active region may result in the formation of cold dense loops or filaments surrounded by hot rarified plasma (see Somov, 1992). The effect results from the great difference of electric conductivities outside and inside the filaments. The force can generate vortex flows (see Section 20.4) inside and in the vicinity of the filaments as well as initiate the non-equilibrium responsible for transient activities: flares, CMEs etc.

The virial theorem confirms this possibility and clarifies the role of preflare reconnecting current sheets in MHD equilibrium and non-equilibrium of an active region. Correct use of the virial theorem confirms the applicability of reconnection in current sheets for explaining the energetics of flares (Litvinenko and Somov, 1991a, 2001) and other non-steady phenomena in the solar atmosphere.

19.6 Practice: Exercises and Answers

Exercise 19.1. Show that, apart from the trivial case of a potential field, the magnetic fields for which

$$\text{curl}\, \mathbf{B} = \alpha \mathbf{B} \tag{19.66}$$

will be force-free. In the most general case, α will be spatially dependent.

Answer. Just substitute formula (19.66) in Equation (19.29).

Exercise 19.2. Show that the force-free fields with α = const represent the state of *minimal* magnetic energy in a closed system (Woltjer, 1958).

Hint. First, assume perfect conductivity and rewrite the freezing-in equation (12.71) by using $\mathbf{B} = \text{curl}\, \mathbf{A}$ as follows

$$\frac{\partial \mathbf{A}}{\partial t} = \mathbf{v} \times (\nabla \times \mathbf{A})\,. \tag{19.67}$$

Here $\mathbf{A}$ is the vector potential. Using Equation (19.67), show that

$$\boxed{\mathcal{H} = \int_V \mathbf{A} \cdot (\nabla \times \mathbf{A})\, dV = \text{const}} \tag{19.68}$$

for all $\mathbf{A}$ which are constant on the boundary S of the region V. The integral $\mathcal{H}$ is called the *global magnetic helicity* of the closed system under consideration (for more detail see vol. 2, Section 12.1.1).

Second, examine the stationary values of the magnetic energy

$$\mathcal{M} = \int_V \frac{B^2}{8\pi}\, dV = \int_V \frac{1}{8\pi}\, (\text{curl}\, \mathbf{A})^2\, dV. \tag{19.69}$$

Introduce a Lagrangian multiplier $\alpha/8\pi$ and obtain the following condition for stationary values

$$\delta \int_V \Big[(\text{curl}\, \mathbf{A})^2 - \alpha\, \mathbf{A} \cdot \text{curl}\, \mathbf{A} \Big]\, dV = 0\,. \tag{19.70}$$

Performing the variation, Equation (19.66) follows with α = const. Such fields are called *linear* force-free fields.

Exercise 19.3. The highly-conductive plasma in the solar corona can support an electric field $\mathbf{E}_{\|}$ if $E_{\|} \ll E_{\rm Dr}$ where $E_{\rm Dr}$ is the Dreicer field (8.70). In the corona $E_{\rm Dr} \approx 7 \times 10^{-6}\ \text{V cm}^{-1}$ (Exercise 8.4). Evaluate the characteristic values of the magnetic field B and the velocity v of plasma motions in the corona which allow us to consider an equilibrium of moving plasma in the corona as a force-free one.

Answer. Let us evaluate an electric field as the electric field related to a motion of magnetic field lines in the corona

$$E_{\parallel} \approx E \approx \frac{1}{c}\, vB \approx 10^{-8}\, v\,(\text{cm s}^{-1})\, B\,(\text{G})\,, \ \ \text{V cm}^{-1}\,. \tag{19.71}$$

From the condition that this field must be much smaller than the Dreicer field we find that

$$v\,(\text{cm s}^{-1})\, B\,(\text{G}) \ll 10^{8}\, E_{\text{Dr}} \approx 7 \times 10^{2}\,. \tag{19.72}$$

So, with the magnetic field in the corona $B \sim 100$ G, the plasma motion velocity must be very small: $v \ll 10$ cm s^{-1}. Hence, if the electric fields that are parallel to the magnetic field lines have the same order of magnitude as the perpendicular electric fields, the solar corona hardly can remain force-free with ordinary collisional conductivity because of the motion of magnetic field lines. The electric runaway effects (Section 8.4.2) can become important even at very slow motions of the field lines in the corona. The *minimum current corona* (see vol. 2, Sections 3.3.1 and 3.4.3) seems to be a more realistic approximation everywhere except the strongly-twisted magnetic-flux tubes.

Exercise 19.4. Derive formula (19.50) in Section 19.3.2 for the total electric current flowing through the system of noncrossing partitions between two magnetic surfaces.

Answer. Substitute the electric current density (19.49) in the integral (19.48):

$$\mathbf{j} \cdot (d\mathbf{l}_1 \times d\mathbf{l}_2) = \frac{c}{B^2}\,(\mathbf{B} \times \nabla p) \cdot (d\mathbf{l}_1 \times d\mathbf{l}_2) + f\,\mathbf{B} \cdot (d\mathbf{l}_1 \times d\mathbf{l}_2)\,. \tag{19.73}$$

Let us rearrange the first item, using the well-known Lagrange identity in vector analysis:

$$(\mathbf{a} \times \mathbf{b}) \cdot (\mathbf{c} \times \mathbf{d}) = (\mathbf{a} \cdot \mathbf{c})\,(\mathbf{b} \cdot \mathbf{d}) - (\mathbf{b} \cdot \mathbf{c})\,(\mathbf{a} \cdot \mathbf{d})\,.$$

We get

$$(\mathbf{B} \times \nabla p) \cdot (d\mathbf{l}_1 \times d\mathbf{l}_2) = (\mathbf{B} \cdot d\mathbf{l}_1)\,(\nabla p \cdot d\mathbf{l}_2) - (\mathbf{B} \cdot d\mathbf{l}_2)\,(\nabla p \cdot d\mathbf{l}_1)\,.$$

By virtue of (19.42) and (19.46),

$$(\mathbf{B} \cdot d\mathbf{l}_1) = (\mathbf{B} \cdot \nabla p)\,\frac{dp}{|\,\nabla p\,|^2} = 0\,, \qquad (\nabla p \cdot d\mathbf{l}_1) = dp\,.$$

Hence

$$(\mathbf{B} \times \nabla p) \cdot (d\mathbf{l}_1 \times d\mathbf{l}_2) = -\,(\mathbf{B} \cdot d\mathbf{l}_2)\,dp\,. \tag{19.74}$$

Substitute (19.74) in (19.73):

$$\mathbf{j} \cdot (d\mathbf{l}_1 \times d\mathbf{l}_2) = -c\,\frac{dp}{B^2}\,(\mathbf{B} \cdot d\mathbf{l}_2) + f(\mathbf{r})\,\mathbf{B} \cdot (d\mathbf{l}_1 \times d\mathbf{l}_2)\,.$$

Thus the expression (19.48) for current dJ_3 takes the form

$$dJ_3 = -\,c\,dp \oint\limits_{l_2} \frac{\mathbf{B} \cdot d\mathbf{l}_2}{B^2} + \oint\limits_{l_2} f(\mathbf{r})\,\mathbf{B} \cdot (d\mathbf{l}_1 \times d\mathbf{l}_2)\,, \tag{19.75}$$

q.e.d.

Chapter 20

Stationary Flows in a Magnetic Field

There exist two different sorts of stationary MHD flows depending on whether or not a plasma can be considered as ideal or non-ideal medium. Both cases have interesting applications in modern astrophysics.

20.1 Ideal plasma flows

Stationary motions of an ideal conducting medium in a magnetic field are subject to the following set of MHD equations (cf. (12.67)):

$$(\mathbf{v} \cdot \nabla)\,\mathbf{v} = -\frac{1}{\rho}\,\nabla\left(p + \frac{B^2}{8\pi}\right) + \frac{1}{4\pi\rho}\,(\mathbf{B} \cdot \nabla)\,\mathbf{B}\,, \tag{20.1}$$

$$\operatorname{curl}\,(\mathbf{v} \times \mathbf{B}) = 0\,, \tag{20.2}$$

$$\operatorname{div}\,\rho\mathbf{v} = 0\,, \tag{20.3}$$

$$\operatorname{div}\mathbf{B} = 0\,, \tag{20.4}$$

$$(\mathbf{v} \cdot \nabla)\,s = 0\,, \tag{20.5}$$

$$p = p\,(\rho, s)\,. \tag{20.6}$$

The induction Equation (20.2) is satisfied identically, provided the motion of the medium occurs along the magnetic field lines, i.e.

$$\boxed{\mathbf{v} \parallel \mathbf{B}\,.} \tag{20.7}$$

20.1.1 Incompressible medium

In the case of an incompressible fluid (ρ = const) Equations (20.1)–(20.6) have the general solution (Syrovatskii, 1956, 1957):

$$\mathbf{v} = \pm \frac{\mathbf{B}}{\sqrt{4\pi\rho}}, \tag{20.8}$$

$$\nabla \left(p + \frac{B^2}{8\pi} \right) = 0. \tag{20.9}$$

Here $\mathbf{B}$ is an arbitrary magnetic field: the form of the field lines is unimportant, once condition (20.4) holds. A conducting fluid flows parallel or antiparallel to the magnetic field. We shall learn more about such equilibrium flows later on.

It follows from (20.8) that

$$\frac{\rho v^2}{2} = \frac{B^2}{8\pi}, \tag{20.10}$$

while Equation (20.9) gives

$$p + \frac{B^2}{8\pi} = \text{const}. \tag{20.11}$$

For the considered class of plasma motions along the field lines, the equipartition of energy between that of the magnetic field and the kinetic energy of the medium takes place, whereas the sum of the gas pressure and the magnetic pressure is everywhere constant.

The existence of the indicated solution means that

> an arbitrary magnetic field and an ideal incompressible medium in motion are in equilibrium, provided the motion of the medium occurs with the Alfvén speed along magnetic field lines.

Stationary flows of this type can be continuous in the whole space as well as discontinuous at some surfaces. For example, the solution (20.10) and (20.11) can be realized as a stream or non-relativistic jet of an arbitrary form, flowing in an immovable medium without a magnetic field.

Note that the tangential discontinuity at the boundary of such a jet is *stable*, since, by virtue of (20.10), the condition (16.38) by Syrovatskii is valid:

$$\frac{B^2}{8\pi} > \frac{1}{4} \frac{\rho v^2}{2}. \tag{20.12}$$

Such stable stationary jets of an incompressible fluid can close in *rings* and *loops* of an arbitrary type.

20.1.2 Compressible medium

In a compressible plasma ($\rho \neq$ const) the solution (20.8) is still possible, once the density of the plasma does not change along the field lines:

$$\mathbf{B} \cdot \nabla \rho = 0 \,. \tag{20.13}$$

Obviously, this condition is necessary, but not sufficient. On substituting the solution (20.8) in Equation (20.3), we get

$$\operatorname{div} \rho \mathbf{v} = \pm \frac{1}{\sqrt{4\pi}} \left[\frac{1}{\sqrt{\rho}} \operatorname{div} \mathbf{B} - \frac{1}{2} \rho^{-3/2} \, \mathbf{B} \cdot \nabla \rho \right] = 0$$

by virtue of (20.4) and (20.13). Thus the condition (20.13) is enough for Equation (20.3) to be satisfied identically. However, to ensure the fulfilment of condition (20.9), we must require constancy of the gas and the magnetic pressure or the absolute value of the magnetic field intensity. The latter means that each magnetic flux tube must have a *constant cross-section*. Hence, by virtue of (20.8), the flow velocity along the tube will be constant as well.

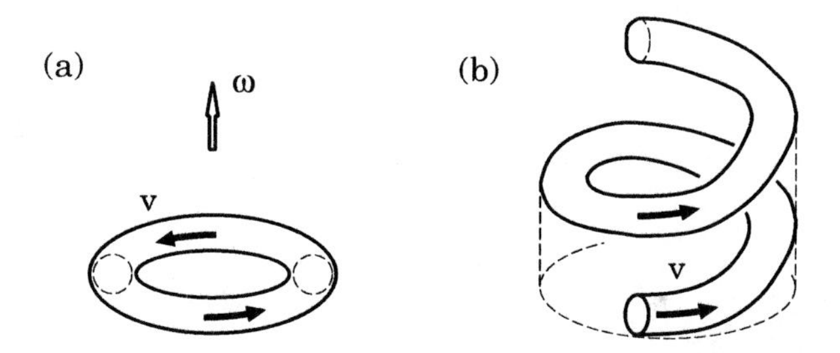

Figure 20.1: Rotational (a) and helical (b) stationary flows of a compressible plasma.

Therefore stationary flows corresponding to the solutions (20.8) and (20.9), which are flows with a *constant velocity* in magnetic tubes of a constant cross-section, are possible in a compressible medium. An example of such a flow is the plasma rotation in a ring tube (Figure 20.1a). We can envisage spiral motions of the plasma, belonging to the same type of stationary solutions in MHD (Figure 20.1b). This may be, for example, the case of an astrophysical jet when plasma presumably moves along a spiral trajectory.

20.1.3 Astrophysical collimated streams (jets)

Powerful extragalactic radio sources comprise two extended regions containing magnetic field and synchrotron-emitting relativistic electrons, each linked by a jet to a central compact radio source located in the nucleus of the associated active galaxy (Begelman et al., 1984). These jets are well **collimated streams of plasma** that emerge from the nucleus in opposite directions, along which flow mass, momentum, energy, and magnetic flux. The oscillations of jets about their mean directions are observed. The origin of the jet is crucial to understanding all active nuclei (Section 13.3).

The microquasars recently discovered in our Galaxy offer a unique opportunity for a deep insight into the physical processes in relativistic jets observed in different source populations (e.g., Mirabel and Rodriguez, 1998; Atoyan and Aharonian, 1999). Microquasars are stellar-mass black holes in our Galaxy that mimic, on a small scale, many of the phenomena seen in quasars. Their discovery opens the way to study the connection between the accretion of plasma onto the black holes and the origin of the relativistic jets observed in remote quasars (Section 13.3).

In spite of the vast differences in luminosity and the sizes of microquasars in our Galaxy and those in active galaxies both phenomena are believed to be powered by gravitational energy released during the accretion of plasmas onto black holes. Since the accreting plasmas have non-zero angular momentum, they form accreation disks orbiting around black holes. If the accreting plasmas have non-zero poloidal magnetic field, the magnetic flux accumulates in the inner region of the disk to form a global poloidal field penetrating the disk. Such poloidal fields could also be generated by dynamo action inside the accreation disk.

In either case, poloidal fields are twisted by the rotating disk toward the azimuthal direction. Moreover this process extracts angular momentum from the disk, enabling efficient accreation of disk plasmas onto black holes. In addition, magnetic twist generated during this process accelerates plasmas in the surface layer of the disk toward the polar direction by the Lorentz force to form bi-directional relativistic jets which are also collimated by the magnetic force (Lovelace, 1976).

20.1.4 MHD waves of arbitrary amplitude

Let us return to the case of an incompressible medium. Consider a steady flow of the type (20.8) and (20.9) in the magnetic field shown in Figure 20.2.

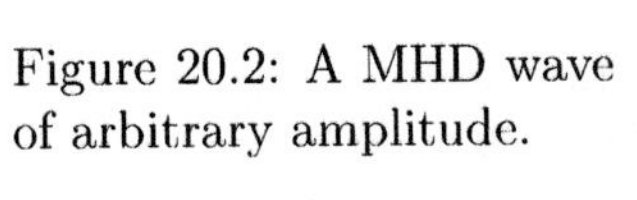
Figure 20.2: A MHD wave of arbitrary amplitude.

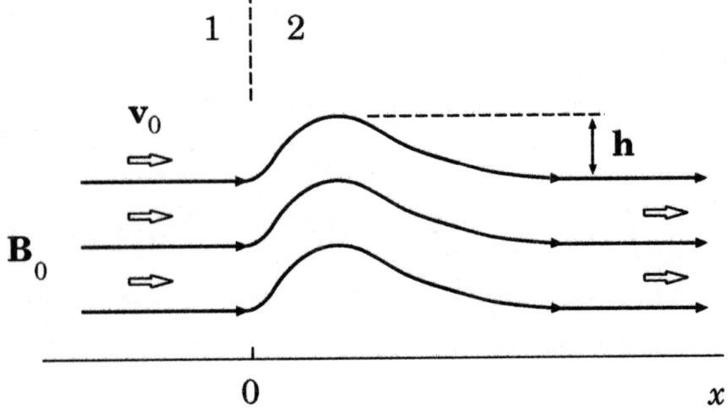

In the region 1, transformed to the frame of reference, the wave front of an arbitrary amplitude $\mathbf{h} = \mathbf{h}(x)$ runs against the immovable plasma in the uniform magnetic field $\mathbf{B}_0$, the front velocity being the Alfvén one:

$$\mathbf{v}_0 = \frac{\mathbf{B}_0}{\sqrt{4\pi\rho}} \,. \tag{20.14}$$

On the strength of condition (20.11), in such a wave

$$p + \frac{(\mathbf{B}_0 + \mathbf{h})^2}{8\pi} = p_0 + \frac{B_0^2}{8\pi} = \text{const}\,, \tag{20.15}$$

i.e. the gas pressure is balanced everywhere by the magnetic pressure.

> The non-compensated magnetic tension, $(\mathbf{B} \cdot \nabla)\, \mathbf{B}/4\pi$, provides the wave motion of arbitrary amplitude

(cf. Section 15.2.2). In this sense, the MHD waves are analogous to elastic waves in a string. MHD waves of an arbitrary amplitude were found for the first time by Alfvén (1950) as non-stationary solutions of the MHD equations for an incompressible medium (see also Alfvén, 1981).

The Alfvén or rotational discontinuity considered in Section 16.2 is a particular case of the solutions (20.8) and (20.9), corresponding to a discontinuous velocity profile. Behaviour of Alfvén waves in the isotropic and anisotropic astrophysical plasmas can be essentially different (see Section 7.3).

20.1.5 Differential rotation and isorotation

Now we consider another exact solution to the stationary equations of ideal MHD. Let us suppose that an equilibrium configuration (for example, a star) rigidly rotates about the symmetry axis of the cylindrically symmetric ($\partial/\partial\varphi = 0$) magnetic field. The angular velocity $\boldsymbol{\omega}$ is a constant vector. Then

$$\mathbf{v} = \mathbf{r} \times \boldsymbol{\omega} = \{\, 0,\, 0,\, v_{\varphi} \}\,, \tag{20.16}$$

where

$$v_{\varphi} = \omega\, r\,.$$

The induction Equation (20.2) is satisfied identically in this case.

Now we relax the assumption that $\boldsymbol{\omega}$ is a constant. Consider the case of the so-called *differential rotation*. Let the vector $\boldsymbol{\omega}$ be everywhere parallel to the z axis, i.e. the symmetry axis of the field $\mathbf{B}$, but the quantity $|\,\boldsymbol{\omega}\,| = \omega$ be dependent on the coordinates r and z, where r is the cylindrical radius:

$$\omega = \omega\,(r, z)\,.$$

Hence

$$v_{\varphi} = \omega\,(r, z)\, r\,. \tag{20.17}$$

Substitution of (20.17) in the induction Equation (20.2), with allowance being made for $\partial/\partial\varphi = 0$ and (20.4), gives

$$\operatorname{curl}(\mathbf{v}\times\mathbf{B}) = \mathbf{e}_\varphi\, r\,(\mathbf{B}\cdot\nabla\omega) = 0\,.$$

Therefore

$$\boxed{\mathbf{B}\cdot\nabla\omega = 0\,,} \tag{20.18}$$

i.e. the magnetic field lines are situated at $\omega =$ const surfaces. When treated in astrophysics, this case is called *isorotation*.

As a consequence of cylindrical symmetry, the $\omega =$ const surfaces are those of rotation, hence **isorotation does not change the magnetic field**.

On the other hand, if the condition for isorotation (20.18) is not valid, **differential rotation twists the field lines**, for example as shown in Figure 20.3, creating a *toroidal* field B_φ. The magnetic field is amplified.

Rigid rotation and isorotation are widely discussed, when applied to stellar physics, because

> rotation is an inherent property of the majority of the stars having strong magnetic fields

(Schrijver and Zwaan, 1999). What is the actual motion of the plasma in the interior of stars?

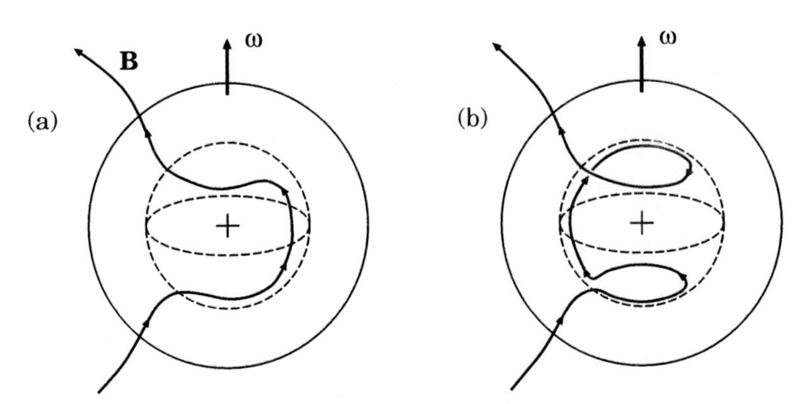

Figure 20.3: Differential rotation creates the toroidal (B_φ) component of a magnetic field inside a star.

Suppose there is no tangential stress at the surface of a star. The rigid rotation must be gradually established owing to viscosity in the star. However the observed motion of the Sun, as a well studied example, is by no means rigid: **the equator rotates faster than the poles**. This effect cannot be explained by surface rotation. Deep layers of the Sun and fast-rotating solar-type stars participate in complex motions: differential rotation, convection, and meridional circulation (see Rüdiger and von Rekowski, 1998). Such motions ensure mixing of deep solar layers down to the solar core. The circumstantial evidence for this comes from observations of the solar neutrino flux as well as helioseismological data. The latter show, in particular, that **the solar core rotates faster than the surface**. The results of the *SOHO* helioseismology enable us to know the structure of the solar internal differential rotation (Schou et al., 1998).

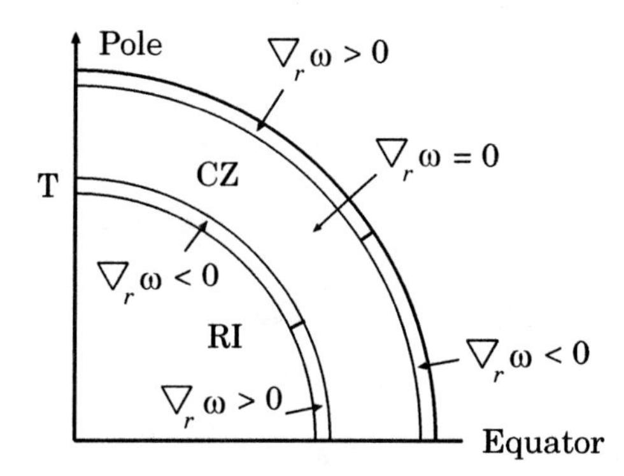

Figure 20.4: Schematic summary of a radial gradient in rotation that have been inferred from helioseismic measurements.

Roughly speaking, in the convective zone (see CZ in Figure 20.4) the angular velocity ω is independent of radius r. The radiative interior (RI) appears to rotate almost uniformly, and is separated from the differentially rotating convective zone by a thin shear layer called the tachocline (shown by T in Figure 20.4). The last is, in fact, too thin to be convincingly resolved by the *SOHO* data.

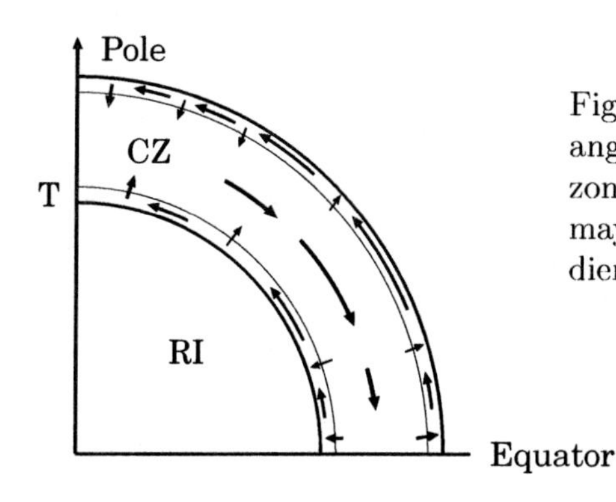

Figure 20.5: Schematic of the flow of angular momentum in the convective zone, tachocline, and photosphere, that may be responsible for the rotation gradients summarized in previous Figure.

Numerical simulations are still rather far from producing a radius-independent differential rotation in the convective zone. A qualitative perspective, which probably will define a context for progress in the future, invoke the concepts of angular momentum balance and transport, and angular momentum cycles in the Sun. With this perspective, it is possible to consider all the angular velocity domains in the outer part of the Sun in a unified way (Gilman, 2000). Figure 20.5 illustrates how angular momentum could be continually cycling in the convective zone and adjacent layers.

If we accept that some process dominates in the cycle by transporting angular momentum from high latitudes to low in the bulk of the convective zone, then everything else follows. All that is required is that some of this momentum 'leak' into the tachocline below and the granulation and supergranulation layers above. Then, to complete the cycle, there is transport of angular momentum back toward the pole in both layers. There the momentum reenters the bulk of the convective zone to be recycled again.

Recommended Reading: Elsasser (1956), Parker (1979), Moreau (1990).

20.2 Flows at small magnetic Reynolds numbers

While investigating MHD flows in a laboratory, the finite conductivity being significant, one has to account for the magnetic field dissipation. Furthermore one has to take account of the fact that the freezing-in condition breaks down owing to the smallness of the magnetic Reynolds number (12.62):

$$\mathrm{Re_m} = \frac{vL}{\nu_{\mathrm{m}}} \ll 1\,. \tag{20.19}$$

The analogous situation takes place, for example, in deep layers of the solar atmosphere near the temperature minimum. The conductivity is small here, since the number of neutral atoms is relatively large (e.g., Hénoux and Somov, 1987, 1991).

Stationary flows are possible in the case of finite conductivity. However they differ greatly from the ideal medium flows considered in the previous Section. The difference manifests itself in the fact that, given dissipative processes, steady flows are realized only under action of some external force, a pressure gradient, for instance. A second difference is that **the plasma of finite conductivity can flow across the field lines**.

20.2.1 Stationary flows inside a duct

We shall examine a flow which has been well studied for reasons of practical importance. Let us consider the steady flow of a viscous conducting fluid along a duct with a transversal magnetic field. Let the x axis of the Cartesian system (Figure 20.6) be chosen in the flow direction, the external uniform field $\mathbf{B}_0$ coinciding with the z axis:

$$\mathbf{v} = \{\, v(z),\, 0,\, 0 \,\}\,, \qquad \mathbf{B}_0 = \{\, 0,\, 0,\, B_0 \,\}\,. \tag{20.20}$$

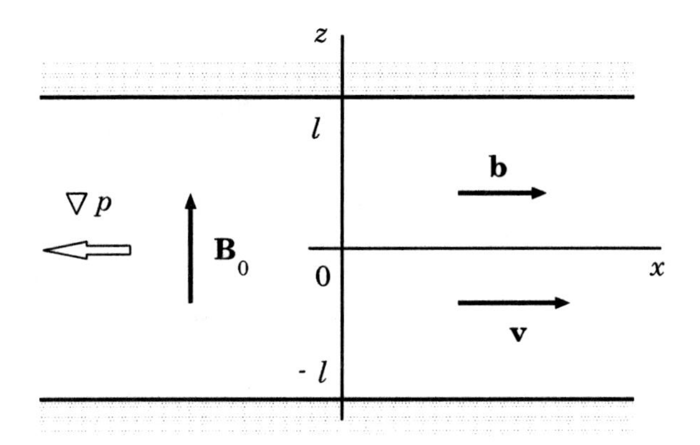

Figure 20.6: Formulation of the problem on the finite conductivity plasma flow in a duct.

Let the width of the duct be $2l$.

We start from the set of Equations (12.42)–(12.47) for a steady flow of an incompressible medium:

$$\rho = \text{const} \, . \tag{20.21}$$

Consider two equations:

$$\operatorname{curl}\,(\,\mathbf{v} \times \mathbf{B}\,) + \nu_{\text{m}}\, \Delta\, \mathbf{B} = 0 \, , \tag{20.22}$$

$$(\,\mathbf{v} \cdot \nabla\,)\, \mathbf{v} = -\, \frac{\nabla p}{\rho} - \frac{\mathbf{B} \times \operatorname{curl} \mathbf{B}}{4\pi\rho} + \nu\, \Delta\, \mathbf{v} \, . \tag{20.23}$$

The pressure gradient $\partial p / \partial x$ along the x axis, which is independent of x, is assumed to be the cause of the motion. Supposing the flow to be relatively slow, neglect the term on the left-hand side of Equation (20.23).

Let $b = b\,(z)$ be the magnetic field component along the velocity. In the coordinate form, Equations (20.22) and (20.23) are reduced to the following three equations:

$$B_0 \, \frac{\partial v}{\partial z} + \nu_{\text{m}} \, \frac{\partial^2 b}{\partial z^2} = 0 \, , \tag{20.24}$$

$$\rho\, \nu \, \frac{\partial^2 v}{\partial z^2} + \frac{B_0}{4\pi} \, \frac{\partial b}{\partial z} - \frac{\partial p}{\partial x} = 0 \, , \tag{20.25}$$

$$\frac{\partial}{\partial z} \left(p + \frac{b^2}{8\pi} \right) = 0 \, . \tag{20.26}$$

Differentiating Equation (20.26) with respect to x gives

$$\frac{\partial^2 p}{\partial x \, \partial z} = 0 \, . \tag{20.27}$$

Differentiating (20.25) with respect to z, with care taken of (20.27), gives

$$\rho\nu\frac{\partial^3 v}{\partial z^3}+\frac{B_0}{4\pi}\frac{\partial^2 b}{\partial z^2}=0\,. \tag{20.28}$$

Eliminate $\partial^2 b/\partial z^2$ between Equations (20.24) and (20.28). The result is

$$\frac{d^3 v}{dz^3}-\frac{B_0^2}{4\pi\rho\,\nu\nu_{\rm m}}\frac{dv}{dz}=0\,. \tag{20.29}$$

This equation is completed by the boundary conditions on the duct walls

$$v\,(l)=v\,(-l)=0\,. \tag{20.30}$$

The corresponding solution is of the form

$$v(z)=v_0\,\frac{\cosh \mathrm{Ha}-\cosh\left(\mathrm{Ha}\,z/l\,\right)}{\cosh \mathrm{Ha}-1}\,. \tag{20.31}$$

Here $v_0=v\,(0)$ is the flow velocity at the centre of the duct, the dimensionless parameter characterizing the flow is

$$\boxed{\mathrm{Ha}=\frac{l\,B_0}{\sqrt{4\pi\rho\,\nu\nu_{\rm m}}}\,.} \tag{20.32}$$

It is called the *Hartmann number*, the flow (20.31) being the Hartmann flow. As $\mathrm{Ha}\to 0$, formula (20.31) converts to the usual parabolic velocity profile which is typical of viscous flows in a duct without a magnetic field:

$$v(z)=v_0\left(1-\frac{z^2}{l^2}\right). \tag{20.33}$$

The influence of a transversal magnetic field shows itself as the appearance of **an additional drag to the plasma flow** and the change of the velocity profile which becomes *flatter* in the central part of the duct (Figure 20.7).

In the limit $\mathrm{Ha}\to\infty$, the Hartmann formula (20.31) gives

$$v(z)=v_0\left\{1-\exp\left[-\mathrm{Ha}\left(1-\frac{z}{l}\right)\right]\right\}. \tag{20.34}$$

Such a velocity profile is flat, $v(z)\approx v_0$, the exception being a thin layer near the walls, the *boundary layer* of the thickness l/Ha.

20.2.2 The MHD generator or pump

What factors determine the value of velocity v_0 at the center of the duct? To find them let us calculate the electric current density in the duct

$$j_y=\frac{c}{4\pi}\frac{\partial b}{\partial z}=\frac{c}{4\pi}\left(\frac{4\pi}{B_0}\frac{\partial p}{\partial x}-\rho\,\nu\,\frac{4\pi}{B_0}\frac{\partial^2 v}{\partial z^2}\right)=$$

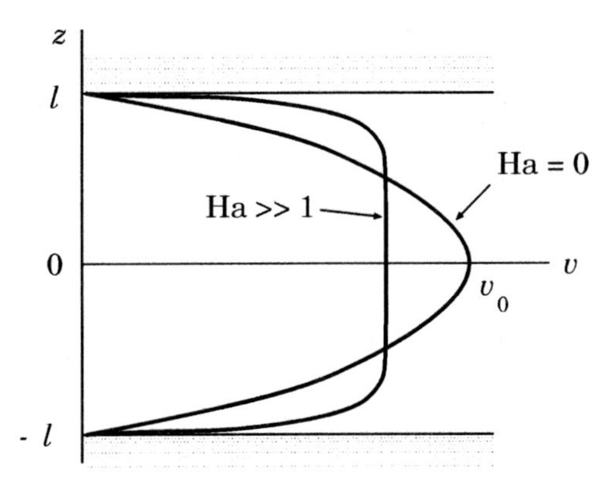

Figure 20.7: Usual parabolic (Ha = 0) and Hartmann profiles of the viscous flow velocity in a duct with a transverse magnetic field.

$$= \frac{c}{B_0} \left(\frac{\partial p}{\partial x} - \rho \nu \frac{\partial^2 v}{\partial z^2} \right) . \tag{20.35}$$

Here the use is made of formula (20.25) to find the derivative $\partial b / \partial z$. Let us substitute in (20.35) an expression for velocity of the type (20.31), i.e.

$$v(z) = A \left(\cosh \mathrm{Ha} - \cosh \frac{\mathrm{Ha}\, z}{l} \right) . \tag{20.36}$$

We get the following equation

$$\frac{j_y B_0}{c} = \frac{\partial p}{\partial x} - \rho \nu A \left(\frac{\mathrm{Ha}}{l} \right)^2 \cosh \frac{\mathrm{Ha}\, z}{l} . \tag{20.37}$$

Let us integrate Equation (20.37) over z from $-l$ to $+l$. The result is

$$\frac{I B_0}{c} = 2l \frac{\partial p}{\partial x} - A\, 2\rho \nu \left(\frac{\mathrm{Ha}}{l} \right) \sinh \mathrm{Ha} , \tag{20.38}$$

where

$$I = \int_{-l}^{l} j_y \, dz \tag{20.39}$$

is the total current per unit length of the duct. We shall assume that there is an electrical circuit for this current to flow outside the duct. The opposite case is considered in Landau and Lifshitz, *Fluid Mechanics*, 1959a, Chapter 8, § 67.

Finally it follows from Equation (20.38) that the sought-after coefficient in formula (20.36) is

$$\boxed{A = \frac{\partial p / \partial x - (1/2lc)\, I B_0}{(\rho \nu / l^2)\; \mathrm{Ha} \sinh \mathrm{Ha}} .} \tag{20.40}$$

Thus

> the velocity of the plasma flow in the duct is proportional to the gas pressure gradient and the magnetic Lorentz force.

This is why two different operational regimes are possible for the duct.

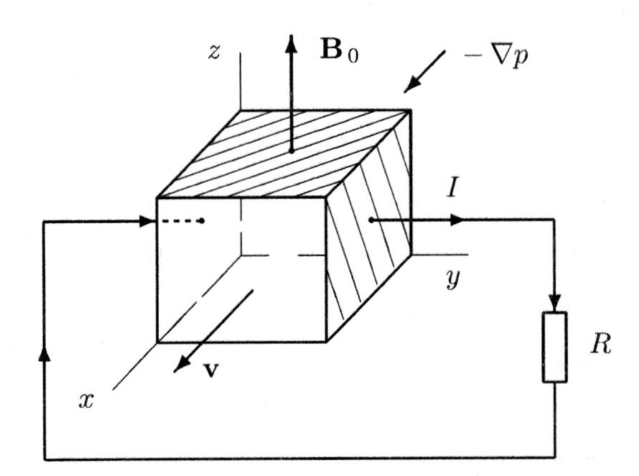

Figure 20.8: Utilization of the MHD duct as the generator of the current I; R is an external load.

If the flow in the duct is realized under the action of an external pressure gradient, the duct operates as the MHD generator shown in Figure 20.8. The same principle explains the action of flowmeters (for more detail see Shercliff, 1965, § 6.5; Sutton and Sherman, 1965, § 10.2) which are important, for example, in controling the flow of the metallic heat conductor in reactors.

The second operating mode of the duct occurs when an external electromagnetic force (instead of a passive load R in Figure 20.8) creates the electric current I between the walls of the duct. Interaction of the current with the external magnetic field $\mathbf{B}_0$ gives rise to the Lorentz force that makes the plasma move along the duct, i.e. in the direction of the x axis. Hence the duct operates as the MHD pump, and this is also used in some technical applications.

20.2.3 Weakly-ionized plasma in astrophysics

Under astrophysical conditions, both operating modes of the MHD duct are realized, once the plasma resistivity is high due, for instance, to its low temperature. In the solar atmosphere, in the minimum temperature region, neutral atoms move in the directions of convective flows and collide with ions, thus setting them in motion. At the same time, electrons remain 'frozen' in the magnetic field. This effect (termed the *photospheric dynamo*) can generate electric currents and amplify the magnetic field in the photosphere and the low chromosphere (see vol. 2, Section 12.4).

A violent outflow of high-velocity weakly-ionized plasma is one of the first manifestations of the formation of a new stars (Bachiller, 1996; Bontemps et al., 1996). Such outflows emerge bipolarly from the young object and involve amounts of energy similar to those involved in accretion processes. The youngest proto-stellar low-mass objects known to date (the class 0 protostars) present a particularly efficient outflow activity, indicating that outflow and infall motions happen simultaneously and are closely linked since the very first stages of the star formation processes.

The idea of a new star forming from relatively simple hydrodynamic infall of weakly-ionized plasma is giving place to a picture in which magnetic fields play a crucial role and stars are born through the formation of complex engines of accreation/ejection. It seems inevitable that future theories of star formation will have to take into account, together with the structure of the protostar and its surrounding accretion disk, the processes related to **multi-fluid hydrodynamics of weakly-ionized plasma**. These are the effects similar to the photospheric dynamo and magnetic reconnection in weakly-ionized plasma (vol. 2, Section 12.3).

Recommended Reading: Sutton and Sherman (1965), Ramos and Winowich (1986).

20.3 The σ-dependent force and vortex flows

20.3.1 Simplifications and problem formulation

As was shown in Section 19.4, a body plunged into a conducting fluid with magnetic and electric fields is acted upon by an *expulsion* force or, more exactly, by the magnetic σ-*dependent* force. As this takes place, the electric field $\mathbf{E}$ and current density $\mathbf{j}$ are non-uniform, and the volume **Lorentz force inside the fluid is non-potential**. The force generates vortex flows of the fluid in the vicinity of the body.

(a) Let us consider the stationary problem for an incompressible fluid having uniform constant viscosity ν and magnetic diffusivity $\nu_{\rm m}$ (Syrovatskii and Chesalin, 1963; Marty and Alemany, 1983; Gerbeth et al., 1990). Let, at first, both the usual and magnetic Reynolds numbers be small:

$$\mathrm{Re} = \frac{vL}{\nu} \ll 1\,, \tag{20.41}$$

$$\mathrm{Re}_{\rm m} = \frac{vL}{\nu_{\rm m}} \ll 1\,. \tag{20.42}$$

The freezing-in condition (12.63) can be rewritten in the form

$$\Delta \mathbf{B} + \mathrm{Re}_{\rm m}\, \mathrm{curl}\,(\, \mathbf{v} \times \mathbf{B}\,) = 0\,, \tag{20.43}$$

where, in view of (20.42), $\mathrm{Re_m}$ is a small parameter. In a zeroth approximation in this parameter, the magnetic field is potential:

$$\Delta \mathbf{B} = 0 .$$

Moreover the magnetic field will be assumed to be uniform, in accordance with the formulation of the problem discussed in Section 19.4. Strictly speaking, the assumption of **a uniform magnetic field** implies the inequality

$$B \gg \frac{4\pi}{c} Lj . \tag{20.44}$$

Its applicability will be discussed later on, in connection with the simplified form of Ohm's law to be used while solving the problem.

(b) Assuming the stationary flows occurring in the fluid to be slow, the inertial force (proportional to v^2) will be ignored in the equation of motion (20.23) as compared to the other forces: **pressure gradient, Lorentz force, viscous force**. The term describing the gravity force will be dropped, since its effect has already been studied in Section 19.4. Finally, on multiplying the equation

$$0 = -\frac{\nabla p}{\rho} - \frac{\mathbf{B} \times \mathrm{curl}\, \mathbf{B}}{4\pi\rho} + \nu \, \Delta \mathbf{v}$$

by the fluid density $\rho = \rho_0$, it is rewritten in the form

$$\eta \, \Delta \mathbf{v} = \nabla p - \mathbf{f} . \tag{20.45}$$

Here $\eta = \rho_0 \, \nu$ is the *dynamic* viscosity coefficient, and

$$\mathbf{f} = \frac{1}{c} \, \mathbf{j} \times \mathbf{B}_0 \tag{20.46}$$

is the Lorentz force in the same approximation.

Recall that, in view of the assumed incompressibility of the fluid, the velocity field obeys the equation

$$\mathrm{div}\, \mathbf{v} = 0 . \tag{20.47}$$

(c) The electric field $\mathbf{E}$ is assumed to be uniform at infinity

$$\mathbf{E} \to \mathbf{E}_0 , \qquad r \to \infty . \tag{20.48}$$

Given the conductivities of the fluid σ_0 and of the submerged body σ_1, we can find the current $\mathbf{j}$ in the whole space using the following conditions:

$$\mathrm{div}\, \mathbf{j} = 0 , \tag{20.49}$$

$$\mathbf{j} = \sigma \mathbf{E} , \tag{20.50}$$

$$\mathrm{curl}\, \mathbf{E} = 0 . \tag{20.51}$$

The current $(\sigma/c)\,\mathbf{v}\times\mathbf{B}$ has been ignored in Ohm's law (20.50). This may be done, once the velocity of engendered vortex flows is much less than the drift velocity, i.e. once the inequality

$$v \ll v_{\rm d} = c\,\frac{E}{B} \tag{20.52}$$

holds. Note that substituting (20.44) in (20.52) results in the inequality

$$\frac{vL}{(c^2/4\pi\sigma)} \ll 1\,, \tag{20.53}$$

which coincides with the initial assumption (20.42).

20.3.2 The solution for a spherical ball

Let us solve the problem for a ball of radius a. We choose the Cartesian frame of reference, in which the direction of the x axis is parallel to $\mathbf{E}_0$, and the origin of coordinates coincides with the center of the ball as shown in Figure 20.9.

By virtue of Ohm's law (20.50), the electric current at infinity

$$\mathbf{j}_0 = \sigma_0\,\mathbf{E}_0 \tag{20.54}$$

is also parallel to the x axis.

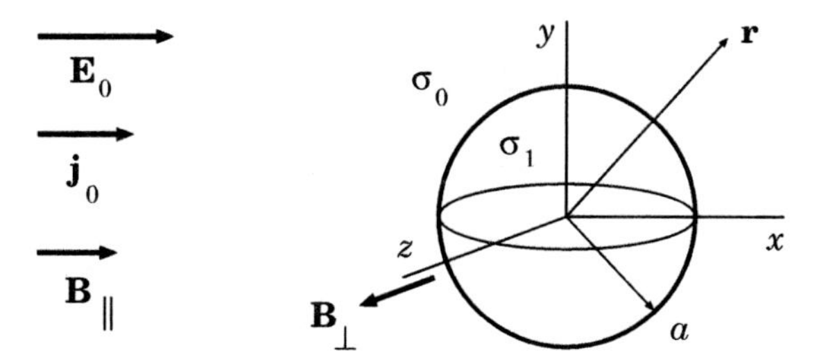

Figure 20.9: An uniform conducting ball of radius a, submerged in a conducting fluid with electric and magnetic fields.

It follows from Equation (20.51) that the current can be represented in the form

$$\mathbf{j} = \nabla\psi\,. \tag{20.55}$$

Here a scalar function ψ, in view of Equation (20.49), satisfies the Laplace equation

$$\Delta\psi = 0\,. \tag{20.56}$$

Let us try to find the solution to the problem in the form of uniform and dipole components:

$$\psi = \mathbf{j}_0\cdot\mathbf{r} + c_0\,\mathbf{j}_0\cdot\nabla\frac{1}{r}\,, \qquad r \geq a\,, \tag{20.57}$$

and

$$\psi = \mathbf{j}_1 \cdot \mathbf{r}, \qquad r < a. \tag{20.58}$$

Here c_0 is an unknown constant, $\mathbf{j}_1 = \{j_1, 0, 0\}$ is an unknown current density inside the ball. Both unknowns are to be found from the matching conditions at the surface of the ball:

$$\{j_r\} = 0 \quad \text{and} \quad \{\mathbf{E}_\tau\} = 0.$$

These conditions can be rewritten as follows

$$\frac{\mathbf{j} \cdot \mathbf{r}}{r} = \frac{\mathbf{j}_1 \cdot \mathbf{r}}{r} \quad \text{at} \quad r = a, \tag{20.59}$$

and

$$\frac{\mathbf{j}_\tau}{\sigma_0} = \frac{\mathbf{j}_{\tau 1}}{\sigma_1} \quad \text{at} \quad r = a. \tag{20.60}$$

On substituting (20.57) and (20.58) in (20.59) and (20.60), the constants c_0 and j_1 are found. The result is

$$\psi = \left[1 + \beta \left(\frac{a}{r} \right)^3 \right] \mathbf{j}_0 \cdot \mathbf{r} \quad \text{for} \quad r \geq a, \tag{20.61}$$

and

$$\psi = (1 - 2\beta)\, \mathbf{j}_0 \cdot \mathbf{r} \quad \text{for} \quad r < a. \tag{20.62}$$

Here the constant

$$\beta = \frac{\sigma_0 - \sigma_1}{2\sigma_0 + \sigma_1}. \tag{20.63}$$

Specifically, inside the ball

$$\mathbf{j}_1 = (1 - 2\beta)\, \mathbf{j}_0, \tag{20.64}$$

and $\mathbf{j}_1 = \mathbf{j}_0$, once $\sigma_1 = \sigma_0$.

20.3.3 Forces and flows near a spherical ball

Knowing the current in the whole space, we can find the Lorentz force (20.46)

$$\mathbf{f} = \frac{1}{c}\, \nabla\psi \times \mathbf{B}_0 = \text{curl}\, \frac{\psi \mathbf{B}_0}{c}. \tag{20.65}$$

In the case at hand,

> the volume Lorentz force has a rotational character and hence generates vortex flows in the conducting fluid.

Let us operate with curl curl on Equation (20.45). Using the known vector identity

$$\text{curl curl}\,\mathbf{a} = \nabla(\nabla\mathbf{a}) - \Delta\,\mathbf{a}$$

and taking account of relations (20.49)–(20.51), a biharmonic equation for the velocity field is obtained

$$\Delta\Delta\,\mathbf{v} = 0\,. \tag{20.66}$$

Operating with divergence on (20.45) and taking account of (20.49)–(20.51), we get

$$\Delta\,p = 0\,. \tag{20.67}$$

Equations (20.66) and (20.67) are to be solved together with Equations (20.45) and (20.47). For bodies with spherical or cylindrical symmetry, it is convenient to make use of the identity

$$\mathbf{r}\cdot\Delta\,\mathbf{q} = \Delta\,(\mathbf{q}\cdot\mathbf{r})\,, \tag{20.68}$$

where $\mathbf{q}$ is any vector satisfying the condition $\operatorname{div}\mathbf{q} = 0$. Then from Equation (20.66) subject to the condition (20.47) we find

$$\Delta\Delta\,(v_r\,r) = 0\,. \tag{20.69}$$

The boundary conditions are taken to be

$$v\,\Big|_S = 0\,, \quad v\,\Big|_\infty = 0\,. \tag{20.70}$$

Here S is the surface of the submerged body which is assumed to be a ball of radius a (cf. Figure 20.7). At its surface $r = a = \text{const}$, Equation (20.47) and the first of conditions (20.70) give

$$\left.\frac{\partial v_r}{\partial r}\right|_S = 0\,. \tag{20.71}$$

The solution of Equation (20.69), satisfying the boundary condition (20.71) and the second of conditions (20.70), is clearly seen to be

$$v_r \equiv 0\,. \tag{20.72}$$

Thus

> in the case of a spherical ball, the flow lines of a conducting incompressible fluid are situated at $r = \text{const}$ surfaces.

Next an equation for the pressure is found using Equation (20.45) and taking into account that, by virtue of (20.68),

$$\mathbf{r}\cdot\Delta\mathbf{v} = \Delta\,(v_r r) = 0\,.$$

The resulting equation is

$$\frac{\partial p}{\partial r} = f_r \, . \tag{20.73}$$

The function f_r occuring on the right-hand side is the radial component of the above mentioned Lorentz force (20.65).

Once the plasma pressure has been found by integrating Equations (20.73) and (20.67), the velocity is determined from Equation (20.45) with the known right-hand side.

Choose the Cartesian frame of reference in which

$$\mathbf{B}_0 = \{ B_{0x}, \; 0, \; B_{0z} \} \, ,$$

$B_{0x} = B_{\parallel}$ and $B_{0z} = B_{\perp}$ being the magnetic field components parallel and perpendicular to $\mathbf{j}_0$, respectively (see Figure 20.9). The current in the conducting fluid (cf. formula (20.61)) is

$$\mathbf{j} = \nabla \psi \, , \qquad \psi = j_0 \, x + j_0 \, \frac{\beta \, a^3 x}{r^3} \, , \tag{20.74}$$

the current inside the ball being defined by formula (20.64). The pressure in the fluid

$$p = \frac{1}{c} \, j_0 B_{\perp} \, y \left(\frac{\beta a^3}{2r^3} - 1 \right) + \text{const} \, . \tag{20.75}$$

It is convenient to rewrite the velocity distribution in spherical coordinates

$$\mathbf{v} = \{ v_r, \; v_\theta, \; v_\varphi \} \tag{20.76}$$

(cf. Syrovatskii and Chesalin, 1963):

$$\begin{aligned} v_r &= 0 \, , \\ v_\theta &= \frac{\beta j_0 a^2}{4c\eta} \, \frac{a}{r} \left(1 - \frac{a^2}{r^2} \right) \left(- B_{\perp} \cos\theta \, \sin\varphi + B_{\parallel} \, \sin\theta \, \sin 2\varphi \right) , \\ v_\varphi &= \frac{\beta j_0 a^2}{4c\eta} \, \frac{a}{r} \left(1 - \frac{a^2}{r^2} \right) \left(B_{\perp} \cos 2\theta \, \cos\varphi + B_{\parallel} \, \sin 2\theta \, \cos^2 \varphi \right) . \end{aligned}$$

This velocity field pattern is shown in Figure 20.10.

The force acting on the body is defined to be (cf. formula (19.60))

$$\mathbf{F} = \frac{1}{c} \int_V \mathbf{j} \times \mathbf{B}_0 \, dV + \oint_S p \, \mathbf{n} \, dS \; - \oint_S \sigma'_{\mathrm{n}} \, dS \, , \tag{20.77}$$

where $\mathbf{n}$ is the inward normal to the sphere;

$$\sigma'_{\mathrm{n}} = (\sigma'_{\alpha\beta} \, n_\beta)_{\mathrm{n}} \, , \tag{20.78}$$

$\sigma'_{\alpha\beta}$ being the viscous stress tensor, see definition (12.53).

On substituting the velocity distribution (20.76) in the viscous force formula (20.78) and integrating (20.77) over the surface S of the ball,

Figure 20.10: Vortex flows near the conducting ball submerged in a conducting fluid with electric and magnetic fields.

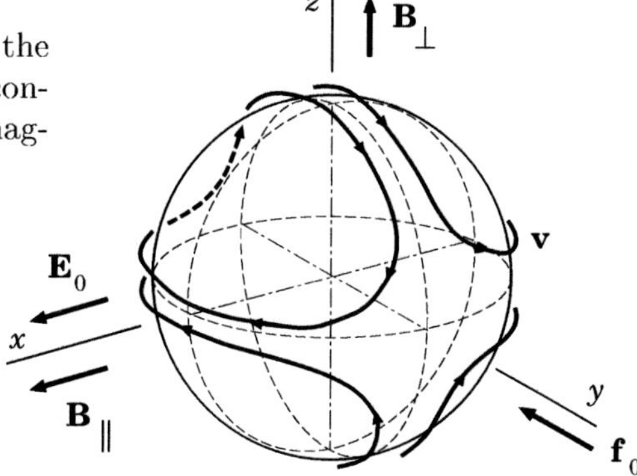

the sum of the viscous forces is concluded to be zero. The moment of the viscous forces acting on the ball is also zero.

The remaining force determined by (20.77) is directed along the y axis and is equal to

$$F = \frac{4\pi a^3}{3} \frac{j_0 B_\perp}{c} \left\{ -(1 - 2\beta) + \left(1 - \frac{\beta}{2}\right) \right\}. \tag{20.79}$$

The constant β is defined by formula (20.63):

$$\beta = \frac{\sigma_0 - \sigma_1}{2\sigma_0 + \sigma_1}.$$

The first term in the curly brackets corresponds to the force $\mathbf{j}_1 \times \mathbf{B}_0 / c$ which immediately acts on the current $\mathbf{j}_1$ inside the ball. Note that

$$1 - 2\beta = \frac{3\sigma_1}{2\sigma_0 + \sigma_1} > 0,$$

in agreement with the direction of the vector product $\mathbf{j}_1 \times \mathbf{B}_0$ or $\mathbf{j}_0 \times \mathbf{B}_0$ (Figure 20.9). Moreover, provided $\sigma_1 = 0$, the term $(1 - 2\beta) = 0$ as it should be the case for a non-conducting ball, since there is no current inside it.

The second term in the curly brackets of formula (20.79) expresses the sum of the forces of the pressure on the surface of the ball. The coefficient

$$1 - \frac{\beta}{2} = \frac{3(\sigma_0 + \sigma_1)}{2(2\sigma_0 + \sigma_1)} > 0,$$

signifying that

the actual σ-dependent force is always somewhat less than the force owing to the interaction of the current $\mathbf{j}_1$ and the magnetic field $\mathbf{B}_0$. Moreover the total force can be opposite in sign.

In the particular case $\sigma_1 = 0$, when the current $j_1 = 0$

$$1 - \frac{\beta}{2} = \frac{3}{4}\,.$$

Hence $F > 0$. The non-conducting ball is expelled in the direction opposite to that of the vector product $\mathbf{j}_0 \times \mathbf{B}_0$ (Figure 20.11).

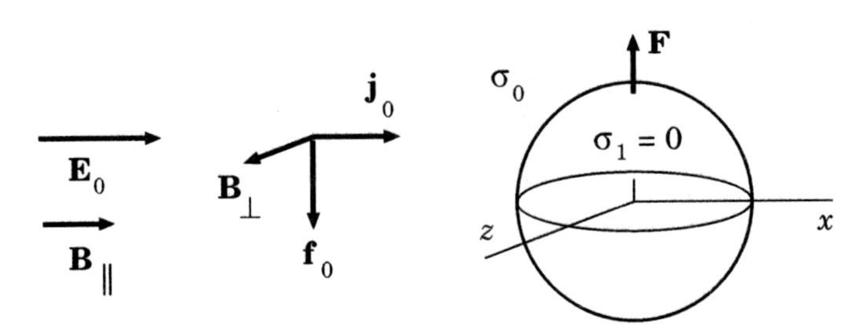

Figure 20.11: The expulsion force **F** acting on the non-conducting ball submerged in a conducting fluid with electric and magnetic fields.

The above properties of the magnetic σ-dependent force are used in technical MHD. They constitute the principle of action for magnetic separators which are intended for dividing mechanical mixtures having different conductivities.

Having the physical sense of the two terms determining the magnetic σ-dependent force (20.79), let us combine them in the following descriptive formula:

$$\boxed{\mathbf{F} = -\,\mathbf{f}_0\, V \times \frac{3}{2}\,\beta\,.} \tag{20.80}$$

Here $V = 4\pi a^3/3$ is the volume of the ball, $\mathbf{f}_0 = \mathbf{j}_0 \times \mathbf{B}_0 / c$ is the Lorentz force in the conducting fluid with uniform magnetic $\mathbf{B}_0$ and electric $\mathbf{E}_0$ fields (cf. (19.59)), the coefficient β being determined by formula (20.63).

20.4 Large magnetic Reynolds numbers

In the previous section we have considered the solution to the MHD problem concerning the magnetic σ-dependent force in the limit of small (usual and magnetic) Reynolds numbers. Leenov and Kolin (1954) were the first to obtain similar solutions in connection with the problem of electromagnetophoresis.

As a rule the opposite limiting case is applicable for astrophysical use. In this case, the problem of the magnetic σ-dependent force is difficult and can hardly be solved completely, especially given

$$\mathrm{Re} \ll 1\,, \qquad \mathrm{Re}_{\mathrm{m}} \gg 1\,. \tag{20.81}$$

A situation of this kind occurs, for example, in solar prominences (Section 20.4.2). In what follows we will show (Litvinenko and Somov, 1994; Somov, 1994b) that an expression for the magnetic σ-dependent force can be found for large magnetic Reynolds numbers, without rigorous calculations of the characteristics of the plasma flow near a body.

20.4.1 The general formula for the σ-dependent force

The equations of stationary MHD for flows of an incompressible fluid with density ρ_0 and dynamic viscosity $\eta = \rho_0 \nu$ are of the form:

$$\rho\,(\mathbf{v} \cdot \nabla)\,\mathbf{v} = -\nabla p + \frac{1}{c}\,\mathbf{j} \times \mathbf{B} + \eta\,\Delta \mathbf{v}\,,$$

$$\operatorname{curl}\,(\mathbf{v} \times \mathbf{B}) + \nu_{\rm m}\,\Delta \mathbf{B} = 0\,, \tag{20.82}$$

$$\operatorname{curl}\mathbf{B} = \frac{4\pi}{c}\,\mathbf{j}\,, \quad \operatorname{div}\mathbf{v} = 0\,, \quad \operatorname{div}\mathbf{B} = 0\,.$$

Let us find the σ-dependent force density $\mathbf{f}$ on the basis of *similarity* considerations. The given set of equations implies that five quantities are the determining parameters of the problem: ν, $\nu_{\rm m}$, a, ρ_0, and $\mathbf{f}_0$. By way of example, velocity v_0 depends on these parameters. Hence v_0 rather than ρ_0 may be treated as a determining parameter. The standard procedure of *dimensional analysis*, described by Bridgman (1931), gives us the formula

$$\mathbf{f} = -\,\mathbf{f}_0\,\Phi\,({\rm Re}, {\rm Re}_{\rm m})\,. \tag{20.83}$$

In the limit ${\rm Re}_{\rm m} = 0$ it reproduces (in a slightly different notation) the result presented in the theoretical part of the paper by Andres et al. (1963). Experimental data, which are stated in the same paper for ${\rm Re} < 10^2$, allow one to conclude that, with an accuracy which is completely sufficient for astrophysical applications,

$$\Phi\,({\rm Re}, {\rm Re}_{\rm m}) \approx \Phi_1({\rm Re}_{\rm m})\,, \tag{20.84}$$

where $\Phi_1(0) \approx 1$.

Generally, the behaviour of the magnetic field lines near the body for ${\rm Re}_{\rm m} \neq 0$ can become nonregular and intricate, as a consequence of the electric current redistribution and vortex flow generation. For example, if ${\rm Re}_{\rm m} < 1$, then the value of the nonregular field component $\delta B \approx {\rm Re}_{\rm m}\,B_0$. The effective magnitude of the field and the magnetic σ-dependent force decrease as compared to the case ${\rm Re}_{\rm m} = 0$.

The form of the decreasing function Φ_1 for ${\rm Re}_{\rm m} \gg 1$ can be determined as follows. Far from the body, at infinity, the electromagnetic energy flux is equal to

$$\mathbf{G}_0 = \frac{c}{4\pi}\,\mathbf{E}_0 \times \mathbf{B}_0\,. \tag{20.85}$$

In close proximity to the body, the magnitude of the Poynting vector must diminish once the disordered behaviour of lines of force is assumed. The difference $(G_0 - G)$ is equal to the power of engendered vortex flows, hence generally we get

$$f a^3 v_0 \leq G_0 \, a^2 . \tag{20.86}$$

The equality (20.86) is achieved in the limit $\mathrm{Re_m} \to \infty$. Here the characteristic velocity v_0 is determined from the equation of motion in the set (20.82):

$$v_0 = f a^2 / \eta \qquad \text{for} \quad \mathrm{Re} \ll 1 \, , \tag{20.87}$$

$$v_0 = (f a / \rho_0)^{1/2} \quad \text{for} \quad \mathrm{Re} \gg 1 \, . \tag{20.88}$$

When $\mathrm{Re_m} \to \infty$, relations (20.84)–(20.88) allow us to obtain the sought-after function appearing in formula (20.83):

$$\Phi\,(\mathrm{Re}, \mathrm{Re_m}) = \begin{cases} 1 & \text{for} \quad \mathrm{Re_m} < 1 \, , \\ \mathrm{Re_m^{-1}} & \text{for} \quad \mathrm{Re_m} > 1 \, . \end{cases} \tag{20.89}$$

The case $\mathrm{Re_m} < 1$ was treated by Leenov and Kolin (1954).

Strictly speaking, we could take also into account the dependence of the function Φ on the usual Reynolds number Re. We could obtain

$$\Phi\,(\mathrm{Re}, \mathrm{Re_m}) = \frac{1}{\mathrm{Re_m}} \, \Phi_2\,(\mathrm{Re}) \, , \tag{20.90}$$

where the function $\Phi_2\,(\mathrm{Re})$ is practically constant.

Note that formula (20.90) can be interpreted as a manifestation of an *incomplete self-similarity* of the function Φ relative to the similarity parameter $\mathrm{Re_m}$ (Barenblatt, 1979). The point is that, from the viewpoint of a 'naive' analysis, the function Φ does not depend on a dimensionless parameter whose magnitude is much greater (or less) than unity. This statement is true only if there exists a final non-zero limit of the function Φ as the parameter at hand tends to infinity (or zero). However, in general, this is not the case, as is clearly demonstrated by (20.90). In fact, $\Phi \to 0$ when $\mathrm{Re_m} \to \infty$. At the same time the function Φ is a power-law one in $\mathrm{Re_m}$; that allows us to write down an expression for the force density **f** in a self-similar form. As this takes place, the exact form of dimensionless combinations cannot be determined from the formal dimensional analysis alone.

Therefore an order-of magnitude expression is obtained for the density of the magnetic σ-dependent force acting on a body submerged into a conducting fluid or plasma (Litvinenko and Somov, 1994; Somov, 1994b):

$$\mathbf{f} = - \frac{c}{4 \pi v_0 \, a} \, \mathbf{E}_0 \times \mathbf{B}_0 \, . \tag{20.91}$$

The expression (20.91) is valid in the limit of large magnetic Reynolds numbers. For a body with a non-zero conductivity σ_1, the electric current flowing inside the body must be taken care of in formula (20.91). The corresponding treatment was presented in Section 20.3.

The physical sense of formula (20.91) is obvious. Comparison of (20.91) with formula (19.59) for the σ-dependent force, which then holds a uniform current flow in the plasma, shows that for $\mathrm{Re_m} \to \infty$ ($\sigma \to \infty$) the plasma in the vicinity of the body possesses, as it were, an *effective* conductivity

$$\boxed{\sigma_{\mathrm{ef}} \approx \frac{c^2}{v_0 a}\,.} \tag{20.92}$$

This finite conductivity of a plasma is a result of the electromagnetic energy losses to generation of *macroscopic* vortex flows.

This mechanism of conductivity of a plasma is different from the usual microscopic one, in which energy losses result from Coulomb collisions of current-carrying electrons with thermal electrons and ions of the plasma. It is no accident that an expression for conductivity, which is equivalent to (20.92), has emerged in quite another problem – while calculating the electrical resistivity of necks in Z-pinches appearing in a highly conductive plasma (Chernov and Yan'kov, 1982).

Note in this context that the σ-dependent force, as well as the characteristic velocity of the plasma flow, depends in a *non-linear* way on the quantity $E_0 B_0$. Using (20.88), (20.88) and (20.91), we see that

$$f \sim \begin{cases} (E_0 B_0)^{1/2}\,, & \mathrm{Re} \ll 1\,, \\ (E_0 B_0)^{2/3}\,, & \mathrm{Re} \gg 1\,. \end{cases} \tag{20.93}$$

Litvinenko and Somov (1994) have supposed that

the magnetic σ-dependent force may play an important part in the dynamics of astrophysical plasma with a non-uniform distribution of temperature and, hence, electric conductivity.

It is this force that can generate large-scale vortex flows of plasma in space. This possibility is illustrated in the next Section.

20.4.2 The σ-dependent force in solar prominences

The solar corona is a natural 'plasma physics laboratory' where formula (20.91), which is applicable at large magnetic Reynolds numbers, can be tested. Recall several of its characteristics: low density $\rho_0 \approx 10^{-16}\,\mathrm{g\,cm^{-3}}$, high temperature $T_0 \approx 10^6\,\mathrm{K}$, dynamic viscosity $\eta \approx 1\,\mathrm{g\,cm^{-1}\,s^{-1}}$, magnetic field $B_0 \approx 10 - 100$ G, electric field $E_0 \approx 10^{-5}$ CGSE units.

On the other hand, according to observational data (Tandberg-Hanssen, 1995), prominences consist of numerous fine threads – *cold dense* formations having a transversal scale $a \approx 10^7$ cm and temperature $T_1 \approx 10^4$ K. Hence the ratio

$$\sigma_1 / \sigma_0 \approx 10^{-3} \ll 1 ,$$

as applied to prominences in the corona. In the vicinity of the threads, as well as near a prominence as a whole, rather fast plasma flows are actually observed.

According to the model under discussion, these flows can be generated by the vortex component of the magnetic σ-dependent force. For Re $\ll 1$, their maximum velocity, as follows from relations (20.88) and (20.91), is determined by the expression

$$v_0 \approx \left(\frac{c E_0 B_0 a}{4 \pi \eta} \right)^{1/2} \approx 10 - 30 \text{ km s}^{-1}, \tag{20.94}$$

that, generally speaking, corresponds to the characteristic values of observed velocities. However the spatial resolution of modern optical, EUV and soft X-ray observations is smaller than is necessary for the model to be confirmed or refuted. Let us consider another possibility.

The symmetric distribution of velocities on the line-of-sight projection (i.e., in the direction towards the observer) is a distinguishing feature of the model since it predicts the presence of **a large number of vortex flows of plasma** inside the prominence. Such a distribution can be observed as a symmetric broadening of spectral lines, which it will be necessary to study if one wishes to study the effect quantitatevly. A similar observational effect can be related to the existence of reconnecting current layers in the same region (Antonucci and Somov, 1992; Antonucci et al., 1996).

The gravity force acting on the prominences is supposed to be balanced by the σ-dependent expulsion. The equilibrium condition makes it possible to evaluate the characteristic value of the plasma density related to the fine threads forming the prominence

$$(\rho_1 - \rho_0) \, g_{\odot} \approx f \, . \tag{20.95}$$

Here the specific gravity of the Sun $g_{\odot} \approx 3 \times 10^4 \text{ cm s}^{-2}$. Formulae (20.91), (20.92), and (20.95) result in

$$\rho_1 \approx \left(\frac{c E_0 B_0 \, \eta}{4 \pi \, g_{\odot}^2 a^3} \right)^{1/2} \approx 3 \times 10^{-13} \text{ g cm}^{-3}, \tag{20.96}$$

in accordance with observational data.

Even faster flows with characteristic velocities $10^2 - 10^3 \text{ km s}^{-1}$ in so-called *eruptive* prominences are probably a consequence of the fact that the coronal fields $\mathbf{E}_0$ and $\mathbf{B}_0$ can change (in magnitude or direction) during the course of evolution. As this takes place, the equilibrium described by equation (20.95) can be violated.

Observations with high spectral resolution in EUV and soft X-ray ranges are necessary to study the effect of the magnetic force stimulated by the presence of plasma regions with considerably different conductivity in the solar atmosphere.

20.5 Practice: Exercises and Answers

Exercise 20.1. Discuss a possible behavior of electrically conducting spheres in an insulating bounded fluid placed in a vertical traveling magnetic field.

Hint. The spheres move in response to the induced electromagnetic forces, the motion being influenced by gravity, viscous drag, vessel boundary reaction, and collisions. The range of possible behaviors, stable, unstable, and chaotic, is very wide. The term 'electromagnetic billiards' seems appropriate to describe this phenomenon (Bolcato et al., 1993).

Appendix 1. Notation

Latin alphabet

Symbol	*Description*	*Introduced in Section (Formula)*
a	current layer half-thickness	8.3
$\mathbf{A}$	vector potential of a magnetic field	6.2
b	half-width of a reconnecting current layer (RCL)	8.3
$\mathbf{b}$	perturbation of a magnetic field	20.2.1
$\mathbf{B}$	magnetic field	1.2
$\mathbf{B}_\tau$	tangential magnetic field	16.2
e, e_a	electric charge	1.2
$\mathbf{e}_c$	unit vector from the curvature centre	5.2
$\mathcal{E}$	energy of a particle	5.1
$\mathbf{E}$	electric field	1.2
$\mathbf{E}_u$	electric field in the plasma rest-frame	11.1
f_k	averaged distribution function for particles of kind k	1.1
f_{kl}	binary correlation function	2.2
f_{kln}	triple correlation function	2.3
$\hat{f}_k$	exact distribution function for particles of kind k	2.2
F	complex potential	14.2
$\mathbf{F}, \mathbf{F}_k$	force	1.1
$\langle \mathbf{F}_k \rangle_v$	mean force per unit volume	9.1
$\mathbf{F}_{\mathbf{kl}}$	force density in the phase space	2.2
$\mathbf{F}'$	fluctuating force	2.1
g	velocity-integrated correlation function	3.2

G	gravitational constant	1.2
$\mathbf{G}$	energy flux density	(1.52)
$\mathbf{h}$	magnetic field at a wave front	20.1
Ha	Hartmann number	20.2
$\mathbf{j}$	electric current density	1.2
$\mathbf{j}'$	current density in the plasma rest-frame	11.1
$\mathbf{j}_k^{\rm q}$	current density due to particles of kind k	9.1
$\mathbf{j}_k$	particle flux density in the phase space	3.1
J	electric current	19.3
k	friction coefficient	1.1
$\mathbf{k}$	wave vector	15.1
$\mathcal{K}$	kinetic energy of a particle	(5.58)
m	magnetic dipole moment	14.4
m, m_a	particle mass	1.2
M	mass of star	19.1
$\mathcal{M}$	magnetic moment of a particle	5.2
	magnetic energy of a system	19.1
n, n_k	number density	8.1
$\mathbf{n}$	unit vector along a magnetic field	5.1
N_k	number of particles of kind k	1.1
p_k	gas pressure of particles of kind k	9.1
$p_{\rm m}$	magnetic pressure	15.1
$p_{\alpha\beta}$	pressure tensor	9.1
$\mathbf{p}$	particle momentum	5.1
$\mathbf{P}$	generalized momentum	6.2
q	generalized coordinate	6.2
$\mathbf{q}$	heat flux density	12.1
$\mathbf{q}_k$	heat flux density due to particles of kind k	9.1
Q_k	rate of energy release in a gas of particles of kind k	9.1
$r_{\rm D}$	Debye radius	8.2
$r_{\rm L}$	Larmor radius	5.1
$\mathbf{r}_a$	coordinates of ath particle	1.2
R	radius of star	14.4
$R_\perp$	guiding centre spiral radius	5.2
$\mathcal{R}$	rigidity of a particle	5.1
$\mathbf{R}$	guiding centre vector	5.2
Re	Reynolds number	12.3
$\mathrm{Re}_{\rm m}$	magnetic Reynolds number	12.3
s	entropy per unit mass	12.2
T	temperature	12.2
	kinetic energy of a macroscopic motion	19.1
$T_{\rm B}$	period of the Larmor rotation	5.2
$T_{\alpha\beta}$	Maxwellian stress tensor	12.1

$\mathbf{u}$	relative velocity	5.1
	velocity of the centre-of-mass system	11.1
$\mathbf{u}_e$	mean electron velocity	11.1
$\mathbf{u}_i$	mean ion velocity	11.1
$\mathbf{u}_k$	mean velocity of particles of kind k	9.1
U	interaction potential	8.1
	volume of a fluid particle	14.2
	specific volume of a magnetic tube	19.3
U_{th}	thermal energy	19.1
$\mathbf{U}$	velocity of the moving reference frame	16.2
	shock speed	17.1
$\mathbf{v}$	macroscopic velocity of a plasma	12.2
$\mathbf{v}, \mathbf{v}_a$	particle velocity	1.2
$\mathbf{v}_{\mathrm{d}}$	drift velocity	5.1
v_{n}	normal component of the velocity	16.2
v_x	velocity orthogonal to a discontinuity surface	16.1
$\mathbf{v}'$	deviation of particle velocity from its mean value	9.1
$\mathbf{v}_\tau$	tangential velocity	16.1
$v_{\parallel}$	velocity component along the magnetic field lines	5.1
$v_{\perp}$	transversal velocity	5.1
V_{A}	Alfvén speed	13.1
$\mathbf{V}_{\mathrm{gr}}$	group velocity of a wave	15.1
$\mathbf{V}_{\mathrm{ph}}$	phase velocity of a wave	15.1
V_{s}	sound speed velocity	15.1
V_{Te}	mean thermal velocity of electons	(5.54)
V_{Ti}	mean thermal velocity of ions	(5.53)
V_{Tp}	mean thermal velocity of protons	(5.55)
$V_{\pm}$	speed of a fast (slow) magnetoacoustic wave	15.1
w	probability density	2.1
w, w_k	heat function per unit mass	9.1
W	energy density of an electromagnetic field	(1.51)
X	phase space	1.1
Z	ion charge number	8.2

Greek alphabet

Symbol	*Description*	*Introduced in Section (Formula)*
α_{B}	parameter of the magnetic field inhomogeneity	5.2
α_{E}	parameter of the electric field inhomogeneity	5.2
β	coefficient in an expulsion force	20.3
γ	dimensionless parameter of ideal MHD	13.1
γ_g	ratio of specific heats	16.1
Γ	$6N$-dimensional phase space	2.1
δ	dimensionless parameter of ideal MHD	12.3
ε	mean kinetic energy of a chaotic motion	12.1
	dimensionless parameter of ideal MHD	13.1
ζ	second viscosity coefficient	12.2
ζ_{i}	interaction parameter	3.1
ζ_{p}	plasma parameter	3.1
η	first viscosity coefficient (dynamic viscosity)	12.2
θ	pitch-angle	5.1
	angle between a wave vector and the magnetic field	15.1
κ_{e}	classical electron conductivity	8.3
λ	mean free path	8.1
$\ln \Lambda$	Coulomb logarithm	8.1
ν	collisional frequency	8.1
ν	kinematic viscosity	12.2
ν_{ei}	electron-ion mean collisional frequency	11.1
ν_{kl}	mean collisional frequency	9.1
ν_{m}	magnetic diffusivity	12.2
ξ	column depth	8.3
$\pi_{\alpha\beta}^{(k)}$	viscous stress tensor	9.1
$\Pi_{\alpha\beta}^{*}$	total momentum flux density tensor	12.2
ρ	plasma mass density	9.1
ρ_k	mass density for particles of kind k	9.1
ρ^{q}	electric charge density	1.2
ρ_k^{q}	charge density due to particles of kind k	9.1

$\boldsymbol{\rho}$	rotational motion vector	5.2
σ	isotropic electric conductivity	11.1
σ_{H}	Hall conductivity	11.1
$\sigma_{\parallel}$	conductivity parallel to the magnetic field	11.1
$\sigma_{\perp}$	conductivity perpendicular to the magnetic field	11.1
$\sigma_{\alpha\beta}^{\mathrm{v}}$	viscous stress tensor	12.2
τ	characteristic time scale	5.2
τ_{ee}	electron collisional time	8.3
τ_{ei}	electron-ion collisional time	8.3
τ_{ii}	ion collisional time	8.3
ϕ	gravitational potential	1.2
φ	electrostatic potential	8.2
φ	angle in the spherical frame	14.4
ϕ, φ	angle in the cylindrical frame	19.2
$\hat{\varphi}_k$	deviation of the exact distribution function from an averaged distribution function	2.2
Φ	magnetic flux	14.2
	stream function	14.4
χ	deflection angle	8.1
ψ	angle to the x axis	14.4
	potential of an electric current	20.3
Ψ	potential of a current-free magnetic field	13.1
ω	wave frequency	15.1
ω_0	wave frequency in a moving frame of reference	15.1
ω_{B}	cyclotron or Larmor frequency	5.1
ω_{pl}	electron plasma frequency	8.2
Ω	gravitational energy	19.1
$\boldsymbol{\omega}$	vector of angular velocity	20.1

Appendix 2 Useful Expressions

Source formulae

Larmor frequency of a non-relativistic electron (5.11), (5.51)

$$\omega_{\mathrm{B}}^{(\mathrm{e})} = \frac{eB}{m_{\mathrm{e}}c} \approx 1.76 \times 10^{7}\, B\,(\mathrm{G})\,, \quad \mathrm{rad\ s^{-1}}\,.$$

Larmor frequency of a non-relativistic proton (5.52)

$$\omega_{\mathrm{B}}^{(\mathrm{p})} \approx 9.58 \times 10^{3}\; B\,(\mathrm{G})\,, \quad \mathrm{rad\ s^{-1}}\,.$$

Larmor radius of a non-relativistic electron (5.14), (5.59)

$$r_{\mathrm{L}}^{(\mathrm{e})} = \frac{c p_{\perp}}{eB} \approx 5.69 \times 10^{-8}\, \frac{v\,(\mathrm{cm\,s^{-1}})}{B\,(\mathrm{G})}\,, \quad \mathrm{cm}\,.$$

Larmor radius of a non-relativistic proton (5.14), (5.61)

$$r_{\mathrm{L}}^{(\mathrm{p})} \approx 1.04 \times 10^{-4}\, \frac{v\,(\mathrm{cm\,s^{-1}})}{B\,(\mathrm{G})}\,, \quad \mathrm{cm}\,.$$

Mean thermal velocity of electrons (5.54)

$$V_{\mathrm{Te}} = \left(\frac{3 k_{\mathrm{B}}\, T_{\mathrm{e}}}{m_{\mathrm{e}}} \right)^{1/2} \approx 6.74 \times 10^{5} \sqrt{T_{\mathrm{e}}\,(\mathrm{K})}\,, \quad \mathrm{cm\ s^{-1}}\,.$$

Mean thermal velocity of protons (5.55)

$$V_{\mathrm{Tp}} \approx 1.57 \times 10^{4} \sqrt{T_{\mathrm{p}}\,(\mathrm{K})}\,, \quad \mathrm{cm\ s^{-1}}\,.$$

Larmor radius of non-relativistic *thermal* electrons (5.56)

$$r_{\mathrm{L}}^{(\mathrm{e})} = \frac{V_{\mathrm{Te}}}{\omega_{\mathrm{B}}^{(\mathrm{e})}} \approx 3.83 \times 10^{-2}\, \frac{\sqrt{T_{\mathrm{e}}\,(\mathrm{K})}}{B\,(\mathrm{G})}\,, \quad \mathrm{cm}\,.$$

Larmor radius of non-relativistic *thermal* protons (5.57)

$$r_{\rm L}^{(\rm p)} = \frac{V_{\rm Tp}}{\omega_{\rm B}^{(\rm p)}} \approx 1.64\, \frac{\sqrt{T_{\rm p}\,({\rm K})}}{B\,({\rm G})}\,, \quad {\rm cm}\,.$$

Drift velocity (5.20)

$$\mathbf{v}_{\rm d} = \frac{c}{e}\, \frac{\mathbf{F} \times \mathbf{B}}{B^2}\,.$$

Magnetic moment of a particle on the Larmor orbit (6.6)

$$\mathcal{M} = \frac{1}{c}\, JS = \frac{e\, \omega_{\rm B}\, r_{\rm L}^2}{2c} = \frac{p_{\perp}^2}{2mB} = \frac{\mathcal{E}_{\perp}}{B}\,.$$

Debye radius ($T_{\rm e} = T\,,\ T_{\rm i} = 0$ or $T_{\rm e} \gg T_{\rm i}$) (8.33)

$$r_{\rm D} = \left(\frac{k_{\rm B} T}{4\pi\, n e^2} \right)^{1/2}.$$

Debye radius in electron-proton thermal plasma ($T_{\rm e} = T_{\rm p} = T$) (8.77)

$$r_{\rm D} = \left(\frac{k_{\rm B} T}{8\pi\, e^2\, n} \right)^{1/2} \approx 4.9 \left(\frac{T}{n} \right)^{1/2}, \quad {\rm cm}\,.$$

Coulomb logarithm (8.75)

$$\ln \Lambda = \ln \left[\left(\frac{3 k_{\rm B}^{3/2}}{2\pi^{1/2}\, e^3} \right) \left(\frac{T_{\rm e}^3}{n_{\rm e}} \right)^{1/2} \right] \approx \ln \left[1.25 \times 10^4 \left(\frac{T_{\rm e}^3}{n_{\rm e}} \right)^{1/2} \right].$$

Electron plasma frequency (8.78)

$$\omega_{pl}^{(\rm e)} = \left(\frac{4\pi\, e^2\, n_{\rm e}}{m_{\rm e}} \right)^{1/2} \approx 5.64 \times 10^4\, \sqrt{n_{\rm e}}\,, \quad {\rm rad\ s^{-1}}\,.$$

Thermal electron collisional time (8.80)

$$\tau_{\rm ee} = \frac{m_{\rm e}^2}{0.714\, e^4\, 8\pi\, \ln \Lambda}\, \frac{V_{\rm Te}^3}{n_{\rm e}} \approx 4.04 \times 10^{-20}\, \frac{V_{\rm Te}^3}{n_{\rm e}}\,, \quad {\rm s}\,.$$

Thermal proton collisional time (8.81)

$$\tau_{\rm pp} = \frac{m_{\rm p}^2}{0.714\, e^4\, 8\pi\, \ln \Lambda}\, \frac{V_{\rm Tp}^3}{n_{\rm p}} \approx 1.36 \times 10^{-13}\, \frac{V_{\rm Tp}^3}{n_{\rm p}}\,, \quad {\rm s}\,.$$

Electron-ion collision (energy exchange) time Section 8.3

$$\tau_{\rm ei}\,(\mathcal{E}) = \frac{m_{\rm e} m_{\rm i}\, [\, 3 k_{\rm B}\, (T_{\rm e}/m_{\rm e} + T_{\rm i}/m_{\rm i})\,]^{3/2}}{e_{\rm e}^2\, e_{\rm i}^2\, (6\pi)^{1/2}\, 8\, \ln \Lambda}\,.$$

Time of energy exchange between electrons and protons (8.44)

$$\tau_{\rm ep}\,(\mathcal{E}) \approx 22\,\tau_{\rm pp} \approx 950\,\tau_{\rm ee}\,.$$

Dreicer field (8.83)

$$E_{\rm Dr} = \frac{4\pi e^3 \ln\Lambda}{k_{\rm B}}\,\frac{n_{\rm e}}{T_{\rm e}} \approx 6.54\times 10^{-8}\,\frac{n_{\rm e}}{T_{\rm e}}\,, \;\; {\rm V\,cm^{-1}}\,.$$

Conductivity of magnetized plasma Section 11.3

$$\sigma_{\parallel} = \sigma = \frac{e^2 n}{m_{\rm e}}\,\tau_{\rm ei} \approx 2.53\times 10^{8}\,n\,({\rm cm^{-3}})\,\tau_{\rm ei}\,({\rm s})\,, \;\; {\rm s^{-1}}\,,$$

$$\sigma_{\perp} = \sigma\,\frac{1}{1+\left(\omega_{\rm B}^{\rm (e)}\,\tau_{\rm ei}\right)^2}\,, \qquad \sigma_{\rm H} = \sigma\,\frac{\omega_{\rm B}^{\rm (e)}\,\tau_{\rm ei}}{1+\left(\omega_{\rm B}^{\rm (e)}\,\tau_{\rm ei}\right)^2}\,.$$

Magnetic diffusivity (or viscosity) (12.49)

$$\nu_{\rm m} = \frac{c^2}{4\pi\sigma} \approx 7.2\times 10^{19}\,\frac{1}{\sigma}\,, \;\; {\rm cm^2\,s^{-1}}\,.$$

Magnetic Reynolds number (12.62)

$${\rm Re_m} = \frac{L^2}{\nu_{\rm m}\,\tau} = \frac{vL}{\nu_{\rm m}}$$

Alfvén speed (13.14), (13.34)

$$V_{\rm A} = \frac{B}{\sqrt{4\pi\rho}} \approx 2.18\times 10^{11}\,\frac{B}{\sqrt{n}}\,, \;\; {\rm cm\ s^{-1}}\,.$$

Sound speed in electron-proton plasma (16.98)

$$V_s = \left(\gamma_g\,\frac{p}{\rho}\right)^{1/2} \approx 1.66\times 10^4\,\sqrt{T({\rm K})}\,, \;\; {\rm cm\ s^{-1}}\,.$$

Electric field in magnetized plasma (19.71)

$$E \approx \frac{1}{c}\,vB \approx 10^{-8}\,v\,({\rm cm\ s^{-1}})\,B\,({\rm G})\,, \;\; {\rm V\ cm^{-1}}\,.$$

Appendix 3. Constants

Fundamental physical constants

Speed of light	c	$2.998 \times 10^{10}\,\mathrm{cm\ s^{-1}}$
Electron charge	e	$4.802 \times 10^{-10}\,\mathrm{CGSE}$
Electron mass	m_e	$9.109 \times 10^{-28}\,\mathrm{g}$
Proton mass	m_p	$1.673 \times 10^{-24}\,\mathrm{g}$
Boltzmann constant	k_{B}	$1.381 \times 10^{-16}\,\mathrm{erg\ K^{-1}}$
Gravitational constant	G	$6.673 \times 10^{-8}\,\mathrm{dyne\ cm^2\ g^{-2}}$
Planck's constant	h	$6.625 \times 10^{-27}\,\mathrm{erg\ s}$

Some useful constants and units

Ampere (current)	A	3×10^9 CGSE
Angström (length)	A	10^{-8} cm
Electron Volt (energy)	eV	1.602×10^{-12} erg
	eV	11605 K
Gauss (magnetic induction)	G	3×10^{10} CGSE
Henry (inductance)	H	$1.111 \times 10^{-12}\,\mathrm{s^2\ cm^{-1}}$
Ionization potential of hydrogen		13.60 eV
Joule (energy)	J	10^7 erg
Maxwell (magnetic flux)	M	3×10^{10} CGSE
Ohm (resistance)	Ω	$1.111 \times 10^{-12}\,\mathrm{s\ cm^{-1}}$
Tesla (magnetic induction)		10^4 Gauss
Volt (potential)	V	3.333×10^{-3} CGSE
Watt (power)	W	$10^7\,\mathrm{erg\ s^{-1}}$
Weber (magnetic flux)	Wb	10^8 Maxwell

Some astrophysical constants

Astronomical unit	AU	1.496×10^{13} cm
Mass of the Sun	$M_\odot$	1.989×10^{33} g

Mass of the Earth	M_E	5.98×10^{27} g
Solar radius	$R_\odot$	6.960×10^{10} cm
Solar surface gravity	$g_\odot$	2.740×10^{4} cm s^{-2}
Solar luminosity	$L_\odot$	3.827×10^{33} erg s^{-1}
Mass loss rate	$\dot{M}_\odot$	10^{12} g s^{-1}
Rotation period of the Sun	$T_\odot$	26 days (at equator)

Bibliography

Each reference is cited in the Sections of the book indicated within square brackets.

Acton, L.: 1996, Coronal structures, local and global, in *Magnetohydrodynamic Phenomena in the Solar Atmosphere: Prototypes of Stellar Magnetic Activity*, eds Y. Uchida, T. Kosugi, and H. Hudson, Dordrecht, Kluwer Academic Publ., p. 3–11. [§ 19.3]

Akhiezer, A.I., Lyubarskii, G.Ya., and Polovin, R.V.: 1959, On the stability of shock waves in MHD, *Soviet Physics–JETP*, v. 8, No. 3, 507–512. [§ 17.2]

Alekseyev, I.I. and Kropotkin, A.P.: 1970, Passage of energetic particles through a MHD discontinuity, *Geomagn. Aeron.*, v. 10, No. 6, 755–758. [§ 18.3]

Alfvén, H.: 1949, On the solar origin of cosmic radiation, *Phys. Rev.*, v. 75, No. 11, 1732–1735. [§ 7.2]

Alfvén, H.: 1950, *Cosmic Electrodynamics*, Oxford, Clarendon Press, p. 228. [Intr., § 12.2, § 13.4, § 15.2, § 20.1]

Alfvén, H.: 1981, *Cosmic Plasma*, Dordrecht, D. Reidel Publ., p. 164. [§ 20.1]

Alfvén, H. and Fälthammar, C.-G.: 1963, *Cosmic Electrodynamics*, Oxford, Clarendon Press, p. 228. [§ 8.1, § 8.2, § 11.1]

Altyntsev, A.T., Krasov, V.I., and Tomozov V.M.: 1977, Magnetic field dissipation in neutral current sheets, *Solar Phys.*, v. 55, No. 1, 69–81. [§ 12.3]

Anderson, J.E.: 1963, *Magnetohydrodynamic Shock Waves*, Cambridge, Massachusetts; M.I.T. Press, p. 226. [§16.2, § 17.4]

Andres, U.T., Polak, L.S., and Syrovatskii, S.I.: 1963, Electromagnetic expulsion of spherical bodies from a conductive fluid, *Soviet Phys.–Technical Physics*, v. 8, No. 3, 193–196. [§ 19.4, § 20.4]

Anile, A.M.: 1989, *Relativistic Fluids and Magneto-Fluids*, Cambridge Univ. Press, p. 336. [§ 12.2]

Antonucci, E. and Somov, B.V.: 1992, A diagnostic method for reconnecting magnetic fields in the solar corona, in *Coronal Streamers, Coronal Loops, and Coronal and Solar Wind Composition*, Proc. First SOHO Workshop, ESA SP-348, p. 293–294. [§ 8.3, § 20.4]

Antonucci, E., Benna, C., and Somov, B.V.: 1996, Interpretation of the observed plasma 'turbulent' velocities as a result of reconnection in solar flares, *Astrophys. J.*, v. 456, No. 2, 833–839. [§ 8.3, § 20.4]

Aschwanden, M.J.: 2002, *Particle Acceleration and Kinematics in Solar Flares: A Synthesis of Recent Observations and Theoretical Concepts*, Dordrecht, Boston, London; Kluwer Academic Publ., 227 p. [§ 4.5]

Aschwanden, M.J., Kliem, B., Schwarz, U., et al.: 1998, Wavelet analysis of solar flare hard X-rays, *Astrophys. J.*, v. 505, No. 2, 941–956. [§ 4.5]

Atoyan, A.M. and Aharonian, F.A.: 1999, Modelling of the non-thermal flares in the Galactic microquasar GRS 1915+105, *Mon. Not. Royal Astron. Soc.*, v. 302, No. 1, 253–276. [§ 20.1]

Axford, W.I., Leer, E., and Skadron, G.: 1977, The acceleration of cosmic rays by shock waves, *Proc. 15th Int. Cosmic Ray Conf.*, Plovdiv, v. 11, p. 132–137. [§ 18.2]

Bachiller, R.: 1996, Bipolar molecular outflows from young stars and protostars, *Ann. Rev. Astron. Astrophys.*, v. 34, 111–154. [§ 20.2]

Bagalá, L.G., Mandrini, C.H., Rovira, M.G., et al.: 1995, A topological approach to understand a multi-loop flare, *Solar Phys.*, v. 161, No. 1, 103–121. [Intr.]

Bai, T., Hudson, H.S., Pelling, R.M., et al.: 1983, First-order Fermi acceleration in solar flares as a mechanism for the second-step acceleration of protons and electrons, *Astrophys. J.*, v. 267, No. 1, 433–441. [§ 6.2.4]

Balbus, S.A. and Papaloizou, J.C.B.: 1999, On the dynamical foundations of α disks, *Astrophys. J.*, v. 521, No. 2, 650–658. [§ 13.2]

Balescu, R.: 1963, *Statistical Mechanics of Charged Particles*, London, New York, Sydney; Interscience Publ., John Wiley and Sons, Ltd., p. 477. [§ 4.1]

Balescu, R.: 1975, *Equilibrium and Nonequilibrium Statistical Mechanics*, New York, London, Sydney, Toronto; A Wiley-Interscience Publ., John Wiley and Sons, Ltd. [§ 3.1]

Balescu, R.: 1988, *Transport Processes in Plasmas*, Amsterdam, [§ 9.4]

Balikhin, M., Gedalin, M., and Petrukovich, A.: 1993, New mechanism for electron heating in shocks, *Phys. Rev. Lett.*, v. 70, 1259–1262. [§ 18.3]

Balogh, A. and Erdös, G.: 1991, Fast acceleration of ions at quasi-perpendicular shocks, *J. Geophys. Res.*, v. 96, No. A9, 15853–15862. [§ 18.3]

Barenblatt, G.I.: 1979, *Similarity, Self-Similarity, and Intermediate Asymptotics*, New York, Consultants Bureau, Plenum. [§ 20.4]

Bednarek, W. and Protheroe, R.J.: 1999, Gamma-ray and neutrino flares produced by protons accelerated on an accretion disc surface in active galactic nuclei, *Mon. Not. Royal Astron. Soc.*, v. 302, 373–380. [§ 13.2]

Begelman, M.C., Blandford, R.D., and Rees, M.J.: 1984, Theory of extragalactic radio sorces, *Rev. Mod. Phys.*, v. 56, No. 2, 255–351. [§ 7.3, § 13.3, § 20.1]

Beloborodov, A.M.: 1999, Plasma ejection from magnetic flares and the X-ray spectrum of Cygnus X-1, *Astrophys. J.*, v. 510, L123–L126. [§ 13.2]

Benz, A.: 2002, *Plasma Astrophysics: Kinetic Processes in Solar and Stellar Coronae, Second Edition*, Dordrecht, Kluwer Academic Publ., p. 299.

[§ 3.1, § 7.1]

Bernstein, I.B., Frieman, E.A., Kruskal, M.D., et al.: 1958, An energy principle for hydromagnetic stability problems, *Proc. Royal Soc.*, v. 244, No. A1, 17–40. [§ 19.3]

Bertin, G.: 1999, *The Dynamics of Galaxies*, Cambridge Univ. Press, p. 448. [§ 1.3, § 9.6]

Bethe, H.A.: 1942, Office of Scientific Research and Development, Rep. No. 445. [§ 17.1]

Bezrodnykh, S.I. and Vlasov, V.I.: 2002, The Riemann-Hilbert problem in a complicated domain for the model of magnetic reconnection in plasma, *Computational Mathematics and Mathematical Physics*, v. 42, No. 3, 263-298. [§ 14.2.2]

Bhatnagar, P.L., Gross, E.P., and Krook, M.: 1954, A model for collision processes in gases. 1. Small amplitude processes in charged and neutral one-component systems, *Phys. Rev.*, v. 94, No. 3, 511-525. [§ 9.7]

Bhattacharjee, A.: 2004, Impulsive magnetic reconnection in the Earth's magnetotail and the solar corona, *Ann. Rev. Astron. Astrophys.*, v. 42, 365-384. [§ 11.4.2]

Bianchini, A., Della Valle, M., and Orio, M. (eds): 1995, *Cataclysmic Variables*, Dordrecht, Boston, London; Kluwer Academic Publ., p. 540. [§ 13.2.2]

Binney, J. and Tremaine, S.: 1987, *Galactic Dynamics*, Princeton, New Jersey; Princeton Univ. Press. [§ 3.3, § 8.5]

Birkinshaw, M.: 1997, Instabilities in astrophysical jets, in *Advanced Topics on Astrophysical and Space Plasmas*, eds E.M. de Gouveia Dal Pino *et al.*, Dordrecht, Kluwer Academic Publ., p. 17–91. [§ 13.3.1]

Biskamp, D. and Welter, H.: 1989, Magnetic arcade evolution and instability, *Solar Phys.*, v. 120, No. 1, 49–77. [§ 19.5]

Blackman, E.G.: 1999, On particle energization in accretion flow, *Mon. Not. Royal Astron. Soc.*, v. 302, No. 4, 723–730. [§ 8.3]

Blackman, E.G. and Field, G.B.: 2000, Constraints on the magnitude of α in dynamo theory, *Astrophys. J.*, v. 534, No. 2, 984–988. [§ 13.1]

Blandford, R.D.: 1994, Particle acceleration mechanisms, *Astrophys. J., Suppl.*, v. 90, No. 2, 515–520. [§ 18.1, § 18.2]

Bliokh, P., Sinitsin, V., and Yaroshenko, V.: 1995, *Dusty and Self-Gravitational Plasmas in Space*, Dordrecht, Kluwer Academic Publ., p. 250. [§ 1.2]

Blokhintsev, D.I.: 1945, Moving receiver of sound, *Doklady Akademii Nauk SSSR (Soviet Physics Doklady)*, v. 47, No. 1, 22–25 (in Russian). [§ 15.2]

Bobrova, N.A. and Syrovatskii, S.I.: 1979, Singular lines of 1D force-free field, *Solar Phys.*, v. 61, No. 2, 379–387. [§ 19.2]

Bocquet, M., Bonazzola, S., Gourgoulhon, E., et al.: 1995, Rotating neutron star models with a magnetic field, *Astron. Astrophys.*, v. 301, No. 3, 757-775. [§ 19.1]

Bodmer, R. and Bochsler, P.: 2000, Influence of Coulomb collisions on isotopic and elemental fractionation in the solar wind, *J. Geophys. Res.*, v. 105, No. A1, 47–60. [§ 8.4, § 10.1]

Bogdanov, S.Yu., Frank, A.G., Kyrei, N.P., et al.: 1986, Magnetic reconnection, generation of plasma fluxes and accelerated particles in laboratory experiments, in *Plasma Astrophys.*, ESA SP-251, 177-183. [§ 12.3]

Bogdanov, S.Yu., Kyrei, N.P., Markov, V.S., et al.: 2000, Current sheets in magnetic configurations with singular X-lines, *JETP Letters*, v. 71, No. 2, 78–84. [§ 12.3]

Bogoliubov, N.N.: 1946, *Problems of a Dynamical Theory in Statistical Physics*, Moscow, State Technical Press (in Russian). [§ 2.4]

Bolcato, R., Etay, J., Fautrelle, Y., et al.: 1993, Electromagnetic billiards, *Phys. Fluids*, v. 5, No. A7, 1852–1853. [§ 20.5]

Boltzmann, L.: 1872, *Sitzungsber. Kaiserl. Akad. Wiss. Wien*, v. 66, 275–284. [§ 3.5]

Boltzmann, L.: 1956, *Lectures on the Theory of Gases*, Moscow, Gostehizdat (in Russian). [§ 3.5]

Bondi, H.: 1952, On spherical symmetrical accretion, *Mon. Not. Royal Astron. Soc.*, v. 112, No. 1, 195–204. [§ 13.2]

Bontemps, S., André, P., Terebey, S., et al.: 1996, Evolution of outflow activity around low-mass embedded young stellar objects, *Astron. Astrophys.*, v. 311, 858–875. [§ 20.2]

Born, M. and Green, H.S.: 1949, *A General Kinetic Theory of Liquids*, Cambridge, Cambridge Univ. Press. [§ 2.4]

Braginskii, S.I.: 1965, Transport processes in plasma, in *Reviews of Plasma Physics*, ed. M. Leontovich, New York, Consultants Bureau, v. 1, 205–311. [§ 9.5, § 10.5, § 11.4.2]

Bridgman, P.W.: 1931, *Dimensional Analysis*, New Haven, Yale Univ. Press, p. 113. [§ 20.4]

Broderick, A., Prakash, M., and Lattimer, J.M.: 2000, The equation of state of neutron star matter in strong magnetic fields, *Astrophys. J.*, v. 537, No. 1, 351–367. [§ 19.1]

Brown, J.C.: 1971, The deduction of energy spectra of non-thermal electrons in flares from the observed dynamic spectra of hard X-ray bursts, *Solar Phys.*, v. 18, No. 2, 489–502. [§ 4.3, § 8.1]

Brown, J.C.: 1972, The directivity and polarization of thick target X-ray bremsstrahlung from flares, *Solar Phys.*, v. 26, No. 2, 441–459. [§ 4.4]

Brown, J.C., McArthur, G.K., Barrett, R.K., et al.: 1998a, Inversion of the thick-target bremsstrahlung spectra from non-uniformly ionised plasmas, *Solar Phys.*, v. 179, No. 2, 379–404. [§ 4.5]

Brown, J.C., Conway, A.J., and Aschwanden, M.J.: 1998b, The electron injection function and energy-dependent delays in thick-target hard X-rays, *Astrophys. J.*, v. 509, No. 2, 911–917. [§ 4.5]

Brown, J.C., Emslie, A.G., and Kontar, E.P.: 2003, The determination and use of mean electron flux spectra in solar flares, *Astrophys. J.*, v. 595,

No. 2, L115–L117. [§ 4.5]

Bykov, A.M., Chevalier, R.A., Ellison, D.C., et al.: 2000, Nonthermal emission from a supernova remnant in a molecular cloud, *Astrophys. J.*, v. 538, No. 1, 203–216. [§ 8.4.1]

Cadjan, M.G. and Ivanov, M.F.: 1999, Langevin approach to plasma kinetics with collisions, *J. Plasma Phys.*, v. 61, No. 1, 89–106. [§ 3.4]

Cai, H.J. and Lee, L.C.: 1997, The generalized Ohm's law in collisionless reconnection, *Phys. Plasmas*, v. 4, No. 3, 509–520. [§ 1.2]

Camenzind, M.: 1995, Magnetic fields and the physics of active galactic nuclei, *Rev. Mod. Astron.*, v. 8, 201–233. [§ 13.3]

Campbell, C.G.: 1997, *Magnetohydrodynamics of Binary Stars*, Dordrecht, Kluwer Academic Publ., p. 306. [§ 13.2]

Cercignani, C.: 1969, *Mathematical Methods in Kinetic Theory*, MacMillan. [§ 3.5]

Chakrabarti, S.K. (ed.): 1999, *Observational Evidence for Black Holes in the Universe*, Dordrecht, Kluwer Academic Publ., p. 399. [§ 8.3]

Chandrasekhar, S.: 1943a, Stochastic problems in physics and astronomy, *Rev. Mod. Phys.*, v. 15, No. 1, 1–89. [§ 3.1, § 8.1, § 8.3]

Chandrasekhar, S.: 1943b, Dynamical friction. 1. General considerations, *Astrophys. J.*, v. 97, No. 1, 255–262. [§ 8.3, § 8.5]

Chandrasekhar, S.: 1943c, Dynamical friction. 2. The rate of escape of stars from clusters and the evidence for the operation of dynamic friction, *Astrophys. J.*, v. 97, No. 1, 263–273. [§ 8.3, § 8.5]

Chandrasekhar, S.: 1981, *Hydrodynamic and Hydromagnetic Stability*, New York, Dover Publ., p. 654. [§ 19.1, § 19.3]

Chandrasekhar, S. and Fermi, E.: 1953, Problems of gravitational stability in the presence of a magnetic field, *Astrophys. J.*, v. 118, No. 1, 116–141. [§ 19.1]

Cherenkov, P.A.: 1934, *C. R. Ac. Sci. U.S.S.R.*, v. 8, 451 (in Russian). [§ 7.4]

Cherenkov, P.A.: 1937, Visible radiation produced by electrons moving in a medium with velocities exceeding that of light, *Phys. Rev.*, v. 52, 378–379. [§ 7.4]

Chernov, A.A. and Yan'kov, V.V.: 1982, Electron flow in low-density pinches, *Soviet J. Plasma Phys.*, v. 8, No. 5, 522–528. [§ 20.4]

Chew, G.F., Goldberger, M.L., and Low, F.E.: 1956, The Boltzmann equation and the one-fluid hydromagnetic equations in the absence of particle collisions, *Proc. Royal Soc. London*, v. A236, No. 1, 112–118. [§ 11.5.1, § 16.4]

Choudhuri, A.R.: 1998, *The Physics of Fluids and Plasmas: An Introduction for Astrophysicists*, Cambridge, Cambridge Univ. Press, p. 427. [Intr., § 19.1]

Clausius, R.: 1870, On a mechanical theorem applicable to heat, *Philosophical Magazine* (Series 4), v. 40, No. 1, 122–127. [§ 19.1]

Cole, J.D. and Huth, J.H.: 1959, Some interior problems of hydromagnetics, *Phys. Fluids*, v. 2, No. 6, 624–626. [§ 14.5]

Collins, G.W.: 1978, *The Virial Theorem in Stellar Astrophysics*, Tucson, Pachart. [§ 19.1]

Courant, R. and Friedrichs, K.O.: 1985, *Supersonic Flow and Shock Waves*, New York, Berlin, Heidelberg, Tokyo; Springer-Verlag, p. 464. [§ 17.1]

Cowling, T.G.: 1976, *Magnetohydrodynamics*, Bristol, Adam Hilger, p. 135. [§ 11.6]

Crooker, N., Joselyn, J.A., and Feynman, J. (eds): 1997, *Coronal Mass Ejections*, Washington, Amer. Geophys. Un., p. 299. [Intr.]

Cuperman, S. and Dryer, M.: 1985, On the heat conduction in multicomponent, non-Maxwellian spherically symmetric solar wind plasmas, *Astrophys. J.*, v. 298, 414–420. [§ 9.5]

Dadhich, N. and Kembhavi, A. (eds): 2000, *The Universe: Visions and Perspectives*, Dordrecht, Kluwer Academic Publ., p. 346. [§ 1.3]

Darwin, C.: 1949, Source of the cosmic rays, *Nature*, v. 164, 1112–1114. [§ 18.1]

Davis, L.Jr.: 1956, Modified Fermi mechanism for the acceleration of cosmic rays, *Phys. Rev.*, v. 101, 351–358. [§ 6.2.4]

de Hoffmann, F. and Teller, E.: 1950, MHD shocks, *Phys. Rev.*, v. 80, No. 4, 692–703. [§ 16.2, § 16.5]

de Martino, D., Silvotti, R., Solheim, J.-E., et al. (eds): 2003, *White Dwarfs*, Dordrecht, Boston, London; Kluwer Academic Publ., p. 429. [§ 1.4, § 3.5]

Debye, P. and Hückel, E.: 1923, *Phys. Zs.*, v. 24, 185. [§ 8.2]

Decker, R.B.: 1983, Formation of shock-spike events in quasi-perpendicular shocks, *J. Geophys. Res.*, v. 88, No. A12, 9959–9973. [§ 18.3]

Decker, R.B.: 1993, The role of magnetic loops in particle acceleration at nearly perpendicular shocks, *J. Geophys. Res.*, v. 98, No. A1, 33–46. [§ 18.3]

Decker, R.B. and Vlahos, L.: 1986, Numerical studies of particle acceleration at turbulent, oblique shocks with an application to prompt ion acceleration during solar flares, *Astrophys. J.*, v. 306, No. 2, 710–729. [§ 18.3.3]

Diakonov, S.V. and Somov, B.V.: 1988, Thermal electrons runaway from a hot plasma during a flare in the reverse-current model and their X-ray bremsstrahlung, *Solar Phys.*, v. 116, No. 1, 119–139. [§ 4.5, § 8.4]

Di Matteo, T., Celotti, A., and Fabian, A.C.: 1999, Magnetic flares in accretion disc coronae and the spectral states of black hole candidates: the case of GX339-4, *Mon. Not. Royal Astron. Soc.*, v. 304, 809–820. [§ 13.2]

Di Matteo, T., Quataert, E., Allen, S.W., et al.: 2000, Low-radiative-efficiency accretion in the nuclei of elliptic galaxies, *Mon. Not. Royal Astron. Soc.*, v. 311, No. 3, 507–521. [§ 13.2]

Di Matteo, T., Johnstone, R.M., Allen, S.W., et al.: 2001, Accretion onto nearby supermassive black holes: *Chandra* constraints on the dominant cluster galaxy NGC 6166, *Astrophys. J.*, v. 550, No. 1, L19–L23. [§ 13.2]

Dokuchaev, V.P.: 1964, Emission of magnetoacoustic waves in the motion of stars in cosmic space, *Soviet Astronomy–AJ*, v. 8, No. 1, 23–31. [§ 15.4]

Drake, J.F. and Kleva R.G.: 1991, Collisionless reconnection and the sawtooth crash, *Phys. Rev. Lett.*, v. 66, No. 11, 1458–1461. [§ 11.2]

Dreicer, H.: 1959, Electron and ion runaway in a fully ionized gas, *Phys. Rev.*, v. 115, No. 2, 238–249. [§ 8.4, § 10.1]

Duijveman, A., Somov, B.V., and Spektor, A.R.: 1983, Evolution of a flaring loop after injection of fast electrons, *Solar Phys.*, v. 88, No. 1, 257–273. [§ 8.3]

Dyer, K.K., Reynolds, S.R., Borkowski, K.J., et al.: 2001, Separating thermal and nonthermal X-rays in supernova remnants. I. Total fits to SN 1006 AD, *Astrophys. J.*, v. 551, No. 1, 439–453. [§ 18.2]

D'yakov, S.P.: 1954, *Zhurnal Exper. Teor. Fiz.*, v. 27, 288–297 (in Russian). [§ 17.5]

Eichler, D.: 1979, Particle acceleration in solar flares by cyclotron damping of cascading turbulence, *Astrophys. J.*, v. 229, No. 1, 413–418. [§ 6.2.4]

Elsasser, W.M.: 1956, Hydromagnetic dynamo theory, *Rev. Mod. Phys.*, v. 28, No. 2, 135–163. [§ 13.1, § 20.1]

Erdös, G. and Balogh, A.: 1994, Drift acceleration at interplanetary shocks, *Astrophys. J., Suppl.*, v. 90, No. 2, 553–559. [§ 18.3]

Falle, S.A. and Komissarov, S.S.: 2001, On the inadmissibility of non-evolutionary shocks, *J. Plasma Phys.*, v. 65, No. 1, 29–58. [§ 16.3]

Fedoryuk, V.M.: 1985, *Ordinary Differential Equations*, Moscow, Nauka (in Russian). [§ 17.4.1]

Feldman, W.C., Bame, S.J., Gary, S.P., et al.: 1982, Electron heating within the Earth's bow shock, *Phys. Rev. Lett.*, v. 49, 199–202. [§ 18.3]

Fermi, E.: 1949, On the origin of cosmic radiation, *Phys. Rev.*, v. 75, 1169–1174. [§ 6.2.4]

Fermi, E.: 1954, Galactic magnetic fields and the origin of cosmic radiation, *Astrophys. J.*, v. 119, No. 1, 1–6. [§ 6.2.4]

Feroci, M., Hurley, K., Duncan, R.C., et al.: 2001, The giant flare of 1998 August 27 from SGR 1900+14. 1. An interpretive study of *Bepposax* and *Ulysses* observations, *Astrophys. J.*, v. 549, No. ?, 1021–1038. [§ 19.1]

Field, G.B.: 1965, Thermal instability, *Astrophys. J.*, v. 142, No. 2, 531–567. [§ 8.3]

Fokker, A.D.: 1914, Die mittlere Energie rotieren der elektrischer Dipole im Strahlungsfeld, *Ann. der Physik*, v. 43, No. 5, 810–820. [§ 3.1]

Fox, D.C. and Loeb, A.: 1997, Do the electrons and ions in X-ray clusters share the same temperature? *Astrophys. J.*, v. 491, No. 2, 459–466. [§ 8.3]

Galeev, A.A., Rosner, R., and Vaiana, G.S.: 1979, Structured coronae of accretion discs, *Astrophys. J.*, v. 229, No. 1, 318–326. [§ 13.2]

Gedalin, M. and Griv, E.: 1999, Collisionless electrons in a thin high Much number shocks: dependence on angle and β, *Ann. Geophysicae*, v. 17, No. 10, p. 1251–1259. [§ 16.4, § 18.3]

Gel'fand, I.M.: 1959, Some problems of the theory of quasilinear equations, *Usp. Mat. Nauk*, v. 14, No. 2, 87–158 (in Russian). [§ 17.1]

Gerbeth, G., Thess, A., and Marty, P.: 1990, Theoretical study of the MHD flow around a cylinder in crossed electric and magnetic fields, *Eur. J. Mech., B/Fluids*, v. 9, No. 3, 239–257. [§ 19.4, § 20.3]

Germain, P.: 1960, Shock waves and shock-wave structure in MHD, *Rev. Mod. Phys.*, v. 32, No. 4, 951–958. [§ 17.4]

Giacalone, J. and Ellison, D.C.: 2000, Three-dimensional numerical simulations of particle injection and acceleration at quasi-perpendicular shocks, *J. Geophys. Res.*, v. 105, No. A6, 12541–12556. [§ 18.1, § 18.3]

Gieseler, U.D.J., Kirk, J.G., Gallant, Y.A., et al.: 1999, Particle acceleration at oblique shocks and discontinuities of the density profile, *Astron. Astrophys.*, v. 435, No. 1, 298–306. [§ 18.2]

Gilman, P.A.: 2000, Fluid dynamics and MHD of the solar convection zone and tachocline, *Solar Phys.*, v. 192, No. 1, 27–48. [§ 20.1]

Ginzburg, V.L. and Syrovatskii, S.I.: 1964, *The Origin of Cosmic Rays*, Oxford, Pergamon Press. [§ 5.1]

Ginzburg, V.L. and Syrovatskii, S.I.: 1965, Cosmic magneto-bremsstrahlung (synchrotron) radiation, *Annual Rev. Astron. Astrophys.*, v. 3, 297–350. [§ 5.3]

Ginzburg, V.L. and Zheleznyakov, V.V.: 1958, On the possible mechanisms of sporadic solar radio emission, *Soviet Astonomy–AJ*, v. 2, No. 5, 653–668. [§ 7.1]

Ginzburg, V., Landau, L., Leontovich, M., et al.: 1946, On the insolvency of the A.A. Vlasov works on general theory of plasma and solid-state matter, *Zhur. Eksp. Teor. Fiz.*, v. 16, No. 3, 246-252 (in Russian). [§ 3.1]

Giovanelli, R.G.: 1949, Electron energies resulting from an electric field in a highly ionized gas, *Phil. Mag.*, Seventh Series, v. 40, No. 301, 206–214. [§ 8.4]

Gisler, G. and Lemons, D.: 1990, Electron Fermi acceleration in collapsing magnetic traps: Computational and analytical models, *J. Geophys. Res.*, v. 95, No. A9, 14925–14938. [§ 18.3]

Glasstone, S. and Loveberg, R.H.: 1960, *Controlled Thermonuclear Reactions*, Princeton, Van Nostrand, p. 523. [Intr.]

Goldreich, P. and Sridhar, S.: 1997, Magnethydrodynamic turbulence revisited, *Astrophys. J.*, v, 485, No. 2, 680–688. [§ 7.2]

Goldston, R.J. and Rutherford, P.H.: 1995, *Introduction to Plasma Physics*, Bristol, Inst. of Phys. Publ., p. 492. [Intr.]

Gombosi, T.I.: 1999, *Physics of the Space Environment*, Cambridge Univ. Press, p. 339. [§ 18.2]

Gorbachev, V.S. and Kel'ner, S.R.: 1988, Formation of plasma condensations in fluctuating strong magnetic field, *Soviet Physics–JETP*, v. 67, No. 9, 1785–1790. [§ 14.4]

Grant, H.L., Stewart, R.W., and Moilliet, A.: 1962, *J. Fluid Mech.*, v. 12, 241–248. [§ 7.2]

Gurevich, A.V.: 1961, On the theory of runaway electrons, *Soviet Physics–JETP*, v. 12, No. 5, 904–912. [§ 8.4]

Gurevich, A.V. and Istomin, Y.N.: 1979, Thermal runaway and convective heat transport by fast electrons in a plasma, *Soviet Physics–JETP*, v. 50, No. 3, 470–475. [§ 8.4]

Gurevich, A.V. and Zhivlyuk, Y.N.: 1966, Runaway electrons in a non-equilibrium plasma, *Soviet Physics–JETP*, v. 22, No. 1, 153–159. [§ 4.5]

Hattori, M. and Umetsu, K.: 2000, A possible route to spontaneous reduction of the heat conductivity by a temperature gradient-driven instability in electron-ion plasmas, *Astrophys. J.*, v. 533, No. 1, 84–94. [§ 8.3]

Hawley, J.F. and Balbus, S.A.: 1999, Instability and turbulence in accretion discs, in *Numerical Astrophysics*, eds S.M. Miyama et al., Dordrecht, Kluwer Academic Publ., p. 187–194. [§ 13.2]

Hawley, J.F., Gammie, C.F., and Balbus, S.A.: 1995, Local three-dimensional magnetohydrodynamic simulations of accretion disks, *Astrophys. J.*, v. 440, No. 2, 742–763. [§ 13.2]

Hénoux, J.-C. and Somov, B.V.: 1987, Generation and structure of the electric currents in a flaring activity complex, *Astron. Astrophys.*, v. 185, No. 1, 306–314. [§ 20.2]

Hénoux, J.-C. and Somov, B.V.: 1991, The photospheric dynamo. 1. Magnetic flux-tube generation, *Astron. Astrophys.*, v. 241, No. 2, 613–617. [§ 11.1, § 20.2]

Hénoux, J.-C. and Somov, B.V.: 1997, The photospheric dynamo. 2. Physics of thin magnetic flux tubes, *Astron. Astrophys.*, v. 318, No. 3, 947–956. [§ 11.1]

Hirotani, K. and Okamoto, I.: 1998, Pair plasma production in a force-free magnetosphere around a supermassive black hole, *Astrophys. J.*, v. 497, No. 2, 563–572. [§ 7.3, § 11.5.2]

Hollweg, J.V.: 1986, Viscosity and the Chew-Goldberger-Low equations in the solar corona, *Astrophys. J.*, v. 306, No. 2, 730–739. [§ 9.5, § 10.5]

Holman, G.D.: 1995, DC electric field acceleration of ions in solar flares, *Astrophys. J.*, v. 452, No. 2, 451–456. [§ 8.4]

Horiuchi, R. and Sato, T.: 1994, Particle simulation study of driven reconnection in a collisionless plasma, *Phys. Plasmas*, v. 1, No. 11, 3587–3597. [§ 1.2, § 11.2]

Hoshino, M., Stenzel, R.L., and Shibata, K. (eds): 2001, *Magnetic Reconnection in Space and Laboratory Plasmas*, Tokyo, Terra Scientific Publ. Co., p. 693. [§ 13.1.3]

Hoyng, P., Brown, J.C., and van Beek, H.F.: 1976, High time resolution analysis of solar hard X-ray flares observed on board the ESRO TD-1A satellite, *Solar Phys.*, v. 48, No. 2, 197–254. [§ 4.5]

Hubrig, S., North, P., and Mathys, G.: 2000, Magnetic Ap stars in the Hertzsprung-Russell diagram, *Astrophys. J.*, v. 539, No. 1, 352–363. [§ 19.1]

Hudson, P.D.: 1965, Reflection of charged particles by plasma shocks, *Mon. Not. Royal Astron. Soc.*, v. 131, No. 1, 23–50. [§ 18.3]

Innes, D.E., Inhester, B., Axford, W.I., et al.: 1997, Bi-directional jets produced by reconnection on the Sun, *Nature*, v. 386, 811–813. [§ 8.3]

Iordanskii, S.V.: 1958, On compression waves in MHD, *Soviet Physics–Doklady*, v. 3, No. 4, 736–738. [§ 16.2]

Iroshnikov, P.S.: 1964, Turbulence of a conducting fluid in a strong magnetic field, *Soviet Astronomy–AJ.*, v. 7, No. 4, 566–571. [§ 7.2]

Jaroschek, C.H., Treumann, R.A., Lesch, H., et al.: 2004, Fast reconnection in relativistic pair plasmas: Analysis of particle acceleration in self-consistent full particle simulations, *Phys. Plasm.*, v. 11, No. 3, 1151–1163. [§ 7.3]

Jeans, J.: 1929, *Astronomy and Cosmogony*, Cambridge Univ. Press. [§ 8.1]

Jones, F.C. and Ellison D.C.: 1991, The plasma physics of shock acceleration, *Space Sci. Rev.*, v. 58, No. 3, 259–346. [§ 18.1, § 18.2, § 18.3]

Jones, M.E., Lemons, D.S., Mason, R.J., et al.: 1996, A grid-based Coulomb collision model for PIC codes, *J. Comput. Phys.*, v. 123, No. 1, 169–181. [§ 3.4]

Kadomtsev, B.B.: 1960, Convective instability of a plasma, in *Plasma Physics and the Problem of Controlled Thermonuclear Reactions*, ed. M.A. Leontovich, London, Oxford; Pergamon Press, v. 4, p. 450–453. [§ 19.3]

Kadomtsev, B.B.: 1966, Hydrodynamic stability of a plasma, in *Reviews of Plasma Physics*, ed. M.A. Leontovich, New York, Consultants Bureau, v. 2, p. 153–198. [§ 19.3]

Kadomtsev, B.B.: 1976, *Collective Phenomena in Plasma*, Moscow, Nauka, p. 238 (in Russian). [§ 7.1]

Kandrup, H.E.: 1998, Collisionless relaxation in galactic dynamics and the evolution of long-range order, *Annals of the New York Acad. of Sci.*, v. 848, 28–47. [§ 3.3]

Kikuchi, H.: 2001, *Electrohydrodynamics in Dusty and Dirty Plasmas*, Dordrecht, Kluwer Academic Publ., p. 207. [§ 1.2]

Kirkwood, J.G.: 1946, The statistical mechanical theory of transport processes. I. General theory, *J. Chem. Phys.*, v. 14, 180–201. [§ 2.4]

Kittel, C.: 1995, *Introduction to Solid State Physics*, 7th edition, John Wiley. [§ 1.4, § 3.5]

Kivelson, M.G. and Russell, C.T. (eds): 1995, *Introduction to Space Physics*, Cambridge, Cambridge Univ. Press, p. 568. [§ 4.1, § 6.2.4]

Klimontovich, Yu.L.: 1975, *Kinetic Theory of Non-ideal Gas and Non-ideal Plasma*, Moscow, Nauka, p. 352 (in Russian). [§ 2.4]

Klimontovich, Yu.L.: 1986, *Statistical Physics*, New York, Harwood Academic. [Intr., § 2.4, § 3.1]

Klimontovich, Yu.L.: 1998, Two alternative approaches in the kinetic theory of a fully ionized plasma, *J. Plasma Phys.*, v. 59, No. 4, 647–656. [§ 3.1]

Klimontovich, Yu.L. and Silin, V.P.: 1961, On magnetic hydrodynamics for a non-isothermal plasma without collisions, *Soviet Physics–JETP*, v. 40, 1213–1223. [§ 11.5, § 16.4]

Kogan, M.N.: 1967, *Dynamics of a Dilute Gas*, Moscow, Nauka (in Russian). [§ 3.5]

Koide, S., Shibata, K., and Kudoh, T.: 1999, Relativistic jet formation from black hole magnetized accreation discs, *Astrophys. J.*, v. 522, 727–752. [§ 12.2, § 13.3]

Kolmogorov, A.N.: 1941, The local structure of turbulence in incompressble viscous fluid for very large Reynolds numbers, *C.R. Acad. Sci. USSR*, v. 30, 201–206. [§ 7.2]

Korchak, A.A.: 1971, On the origin of solar flare X-rays, *Solar Phys.*, v. 18, No. 2, 284–304. [§ 8.1]

Korchak, A.A.: 1980, Coulomb losses and the nuclear composion of the solar flare accelerated particles, *Solar Phys.*, v. 66, No. 1, 149–158. [§ 8.4]

Kotchine, N.E.: 1926, Rendiconti del Circolo Matematico di Palermo, v. 50, 305–314. [§ 17.1]

Kovalev, V.A. and Somov, B.V.: 2002, On the acceleration of solar-flare charged particles in a collapsing magnetic trap with an electric potential, *Astronomy Letters*, v. 28, No. 7, 488–493. [§ 8.1]

Kraichnan, R.H.: 1965, Inertial-range spectrum of hydromagnetic turbulence, *Phys. Fluids*, v. 8, No. 7, 1385–1389. [§ 7.2]

Krall, N.A. and Trivelpiece, A.W.: 1973, *Principles of Plasma Physics*, New York, McGraw-Hill Book Co. [§ 9.5]

Krymskii, G.F.: 1977, A regular mechanism for the acceleration of charged particles on the front of a shock wave, *Soviet Phys. Dokl.*, v. 22, No. 6, 327–328. [§ 18.2]

Kudriavtsev, V.S.: 1958, Energetic diffusion of fast ions in equalibrium plasma, *Soviet Physics–JETP*, v. 7, No. 6, 1075–1079. [§ 4.1]

Kulikovskii, A.G. and Liubimov, G.A.: 1961, On the structure of an inclined MHD shock wave, *Appl. Math. Mech.*, v. 25, No. 1, 171–179. [§ 17.4]

Kunkel, W.B.: 1984, Generalized Ohm's law for plasma including neutral particles, *Phys. Fluids*, v. 27, No. 9, 2369–2371. [§ 11.1]

Lahav, O., Terlevich, E., and Terlevich, R.J. (eds): 1996, *Gravitational Dynamics*, Cambridge, UK; Cambridge Univ. Press, p. 270. [§ 1.3]

Lancellotti, C. and Kiessling, M.: 2001, Self-similar gravitational collapse in stellar dynamics, *Astrophys. J.*, v. 549, L93–L96. [§ 3.5]

Landau, L.D.: 1937, Kinetic equation in the case of Coulomb interaction, *Zh. Exp. Teor. Fiz.*, v. 7, No. 1, 203–212 (in Russian). [§ 3.1]

Landau, L.D.: 1946, On the vibrations of the electron plasma, *J. Phys. USSR*, v. 10, No. 1, 25–30. [§ 3.1, § 7.1]

Landau, L.D. and Lifshitz, E.M.: 1959a, *Fluid Mechanics*, Oxford, London; Pergamon Press, p. 536. [§ 12.2, § 15.4, § 16.1, § 16.2.2, § 20.2]

Landau, L.D. and Lifshitz, E.M.: 1959b, *Statistical Physics*, London, Paris; Pergamon Press, p. 478. [§ 1.1.5, § 1.4, § 3.5, § 16.5]

Landau, L.D. and Lifshitz, E.M.: 1975, *Classical Theory of Field*, 4th edition, Oxford, Pergamon Press, p. 374. [§ 1.2, § 5.1, § 5.3, § 6.2.1, § 7.4, § 13.4, § 18.4, § 19.1]

Landau, L.D. and Lifshitz, E.M.: 1976, *Mechanics*, 3rd edition, Oxford, Pergamon Press, p. 165. [§ 1.1.5, § 1.4, § 6.1, § 8.1, § 19.1]

Landau, L.D., Lifshitz, E.M., and Pitaevskii, L.P.: 1984, *Electrodynamics of Continuous Media*, Oxford, New York; Pergamon Press, p. 460. [§ 11.4.2, § 16.2, § 17.3]

Langmuir, I.: 1928, *Proc. Nat. Acad. Sci. U.S.A.*, v. 14, 627. [§ 3.2.2]

Larrabee, D.A., Lovelace, R.V.E., and Romanova, M.M.: 2003, Lepton acceleration by relativistic collisionless magnetic reconnection, *Astrophys. J.*, v. 586, No. 1, 72–78. [§ 7.3]

Lavrent'ev, M.A. and Shabat, B.V.: 1973, *Methods of the Theory of Complex Variable Functions*, Moscow, Nauka, p. 736 (in Russian). [§ 14.2]

Lax, P.: 1957, Hyperbolic systems of conservation laws, *Comm. Pure Appl. Math.*, v. 10, No. 4, 537–566. [§ 17.1]

Lax, P.: 1973, Hyperbolic Systems of Conservation Laws and the Mathematical Theory of Shock Waves, *SIAM*. [§ 17.1]

Leenov, D. and Kolin, A.: 1954, Theory of electromagnetophoresis. 1. MHD forces experienced by spheric and cylindrical particles, *J. Chemical Phys.*, v. 22, No. 4, 683–688. [§ 20.4]

Leith, C.E.: 1967, Diffusion approximation to inertial energy transfer in isotropic turbulence, *Phys. Fluids*, v. 10, No. 7, 1409–1416. [§ 7.2]

Leontovich, M.A. (ed.): 1960, *Plasma Physics and the Problem of Controlled Thermonuclear Reactions*, London, Oxford, New York, Paris; Pergamon Press, v. 1–4. [Intr.]

Lesch, H. and Pohl, M.: 1992, A possible explanation for intraday variability in active galactic nuclei, *Astron. Astrophys.*, v. 254, No. 1, 29–38. [§ 13.2]

Liberman, M.A.: 1978, On actuating shock waves in a completely ionized plasma, *Soviet Physics–JETP*, v. 48, No. 5, 832–840. [§ 16.2, § 17.4]

Liboff, R.: 2003, *Kinetic Theory: Classical, Quantum, and Relativistic Descriptions*, Heidelberg, New York; Springer, p. 571. [Intr.]

Lichnerowicz, A.: 1967, *Relativistic Hydrodynamics and Magnetohydrodynamics*, New York, Amsterdam, Benjamin, p. 196. [§ 12.2]

Lifshitz, E.M. and Pitaevskii, L.P.: 1981, *Physical Kinetics*, Oxford, New York, Beijing, Frankfurt; Pergamon Press, p. 452. [§ 3.5, § 7.3, § 8.3]

Lin, R.P. and Hudson, H.S.: 1971, 10-100 keV electron acceleration and emission from solar flares, *Solar Phys.*, v. 17, No. 2, 412–435. [§ 4.3]

Lin, R.P., Dennis, B.R., Hurford, G.J., et al.: 2002, The Reuven Ramaty High-Energy Solar Spectroscopic Imager (RHESSI), *Solar Phys.*, v. 210, No. 1, 3-32. [§ 4.5]

Lin, R.P., Krucker, S., Hurford, G.J., et al.: 2003, *RHESSI* observations of particle acceleration and energy release in an intense solar gamma-ray line flare, *Astrophys. J.*, v. 595, No. 2, L69–L76. [§ 4.5]

Litvinenko, Y.E. and Somov, B.V.: 1991a, Solar flares and virial theorem, *Soviet Astronomy–AJ*, v. 35, No. 2, 183–188. [§ 19.1, § 19.2, § 19.5]

Litvinenko, Y.E. and Somov, B.V.: 1991b, Nonthermal electrons in the thick-target reverse-current model for hard X-ray bremsstrahlung, *Solar Phys.*,

v. 131, No. 2, 319–336. [§ 4.5]

Litvinenko, Y.E. and Somov, B.V.: 1994, Electromagnetic expulsion force in cosmic plasma, *Astron. Astrophys.*, v. 287, No. 1, L37–L40. [§ 20.4]

Litvinenko, Y.E. and Somov, B.V.: 2001, Aspects of the global MHD equilibria and filament eruptions in the solar corona, *Space Sci. Rev.*, v. 95, No. 1, 67–77. [§ 19.1, § 19.5]

Liubarskii, G.Ya. and Polovin, R.V.: 1958, Simple magnetoacoustic waves, *Soviet Physics–JETP*, v. 8, No. 2, 351. [§ 16.2]

Lovelace, R.V.E.: 1976, Dynamo model of double radio sources, *Nature*, v. 262, 649–652. [§ 20.1]

Lundquist, S.: 1951, Magneto-hydrostatic fields, *Ark. Fys.*, v. 2, No. 35, 361–365. [§ 19.2]

Macdonald, D.A., Thorne, K.S., Price, R.H., et al.: 1986, Astrophysical applications of black-hole electrodynamics, in *Black Holes: The Membrane Paradigm*, eds. K.S. Thorne, R.H. Price, and D.A. Macdonald, New Haven, London; Yale Univ. Press, p. 121–137. [§ 13.3]

MacDonald, W.M., Rothenbluth, M.N., and Chuck, W.: 1957, Relaxation of a system of particles with Coulomb interactions, *Phys. Rev.*, v. 107, No. 2, 350–353. [§ 4.1]

MacNeice, P., McWhirter, R.W.P., Spicer, D.S., et al.: 1984, A numerical model of a solar flare based on electron beam heating of the chromosphere, *Solar Phys.*, v. 90, No. 2, 357–353. [§ 8.3]

Manmoto, T.: 2000, Advection-dominated accretion flow around a Kerr black hole, *Astrophys. J.*, v. 534, No. 2, 734–746. [§ 8.3, § 13.2]

Markovskii, S.A. and Somov, B.V.: 1989, A model of magnetic reconnection in a current sheet with shock waves, *Fizika Solnechnoi Plasmy* (Physics of Solar Plasma), Moscow, Nauka, 456–472 (in Russian). [§ 14.2.2]

Markovskii, S.A. and Somov, B.V.: 1996, MHD discontinuities in space plasmas: Interrelation between stability and structure, *Space Sci. Rev.*, v. 78, No. 3–4, 443–506. [§ 17.5]

Marty, P. et Alemany, A.: 1983, Écoulement dû à des champs magnétique et électrique croisés autour d'un cylindre de conductivité quelconque, *Journal de Mécanique Théorique et Appliquée*, v. 2, No. 2, 227–243. [§ 19.4, § 20.3]

McDonald, L., Harra-Murnion, L.K., and Culhane, J.L.: 1999, Nonthermal electron energy deposition in the chromosphere and the accompanying soft X-ray flare emission, *Solar Phys.*, v. 185, No. 2, 323–350. [§ 8.3]

Michel, F.C.: 1991, *Theory of Neutron Star Magnetospheres*, Chicago, London; Chicago Univ. Press, p. 456. [§ 7.3, § 11.5.2, § 12.2]

Mikhailovskii, A.B.: 1979, Nonlinear excitation of electromagnetic waves in a relativistic electron-positron plasma, *Soviet J. Plasma Phys.*, v. 6, No. 3, 336–340. [§ 7.3]

Mikhailovskii, A.B., Onishchenko, O.G., and Tatarinov, E.G.: 1985, Alfvén solitons in a relativistic electron-positron plasma, *Plasma Physics and Controlled Fusion*, v. 27, No. 5, 539–556. [§ 7.3]

Mirabel, I.F. and Rodriguez, L.F.: 1998, Microquasars in our Galaxy, *Nature*, v. 392, 673–676. [§ 20.1]

Moffatt, H.K.: 1978, *Magnetic Field Generation in Electrically Conducting Fluids*, London, New York, Melbourne; Cambridge Univ. Press, p. 343. [§ 13.1]

Moreau, R.: 1990, *Magnetohydrodynamics*, Dordrecht, Kluwer Academic Publ., p. 328. [§ 20.1]

Morozov, A.I. and Solov'ev, L.S.: 1966a, The structure of magnetic fields, in: Leontovich M.A. (ed.), *Reviews of Plasma Physics*, New York, Consultans Bureau, v. 2, 1–101. [§ 19.3]

Morozov, A.I. and Solov'ev, L.S.: 1966b, Motion of particles in electromagnetic fields, in: Leontovich M.A. (ed.), *Reviews of Plasma Physics*, New York, Consultants Bureau, v. 2, 201–297. [§ 5.2]

Murata, H.: 1991, Magnetic field intensification and formation of field-aligned current in a non-uniform field, *J. Plasma Physics*, v. 46, No. 1, 29–48. [§ 11.1]

Nakano, T.: 1998, Star formation in magnetic clouds, *Astrophys. J.*, v. 494, No. 2, 587–604. [§ 19.1]

Narayan, R., Garcia, M.R., and McClintock, J.E.: 1997, Advection-dominated accretion and black hole horizons, *Astrophys. J.*, v. 478, No. 2, L79-L82. [§ 8.3]

Negoro, H., Kitamoto, S., Takeuchi, M., et al.: 1995, Statistics of X-ray fluctuations from Cygnus X-1: Reservoirs in the disk? *Astrophys. J.*, v. 452, No. 1, L49–L52. [§ 13.2]

Nishikawa, K.I., Frank, J., Christodoulou, D.M., et al.: 1999, 3D relativistic MHD simulations of extragalactic jets, in *Numerical Astrophysics*, eds S.M. Miyama et al., Dordrecht, Kluwer Academic Publ., p. 217–218. [§ 13.3]

Northrop, T.G.: 1963, *The Adiabatic Motion of Charged Particles*, New York, John Wiley, Interscience. [§ 6.4]

Novikov, I.D. and Frolov, V.P.: 1989, *Physics of Black Holes*, Dordrecht, Boston, London; Kluwer Academic Publ., p. 341. [§ 11.5.2, § 12.2, § 13.3]

Novikov, I.D. and Thorne, K.S.: 1973, in *Black Holes*, eds C.D. Dewitt and B. Dewitt, New York, Gordon and Breach, p. 345–354. [§ 8.3, § 13.2]

Obertz, P.: 1973, Two-dimensional problem of the shape of the magnetosphere, *Geomagn. Aeron.*, v. 13, No. 5, 758–766. [§ 14.2]

Oreshina, I.V. and Somov, B.V.: 1999, Conformal mapping for solving problems of space electrodynamics, *Bull. Russ. Acad. Sci., Physics*, v. 63, No. 8, 1209–1212. [§ 14.5]

Ostriker, E.C.: 1999, Dynamical friction in a gaseous medium, *Astrophys. J.*, v. 513, No. 1, 252–258. [§ 8.5]

Paesold, G. and Benz, A.O.: 1999, Electron firehose instability and acceleration of electrons in solar flares, *Astron. Astrophys.*, v. 351, 741–746. [§ 7.2]

Palmer, P.L.: 1994, *Stability of Collisionless Stellar Systems*, Dordrecht, Kluwer Academic Publ., p. 349. [§ 9.6]

Parker, E.N.: 1979, *Cosmic Magnetic Fields. Their Origin and Their Activity*, Oxford, Clarendon Press, p. 841. [§ 13.1, § 19.3, § 19.4, § 20.1]

Parks, G.K.: 2004, *Physics of Space Plasmas, An Introduction, Second Edition*, Boulder, Oxford; Westview Press, p. 597. [Intr., § 14.5, § 18.1, § 18.2]

Peacock, J.A.: 1999, *Cosmological Physics*, Cambridge, UK; Cambridge Univ. Press, p. 682. [§ 7.3, § 9.6]

Persson, H.: 1963, Electric field along a magnetic line of force in a low-density plasma, *Phys. Fluids*, v. 6, No. 12, 1756–1759. [§ 8.1]

Pfaffelmoser, K.: 1992, Global classic solutions of the Vlasov-Poisson system in three dimensions for general initial data, *J. Diff. Equations*, v. 95, 281–303. [§ 16.5]

Planck, M.: 1917, *Sitz. der Preuss. Akad.*, 324. [§ 3.1]

Polovin, R.V.: 1961, Shock waves in MHD, *Soviet Phys. Usp.*, v. 3, No. 5, 677–688. [§ 16.2, § 17.2]

Polovin, R.V. and Demutskii, V.P.: 1990, *Fundamentals of Magnetohydrodynamics*, New York, Consultants Bureau. [§ 17.4]

Polovin, R.V. and Liubarskii, G.Ya.: 1958, Impossibility of rarefaction shock waves in MHD, *Soviet Physics–JETP*, v. 8, No. 2, 351–352. [§ 16.2]

Priest, E.R.: 1982, *Solar Magnetohydrodynamics*, Dordrecht, Boston, London; D. Reidel Publ. Co., p. 472. [§ 16.2, § 19.3]

Punsly, B.: 2001, *Black Hole Gravitohydromagnetics*, New York, Berlin, Heidelberg, Tokyo; Springer-Verlag, p. 400. [§ 12.2.3]

Ramos, J.I. and Winowich, N.S.: 1986, Magnetohydrodynamic channel flow study, *Phys. Fluids*, v. 29, No. 4, 992–997. [§ 20.2]

Reid, I.N., Liebert, J., and Schmidt, G.D.: 2001, Discovery of a magnetic DZ white dwarf with Zeeman-split lines of heavy elements, *Astrophys. J.*, v. 550, No. 1, L61–L63. [§ 13.2]

Rodrigues-Pacheco, J., Sequeiros, J., del Peral, L., et al.: 1998, Diffusive-shock-accelerated interplanetary ions during the solar cycle 21 maximum, *Solar Phys.*, v. 181, No. 1, 185–200. [§ 18.2]

Roikhvarger, Z.B. and Syrovatskii, S.I.: 1974, Evolutionarity of MHD discontinuities with allowance for dissipative waves, *Soviet Physics–JETP*, v. 39, No. 4, 654–656. [§ 17.1, § 17.3]

Rose, W.K.: 1998, *Advanced Stellar Astrophysics*, Cambridge, UK; Cambridge Univ. Press, p. 494. [§ 1.3, § 5.3, § 7.3, § 12.2, § 13.2]

Rosenbluth, M. and Longmire, C.: 1957, Stability of plasmas confined by magnetic fields, *Ann. Phys.*, v. 1, No. 1, 120–140. [§ 19.3]

Ruderman, M.: 1971, Matter in superstrong magnetic fields: The surface of a neutron star, *Phys. Rev. Lett.*, v. 27, No. 19, 1306–1308. [§ 5.3]

Ruderman, M.A. and Sutherland, P.G.: 1975, Theory of pulsars: Polar gaps, sparks, and coherent radiation, *Astrophys. J.*, v. 196, No. 1, 51–72. [§ 7.3]

Rüdiger, G. and von Rekowski, B.: 1998, Differential rotation and meridional flow for fast-rotating solar-type stars, *Astrophys. J.*, v. 494, No. 2, 691–699. [§ 13.1, § 20.1]

Ruffolo, D.: 1999, Transport and acceleration of energetic particles near an oblique shock, *Astrophys. J.*, v. 515, No. 2, 787–800. [§ 18.2]

Sarazin, C.L. and Kempner, J.C.: 2000, Nonthermal bremsstrahlung and hard X-ray emission from clusters of galaxies, *Astrophys. J.*, v. 533, No. 1, 73–83. [§ 8.3]

Sarris, E.T. and Van Allen, J.A.: 1974, Effects of interplanetary shocks on energetic particles, *J. Geophys. Res.*, v. 79, No. 28, 4157–4173. [§ 18.3]

Schabansky, V.P.: 1971, Some processes in the magnetosphere, *Space Sci. Rev.*, v. 12, No. 3, 299–418. [§ 11.1]

Schlickeiser, R.: 2002, *Cosmic Ray Astrophysics*, New York, Berlin, Heidelberg, Tokyo; Springer-Verlag, p. 519. [§ 5.1]

Schlüter, A.: 1951, Dynamic des Plasmas, *Zeitschrift für Naturforschung*, v. 6A, No. 2, 73–78. [§ 11.1]

Schmidt, G.: 1979, *Physics of High Temperature Plasmas*, New York, London; Academic Press, p. 408. [§ 3.1]

Schou, J., Antia, H.M., Basu, S., et al.: 1998, Helioseismic studies of differential rotation in the solar envelope by the solar oscillations investigation using the Michelson Doppler Imager, *Astrophys. J.*, v. 505, No. 1, 390–417. [§ 20.1]

Schram, P.P.J.: 1991, *Kinetic Theory of Gases and Plasmas*, Dordrecht, Boston, London; Kluwer Academic Publ., p. 426. [Intr., § 6.2.2]

Schrijver, C.J. and Zwaan, C.: 1999, *Solar and Stellar Magnetic Activity*, Cambridge, UK; Cambridge Univ. Press, p. 400. [§ 20.1]

Sedov, L.I.: 1973, *Mechanics of Continuous Medium*, Moscow, Nauka, v. 1, p. 536; v. 2, p. 584 (in Russian). [§ 13.1]

Shafranov, V.D.: 1966, Plasma equilibrium in a magnetic field, in *Reviews of Plasma Physics*, ed. M.A. Leontovich, New York, Consultants Bureau, v. 2, 103–151. [§ 19.2, § 19.3]

Shakura, N.I. and Sunyaev, R.A.: 1973, Black holes in binary systems, Observational appearance, *Astron. Astrophys.*, v. 24, No. 2, 337–355. [§ 8.3, § 13.2]

Shercliff, A.J.: 1965, *A Textbook of Magnetohydrodynamics*, Oxford, London, New York; Pergamon Press, p. 265. [§ 13.1, § 16.2, § 17.4, § 20.2.2]

Shkarofsky, I.P., Johnston, T.W., and Bachynski, M.P.: 1966, *The Particle Kinetics of Plasma*, Reading, Massachusetts; Addison-Wesley Publ., p. 518. [§ 1.1, § 9.3, § 9.5, § 11.5, § 12.2.2]

Shoub, E.C.: 1983, Invalidity of local thermodynamic equilibrium for electrons in solar transition region, *Astrophys. J.*, v. 266, No. 1, 339–369. [§ 8.4]

Shoub, E.C.: 1987, Failure of the Fokker-Planck approximation to the Boltzmann integral for (1/r) potentials, *Phys. Fluids.*, v. 30, No. 5, 1340–1352. [§ 3.1, § 3.5]

Shu, F.H.: 1992, *The Physics of Astrophysics, v. 2, Gas Dynamics*, Mill Valley, California; Univ. Science Books, p. 476. [§ 6.2.2, § 19.3]

Silin, V.P.: 1971, *Intoduction to the Kinetic Theory of Gases*, Moscow, Nauka, p. 332 (in Russian). [§ 3.1, § 3.5, § 6.2.2]

Simon, A.L.: 1959, *An Introduction to Thermonuclear Research*, London, Pergamon Press, p. 182. [Intr.]

Sirotina, E.P. and Syrovatskii, S.I.: 1960, Structure of low intensity shock waves in MHD, *Soviet Physics–JETP*, v. 12, No. 3, 521–526. [§ 16.4, § 17.3]

Sivukhin, D.V.: 1965, Motion of charged particles in electromagnetic fields in the drift approximation, in *Reviews of Plasma Physics*, ed. M.A. Leontovich, New York, Consultants Bureau, v. 1, 1–104. [§ 5.2]

Sivukhin, D.V.: 1966, Coulomb collisions in a fully ionized plasma, in *Reviews of Plasma Physics*, ed. M.A. Leontovich, New York, Consultants Bureau, v. 4, 93–341. [§ 8.3, § 8.4]

Sivukhin, D.V.: 1996, *A Course of General Physics. Electricity*, 3rd Edition, Moscow, Nauka, (in Russian). [§ 11.4.2, § 19.2]

Smirnov, B.M.: 1981, *Physics of Weakly Ionized Gases: Problems and Solutions*, Moscow, Mir Publ., p. 432. [§ 3.5]

Smirnov, V.I.: 1965, *A Course of Higher Mathematics*, v. 2, Oxford, New York; Pergamon Press. [§ 1.1, § 12.3]

Somov, B.V.: 1981, Fast reconnection and transient phenomena with particle acceleration in the solar corona, *Bull. Acad. Sci. USSR, Phys. Ser.*, v. 45, No. 4, 114–116. [§ 8.3.2]

Somov, B.V.: 1982, Accumulation and release of flare energy, in *Proc. 12th Leningrad Seminar on Space Physics: Complex Study of the Sun*, Leningrad, LIYaF, p. 6–49 (in Russian). [§ 4.1, § 4.4]

Somov, B.V.: 1986, Non-neutral current sheets and solar flare energetics, *Astron. Astrophys.*, v. 163, No. 1, 210–218. [§ 8.3]

Somov, B.V.: 1992, *Physical Processes in Solar Flares*, Dordrecht, Boston, London; Kluwer Academic Publ., p. 248. [§ 8.3, § 8.4, § 19.5]

Somov, B.V.: 1994a, *Fundamentals of Cosmic Electrodynamics*, Dordrecht, Boston, London; Kluwer Academic Publ., p. 364. [§ 14.2]

Somov, B.V.: 1994b, Features of mass supply and flows related with reconnection in the solar corona, *Space Sci. Rev.*, v. 70, No. 1, 161–166. [§ 19.4, § 20.4]

Somov, B.V. and Syrovatskii, S.I.: 1972a, Plasma motion in an increasing strong dipolar field, *Soviet Phys.–JETP*, v. 34, No. 2, 332–335. [§ 14.4]

Somov, B.V. and Syrovatskii, S.I.: 1972b, Appearance of a current sheet in a plasma moving in the field of a two-dimensional magnetic dipole, *Soviet Phys.–JETP*, v. 34, No. 5, 992–997. [§ 14.2]

Somov, B.V. and Syrovatskii, S.I.: 1976a, Physical processes in the solar atmosphere associated with flares, *Soviet Physics Usp.*, v. 19, No. 10, 813–835. [§ 8.3]

Somov, B.V. and Syrovatskii, S.I.: 1976b, Hydrodynamic plasma flows in a strong magnetic field, in *Neutral Current Sheets in Plasma*, Proc. P.N. Lebedev Phys. Inst., v. 74, ed. N.G. Basov, New York and London, Consultants Bureau, p. 13–71. [§ 13.1, § 14.1, § 14.2, § 14.4]

Somov, B.V. and Tindo, I.P.: 1978, Polarization of hard X-rays from solar flares, *Cosmic Research*, v. 16, No. 5, 555-564. [§ 4.5]

Somov, B.V. and Titov, V.S.: 1983, Magnetic reconnection as a mechanism for heating the coronal loops, *Soviet Astronomy Letters*, v. 9, No. 1, 26–28. [§ 8.3]

Somov, B.V., Syrovatskii, S.I., and Spektor, A.R.: 1981, Hydrodynamic response of the solar chromosphere to elementary flare burst. 1. Heating by accelerated electrons, *Solar Phys.*, v. 73, No. 1, 145–155. [§ 8.3]

Spicer, D.S. and Emslie, A.G.: 1988, A new quasi-thermal trap model for solar hard X-ray bursts: An electrostatic trap model, *Astrophys. J.*, v. 330, No. 2, 997–1007. [§ 8.1]

Spitzer, L.: 1940, The stability of isolated clusters, *Mon. Not. Royal Astron. Soc.*, v. 100, No. 5, 396–413. [§ 8.3]

Spitzer, L.: 1962, *Physics of Fully Ionized Gases*, New York, Wiley Interscience, p. 170. [§ 8.3, § 8.4, § 9.5]

Stewart, R.W. and Grant, H.L.: 1969, Determination of the rate of dissipation of turbulent energy near the sea surface in the presence of waves, *J. Geophys. Res.*, v. 67, 3177–3184. [§ 7.2.2]

Stix, T.H.: 1992, *Waves in Plasmas*, American Inst. of Physics. [§ 10.4]

Störmer, C.: 1955, *The Polar Aurora*, Oxford, Clarendon Press. [§ 6.4]

Strittmatter, P.A.: 1966, Gravitational collapse in the presence of a magnetic field, *Monthly Not. Royal Astron. Soc.*, v. 132, No. 3, 359–378. [§ 19.1]

Strong, K.T., Saba, J.L.R., Haisch, B.M., et al. (eds): 1999, *The Many Faces of the Sun*, New York, Berlin, Heidelberg, Tokyo; Springer, p. 610. [§ 4.3]

Subramanian, P., Becker, P.A., and Kazanas, D.: 1999, Formation of relativistic outflows in shearing black hole accretion coronae, *Astrophys. J.*, v. 523, No. 1, 203–222. [§ 13.3]

Suh, I.S. and Mathews, G.J.: 2001, Cold ideal equation of state for strongly magnetized neutron star matter: Effects on muon production and pion condensation, *Astrophys. J.*, v. 546, No. 3, 1126–1136. [§ 19.1]

Sutton, G.W. and Sherman, A.: 1965, *Engineering Magnetohydrodynamics*, New York, McGraw-Hill Book Co., p. 548. [§ 13.1, § 20.2]

Syrovatskii, S.I.: 1953, On the stability of tangential discontinuities in MHD medium, *Zhur. Exper. Teor. Fiz.*, v. 24, No. 6, 622–630 (in Russian). [§ 16.2]

Syrovatskii, S.I.: 1954, Instability of tangential discontinuities in a compressive medium, *Zhur. Exper. Teor. Fiz.*, v. 27, No. 1, 121–123 (in Russian). [§ 16.2]

Syrovatskii, S.I.: 1956, Some properties of discontinuity surfaces in MHD, *Proc. P.N. Lebedev Phys. Inst.*, v. 8, 13–64 (in Russian). [§ 16.2, § 16.3, § 20.1]

Syrovatskii, S.I.: 1957, Magnetohydrodynamics, *Uspehi Fiz. Nauk*, v. 62, No. 3, 247–303 (in Russian). [§ 12.2, § 16.2, § 19.1, § 20.1]

Syrovatskii, S.I.: 1959, The stability of shock waves in MHD, *Soviet Physics–JETP*, v. 8, No. 6, 1024–1028. [§ 17.1]

Syrovatskii, S.I.: 1971, Formation of current sheets in a plasma with a frozen-in strong field, *Soviet Physics–JETP*, v. 33, No. 5, 933–940. [§ 14.2.2]

Syrovatskii, S.I. and Chesalin, L.S.: 1963, Electromagnetic generation of conductive fluid flows near bodies and expulsive force, in *Questions of Magnetohydrodynamics*, Riga, Zinatne, p. 17–22 (in Russian). [§ 19.4, § 20.3]

Syrovatskii, S.I. and Shmeleva, O.P.: 1972, Heating of plasma by high-energy electrons, and the non-thermal X-ray emission in solar flares, *Soviet Astronomy–AJ*, v. 16, No. 2, 273–283. [§ 4.3]

Syrovatskii, S.I. and Somov, B.V.: 1980, Physical driving forces and models of coronal responses, in *Solar and Interplanetary Dynamics*, eds M. Dryer and E. Tandberg-Hanssen, IAU Symp. **91**, Dordrecht, Reidel, p. 425–441. [§ 14.2]

Takahara, F. and Kusunose, M.: 1985, Electron-positron pair production in a hot accretion plasma around a massive black hole, *Progr. Theor. Phys.*, v. 73, No. 6, 1390–1400. [§ 7.3]

Takizawa, M.: 1998, A two-temperature model of the intracluster medium, *Astrophys. J.*, v. 509, No. 2, 579–584. [§ 8.3]

Tamm, I.E.: 1989, *Basic Theory of Electricity*, 10th edition, Moscow, Nauka, p. 504 (in Russian). [§ 19.3]

Tandberg-Hanssen, E.: 1995, *The Nature of Solar Prominences*, Dordrecht, Boston, London; Kluwer Academic Publ., p. 308. [§ 19.3, § 20.4]

Tidman, D.A. and Krall, N.A.: 1971, *Shock Waves in Collisionless Plasma*, New York, London, Sydney, Toronto; Wiley-Interscience, p. 175. [§ 16.4]

Titov, V.S. and Priest, E.R.: 1993, The collapse of an X-type neutral point to form a reconnecting current sheet. *Geophys. and Astrophys. Fluid Dynamics*, v. 72, 249–276. [§ 14.2]

Toptyghin, I.N.: 1980, Acceleration of particles by shocks in a cosmic plasma, *Space Sci. Rev.*, v. 26, No. 1, 157–213. [§ 18.3]

Treumann, R.A. and Baumjohann, W.: 1997, *Advanced Space Plasma Physics*, London, Imperial College Press, p. 381. [§ 7.1]

Trubnikov, B.A.: 1965, Particle interactions in a fully ionized plasma, in *Reviews of Plasma Physics*, ed. M.A. Leontovich, New York, Consultants Bureau, v. 1, 105–204. [§ 8.4]

Tverskoy B.A.: 1967, Contribution to the theory of Fermi statistical acceleration, *Soviet Phys. JETP.*, v. 25, No. 2, 317–325. [§ 7.2]

Tverskoy B.A.: 1968, Theory of turbulent acceleration of charged particles in a plasma, *Soviet Phys. JETP.*, v. 26, No. 4, 821–828. [§ 7.2]

Tverskoy B.A.: 1969, Main mechanisms in the formation of the Earth's radiation belts, *Rev. Geophys.*, v. 7, No. 1, 219–231. [§ 6.4]

UeNo, S.: 1998, Comparison between statistical features of X-ray fluctuations from the solar corona and accretion disks, in *Observational Plasma Astrophysics: Five Years of Yohkoh and Beyond*, eds T. Watanabe, T. Kosugi, and A.C. Sterling, Dordrecht, Kluwer Academic Publ., p. 45–50. [§ 13.2]

van de Hulst, H.C.: 1951, Interstellar polarization and MHD waves, in *Problems of Cosmical Aerodynamics*, eds J.M. Burgers and H.C. van de Hulst,

p. 45–57. [§ 15.2, § 15.3]

van den Oord, G.H.J.: 1990, The electrodynamics of beam/return current systems in the solar corona, *Astron. Astrophys.*, v. 234, No. 2, 496–518. [§ 4.5]

Vink, J., Laming, J.M., Gu, M.F., et al.: 2003, The slow temperature equilibration behind the shock front of SN 1006, *Astrophys. J.*, v. 587, No. 1, L31–L34. [§ 16.4]

Vladimirov, V.S.: 1971, *Equations of Mathematical Physics*, New York, M. Dekker, p. 418. [§ 1.1.5, § 1.2.2, § 13.1]

Vlasov, A.A.: 1938, On the oscillation properties of an electron gas, *Zhur. Eksp. Teor. Fiz.*, v. 8, No. 1, 29–33 (in Russian). English translation: 1968, The vibrational properties of an electron gas, *Soviet Physics Uspekhi*, v. 10, No. 4, 721–733; see also *Soviet Physics Uspekhi*, v. 19, No. 6, 545–546 [§ 3.1, § 10.2.2]

Vlasov, A.A.: 1945, On the kinetic theory of an ansembly of paricles with collective interactions, *Soviet J. Phys.*, v. 9, No. 1, 25–28. [§ 3.1]

Volkov, T.F.: 1966, Hydrodynamic description of a collisionless plasma, in *Reviews of Plasma Physics*, ed. M.A. Leontovich, New York, Consultant Bureau, v. 4, 1–21. [§ 11.5.1, § 16.4]

Walt, M.: 1994, *Introduction to Geomagnetically Trapped Radiation*, Cambridge, UK; Cambridge Univ. Press, p. 188. [§ 6.4]

Webb, G.M.: 1986, Similarity considerations and conservation laws for magnetostatic atmospheres, *Solar Phys.*, v. 106, No. 2, 287–313. [§ 19.5]

Webb, G.M., Zank, G.P., Ko, C.M., et al.: 1995, Multi-dimensional Green's functions and the statistics of diffusive shock acceleration, *Astrophys. J.*, v. 453, No. 1, 178–189. [§ 18.2]

Wentzel, D.G.: 1963, Fermi acceleration of charged particles, *Astrophys. J.*, v. 137, No. 1, 135–146. [§ 18.3]

Wentzel, D.G.: 1964, Motion across magnetic discontinuities and Fermi acceleration of charged particles, *Astrophys. J.*, v. 140, No. 3, 1013–1024. [§ 6.2.4, § 18.3]

Wiita, P.J.: 1999, Accretion disks around black holes, in *Black Holes, Gravitational Radiation and the Universe*, eds B.R. Iyer and B. Bhawal, Dordrecht, Boston, London; Kluwer Academic Publ., p. 249–263. [§ 8.3]

Woltjer, L.: 1958, A theorem on force-free magnetic fields, *Proc. Nat. Acad. Sci. USA*, v. 44, No. 6, 489–491. [§ 19.6]

Yvon, J.: 1935, *La Theorie des Fluids et l'Equation d'Etat*, Paris, Hermann et Cie. [§ 2.4]

Zank, G.P.: 1991, Weyl's theorem for MHD, *J. Plasma Phys.*, v. 46, No. 1, 11–14. [§ 16.2]

Zel'dovich, Ya.B. and Novikov, I.D.: 1971, *Relativistic Astrophysics. Vol. 1, Stars and Relativity*, Chicago, Univ. of Chicago Press. [§ 12.2, § 19.3]

Zel'dovich, Ya.B. and Raizer, Yu.P.: 1966, *Physics of Shock Waves and High-Temperature Hydrodynamic Phenomena*, New York, San Francisco, London; Academic Press, v. 1, p. 464; v. 2, p. 452. [§ 8.3, § 16.1, § 16.4]

Zel'dovich, Ya.B. and Raizer, Yu.P.: 2002, *Physics of Shock Waves and High-Temperature Hydrodynamic Phenomena*, eds W.D. Hayes and R.F. Probstein, Mineola, Dover. [§ 8.3, § 16.1, § 16.4]

Zel'dovich, Ya.B., Ruzmaikin, A.A., and Sokolov, D.D.: 1983, *Magnetic Fields in Astrophysics*, New York, Gordon and Breach. [§ 13.1]

Zheleznyakov, V.V.: 1996, *Radiation in Astrophysical Plasmas*, Dordrecht, Boston, London; Kluwer Academic Publ., p. 462. [§ 7.1, § 7.4, § 10.4]

Zhou, Y. and Matthaeus, W.H.: 1990, Models of inertial range spectra of MHD turbulence, *J. Geophys. Res.*, v. 95, No. A9, 14881–14892. [§ 7.2]

Index

Astrophysics and Space Science Library

Volume 337: ***Astrophysical Disks,*** edited by A. M. Fridman, M. Y. Marov, I.G. Kovalenko. Hardbound ISBN 1-4020-4347-3, June 2006

Volume 336: ***Scientific Detectors for Astronomy 2005***, edited by J.E. Beletic, J.W. Beletic, P. Amico. Hardbound ISBN 1-4020-4329-5, December 2005

Volume 335: ***Organizations and Strategies in Astronomy 6***, edited by A. Heck. Hardbound ISBN 1-4020-4055-5, November 2005

Volume 334: ***The New Astronomy: Opening the Electromagnetic Window and Expanding our View of Planet Earth,*** edited by W. Orchiston. Hardbound ISBN 1-4020-3723-6, October 2005

Volume 333: ***Planet Mercury***, by P. Clark and S. McKenna-Lawlor. Hardbound ISBN 0-387-26358-6, November 2005

Volume 332: ***White Dwarfs: Cosmological and Galactic Probes,*** edited by E.M. Sion, S. Vennes, H.L. Shipman. Hardbound ISBN 1-4020-3693-0, September 2005

Volume 331: ***Ultraviolet Radiation in the Solar System***, by M. Vázquez and A. Hanslmeier. Hardbound ISBN 1-4020-3726-0, November 2005

Volume 330: ***The Multinational History of Strasbourg Astronomical Observatory,*** edited by A. Heck. Hardbound ISBN 1-4020-3643-4, June 2005

Volume 329: ***Starbursts – From 30 Doradus to Lyman Break Galaxies,*** edited by R. de Grijs, R.M. González Delgado. Hardbound ISBN 1-4020-3538-1, May 2005

Volume 328: ***Comets***, by J.A. Fernández. Hardbound ISBN 1-4020-3490-3, July 2005

Volume 327: ***The Initial Mass Function 50 Years Later***, edited by E. Corbelli, F. Palla, H. Zinnecker. Hardbound ISBN 1-4020-3406-7, June 2005

Volume 325: ***Kristian Birkeland – The First Space Scientist,*** by A. Egeland, W.J. Burke. Hardbound ISBN 1-4020-3293-5, April 2005

Volume 324: ***Cores to Clusters – Star Formation with next Generation Telescopes,*** edited by M.S. Nanda Kumar, M. Tafalla, P. Caselli. Hardbound ISBN 0-387-26322-5, October 2005

Volume 323: ***Recollections of Tucson Operations,*** by M.A. Gordon. Hardbound ISBN 1-4020-3235-8, December 2004

Volume 322: ***Light Pollution Handbook,*** by K. Narisada, D. Schreuder Hardbound ISBN 1-4020-2665-X, November 2004

Volume 321: ***Nonequilibrium Phenomena in Plasmas***, edited by A.S. Shrama, P.K. Kaw. Hardbound ISBN 1-4020-3108-4, December 2004

Volume 320: ***Solar Magnetic Phenomena,*** edited by A. Hanslmeier, A. Veronig, M. Messerotti. Hardbound ISBN 1-4020-2961-6, December 2004

Volume 319: ***Penetrating Bars through Masks of Cosmic Dust***, edited by D.L. Block, I. Puerari, K.C. Freeman, R. Groess, E.K. Block. Hardbound ISBN 1-4020-2861-X, December 2004

Volume 318: ***Transfer of Polarized light in Planetary Atmospheres***, by J.W. Hovenier, C. van der Mee, J.W. Domke. Hardbound ISBN 1-4020-2855-5. Softcover ISBN 1-4020-2889-X, November 2004

Volume 317: ***The Sun and the Heliosphere as an Integrated System,*** edited by G. Poletto, S.T. Suess. Hardbound ISBN 1-4020-2830-X, November 2004

Volume 316: *Civic Astronomy - Albany's Dudley Observatory, 1852-2002,* by G. Wise
Hardbound ISBN 1-4020-2677-3, October 2004

Volume 315: ***How does the Galaxy Work** - A Galactic Tertulia with Don Cox and Ron Reynolds,* edited by E. J. Alfaro, E. Pérez, J. Franco
Hardbound ISBN 1-4020-2619-6, September 2004

Volume 314: ***Solar and Space Weather Radiophysics- Current Status and Future Developments,*** edited by D.E. Gary and C.U. Keller
Hardbound ISBN 1-4020-2813-X, August 2004

Volume 313: ***Adventures in Order and Chaos***, by G. Contopoulos. Hardbound ISBN 1-4020-3039-8, January 2005

Volume 312: ***High-Velocity Clouds***, edited by H. van Woerden, U. Schwarz, B. Wakker
Hardbound ISBN 1-4020-2813-X, September 2004

Volume 311: ***The New ROSETTA Targets- Observations, Simulations and Instrument Performances***, edited by L. Colangeli, E. Mazzotta Epifani, P. Palumbo
Hardbound ISBN 1-4020-2572-6, September 2004

Volume 310: ***Organizations and Strategies in Astronomy 5,*** edited by A. Heck
Hardbound ISBN 1-4020-2570-X, September 2004

Volume 309: ***Soft X-ray Emission from Clusters of Galaxies and Related Phenomena***, edited by R. Lieu and J. Mittaz
Hardbound ISBN 1-4020-2563-7, September 2004

Volume 308: ***Supermassive Black Holes in the Distant Universe,*** edited by A.J. Barger
Hardbound ISBN 1-4020-2470-3, August 2004

Volume 307: ***Polarization in Spectral Lines***, by E. Landi Degl'Innocenti and M. Landolfi
Hardbound ISBN 1-4020-2414-2, August 2004

Volume 306: ***Polytropes – Applications in Astrophysics and Related Fields,*** by G.P. Horedt
Hardbound ISBN 1-4020-2350-2, September 2004

Volume 305: ***Astrobiology: Future Perspectives,*** edited by P. Ehrenfreund, W.M. Irvine, T. Owen, L. Becker, J. Blank, J.R. Brucato, L. Colangeli, S. Derenne, A. Dutrey, D. Despois, A. Lazcano, F. Robert
Hardbound ISBN 1-4020-2304-9, July 2004
Paperback ISBN 1-4020-2587-4, July 2004

Volume 304: ***Cosmic Gammy-ray Sources,*** edited by K.S. Cheng and G.E. Romero
Hardbound ISBN 1-4020-2255-7, September 2004

Volume 303: ***Cosmic rays in the Earth's Atmosphere and Underground***, by L.I, Dorman
Hardbound ISBN 1-4020-2071-6, August 2004